The Birds *of* County Cork

The Birds *of* County Cork

A review of history, status, distribution and migration

Patrick Smiddy, Mark Shorten and Russ Heselden

Illustrations by Russ Heselden
Photography by Richard T. Mills

CORK UNIVERSITY PRESS

First published in 2022 by
Cork University Press
Boole Library
University College Cork
Cork T12 ND89
Ireland

Library of Congress Control Number: 2022932548
ISBN: 9781782055198

Distribution in the USA: Longleaf Services, Chapel Hill, NC, USA

Book design and typesetting by Studio 10 Design, Cork
Printed by Gutenberg Press in Malta

Cover images:
Front cover – Common Whitethroat, by Russ Heselden;
Front flap – top: sedge warbler; bottom: Goldfinch, both by Richard T. Mills

www.corkuniversitypress.com

EIRGRID

Comhairle Contae Chorcaí
Cork County Council

ABOUT THE AUTHORS

Patrick Smiddy was born in 1950 and has had an interest in natural history, especially birds and mammals, for as long as he can remember. He began birdwatching 'proper' in the mid-1960s. He became a bird-ringer in the early 1970s and has studied Dippers, Grey Wagtails, Barn Swallows and *Acrocephalus* warblers, among other species; his subjects of interest include wetland and river birds, breeding biology, feeding ecology (owls, birds of prey and corvids) and migration. He has published widely in national and international peer-reviewed journals, was a member and secretary of the Irish Rare Birds Committee (1985–96), and editor of the journal *Irish Birds* (2010–17). He worked for the National Parks and Wildlife Service (1980–2008) and more recently in wildlife consultancy. He is a member of BirdWatch Ireland, the British Trust for Ornithology, the British Ornithologists' Union and the European Ornithologists' Union. Now retired, he is an honorary research associate with the University College Cork Ornithology Group and was awarded an honorary MSc degree by the National University of Ireland in 2000.

Mark Shorten was born in 1961 and has been active in birdwatching since 1975. He has participated in all major bird surveys since then and has been joint editor of the *Cork Bird Report* (1990–5). He has been county recorder since 1990 and currently works as a bird surveyor. He is a member of BirdWatch Ireland, and his chief claim to fame is finding the first Irish Lesser Crested Tern at Ballycotton in 1996.

Russ Heselden was born in 1963 and completed his PhD at University College Cork during 1987–91, studying the limestone geology of the Cork and Cloyne synclines. He was joint editor and illustrator of the *Cork Bird Report* from 1987 to 1990. Now teaching in Norfolk, he has built a parallel career as a wildlife artist and author. He has illustrated several books, including two with the British Trust for Ornithology, has exhibited with the Society of Wildlife Artists and was elected as a signatory member of Artists for Conservation, based in Vancouver, in 2014. He regularly assists with bird surveys and has covered the same Breeding Bird Survey square in north Norfolk for the past eighteen years. Russ has a special interest in waders, chats and *Phylloscopus* warblers. He has been an active birder for over forty years, and has written a number of notes and identification papers.

This book is dedicated
to the memory of
Clive Desmond Hutchinson
(1949–98)

Clive was the most
influential and inspiring Irish
ornithologist of the 1970s,
1980s and 1990s.

CONTENTS

LIST OF ILLUSTRATIONS

Front cover: Common Whitethroat

Frontispiece: Black-tailed Godwits

LIST OF SCIENTIFIC NAMES OF SPECIES

(other than Birds) Mentioned in the Text

MAMMALS

Brown rat *Rattus norvegicus*

Field mouse *Apodemus sylvaticus*

House mouse *Mus domesticus*

Bank vole *Myodes glareolus*

Greater white-toothed shrew *Crocidura russula*

Pygmy shrew *Sorex minutus*

Fox *Vulpes vulpes*

Dog *Canis familiaris*

Cat *Felis catus*

Mink *Neovison vison*

Rabbit *Oryctolagus cuniculus*

Soprano pipistrelle bat *Pipistrellus pygmaeus*

Cattle *Bos*

Sheep *Ovis*

Pig *Sus*

FISHES

Herring *Clupea harengus*

Perch *Perca fluviatilis*

Pike *Esox lucius*

Roach *Rutilus rutilus*

Salmon *Salmo salar*

Sprat *Sprattus sprattus*

PLANTS

Alder *Alnus glutinosa*

Birch *Betula* species

Noble fir *Abies procera*

Rowan *Sorbus aucuparia*

Scots pine *Pinus sylvestris*

Spruce *Picea* species

Gorse *Ulex europaeus*

Ivy *Hedera helix*

Bell heather *Erica cinerea*

Ling heather *Calluna vulgaris*

Bramble *Rubus* species

Common reed *Phragmites australis*

Cord-grass *Spartina anglica*

Maize *Zea mays*

Silverweed *Potentilla anserina*

Bracken *Pteridium aquilinum*

OTHERS

Common frog *Rana temporaria*

Common lizard *Zootoca vivipara*

Common field grasshopper *Chorthippus brunneus*

Spring hawker dragonfly *Brachytron pratense*

Shrimp *Crangon crangon*

Botulism *Clostridium botulinum*

Myxomatosis *Myxoma virus*

Trichomonosis *Trichomonas gallinae*

ABBREVIATIONS

References to 'CCBO' throughout Chapter 7 and Appendix 1 and 2 are to Cape Clear Bird Observatory or to the island of Cape Clear. Note that the correct name for the island is 'Clear Island', but it is commonly, albeit incorrectly, referred to as 'Cape Clear Island' by visiting birdwatchers. The term 'Cape Clear' is, if correctly used, a place on Clear Island. However, we have retained the terminology to avoid confusion with the ornithological literature. Other abbreviations are announced at first mention, separately, within each chapter.

ACKNOWLEDGEMENTS

Many people have contributed information to this book, answered our questions and read sections in draft, all of which have gone to improving the text; we most sincerely thank the following: Colin Barton, Sam Bayley, Simon Berrow, Hugh Brazier, Kieran Buckley, Brian Caffrey, Mark Carmody, Kendrew Colhoun, Dave Cooke, John Coveney, Ciarán Cronin, Seán Cronin, Olivia Crowe, Chris Cullen, Sinéad Cummins, Gordon D'Arcy, Joe Doolan, Brian Duffy, Aidan Duggan, John Earley, Kieran Fahy, Dare Fekonja, Darío Fernández-Bellon, Izabela Fischer, Renaud Flamant, Seán Fleming, Colm Flynn, Jim Fox, Ernest Garcia, Tom Gittings, Mark Golley, Kieran Grace, Ricard Gutiérrez, Clare Heardman, Iain Hill, Joe Hobbs, Harry Hussey, the late Clive Hutchinson, Rachel Hutchinson-Kelly, Tom Kelly, Seán Kingston, the late Tony Lancaster, John Temple Lang, Lesley Lewis, John Lusby, John Lynch, Alan McCarthy, Declan McGrath, Allan Mee, the late Oscar Merne, Richard Mills, Paul Moore, Killian Mullarney, Rick Mundy, Matt Murphy, Susan Murphy Wickens, Tony Nagle, Richard Nairn, Steve Newton, Barry O'Donoghue, Paul O'Donoghue, the late Liam O'Flynn, John O'Halloran, Tadg O'Keeffe, Barry O'Mahony, Dennis O'Sullivan, Paddy O'Sullivan, Brian Power, Ken Preston, John Quinn, David Rees, the late Margaret and Richard Ridgway, the late Michael Roche, Cyril Saich, Derek Scott, Tim Sharrock, Ralph Sheppard, Paddy Sleeman, Eva Sweeney, Pascal Sweeney, Mark Tasker, David Tierney, Paul Walsh, Jim Wilson, Steve Wing, Peter Wolstenholme, Julian Wyllie, Andreas Zours, and staff at the Royal Irish Academy.

We additionally thank Joe Doolan for allowing the use of data from irishbirding.com, Michael O'Clery for his assistance with the graphics, Dr Tom Kelly for advice, friendship and encouragement over many years, and for writing the foreword, Professor John O'Halloran for advice and much help and Barry O'Mahony for his help with, and knowledge of, bird-ringing and recoveries, and for his commitment to the compilation of Cork ringing reports.

We are grateful to Cork County Council for their support, especially Conor Nelligan and Sharon Casey. This publication has been supported by Cork County Council's Heritage Unit as an action of the County Cork Heritage Plan.

EirGrid also provided financial support, and we gratefully acknowledge the help of Martin Corrigan, David Martin, Brendan Tuohy and Suzanne Collins for their kindness and generosity.

We are also most grateful to Maria O'Donovan and Mike Collins of Cork University Press along with copy editor Aonghus Meaney, proof reader Claire Fitzgerald, indexer Fionbar Lyons and designer Alison Burns of Studio 10 for their help and patience, especially towards the closing stages of the writing process.

Last, but not least, we owe a debt of gratitude to all the birdwatchers who have contributed their records over the years, whether to the *Cork Bird Report*, the *Cape Clear Bird Observatory Report*, the *Sherkin Island Bird Report*, or the *Irish Bird Report* (latterly *Irish Birds*). Without the mass of data gathered voluntarily either as part of the pleasures of birdwatching or during surveys, this book could not have been written.

Russ would like to thank his mother, Patricia Heselden, for tolerating his wildlife obsession during his formative years. He would also like to single out Bob Philpott, someone who started as his Young Ornithologists' Club leader and became a lifelong friend, his teachers Roger and Mary Townsin for much early encouragement and for introducing him to Norfolk, and Robert Greenhalf for many happy hours spent sketching together in the field.

Mark wishes to thank his mother, Freda Shorten, for encouraging him to take up birdwatching as a hobby, his wife Thérèse and children Katie and Ben for their patience, support and love, and O'Sullivan's bar (Crookhaven) for sustenance.

Pat wishes to thank those who influenced him early on, especially Tom Kelly and Kieran O'Brien, and subsequently Barry O'Mahony and John O'Halloran for the many good times together gathering data in the field. He also thanks Professor James Fairley for encouraging him in his early attempts at writing papers, and in proposing his honorary degree, and for his always-early Christmas card. Pádraigín O'Donoghue has been a great encouragement at times when writing was dragging on and offered great support and important advice.

FOREWORD

In terms of landmass, Cork is the largest county in Ireland and its heavily indented coastline is the longest. As would be expected of such a large area, Cork has many habitats including mountains, upland bog, semi-natural oak woods, rivers (lotic habitats), sandy beaches, dunes, estuaries and salt marshes but, curiously, bodies of standing freshwater (lentic habitats) are relatively few – the largest being a series of dams and reservoirs that were constructed in the period 1952–7 as part of the River Lee hydro-electric scheme.

Nevertheless, in terms of breeding species, migrants and vagrants, County Cork has a rich avifauna and its documentation dates to the early 1800s as summarised by Ussher and Warren (1900) – which drew heavily on the work of William Thompson (1849–51). Therefore, data relating to birds in County Cork spans an interval of some 200 years. More and more information has accumulated, particularly since the establishment of a bird observatory at Cape Clear Island in 1959, the formation of the Cork Ornithological Society and the publication of the *Cork Bird Report* in 1963, and the sequence of breeding and wintering bird atlases (1976–2011). In addition, collaborative research between the Department of Zoology at University College Cork and various local research groups has been, and continues to be, published in *Irish Birds* (founded by Clive Hutchinson in 1977) and various other journals.

It is therefore something of a major task to review, synthesise, analyse and summarise the different lines of data on the birds of Cork accumulated over this 200-year interval.

The authors of this book have been studying the avifauna of Cork for some fifty years. The lead author, Patrick Smiddy, MSc (NUI), is a former editor of *Irish Birds*, a former member of the Irish Rare Birds Committee, a former joint editor of the *Irish Bird Report* and one of the most prolific publishers of peer-reviewed scientific papers on different aspects of the biology and ecology of the birds of Ireland. He is therefore something of an oracular voice in relation to the avifauna of Ireland generally and particularly that of County Cork. In addition, he lives on the shores of Ballymacoda estuary within Youghal Bay and close to Ballycotton Bay, areas that have received prominent mention in Irish ornithology since the time of Ussher and Warren.

But the authors of this monograph have also seen changes to the species composition of birds in County Cork. The Corn Bunting has become extinct in Ireland, and the Corncrake, Little Tern and Northern Lapwing have all but disappeared as breeding species in Cork while the Twite is rarely, if ever, now seen in winter. On the other hand, Cork has seen the arrival of the Collared Dove, Little Egret, Common Reed Warbler and Common Buzzard.

The opening of a bird observatory on Cape Clear Island in 1959 initiated a new era in the recording of birds in Cork. The island, which scarcely gets a mention in the monographs of Ussher and Warren, or Kennedy, Ruttledge and Scroope (1954), became the dominant location for sightings of vagrant species in Ireland – particularly of rare western Palearctic and Nearctic passerine and near passerine species. Equally, sea-watching observations from the famous headlands of Cape Clear revealed the somewhat episodic but unprecedented passage of large numbers of seabirds including Great and Cory's Shearwaters and, more recently, occasional sightings of the very rare Fea's-type and Wilson's Storm Petrels. Subsequently, new observation points were explored at Ballycotton, Old Head of Kinsale, Galley Head, Mizen Head and Dursey Island, and these also produced outstanding sightings of rare and scarce seabirds. Sadly, access to some of these seabird and whale watch points – arguably among the best in Europe – is much less free than it was in the past.

Many young ornithologists in Ireland learned their fieldcraft at the observatory on Cape Clear – especially the recording of field notes and the writing of descriptions of rare birds in the field. This was one of the reasons why there was an upsurge of records of rare North American wading birds in Cork in the 1960s when, *inter alia*, Long-billed Dowitcher, Baird's Sandpiper, Semipalmated Sandpiper and American/Pacific Golden Plover were recorded in the county for the first time. This group of young observers were mentored by, among others, Tadg O'Keeffe, the late Commandant Michael Hartnett, Captain Liam O'Flynn, Paddy O'Flynn and Frank King. Without the generous support of these mentors the very striking findings made in the autumn of 1966 might have been much less notable than they proved to be.

Of course, as Killian Mullarney has pointed out in his public lectures, there has been a major revolution since the 1960s and 1970s – both in the know-how, i.e. the availability of first-class identification keys, and in the equipment (binoculars and telescopes) and especially cameras used to detect and record sightings made in the field. High-quality photographs have now, to a certain extent, supplanted the traditional skill of field descriptions that would have been submitted in support of the finding of a rare species. However, Richard T. Mills began to contribute photographs of rare birds in Cork in 1971; in the case of Cork's first Solitary Sandpiper on 5–7 September 1971, Major Ruttledge, then editor of the *Irish Bird Report*, remarked that 'three superb photographs by R.M. left no doubt as to identity'. So, for at least fifty years, Richard Mills' photographs have been one of the mainstays of ornithology in County Cork.

Wildfowl and wader counts began in the 1960s along with the first census of seabirds. There have been four atlas surveys and more recently a Countryside Bird Survey. All of these, together with the ongoing census of wetland birds and other surveys, have led to a large but disparate body of literature on the birds of County Cork.

This outstanding monograph on the *Birds of County Cork*, the most detailed of its kind to be published, represents a major landmark in the 200-year history of ornithological studies in Cork. Over the 1800–2018 interval some inland habitats (e.g. the Annagh Bogs) have completely disappeared and others on the coast (e.g. Ballycotton and Shanagarry marshes) mentioned by Ussher and Warren (1900) have been severely degraded despite being Special Protection Areas (SPA). Hopefully, this book, in addition to its scholarly content, will re-focus efforts on the proper conservation and management of these vulnerable habitats.

Tom C. Kelly
September 2021

Map 1:
Place Names in County Cork

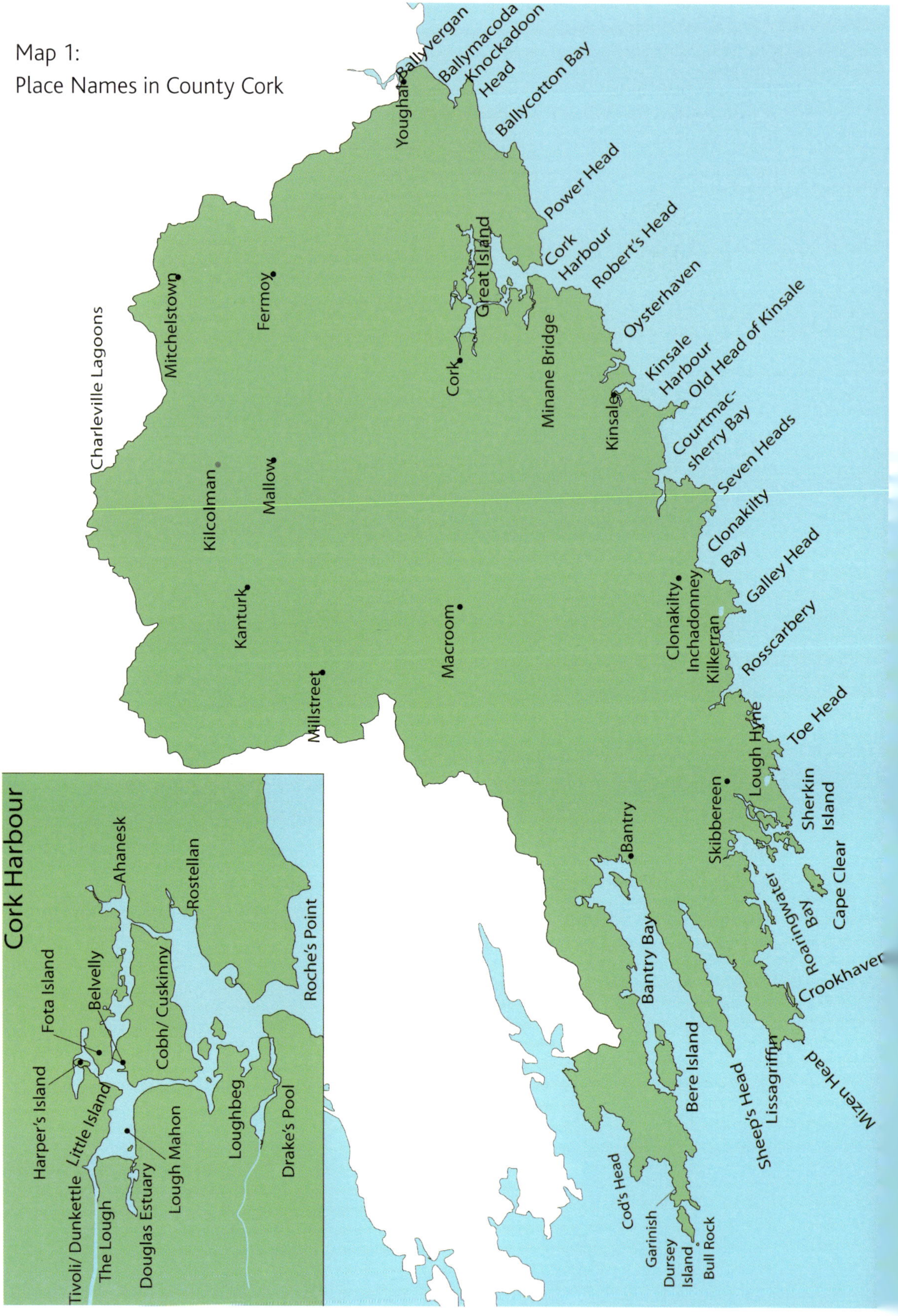

Map 2:
Rivers and Mountains
in County Cork

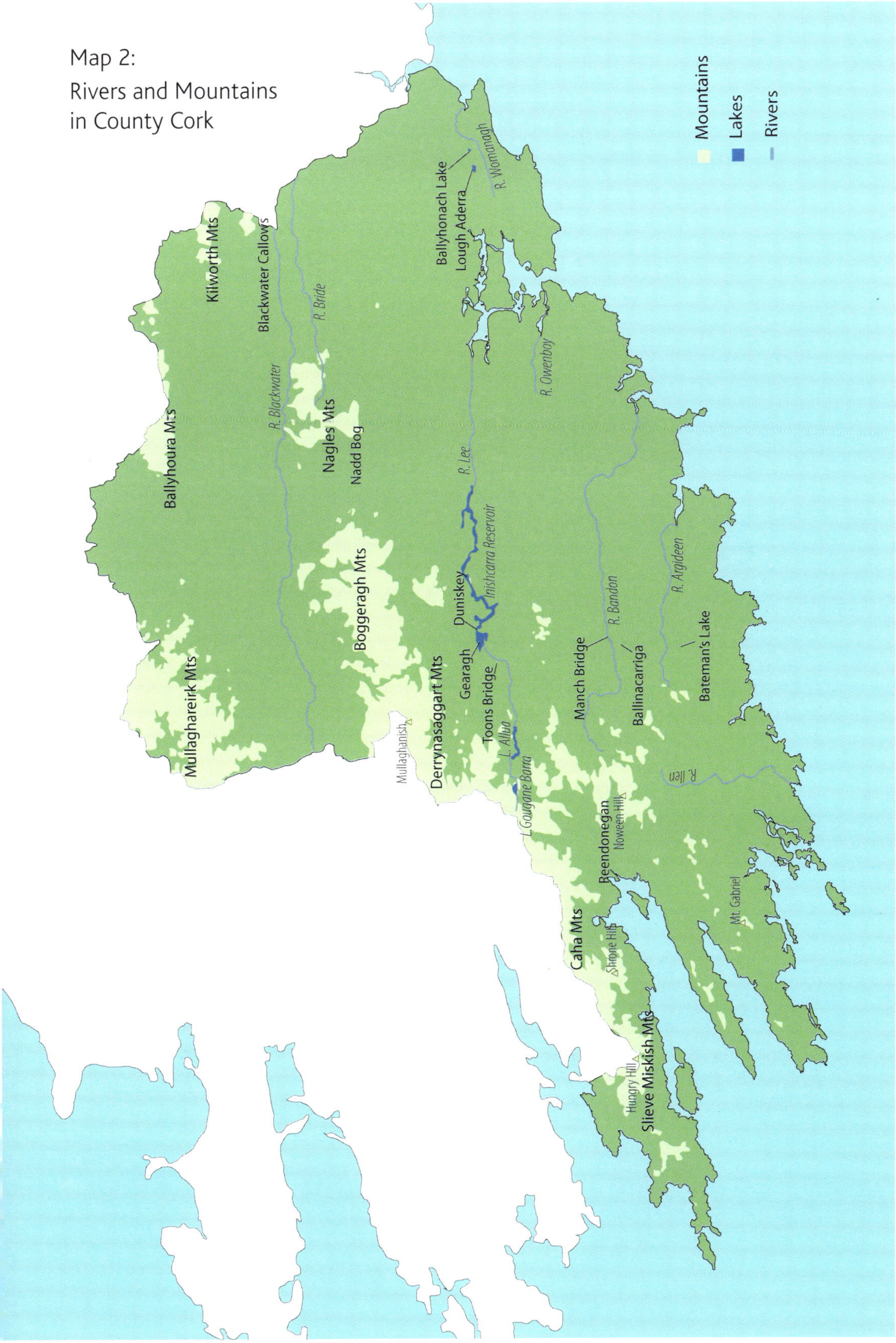
Mountains
Lakes
Rivers
Ballyhonach Lake
Lough Aderra
R. Womanagh
Kilworth Mts
Blackwater Callows
R. Bride
R. Blackwater
R. Owenboy
Ballyhoura Mts
Nagles Mts
Nadd Bog
R. Lee
Inishcarra Reservoir
Boggeragh Mts
Duniskey
Gearagh
Toons Bridge
Mullaghareirk Mts
Derrynasaggart Mts
Mullaghanish
L. Allua
L. Gougane Barra
Manch Bridge
R. Bandon
Ballinacarriga
R. Argideen
Bateman's Lake
R. Ilen
Noween Hill
Beendonegan
Mt. Gabriel
Caha Mts
Shrone Hill
Hungry Hill
Slieve Miskish Mts

Map 3:
Geology of County Cork

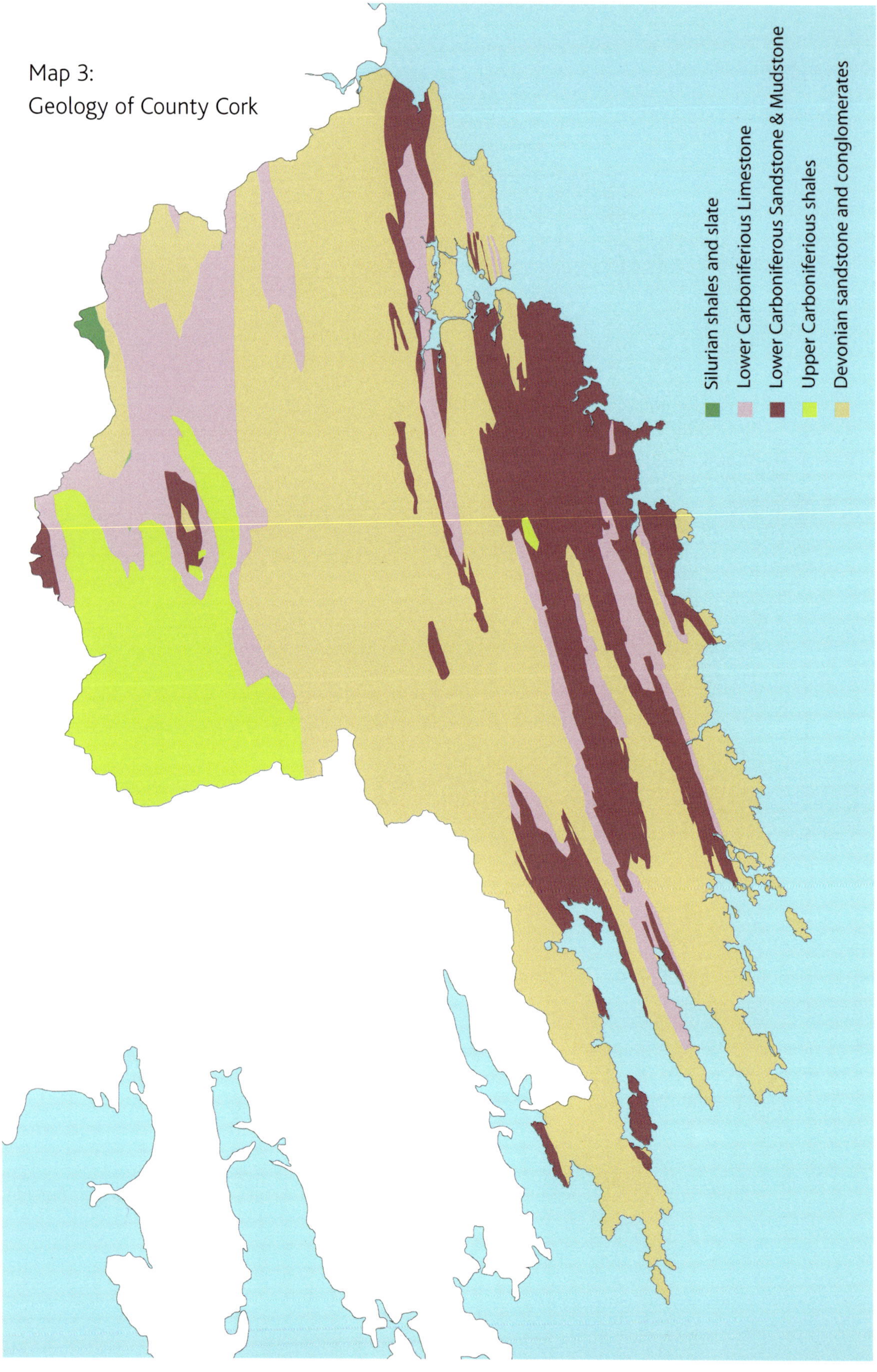

Map 4:
Soils of County Cork

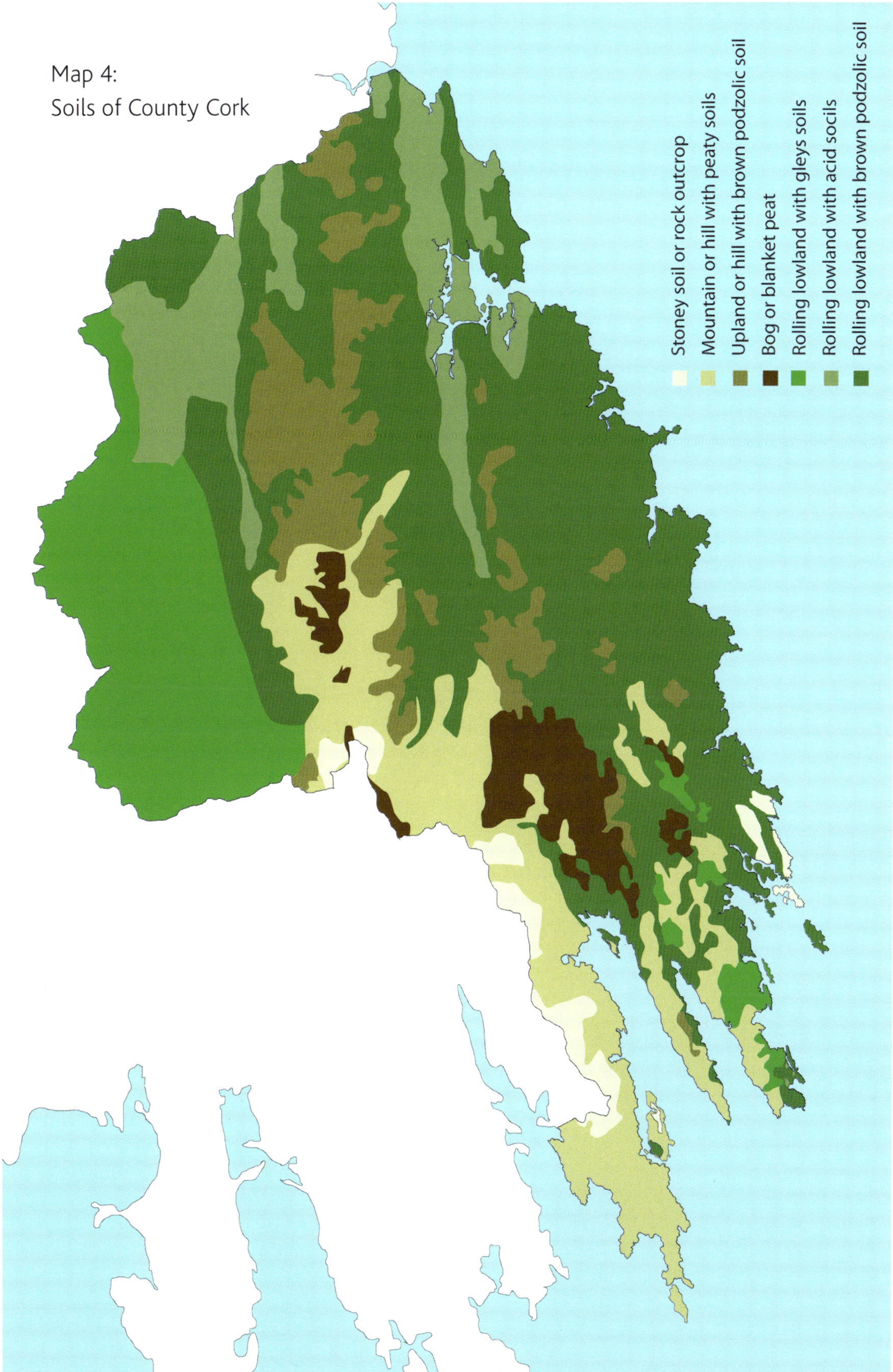
Stoney soil or rock outcrop
Mountain or hill with peaty soils
Upland or hill with brown podzolic soil
Bog or blanket peat
Rolling lowland with gleys soils
Rolling lowland with acid socils
Rolling lowland with brown podzolic soil

Chapter 1

Introduction

Cape Clear Bird Observatory

County Cork has long been known as an important place for birds. Records dating to at least 1750 are scattered through several general works on the birds of Ireland and in various journals, such as the *Zoologist*, *Irish Naturalist*, *Irish Naturalists' Journal* and *British Birds*. Since the late 1950s a growing number of observations have been published in local reports such as the *Cork Bird Report*, *Cape Clear Bird Observatory Report* and *Sherkin Island Bird Report*, as well as in the national *Irish Bird Report*. The journal *Irish Birds* was established in 1977 to provide a specifically Irish outlet for an increasing number of notes and papers, and it has become the place of choice for many observers wishing to publish essentially local, but important, observations and data. The present book represents the first attempt to bring together this information. There is still a great amount to be learned, particularly about numbers, population density, breeding ecology, feeding ecology, distribution and movements of many common species, and the aims of this book include stimulating an interest in research and monitoring which will, hopefully, lead to eventual publication of results.

In summer, County Cork offers the interested observer coastal breeding populations of Peregrine Falcon, Chough, and a variety of seabirds, while in winter large numbers of waterbirds and thrushes arrive to take advantage of the mild conditions (scientific names of birds are given in Chapter 7 and Appendix 1). Cork Harbour is an internationally important site for wintering waterbirds with populations of between 30,000 and 40,000 individuals, and in severe weather over 50,000, although there has been a recent decrease.[1] Coastal sites such as Rosscarbery, Clonakilty Bay, Courtmacsherry Bay, Bandon estuary, Ballycotton, Ballymacoda, River Lee reservoirs, Charleville lagoons and Kilcolman marsh hold nationally important waterbird populations, while smaller sites, such as Lough Aderry, hold nationally important populations of individual species, such as Gadwall. The Little Egret is now a regular feature of estuaries and river valleys, where it has established several breeding populations since the late 1990s, and the White-tailed Eagle has returned to breed following a re-introduction programme in the adjoining county of Kerry.

Through much of the year there are often spectacular seabird movements off the coast, usually easily observed from headlands where the birds often pass very close inshore (Fig. 1.1). In particular, the extreme south-west of the county, especially Cape Clear Island, has an international reputation

Fig. 1.1 Great Shearwater; a regular autumn visitor to south and west coasts, especially off Cape Clear, where numbers vary from year to year.

as the finest place in Europe to watch shearwater migration in autumn, although significant movements can also be observed from Galley Head and Old Head of Kinsale.

Inland, woodland habitats abound with the songs of Blackcap, Goldcrest and Coal Tit, while the less easy to observe Sparrowhawk and Long-eared Owl are also present. There are Jay populations in most woods and the Great Spotted Woodpecker is increasingly seen, but Woodcock are apparently scarce and require special effort to find. In recent years, the Common Buzzard has become the most frequently observed bird of prey, and although absent from few places it is scarcest in the west.

Upland areas hold small populations of breeding Hen Harrier and Common Kestrel, where Skylark and Meadow Pipit are both common

Fig. 1.2 Common Kestrel; breeds widely at low density, and feeds increasingly on the recently introduced Greater White-toothed Shrew.

(Fig. 1.2). However, the Ring Ouzel no longer haunts the cliffs and corries of the mountains, and the Red Grouse has declined significantly.

The Yellowhammer is common on farmland in the east although it has disappeared from the west, but the Corncrake and Grey Partridge have not bred for many years. The Corn Bunting and Twite have both ceased to breed, and the always small Tree Sparrow breeding population is also probably extinct. On the other hand, the Barn Swallow is doing well on farmland, as are several corvid species. Breeding waders, such as Northern Lapwing and Common Snipe, have declined and been pushed to marginal areas and uplands, and the Common Curlew is almost extinct.

Rivers hold good breeding populations of Mute Swan, Mallard, Common Kingfisher, Dipper and Grey Wagtail, while the Reed Bunting haunts the marginal habitats along riverbanks, and in marshes (Fig. 1.3). The Sedge Warbler is common in reedbeds and in rank wetland vegetation, as well as sometimes occurring in drier habitats, often in young conifer plantations. However, the small Common Reed Warbler population is confined to reedbeds in fens and river margins, mainly in the east of the county.

Artificial sites such as sand pits and rock quarries provide nesting habitats for Sand Martin, Peregrine Falcon and Raven, and industrial yards and car parks for Common Ringed Plover. The Common Tern has had mixed fortunes at industrial sites within Cork Harbour, but their numbers

Fig. 1.3 Dipper; common in rivers and streams, often breeding alongside the Grey Wagtail.

Fig. 1.4 Pied Wagtail; a widespread breeder, forming winter roosts in urban areas, especially in Cork city.

have held up in the face of many difficulties, and conservation measures are ongoing. Urban sites are often considered of no importance for birds, but there are significant breeding populations of Common Swift and House Martin, and winter roosts of Pied Wagtail and Common Starling, in many towns and other urban settings, as well as in Cork city, where at least one pair of the Peregrine Falcon breeds (Fig. 1.4).

Perhaps of less significance in conservation terms, though no less fascinating, are the rare and unusual species which regularly reach the

Fig. 1.5 Yellow-rumped Warbler; an autumn vagrant to south-west Cork from North America.

county during migration seasons. Although the numbers of migrants are generally small as Cork lies too far west of the main migration flyways for large arrivals of common species, the extreme westerly location, and the dominance of westerly winds, ensures that these include vagrant waders and land birds blown across the Atlantic Ocean from North America. Coastal headlands and islands are also good places to search for overshooting southern European migrants in spring, and eastern vagrants in autumn. The list of rarities is extensive, and the superb scenery and relaxed atmosphere makes seeking them out an enjoyable experience for the dedicated (Fig. 1.5).

The number of active birdwatchers in Cork has always been small, although growing interest and expertise since the late 1950s has been the

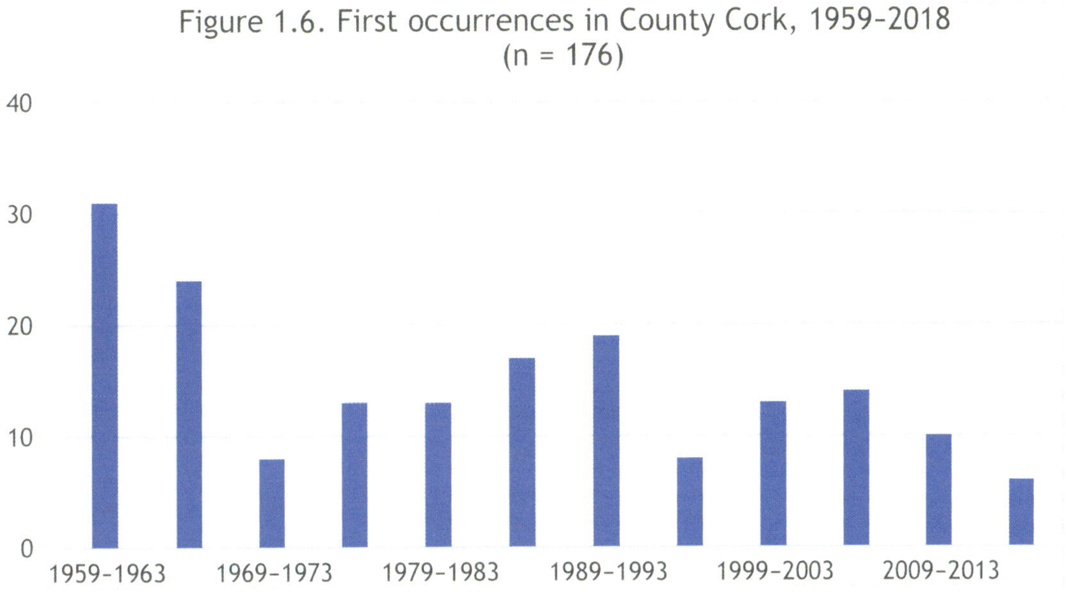

Fig. 1.6 First occurrences in County Cork, 1959–2018.

dominant factor responsible for the increase in first records since that time (Fig. 1.6) (mean per annum = 2.9 species). In the Systematic List (see Chapter 7) the status of several species shows a dramatic change during the period 1959–68, not because there was any real change in their populations, but because that was the first decade of regular observations at Cape Clear Island following the establishment of the bird observatory. In reality, except for extreme rarities, many of the species first recorded during that ten-year period have probably always occurred, but nobody was there to identify them. In the first ten years of existence of Cape Clear Bird Observatory (1959–68), thirty-eight species were added to the Cork List, while only seventeen species were added across the rest of the county. Some of these species were extreme rarities, but many became regular passage migrants there, and elsewhere on the Cork coast, once it was realised there was the probability of recording them during favourable weather conditions, especially in autumn. Many more species, formerly believed to be of very rare occurrence in Ireland, were also shown to occur regularly as a direct result of migration work at the observatory. However, several species have genuinely become more regular, even common, since 1959.

The number of observers contributing records to the *Cork Bird Report*, a rough guide to the level of birdwatching activity in the county, was very low in the 1960s (mean of 29 during 1963–71) but increased in the 1980s (mean of 82 during 1976–85) and remained static in the 1990s (mean of 80 during 1986–95). Birdwatchers now visit many more sites than formerly, and several coastal headlands attract autumn coverage of a standard near that expected at a bird observatory; Dursey Island, the Beara peninsula, Mizen Head, Galley Head, Old Head of Kinsale and Knockadoon Head are cases in point, and each has contributed records of first occurrences in Cork in recent decades (Table 1.1). No fewer than 94 of the 427 species that form the Cork List were also first Irish records, most of them observed at Cape Clear Island (Table 1.2). Many of these were vagrants either from the west (North America) or from the east (Siberia), part of the reason why the Cork coast is visited by so many birdwatchers each autumn (see Appendix 2).

Dursey Island and Beara	15 (1977)
Mizen Head	2 (1992)
Galley Head	5 (1985)
Old Head of Kinsale	4 (1987)
Knockadoon Head	2 (1980)
Cape Clear Island	58 (1959)

Table 1.1 Number of first Cork records of passerines and near-passerines at the most regularly watched coastal headland and island sites since the foundation of Cape Clear Bird Observatory (1959–2018) (years of first records in parentheses).

Site/area	**No.**
Dursey Island and Beara peninsula (includes Bull Rock)	12
Cape Clear Island (includes Fastnet Rock)	30
Galley Head area and Clonakilty Bay	8
Cork Harbour area	6
Ballycotton (includes Garryvoe)	9
Knockadoon Head and Ballymacoda	4
Youghal area	8
Other coastal sites	8
Inland sites	8
County Cork (unknown)	1

Table 1.2 County Cork sites where ninety-four first Irish records have occurred (all records). None of the 'other coastal sites' and 'inland sites' have had more than two records each.

Visiting birdwatchers may spend only a week or two in the county, typically in the autumn and almost exclusively at coastal localities. For many years the regular core of local observers (those submitting observations throughout the year), who are mainly resident in coastal areas, especially in and around Cork city, usually numbered fewer than twenty. Comparison with the annual *Kent Bird Report* (a similar-sized county in south-east England) illustrates the coverage problem, with the list of annual contributors there typically numbering up to 500 individuals. Therefore, it is no surprise that there are still considerable gaps in our knowledge of the Cork avifauna. This is especially true of central and northern parts of the county, which means that, except where special surveys have been organised, coverage has historically been less than adequate and is the main reason why so little is known about so many inland and upland species, with a few exceptions. A feature of bird recording in County Cork over the years has been a surge in records from previously poorly covered areas following residency or regular visitations by new observers, e.g. Dursey Island, Beara peninsula, Mizen Head, Galley Head, Cobh, Ballycotton, Ballymacoda, the Gearagh, Kilcolman and Charleville lagoons. Many observers at these sites have also kept up a string of publications, frequently enhanced by the superb photographs of Richard T. Mills, a Cork resident of long standing. Their data and published papers have helped to form the backbone of this book.

While the main part of this book is given over to individual accounts of the 427 species in the Systematic List (see Chapter 7), we attempt to put these into context by providing summary chapters on the Cork environment (Chapter 2), ornithology and ornithologists (Chapter 3), research and monitoring (Chapter 4), and recent changes of status (Chapter 5), before introducing the Cork List in Chapter 6.

The pressures on habitats in County Cork, and hence on birds and other wildlife, are growing. Pollution has always been a threat to delicate wetland ecosystems, especially rivers, although oil pollution in the sea appears to have declined. The changing agricultural landscape through intensification since Ireland's accession to the European Economic Community (later European Union) in 1973 has made it almost impossible for ground-nesting and specialist species to breed in that habitat, and hedgerow removal along with herbicide and insecticide use has decreased both habitat and food for some species, although some generalists have benefited and increased in farmland habitat.[2] New forest plantations on lowland and upland areas have provided habitat for many species, but the sites selected for some have removed the habitat of other species, including some red-listed on 'birds of conservation concern'.[3]

The growing importance of tourism in this scenic corner of Ireland has introduced further pressures through disturbance and erosion, while uplands have become more accessible than formerly through road building for renewable energy projects. These activities may become a further threat to some vulnerable species, such as Hen Harrier and Red Grouse, through habitat fragmentation. Some sensitive coastal wetlands have been damaged while others are threatened by activities as diverse as industrial development, flood defences, road building, urbanisation, and recreational activities such as golf courses and boating marinas. These activities keep organisations such as BirdWatch Ireland (BWI) and local activists constantly on the alert.

We hope this book will not only be a source of reference to resident and visiting birdwatchers, but that it will provide reliable information on the past and present status of birds to state and voluntary conservation agencies, planners, environmental consultants, service providers, industrialists, the fishing community, farmers, and other land and resource users. Most of all we hope it will provide a record of species lost and gained, what riches we currently have, what changes may take place in the future and what could be lost through insensitive use of our environment (Fig. 1.7).

Fig. 1.7 Chough; occurs along the entire coast, and inland in the south-west, but is sensitive to insecticide use and habitat change.

Chapter 2

Cork Environment

Galley Head

Location

County Cork lies at the south-west corner of Ireland. It is the southernmost of the island's thirty-two counties and is also the largest, with an area of about 7,500 km^2 (8.9 per cent of the area of the whole island of Ireland). It is bounded by the Atlantic Ocean to the south and south-west, Kerry to the north-west, Limerick to the north, and Tipperary and Waterford to the north-east and east. The most westerly, southerly and easterly points are Bull Rock, Fastnet Rock and Youghal, respectively. It measures 171 km from west to east, and 110 km from south to north. However, because of its indented coastline, particularly in the west, the total coastline length, including the islands, runs to about 1,100 km.

Climate

Due to the influence of the Atlantic Ocean the climate of Cork is markedly oceanic, dominated throughout the year by westerly winds winds and rain-bearing frontal depressions. These depressions move across the Atlantic, sometimes in as little as twenty-four hours, and in autumn may carry exhausted migrant passerines and waders from North America. Parts of the west coast may experience forty days of westerly gales per annum. Such conditions are conducive to the appearance close inshore of shearwaters and other seabirds.

The air is clear and unpolluted unless easterly winds bring smog from Britain and the continent, as sometimes happens during established anti-cyclones in spring. The air is always humid, and measurable rainfall occurs on approximately two-thirds of the days. The mean January temperature is generally above 4°C and surface waters seldom freeze for long periods. For this reason, Cork is extremely important as a wintering area for waterbirds, and as a refuge for birds forced westwards by hard weather conditions in Britain and Europe, although this is changing due to climate warming. Soils usually remain unfrozen in winter providing optimal feeding conditions for plovers and thrushes, which may occur in large numbers. Summers are generally rather wet and warm, but are seldom hot, and a mean of 15°C limits the number of insect species on which some birds depend.

Geology and topography

Past geological events have had a fundamental influence in shaping the Cork landscape. The present topography, in turn, plays an important role in determining distribution of animals, none more so than birds. This account is intended as a non-technical summary of how different parts of the county took on their present distinctive characteristics.

Most of Cork's bedrock was laid down in the Devonian and Carboniferous periods (415–360 and 360–300 million years ago, respectively). In Devonian times, Ireland lay on the southern margin of a landmass called Laurasia, about mid-way between European Russia in the east and North America in the west.[1] Much of the area that is now County Cork was then a depression in the Earth's crust, a sedimentary basin termed the Munster Basin by geologists. This was a major east–west feature which probably originated in the Middle Devonian. The landmass of Gondwanaland lay to the south, separated from Laurasia by the proto-Tethys Ocean.[2] Deposition of sedimentary rocks within the Munster Basin, and the relative movements of the two landmasses (Laurasia and Gondwanaland), are the factors responsible for much of the present-day landscape.

In the late Devonian, the southern shoreline and coastal plains of Laurasia were in the southern tropics, and the interior hinterland and its lake basins were arid and hot. To the east, north and south-west of the Munster Basin were mountains. Rivers flowed from these onto the wide plains to the south, and from there to the sea, carrying with them eroded sediment. On occasion, the rivers flooded outwards and left a thin spread of mud and silt on the plains. Considerable depths of non-marine sediment (known as Old Red Sandstone because of its largely red colour) accumulated locally, as much as 6 or 7 km in the west of the county.[3] While this was happening, the basin was continually subsiding. In the late Devonian, the sea broke through from the south-east and flooded the area. At about the same time, the locus of greatest subsidence shifted from west Cork to south Cork, and a sub-basin, termed the South Munster Basin, developed south of a line from the Kenmare River to Cork Harbour.[4]

Since the sea was now covering the area, the distinctively red terrestrial sandstones of the mid–late Devonian were overlain by green and grey shallow marine silts and sandstones, and then in the early Carboniferous period by thin-bedded deep-water mudrocks as water levels in the South Munster Basin deepened. In contrast, as the waters began to encroach on the land to the north (the northern part of the Munster Basin) a shallow warm tropical sea was established in which corals and other marine organisms thrived. This area thus became the site of dominantly limestone sedimentation throughout the Lower Carboniferous.[5]

At the end of the Carboniferous period, a progressive closure of the proto-Tethys Ocean occurred as Laurasia and Gondwanaland collided, causing a major mountain-building episode (the Hercynian Orogeny). This uplifted the siltstones, mudstones, sandstones and limestones and folded them together along east–west axes into a series of anticlines and synclines. The modern landscape features owe virtually everything to the effects of this folding. The Old Red Sandstones are exposed along the crests of east–west ridges, and Carboniferous rocks outcrop in valley bottoms. An overall west–east tectonic tilt, combined with the original geographical distribution of limestones, ensures that in the east of the county these valley bottoms are lined with fertile limestones. Larger rivers are mainly associated with the synclinal axes, and flow eastward in response to the overall west–east tectonic tilting. Only near to their estuaries do the rivers show a marked disregard for the structure.

Over most of the county, but more so in the east, the solid bedrock is now largely concealed by drift, Ice Age deposits of boulder clay, sands and gravels sorted from it by meltwaters flowing from long-vanished glaciers. Westwards, where the land is craggy and mountainous, the bedrock is increasingly exposed. The landscape of west Cork is partly submerged. The deep-sea inlets of Roaringwater, Dunmanus and Bantry Bays are rias, or drowned river valleys. The western islands are what Pochin-Mould called 'the heights of a county lost to the ocean'.[6]

A lyrical overview of the county and its landscapes, worth repeating here, was given by Pochin-Mould. It can readily be related to the geological picture outlined above.

> In the north are heathy uplands, now widely afforested, the Nagles and Boggeragh Mountains and high cold uplands around them; to the south and east is rolling farmland; west, the country roughens, flat pastures and wide cornfields changing to small mountainy crofts, the rocky bones of the land breaking through the thin skin of soil.
>
> To the east, the coast has a certain plainness, a directness of line that is entirely lost in the west. Here the rias, drowned valleys, finger deep inland in long narrow bays, and the sea is set with islands like gems,

green and magical in a shimmering ocean; Carbery's Hundred Isles; the Fastnet Rock like a black swan sailing on a magic sea.[7]

The divisions recognised here form a useful framework for discussion, and in the following account the county is considered under four simple headings: the western peninsulas, the south coast and river valleys, the eastern and central farmlands, and the northern hills.

The western peninsulas

The coast of west Cork consists of a series of peninsulas separated by drowned river valleys, all of which owe their origin to Hercynian folding and all of which are composed of erosion-resistant sandstones of Devonian age. Their position, jutting out into the Atlantic, makes them the first landfall for disoriented migrants in spring and autumn, while their jagged coastlines harbour small seabird colonies and the bulk of Cork's Chough population. The backbone of the Beara peninsula, extending in a rocky ridge to Crow Head and Dursey Island, and beyond to the Bull and Cow Rocks, the most important seabird colonies in the county, is made up of a tightly folded anticlinorium of Old Red Sandstone which forms the summits of the Caha and Slieve Miskish Mountains (Fig. 2.1). The sandstones dip steeply to both

Fig. 2.1 Dursey Island, looking east; an important Chough habitat, and has a great record for vagrants, especially in autumn.

Fig. 2.2 European Stonechat; mostly a coastal species which extends its range inland following a series of mild winters.

north and south away from the main fold axis to pass beneath flanking Carboniferous rocks which interject rocky promontories and islands into the waters of Kenmare and Bantry Bays. These Carboniferous projections (such as Bere Island and Kilcatherine promontory) are of lower elevation and have more easily tilled soils than areas of Old Red Sandstone and it is on these that much of the improved land and most of the settlement is found. Some of the headlands between Dursey Head and Bere Island are formed from resistant volcanic rocks interbedded within the Carboniferous succession. Elsewhere, the landscape is one of furzy heath and rocky knolls, empty fields and ruined cottages, where the coastline consists of sandstone. Less than one-third of the peninsula is under improved land, and European Stonechats, Meadow Pipits and (in summer) Northern Wheatears are typical species (Fig. 2.2). Although rising to more than 600 m, most of the higher summits are featureless peat-covered domes, with few intervening gaps apart from the Healy Pass.

Sheep's Head, with its ice-scrubbed spine of sandstone hills, is (except for its westernmost tip) a simple anticlinal structure. It therefore contrasts sharply with the Mizen peninsula, which is a complicated sequence of folded Devonian and Carboniferous rocks. While its highest summit (Mount Gabriel; 407 m) is composed of Old Red Sandstone, the northern hills are Lower Carboniferous in age. Mizen Head itself has sometimes been called the Land's End of Ireland since it is the most south-west point of the Irish mainland. It is an ideal spot for seawatching in autumn, particularly when south-west winds combine with rain and poor visibility to bring large numbers of seabirds (including the larger shearwaters) close inshore. The cliffs of Mizen Head are formed of hundreds of feet of close-bedded, gently folded sandstones and shales, and have some small seabird colonies. Because of its lower elevation and better soils, the Mizen peninsula appears less barren than its more northerly neighbours, since over half of its area is given over to crops and pasture. Many of the gardens on Mizen Head are worth searching for small passerines in spring, and especially autumn, while Lissagriffin Lake is one of the best sites in Ireland for American waders. The beach at Barley Cove is also the finest for miles around. Farmers from all over west Cork formerly collected sand from here because of its reputed agricultural value, but this ancient privilege has long since been withdrawn. Although there are good agricultural lands, particularly around Skibbereen, farms tend to be rather small. Tourism, not farming, is now prominent in the economy of this part of west Cork.

Roaringwater Bay is the southernmost of the drowned river valleys of west Cork. The bay has been carved from an intricate synclinorium of Devonian and Carboniferous sandstones and slates lying between the major anticlines of Mizen Head and Cape Clear Island, resulting in a tattered coastline and an archipelago of small islands. Most of the smaller islands in Roaringwater Bay are now deserted, and some hold small colonies of terns and gulls. The rocky and windswept eminences of Sherkin and Cape Clear Islands dominate the bay and were formed when the sea breached the sandstone ridge of the Cape Clear peninsula in two places (Fig. 2.3). Further out to

Fig. 2.3 Cape Clear Island; the site of a bird observatory since 1959, and one of the best sites in Europe for large seabird passage movements.

sea, the Fastnet Rock lighthouse stands on the last unconsumed remnant of the slowly disintegrating peninsula. There has been a bird observatory at Cape Clear since 1959. Though the volume of passerine migration is small, rarities are regular and include overshooting European species in spring and Siberian and American vagrants in autumn. Seawatching can be spectacular, and it is arguably the best location in Europe to be during a south-west gale in autumn. Due to its position, Sherkin Island is less ideally situated for seabird migration but should be almost as good for land birds, as has been shown by observations made by workers at Sherkin Island Marine Station.

Due to the sandstone bedrock and peaty soils of much of west Cork, the abundant small lakes in this part of the county tend to be oligotrophic and poor in both breeding and wintering birds. In summer they often hold just the occasional pair of Mute Swans or Moorhens; in winter they may be almost completely without birds. This was graphically demonstrated by a waterbird count of several lakes on the Beara peninsula in January 1994, which produced a total of only twenty-nine individuals of eight species. Only thirteen of these were ducks, and not one bird was found on the Cork side of the border.

The south coast and river valleys

From Cape Clear to Cork Harbour, the southern coast of Cork is roughly aligned with the axes of folding, running approximately west–east. Although there are many headlands and bays, this coast lacks the gigantic rias and peninsulas of the Atlantic coast. From Baltimore to east of Rosscarbery the coastline has been carved out of the southern limb of the Clonakilty anticline, but the remnants of Carboniferous slate which survive on the two projecting promontories of Toe Head and Galley Head (both excellent for small passerines and seabird migration) suggest that the Carboniferous rocks of another anticline lie not far offshore. Along most of this southern coast the bays and estuaries run directly across the strike of the rocks, but this may, in part, have been controlled (or facilitated) by the presence of numerous north–south faults. The muddy estuaries at Rosscarbery and Clonakilty Bay appear to be fault controlled. Both hold important concentrations of waders in winter, Common Ringed Plover at the former, and Common Shelduck, Common Ringed Plover, European

Fig. 2.4 Clonakilty Bay; an important estuary for waders, especially European Golden Plover and Black-tailed Godwit.

Golden Plover, Grey Plover and Black-tailed Godwit at the latter (Fig. 2.4). Beyond Clonakilty Bay and the sandstone promontories of Seven Heads, the broad bay of Courtmacsherry has been carved from Carboniferous slates. Courtmacsherry Bay is important for European Golden Plover and Black-tailed Godwit in winter.

The Old Red Sandstone dies away at Seven Heads, not reappearing on the coast until Cork Harbour is reached to the east. The eastern end of Courtmacsherry Bay is closed by Old Head of Kinsale, a bony finger of slates and grits projecting some 5 km from the main coastline (Fig. 2.5). Because of its easy access from Cork city, Old Head of Kinsale receives reasonably regular coverage by birdwatchers. Seawatching can be very good, though seabirds are rarely seen in such numbers, or as close inshore, as at more westerly points, but the many small gardens provide welcome cover for tired passerine and near-passerine migrants. A small marsh at Garrettstown is also attractive to several species, as are the nearby sandy beaches.

To the north of Old Head of Kinsale is the drowned valley of the River Bandon, a prime example of the way in which Cork's major rivers show a marked disregard for geological structure in their lower reaches. The Bandon, having followed an easterly course along a synclinal axis from Dunmanway to Inishannon, turns abruptly south and flows across the

Fig. 2.5 Old Head of Kinsale; site of an important colony of Kittiwake, Razorbill and Common Guillemot, along with Chough.

Fig. 2.6 Great Island Channel; one of the most important sectors of Cork Harbour for waterbirds, showing the large areas that have been colonised by the invasive Cord-grass.

topographical grain. The Bandon estuary holds important concentrations of Black-tailed Godwit and Common Redshank.

The eastern and central farmlands

For 35 km east of Kinsale (past Barry's Head, Robert's Head and Power Head) the coast is formed of Carboniferous slates, but younger rocks (the Kiltorcan Sandstones) outcrop east of Power Head. The sea has breached the Kiltorcan Sandstones of the Ballycotton anticline at Roche's Point to reach the less resistant Carboniferous limestones and slates of the neighbouring Cloyne syncline, allowing the sea to flood a broad limestone valley and create the magnificent land-locked basin of Cork Harbour, a highly complex estuary which forms the most important site for wintering waterbirds in the county. Islands in the harbour represent hill tops which survived the general submergence (Fig. 2.6). The steep sandstone ridge which now forms Great Island has been broken at two places, east and west of Cobh, by former river valleys now inundated by the sea at East Ferry and Passage West. To the north, the limestones of the Cork to Castlemartyr syncline have been lowered by erosion to form a broad valley reaching from Cork city to the sea at Youghal Bay. Into this lowland the marine submergence has infiltrated its waters to form the picturesque Lough Mahon, thereby isolating Great Island. The limestone valley between Cork city and Castlemartyr broadens eastwards to Youghal Bay and looks as if it ought to be the natural outlet of the River Lee into St George's Channel, but like the Rivers Bandon and Owenboy the Lee swings southwards, ignoring the Castlemartyr limestone corridor and preferring to follow a more difficult course through the sandstone ridge at Passage West and thence to the sea via Cork Harbour. The south-flowing River Owennacurra crosses the Castlemartyr syncline at Midleton and equally perversely follows the East Ferry breach in the sandstone anticline before reaching Cork Harbour.

The variety of habitats provided by the different basins, and the enrichment from the silt and sewage brought down by the River Lee, have made Cork Harbour one of the few sites in the country that regularly supports more than 20,000 waders. In December 1985 Northern Lapwing numbers alone totalled 27,963. Because of these large numbers, and the concentrations of Black-tailed Godwit and Common Redshank, the area is

Fig. 2.7 Black-tailed Godwit; Cork Harbour is the most important site in Cork for this species, but high numbers occur at many estuaries between Rosscarbery and Youghal.

recognised as being of international importance (Fig. 2.7). In addition, up to twenty further species occur here in nationally important numbers. Much of Cork's industry is centred around Cork Harbour, which boasts chemical and pharmaceutical plants, an oil refinery at Whitegate with oil storage tanks, and a jetty at Corkbeg Island. Aghada and Whitegate each have power-generation plants, and some harbour industries use wind turbines to augment their power demand. The area is also an important seaport, with Ringaskiddy and Tivoli being the main sites, while Marino Point will be developed as a port facility in the future.

As the harbour holds such large numbers of waterbirds in winter there is potential for conflict between industrialists and environmentalists. Since the 1950s several large areas of mudflat have been reclaimed for industrial, port-related and road-building purposes. Reclamation, industrial and urban development, pollution and spread of cord-grass have been cited as threats to the area. Bird populations in Cork Harbour are currently carefully monitored by the Cork branch of BWI and the National Parks and Wildlife Service (NPWS), who are prepared to challenge potentially damaging developments and to seek mitigation measures. There are several BWI reserves, and Special Protection Areas (SPA) designated under the European Union Birds Directive (see Appendix 4). Harper's Island, situated within one of the designated areas, was purchased by Cork County Council in the 1990s as part of the N25 road scheme and is managed for wildlife by a steering group representing statutory and conservation agencies as well as a local community association.

The River Lee formerly divided into several branching distributary channels as its waters reached the tidal limits of the estuary. Here, because its velocity was checked, it deposited a large volume of sediment. These channels were covered over to accommodate the spacious streets required by eighteenth-century planners of Cork city. More than one-third of Cork's population now live in Cork city or its suburbs.

Further east lie the important wetland areas of Ballycotton and Ballymacoda. Both areas have a bedrock of Carboniferous limestone and are surrounded by fertile farmland. The habitat at Ballycotton is a series of muddy coastal lagoons flanked in the south-west corner by reedbeds. The lagoons are extremely attractive to waterbirds, both on passage and in winter, and the area holds important concentrations of several species. Because of the limestone bedrock, the water is alkaline and eutrophic, allowing the development of a rich invertebrate fauna which in turn provides sustenance for waders. However, draining of the lagoons has caused a marked decline in the number of species such as Little Grebe, Common Coot, Whooper Swan, Bewick's Swan, Gadwall and Shoveler, and a change in distribution of others within the site (Figs 2.8 and 2.9).

Nearby Ballymacoda, the estuary of the River Womanagh (a broad area of intertidal mudflats surrounded by low-lying marshy fields), has important concentrations of over ten waterbird species. Two of these (European Golden Plover and Black-tailed Godwit) regularly occur in internationally important concentrations. Northern Lapwing numbers can also be very high at times, and a flock of up to 300 Grey Plover is a speciality in spring (Fig. 2.10).

This account has concentrated so far on the coast of the county, but two inland sites, both along the River Lee valley, are also extremely important. The Gearagh was formerly a braided river system rivalling that formed by the River Lee as it entered Lough Mahon. In the 1950s, however, this

Fig. 2.8 Ballycotton; pre-1990 was an important mostly freshwater habitat for many waterbird species, including swans, ducks and Common Coot.

Fig. 2.9 Ballycotton; post-1990 following breaching of the shingle barrier, many waterbirds declined or disappeared, while others changed their distribution within the habitat.

Fig. 2.10 Ballymacoda; the estuary of the River Womanagh, along with Pilmore and Ring Strands, is an important habitat for many waterbird species, especially European Golden Plover and Black-tailed Godwit.

Fig. 2.11 The Gearagh; suffered damage due to the flooding of the River Lee in the 1950s for hydro-electric power generation, one of the most important inland Cork wetlands.

unique landscape of wooded islets was partly lost beneath the waters of a 27 km long lake, dammed to provide water for two modern hydro-electric power stations.[8;9] Upstream of Inishcarra, only Lough Allua is a natural waterway. The combined area of the Gearagh (now a nature reserve) and Inishcarra reservoir holds nationally important numbers of Common Wigeon, Common Teal, Mallard, Common Coot and European Golden Plover (Fig. 2.11).

The northern hills

North of the River Lee valley is a line of sandstone mountains. The Boggeragh (644 m) and Nagles Mountains (428 m) form a barrier between the prosperous farmlands of the River Lee and those of its northern neighbour, the River Blackwater. This line of mountains, their thickly wooded slopes rising darkly above the southern margin of the Blackwater valley, represents the northern limit of severe Hercynian folding (a line that extends from Dingle Bay in the west to Dungarvan Harbour in the east) (Fig. 2.12). On its northern edge, the Blackwater valley is fringed by the Kilworth (283 m) and Ballyhoura Mountains (528 m), forming an area of relatively bleak uplands with poor soils. Comparatively little is known about the birds of

Fig. 2.12 Upland forestry; extensive upland areas, such as at Nagles Mountains, have been planted with non-native conifer trees.

these northern hills. They are the stronghold of such upland or moorland species as Red Grouse and Hen Harrier, both known to be declining. A major cause is likely to be the gradual fragmentation of habitat as many of the slopes are either forested or encroached upon by farming.

Like the River Lee, the Blackwater follows an easterly course for much of its length. Between Mallow and Fermoy, it receives several important tributaries from the north, and together these have lowered the limestone tract which lies between the encircling sandstone uplands of the Nagles, Kilworth and Ballyhoura Mountains. Between Fermoy and Cappoquin this lower ground, though confined to a relatively narrow valley, forms a floodplain known as the Blackwater callows. These callows hold internationally important numbers of Whooper Swan and Black-tailed Godwit, and nationally important numbers of Common Wigeon and Common Teal. The estuary of the River Blackwater holds nationally important concentrations of several species, although these are mostly within the adjoining county of Waterford, as are much of the Blackwater callows.

Kilcolman, lying 5 km north-east of Buttevant on the southern slopes of the Ballyhoura Mountains range, is a fenland occupying a glacially eroded limestone hollow which has been managed as a refuge since 1969 (Fig. 2.13). The regular small flocks of wild swans and Greater

Fig. 2.13 Kilcolman fen; the finest inland fen habitat in Cork, hosting important numbers of Whooper Swan and various duck species.

White-fronted Geese have been intensively studied, and the site also has nationally important numbers of several other duck species. Further north, on the Cork/Limerick border, are Charleville lagoons. Although this site has a nationally important concentration of Shoveler, it is perhaps most notable as one of the best inland sites in the county for waders. Sizeable numbers of Dunlin often occur, and the lagoons are regularly used as a wintering site by Green Sandpiper.

Chapter 3

Ornithology and Ornithologists

Ballymacoda, with Common Curlews

A brief history of Cork ornithology

Although records of birds in County Cork can be traced to the 1700s, it was not until the 1900s that any detail regarding populations and distribution emerged. The small number of resident observers and the lack of a systematic recording system meant that many species were poorly documented, even as recently as the 1950s. It was impossible to detect population changes until they were already well underway. Existing species may have declined significantly, or new colonists arrived undetected, leading to an unknown level of imprecision in assessment of past status.[1] However, careful perusal of the available evidence provides some general information about past events, such as the extinction of the eagles and other large birds of prey as breeding species, the arrival and eventual breeding of species such as the Fulmar and Mistle Thrush, and the ubiquity of the Corncrake and Corn Bunting, now both extinct as breeders. The situation began to change dramatically from the 1950s onwards, and this can be traced to seminal events, such as the establishment of the annual *Irish Bird Report* in 1953, the Cape Clear Bird Observatory in 1959 and the Cork Ornithological Society (COS) in 1961. Here, we broadly chronologically review the annals of Cork ornithology as gleaned from the scattered and sometimes fragmentary published record.

Giraldus Cambrensis, the Welsh monk who visited Ireland in the twelfth century, is credited with being Ireland's first ornithologist. He travelled in Cork and Waterford in 1183 and recorded his observations which included an account of Irish birds.[2] There is, however, little or nothing in this account that can be related directly to Cork. It should be noted that his natural history writing has been criticised as unreliable.[3] However, it is generally considered an important, if somewhat exaggerated, early natural history document.

While the present book is the first on the status of the birds of County Cork, it is not the first attempt at a list of the birds of the county. Charles Smith, from Waterford, published the first list in 1750. Smith's list effectively signals the beginning of scientific ornithology in the county and runs to 100 acceptable species, although it has many omissions and other problems. Hardly anything else of note was published on the birds of the county during the next ninety-five years until Joshua Reuben Harvey produced another list of the birds, containing 164 species, for a meeting of the British Association for the Advancement of Science held at Cork in 1843, the

Fig. 3.1 Mixed flock of small waders; Dunlin, Sanderling, Common Ringed Plover and Turnstone are found along the coast in mudflat and sandy habitats.

proceedings of which were published in 1845. Another edition of Harvey's list was published in 1875 and contained 182 species. These lists are terse and lack detail, but they constitute important publications for their time. The first comprehensive treatment of the ornithology of Ireland appeared in the middle of the nineteenth century. In this work, William Thompson, from Belfast, included records of Cork birds communicated to him by Harvey, the Parker brothers, Robert Ball, William Crawford, Samuel Moss, Joseph Stopford, and others either resident or born in Cork.

The study of Cork's avifauna was patchy throughout the second half of the nineteenth century. Gentlemen shooters added several species to the Cork List by obtaining specimens. Ralph Payne-Gallwey shot extensively in Cork Harbour and added a number of new species from there and elsewhere. Some idea of the scale of shooting at the time can be had from his comment that 'before the era of steam in Cork Harbour, now overrun with shooters, a gentleman fowler, living in that locality, was on one occasion forced to throw birds overboard (though afterwards picked up) to avoid sinking in his single-handed punt'.[4] Some people, like Payne-Gallwey, made their living from shooting and sending birds to the markets in Cork, Dublin, London and elsewhere. Others shot for pleasure, like William Crawford, the wealthy brewer and philanthropist.[5] Crawford shot regularly in Cork Harbour and once killed sixty Curlew Sandpiper and ten Dunlin in one shot from his large strand-gun; he also shot an Osprey at his home at Lakelands on the Mahon peninsula in October 1848 (Fig. 3.1). It should be said that it was normal practice then in ornithology, and well into the twentieth century, to verify records of rare and unusual birds by the preservation of specimens. The saying that 'what was shot was history, what was missed was mystery' was very apt at the time.

This was also an era when the capturing of wildfowl in duck decoys for the market was popular. Two decoys were built in County Cork and details are provided in Appendix 3.

The Royal Cork Institution (RCI) (1807–61) and its offspring, the Cork Cuvierian Society (CCS) (1835–78), were learned societies with scientific libraries in the days before Queen's College Cork (later University College Cork) (UCC). The RCI (and CCS) regularly held public lectures in the city to satisfy a demand for academic learning.[6] It was through the influence of the RCI that the British Association for the Advancement of Science held its meeting at Cork in 1843, and the CCS published the proceedings. There had been a zoology department (natural history to 1909) at UCC since 1849, but there were no published studies on ornithology up to the time when Professor Fergus J. O'Rourke held the chair (1955–81).

Other activities in the late nineteenth century centred on the late Victorian trend of forming and contributing to natural history societies. The Cork Naturalists' Field Club, founded in 1892, never stimulated much interest in birds, and in 1923 handed over its effects to the Cork Camera and Field Club. The Cork Historical and Archaeological Society (founded in 1891) was more influential, its *Journal* was an outlet for published notes on natural history, and it survives to the present day.

Richard John Ussher, also from Waterford, published the next list of the birds of Cork as a series of sixteen articles in the *Journal of the Cork Historical and Archaeological Society* during 1892–4. It is hard to believe, but true, that no comprehensive account of all birds of County Cork has been published in any format since then, a period of 130 years. The articles by Ussher brought the Cork List to 215 species, but sadly documented the loss of the Great Bittern, White-tailed Eagle, Golden Eagle, Marsh Harrier, European Golden Plover, Woodlark and Jay as breeding species through a combination of persecution and land reclamation (Fig. 3.2). The Jay is the only one of these species that continues to breed regularly in Cork in a wild state, having recolonised about 1915 (see Chapter 7).

Richard Ussher, along with Robert Warren, published an important book on the status of birds in Ireland in 1900, and Richard Barrington published the first authoritative work on bird migration in the same year.

Fig. 3.2 White-tailed Eagle; extinct for more than a century, has recently returned as a breeding species to the south-west after a re-introduction programme.

Both publications comprehensively cover the last quarter of the nineteenth century, and include much data relating to Cork. Barrington used data collected by lighthouse-keepers around the Irish coast, including those at Bull Rock, Fastnet Rock, Old Head of Kinsale and Ballycotton. His pioneering work formed the basis of our knowledge of migration during the first half of the twentieth century.

The early decades of the twentieth century saw a decline in the dynamism evident among Irish ornithologists of the nineteenth century; understandable perhaps as several of the key figures died within a few years of each other (Ussher, Warren, Barrington), the world became embroiled in two wars and Ireland suffered much political strife generated during the establishment of an independent state. The period 1900–54 was one of comparative inactivity in Irish ornithology. The *Irish Naturalist* and its successor from 1925, the *Irish Naturalists' Journal*, provided the only Irish outlets for observations apart from occasional notes in the London-based *Zoologist* and *British Birds*, the former incorporated into the latter in 1916. Records of shot rare birds, rather than field observations, formed the basis for most published notes. Several names recur, including those of W. Abbott, J.E. Flynn, J. Glanville, E. O'Donovan, B. O'Regan and F.R. Rohu. Frederick Raynor Rohu was born in Donegal in 1846, became a lighthouse-keeper in 1866 and served at Old Head of Kinsale; following resignation he settled in Cork city and established a furrier and taxidermy business at 72 Grand Parade. He shot Ireland's first Rufous Bush Robin and set up Ireland's only Sandhill Crane.

The 1950s opened with two landmark events in Irish ornithology. The first was the publication of a new standard work fifty-four years after the brilliant tome of Ussher and Warren. The writing of this new work was led by Reverend Patrick Kennedy, and it included much new data on Cork ornithology. The second was the publication of the first *Irish Bird Report*, edited for the next nineteen years by the Major (Robert Ruttledge), before he handed over to Ken Preston of Cork in 1972. The *Irish Bird Report* appeared annually for twenty-three years until replaced in 1977 by a new journal, *Irish Birds*, which today continues to serve Irish ornithologists as a forum for publication of their work. Both events proved important spurs to ornithological activity and signalled the start of regular and systematic observations, recording and publication. This was an exciting time, relatively speaking, although Ireland was very much in the economic doldrums.

There were island bird observatories in three corners of Ireland by the late 1950s, Great Saltee (Wexford), Copeland (Down) and Tory (Donegal). The gap in the south-west was obvious. Cape Clear Bird Observatory was conceived on a stormy day on Great Saltee. In conditions too wet and windy for birdwatching, Ken Williamson (migration research officer with the British Trust for Ornithology) and Major Ruttledge sat down with a map and picked out Cape Clear as a likely spot for the missing observatory. Major Ruttledge spent four days on the island in April 1959 but was discouraged by its size, the vast amount of cover, and the fact that he saw very few birds. The enterprise might have ended there were it not for the fact that in August of that year Tim Sharrock and four other young English birdwatchers (L. Cornwallis, B.H.B. Dickinson, H.M. Dobinson and M.F. Seddon) arrived to work the island. Sharrock remained until November, and later wrote that in the eleven weeks of coverage that autumn '17 species regarded at that time as major rarities in Ireland were seen ... sea passage on a previously unexpected scale was discovered; and massive diurnal migration was observed. The expedition to test the potential of Cape Clear Island as a possible observatory site was a resounding success.'[7]

During January–March 1961 Professor Denis Gwynn contributed a regular column to the *Cork Examiner* newspaper, and many of these articles were concerned with birdwatching in Cork. He repeated several times his opinion that the time was right for the establishment of a club or society devoted to nature study or birdwatching. He was astonished at the volume of correspondence he received as a result, and he eventually persuaded a young Cork man, Tadg O'Keeffe, to attempt establishing a society for birdwatchers. In November 1961, O'Keeffe organised a meeting in Desmond's Hotel, Cork city. Liam O'Flynn, a founder member and later a driving force behind this venture, recounts the following:

> When I arrived there at 8 p.m. I found a crowd of some 30–40 people assembled, of all ages and apparently drawn from many different walks of life. I later discovered that some had travelled from as far away as Skibbereen, Union Hall, Ballycotton, Mallow and Bandon.
>
> There was nobody in the room that I knew, and I found, somewhat to my surprise, that Tadg O'Keeffe, who had convened the meeting, was a young man who appeared to be in his late 'teens. He seemed to be a

> shy diffident person and when the meeting proper began it was another young man, Liam Horgan, and Michael Roche who got things under way. A lively discussion developed with contributions from, among others, John B. Jermyn, Dr Saunders M.O.H., Mr John Kelly F.R.C.S., Paddy O'Flynn and, because I never knew when to stay quiet, myself. One amusing incident occurred. Having decided that a society or association for birdwatchers should be formed, the meeting proceeded to discuss the question of a suitable name. Many would have settled for the *Cork Bird Watching Club* or some such title, but a brash young man whose name I never discovered, declared loftily that this was too mundane and suggested instead the more pretentious title of the *Cork Ornithological Society*. Despite some misgivings this was eventually agreed to. The joke was that the young man who had saddled us with this rather unwieldy name never appeared again at any meeting.[8]

The COS held monthly indoor meetings during September–May, at the Rob Roy pub in Cook Street. The society had a steady membership of 40–50, with an annual subscription of £1 (ten shillings for students). For the first few years of its existence COS contributed little of significance to Irish ornithology. Even its founding members were very much beginners, and a number were happy to remain so. Others, however, developed their knowledge very rapidly and the volume of contributions from Cork to the *Irish Bird Report* increased annually. The first *Cork Bird Report* was published by COS in 1963, under the editorship of Liam O'Flynn. This was Ireland's first ever county bird report and it had a total of seven contributors. O'Flynn continued as editor until 1965, with Tadg O'Keeffe taking over from 1966 to 1968. Over this time the quality and size of the report steadily improved as observations became more numerous. One lasting result of this activity was that it provided a focus for many active and dedicated young birdwatchers 'who were schoolboys in the sixties but who blossomed forth in the seventies as some of the ablest ornithologists in Ireland'.[9] At the same time, news of the massive seabird movements which could be witnessed at Cape Clear was generating international interest, and a wide range of experienced birdwatchers began coming to Cork. Clive Hutchinson later wrote of this period:

> … perhaps the greatest long-term value of the observatory network was the cross-fertilization fostered by the mix of relatively experienced British and novice Irish birdwatchers, and the opportunities provided for Irish birdwatchers to learn the techniques of bird-ringing and the use of mist-nets to catch birds.
>
> Well-known ornithologists … such as Ken Preston, Killian Mullarney and Anthony McGeehan, developed their skills at Cape Clear. The presence of ornithologists at island watchpoints who knew how to keep log-books, record data and make descriptions of birds was a marvellous educating factor. The collection of data and the encouragement to analyse it stimulated many of us in later studies.[10]

Ballycotton found fame during the 1960s as a site for recording American waders, and an observatory operated there in the autumns of 1968 and 1969 (Fig. 3.3). Lissagriffin, Clonakilty and Old Head of Kinsale were also becoming better known as sites worth watching, and Richard and Margaret

Fig. 3.3 Lesser Yellowlegs; a vagrant from North America, occurring mainly in autumn in estuarine and coastal lagoon habitats.

Fig. 3.4 Greenland White-fronted Goose; a wintering flock formerly occurred at Kilcolman fen in north Cork, but it is now seen mainly as an autumn passage visitor.

Ridgway purchased Kilcolman Bog, the first step in the rehabilitation and transformation of the fenland into a nature reserve (Fig. 3.4). The full story of how this was achieved is recounted by Ridgway and Hutchinson.[11]

Fergus O'Rourke (then professor of zoology at UCC), who served as vice-president of COS, published a useful introduction to the fauna of Ireland in 1970, including the birds.[12] This era saw the beginnings of bird study at UCC under lecturers F. O'Gorman and G.A. Walton, and students such as A.J.M. Claassens and T.C. Kelly, both of whom studied bird ectoparasites. Dr Tom Kelly later became a lecturer at UCC, and with Professors Maire Mulcahy, Alan Myers, Paul Giller, John Davenport and John O'Halloran (current president of UCC), and others, enabled generations of students to study ornithology, many now holding teaching and decision-making positions in universities, government and conservation agencies, environmental consultancies, and industry, and in the process making UCC an important ornithological research institute.

In this era of intense activity and discovery several other developments were taking place at national level and were to significantly affect the birdwatching scene in Cork. Two major surveys began in 1968 and 1969, respectively: the first *Breeding Atlas* organised by Tim Sharrock, the founder of Cape Clear Bird Observatory, and the Seabird Group's survey of breeding seabirds, dubbed *Operation Seafarer*, and organised in Ireland by Oscar Merne. These involved a great deal of co-operation and promoted focused scientific enquiry, as well as generating vast amounts of detailed information. While these surveys were still ongoing, the first systematic censusing of Irish coastal and inland wetlands began under what was to become the *Wetlands Enquiry* organised by Clive Hutchinson. It is fair to say that in the late 1960s and early 1970s more was achieved and learned about Irish (and Cork) birds than in the previous 200 years.

Liam O'Flynn represented COS in 1965 at a meeting in Dublin which eventually formed the Irish Wildbird Conservancy (IWC) in December 1968 (now BWI). In April 1971 the COS voted itself out of existence and became the Cork branch of IWC. In the same year the *Cork Bird Report* made its last appearance having been published annually since 1963. However, a meeting organised by Tom Kelly in 1975 revived the *Cork Bird Report* under the editorship of Seán Fleming and a committee; it published regularly between 1976 and 2006 but has not appeared since the latter year (Table 3.1). The Cork branch of IWC published a regular series of articles on birds and birdwatching in the *Cork Examiner* newspaper

Issue(s)	Editor(s)
1963, 1964, 1965	Liam (W.J.) O'Flynn
1966, 1967, 1968	Tadg O'Keeffe
1969, 1970, 1971	Ken Preston
1976, 1977	Seán Fleming
1978, 1979, 1980, 1981	John O'Halloran
1982	Barry O'Mahony
1983	Jim Wilson and Seán Pierce
1984, 1985, 1986	Jim Wilson
1987–1988	Russ Heselden, Peter Leonard, and Jim Wilson
1989	Russ Heselden and Peter Leonard
1990, 1991	Peter Leonard and Mark Shorten
1992, 1993, 1994, 1995	Mark Shorten and Martin Styles
1996–2004, 2005–2006	Ciarán Cronin, Colin Barton, Harry Hussey, and Mark Carmody

Table 3.1 Publication schedule of the *Cork Bird Report* (1963–2006).

in the 1970s, first under the nom-de-plume 'Flemingo' (Seán Fleming) and later 'Redshank' (Pat Smiddy). Liam O'Flynn later served as national chairman of IWC, and another Cork man, Jim Wilson, has since also filled this role.

In the early years the Cork branch of IWC was affiliated to the Federation of County Cork Gun Clubs. This entitled a member of the Cork branch committee to attend meetings of the federation as an observer. This was an invaluable arrangement as it allowed dialogue between two organisations whose aims might appear to be opposed to each other, although some of the founding members of COS were shooting men, as was the case with several of the founding members of IWC, so the arrangement was a sensible one. The current situation is unclear; several persons who recently held office on the Cork branch of BWI, when asked, thought the connection had been severed some years before. This is a loss, since a line of official communication between the two organisations could prove valuable at a time when many species, including gamebirds, are in decline for various reasons.

Sherkin Island Marine Station was founded in 1975 by Matt and Eileen Murphy, and it continues today under the management of their family. While, as the title implies, it is mainly concerned with the marine from the shore seawards, researchers there have conducted much important work on the botany of Roaringwater and Bantry Bays and produced six *Sherkin Island Bird Reports*. These reports are an invaluable addition to the work of the nearby Cape Clear Bird Observatory, and their importance as an ornithological record is strengthened by the fact that the workers there have not confined their observations to Sherkin Island, but to all the islands of Roaringwater Bay. The marine station has also produced a fine book of photographs of Irish birds taken by Richard T. Mills.[13]

The nearby Lough Hyne has been a marine research station of UCC for nearly 100 years, although relatively little work has been done on the birds of the area apart from an important study of shoreline corvid ecology, details of which can be found in Chapter 7.[14]

The establishment of an ornithological research conference at UCC was an important event. Conferences are held at intervals of approximately five years since the first in 1985, and their main function is to bring together researchers for discussion and dissemination of results. Abstracts submitted to the conference have been published in the journal *Irish Birds*, which forms a useful register of past and ongoing research. The most recent conference was the seventh in the series, held in November 2017.[15]

Recording and publication of results has continued from the 1970s to the present day, with Cork (and Cork-based) ornithologists taking part in all major atlas and other surveys during this time. The Cork branch of BWI has remained active, but with some ups and downs, and a second branch has been formed in west Cork.

Cape Clear Bird Observatory continues, currently under the wardenship of Steve Wing. An updated version of the 1973 edition of *The Natural History of Cape Clear Island* was published in 2020, but much data remains to be analysed.[16] However, the *Cape Clear Bird Observatory Report* lapsed after 1996, although an online version has been produced annually since 2016, but its value as a document of record would be enhanced with a detailed systematic list.

Domestic and European Union legislation and directives are now in place with the purpose of conserving birds, other wildlife, their habitats, and biodiversity. Cork County Council and Cork City Council, as well as being the planning authorities for their respective regions, now have new responsibilities of conservation under these laws and directives. The NPWS has overall responsibility for all aspects of wildlife protection and conservation, and conservation rangers work towards this aim throughout the county.

The study of many aspects of natural history in the 1700s and 1800s was generally the preserve of the wealthy landed gentry or the clergy, who pursued their hobby in an amateur capacity but often at a level approaching full-time study. This changed slowly, and it was not until the latter part of the twentieth century that natural history, and ornithology, became popular among a wider spectrum of the population. Nowadays, the term 'citizen scientist' is used to describe a member of the public who collects data on natural history, often collaboratively with professionals; this is aided by the fact that these data can be logged to websites on the internet, for analysis by professionals. Some surveys and monitoring schemes depend to a large extent on citizen scientists to carry out the fieldwork, after appropriate training. Side by side with the changing profile of the data collectors has come the professional ornithologist often working for a government agency, environmental consultancy or a non-government conservation body, and full-time researchers working at a university. Research on birds in Ireland

and in Cork has reached a new level, with many studies taking place at UCC, and by BWI and other voluntary researchers. Bird-ringing, usually carried out by enthusiastic volunteers such as Barry O'Mahony and Sam Bayley, continues apace and is providing new data on movements, mortality and longevity (see Chapter 4).

In common with many other walks of life, there were few women involved in any aspect of ornithology in the early years. An exception was Cynthia Longfield (of Castlemary), the dragonfly expert, who was contributing records to the *Irish Bird Report* before the COS was formed. Happily, over the last two decades almost as many women as men have joined the ranks of professional ornithology, with several coming through UCC with PhDs.

However, it is inevitable that the twenty-first century will bring new opportunities and new threats to Cork's avifauna. Opportunities may come from changes to land use, or climate change due to global warming. For some species, threats may arise from the same quarters or from short-term industrial or leisure developments. Against this background, the careful scientific monitoring of Cork's dynamic bird populations is a very positive development that will allow early detection of changes. As we hope this book shows, Cork's birdlife is an important natural asset, and a stunning legacy for future generations to enjoy. It deserves the greatest care and preservation.

Ornithologists and their work

Many ornithologists, naturalists and other authors have contributed to an understanding of the birds of Cork over nearly three centuries. As in most walks of life, a small number have made outstanding contributions, while most have contributed smaller amounts of data. The purpose of this section is not to grade the ornithologists of the past, but rather to chronologically acknowledge those who have laid the foundations of ornithology in Cork, as well as those who have maintained, improved and expanded the science in more recent times. It follows that such contributors do not necessarily have to be of Cork origin or birth. Perusal of the bibliography at the end of this book will reveal the large number of past and present contributors to the ornithology of Cork, although it has not been possible to cite every reference to the county. This section deals with the small band of ornithologists who, it is considered, have made the greatest historical contribution to Cork ornithology.

Don Philip O'Sullivan Beare (*c.* 1590–1660)

O'Sullivan Beare wrote the manuscript *Zoilomastix* in 1625–6 while exiled in Spain. This manuscript, written in Latin, has been translated, edited and published by Denis O'Sullivan. The O'Sullivan clan came from Knockgraffon (Tipperary), but were driven from their lands by invading Normans. They settled in two Munster branches: the O'Sullivan Beare in the Beara peninsula, and the O'Sullivan Mór in Kerry. Don Philip was born at Dursey about 1590. Following defeat of the Irish at the Battle of Kinsale (1601), the young Don Philip, his parents and members of the family were exiled in Spain from February 1602.

In the *Zoilomastix* Don Philip set out to refute the statements of Giraldus Cambrensis in his *Topographia Hiberniae*, which he considered derogatory of Ireland.[17] However, Don Philip has himself been criticised for 'acting as a propagandist' and for 'going over the top' in his assessment of the work of Cambrensis, and his 'hatred of the English' may have clouded his judgement; he took Cambrensis to task for describing Ireland as 'uneven and mountainous, soft and watery, wooded and boggy, truly a country deserted without roads and wet', surely a good description of Ireland in the twelfth century and for several centuries afterwards, and partly true even to this day.[18] It could be said that Don Philip's glowing views of Ireland, which include no derogatory remarks, might be somewhat embellished. Don Philip listed seven or eight times more species of birds than Cambrensis, and in doing so he may have been trying to prove that his account was more reliable than that of Cambrensis. However, Cambrensis never set out to produce a definitive list of the birds of Ireland. This could leave Don Philip open to the accusation that he may have included some species unjustifiably. Whenever Don Philip gave a description of a bird (e.g. Great Spotted Woodpecker) he was quoting from Pliny the Elder and Aristotle. He was educated at Compostella (Spain) and so would have had access to the literature available at the time, especially that of the ancient writers. Clearly, he could not be using his own experience since he left Ireland when only about twelve years old. It is believed that Don Philip later returned to Ireland, to the Franciscan friary at Kilcrea (Gordon D'Arcy).

To balance this potential bias, it is important to point out that Don Philip used Irish names for species, which he may have remembered from his boyhood in Ireland. He would also have had the benefit of his father's presence in Spain (who was 70 years old when exiled in 1602 and lived to 100), who probably continued to speak Irish throughout his life among the large Irish community there. Don Philip produced an important early work on the birds of Ireland in the *Zoilomastix*, but one in need of further study, and in the meantime should be cautiously approached by ornithologists. However, although born at Dursey, no part of the list can be identified as referring specifically to Cork, although if he was using lore from older family members then much of it must have come from west Cork. The *Zoilomastix* is a unique piece on the birds of Ireland from four centuries ago.

Charles Smith (*c.* 1715–62)

Few details about Smith's life were known until recently. He was born at Waterford about 1715, he took a medical degree at Trinity College (Dublin) in 1738, and he practised pharmacy at Dungarvan (Waterford). He was involved with the Physico-Historical Society (1744–52) and several of his books were published under its patronage. He also served as secretary of the Medico-Philosophical Society (1755–1831) from 1756 until his death.[19]

Smith published the first list of the birds of County Cork in 1750 in a chapter entitled 'A Catalogue of the Birds Observed in this County'.[20] However, Isaac Butler probably collected at least some of the material for this list while employed by the Physico-Historical Society. Butler worked as an inquirer for the society in the 1740s and he travelled widely throughout the country, including Cork, gathering material which was later used by Smith and other authors in their writings. Butler was a Dubliner with an interest in botany and archaeology, and all the indications are that he was a careful observer, and was highly regarded as a botanist, although he published nothing under his own name.[21] Smith was himself employed by the society, and he too travelled widely in Cork, Kerry, Limerick and Waterford.

Smith's list of the birds of Cork (with 100 acceptable species) is clearly incomplete but, taking account of the state of knowledge of natural history in Ireland at the time, it is an important local contribution to ornithology.

However, certain easily identified species are absent from his list, and these were undoubtedly present then as breeding (e.g. Hen Harrier, European Stonechat, Reed Bunting) or wintering birds (e.g. Dunlin, Turnstone). No warbler species are included in the list. Most of the species are recognisable by the English names used by Smith, many being the names (or variants) still in use today, and several of the scientific names used are still recognisable. Indeed, Smith gives a detailed description of some species that could not be bettered today (e.g. Common Shelduck, Chough, Dipper), although others are not described. However, the extent to which these descriptions are merely copied from other works and from the ancient writers is unknown.

A few species are less easy to recognise. Smith clearly referred to some of the ornithological works of Britain and Europe then available, although these were sometimes far from authoritative, and there was little consistency in the use of English or scientific names. This led to considerable confusion between some species (e.g. geese, birds of prey, waders). Smith included a few species for which there was no evidence of their occurrence in Cork at that time (e.g. Black Grouse (*Tetrao tetrix*), Great Bustard, Stone Curlew, Tawny Owl), although some have since been added to the Cork List. The inclusion of certain species (e.g. Capercaillie, Red Kite, Goshawk) was questioned by later authors, although it is now generally accepted that these species did indeed occur in Ireland (and in Cork) in the centuries before Smith wrote.[22] However, despite its shortcomings, Smith's list is a valuable early contribution to the ornithology of Cork. Smith died at Bristol in 1762 and is buried there.

Robert Ball (1802–57)

The Ball family had a long association with Youghal, but Robert was born at Cobh where his father held a post in the customs office. His father returned to Youghal in 1815 when Robert was thirteen years old. Robert was educated at Clonakilty and later at the Quaker school at Ballitore (Kildare), and his interest in natural history was encouraged at both. At home, he developed his interest in natural history further in the company of his sisters Anne and Mary, themselves both important naturalists in fields other than ornithology. Robert Ball went to live in Dublin in 1827, where he worked as a civil servant at Dublin Castle until his retirement in 1852.

Following retirement, he devoted the rest of his life to natural history.

While living at Youghal and during later visits, he made many observations on the birds of the area and was responsible for adding several new species to the Cork and Irish Lists, either through his own observations or reporting the observations of others. Many of the latter would be lost were it not for his diligence. Although he published little on birds, his contribution to Irish ornithology, and to that of Cork in particular, was very important, and he is frequently mentioned as a contributor in William Thompson's *Natural History of Ireland*. Robert Ball wrote papers on varied natural history subjects, but especially on subfossil mammals, seals, cephalopods and corixid water bugs. However, the important fieldwork on water bugs was carried out by his younger sister, Mary. He was in regular communication with like-minded naturalists of his day such as Edward Forbes, Robert Patterson and William Thompson.

Once free of his official duties he took a leading part in the work of the scientific societies in Dublin. He had earlier become honorary secretary of the (Royal) Zoological Society of Ireland (1837) and director of the museum at Trinity College (Dublin) (1844), and he presented his natural history collections to the latter institution. Trinity College also conferred on him the degree of LLD (doctor of laws) in 1850. He became secretary of the newly instituted Queen's University in Ireland in 1851, and a fellow of the Royal Society of London in the year of his death (1857), his candidature having been sponsored by sixteen fellows, including Richard Owen and Thomas Huxley. He was also a member of the Royal Irish Academy and the Royal Dublin Society, and he founded (along with others) the Dublin University Zoological and Botanical Association. It has been said that Robert Ball, along with William Thompson, ranked among the most important naturalists working and living in Ireland in the early nineteenth century. He is buried in Dublin at Mount Jerome Cemetery.[23;24]

Joshua Reuben Harvey (1804–78)

Joshua Reuben Harvey was born at Cork in 1804. He became a medical doctor having studied at Cork, Dublin and Edinburgh, and worked as a physician at several Cork hospitals during his career. He held the chair of midwifery at Queen's College Cork (UCC) from 1849 until his death.

He was president of the Medical Society of Cork and of the CCS and was a fellow of the Botanical Society of Edinburgh and a member of the RCI. Harvey collected marine algae in and around Cork Harbour and his specimens are still in the herbarium at UCC and the National Herbarium (Dublin), although he did not publish his own records. It is believed that his work on the algae was somewhat overshadowed by that of his better-known namesake, William Henry Harvey, who was the foremost phycologist of his day.[25]

As an ornithologist, Joshua Harvey deserves inclusion here because of his list of the birds of Cork prepared for the meeting of the British Association for the Advancement of Science held at Cork in 1843, but not published until 1845, ninety-five years after the first list produced by Smith.[26] However, Harvey's list is terse and very short on detail, and the exclusion (or inclusion) of a species is difficult to interpret in some cases. Another version of his list was published in 1875.[27] The 1845 list has 168 species in a table, although on review it is necessary to reduce this to 164 acceptable species. Similarly, the 1875 list has 197 species in a table, which on review is reduced to 182 acceptable species. One record (Ruddy Shelduck, Clonakilty; an addition to the Cork List) had been overlooked, and not published in the standard literature until recently. It is likely the 1875 list was unknown to ornithologists at the time (such as Richard Ussher) since it is not quoted anywhere, although Ussher was aware of (or had access to) notes by Harvey. Harvey died in 1878, and it appears that he had minimal input into compiling the 1875 revision, although eighteen new species are added (note that errors in the 1845 list are perpetuated in 1875, and a few new errors are introduced). Harvey's name does not head the list, although he is mentioned in the introductory paragraph as the provider of the account. Although the list is a considerable advance on that of Smith, some species known to occur in the county are overlooked.

Harvey amassed an important collection of bird specimens during his lifetime. He was acquainted with many of the naturalists of his day in and around Cork, such as the Parker brothers and Robert Warren. These naturalists and sportsmen provided him with many specimens, some of very rare birds. He presented his collection of more than 350 specimens of native Irish birds to Queen's College Cork (UCC), as well as a large collection of birds' eggs and native Irish mammals. The presentation was made when Harvey left Cork to live with his son in Dublin. Robert Warren described the Harvey collection as at one time being 'the finest collection of native birds in the south of Ireland'.[28;29] Although his publication record is limited, Warren ranked Harvey, among Irish naturalists, as second only to William Thompson.

Richard Dunscombe Parker (1805–81)

Richard Dunscombe Parker deserves inclusion here not only because of his ornithological skills, but because he was an amateur artist who made beautiful paintings of birds, and who has been considered a rival to the American bird artist John James Audubon and English bird artist John Gould. The Parker family home was at Carrigrohane, but Richard later lived a bachelor life at Landscape House, Sunday's Well, with his brothers. He was a gentleman-farmer, naturalist and sportsman who shot over a wide area in Cork and Kerry in the company of his brothers. The Parker brothers are frequently mentioned by William Thompson as providers of information on Cork birds, mainly based on their shooting excursions.

Some of Parker's paintings were exhibited at a meeting of the British Association for the Advancement of Science held at Cork in 1843, and the response was enthusiastic. Parker had been completely forgotten until the rediscovery of his 170 large watercolours of birds in the Ulster Museum (Belfast) in 1976, having been bequeathed to the museum by Parker's niece in 1932, Eleanor Parker of Carrigrohane Lodge. The Ulster Museum put on an exhibition of all 170 watercolours in 1980. A hand-bound limited edition of forty paintings was published four years later, and these were exhibited at the Crawford Municipal Art Gallery (Cork) in 1986.[30;31] Fittingly, these forty paintings were again exhibited at a meeting of the British Association for the Advancement of Science held at Belfast in 1987.

Parker evidently began drawing birds at an early age. Joshua Harvey told William Thompson that Richard Parker made a coloured drawing of a Hobby shot at Carrigrohane in summer 1822 (?) because it 'presented so unusual an appearance'.[32] Although the year has a question mark, if correct, Parker was then only seventeen years old. Based on this information from Harvey, who saw the painting (made from the drawing) about 1848 at the home of Parker, Thompson accepted the occurrence as the first record of a Hobby obtained in Ireland. The Hobby does not appear to have been preserved, but the painting survives in the Ulster Museum to this day.

Robert Warren (1829–1915)

Robert Warren was born at Cork, and when a few years old his parents moved the family to Castle Warren, near Ringaskiddy, where they lived until 1851 (the ruins of the house can be seen to the present day). Here, the young Robert spent his formative years, being educated by a tutor and at a school in Cork. Robert and his brother Edward frequently explored, by boat, the south coast of Cork between Cork Harbour and the Sovereign Islands accompanied by their father, also Robert. He quickly learned the birds of Cork Harbour, especially the gulls, through his interest in shooting. He formed a friendship with Joshua Harvey, to whom he frequently sent specimens for his collection, many of these being preserved to the present day at UCC, having been donated by Harvey. Warren was a correspondent of William Thompson, and he provided him with much information on the birds of Cork for the latter's *Natural History of Ireland*.

Warren left Cork in 1851 to live at Moyview (Sligo), and Cork's loss of a gifted ornithologist was Sligo's and Mayo's gain. There he made many observations on the birds of the River Moy estuary and Killala Bay, where he added several species to the Irish List. He published many papers and notes on birds and other animals during his lifetime in journals such as *Irish Naturalist* and *Zoologist*. He was co-author with Richard Ussher of *Birds of Ireland*. Although Warren himself authored the accounts of only six species, he contributed much new information to Ussher. Ralph Payne-Gallwey considered Warren to be one of the best naturalists and wildfowlers in Ireland in his time. Warren returned to his native Cork on retirement in 1909, where he died in 1915 after an accident at his home on the shores of Cork Harbour, 'Ardnaree', Monkstown. Although most of Warren's active working life was spent away from Cork, his contribution to an understanding of the ornithology of Cork Harbour and the south Cork coast remains an important one.[33;34]

Richard John Ussher (1841–1913)

Richard John Ussher was from Cappagh in west Waterford. He was educated at Portarlington, later at Chester and then at Trinity College (Dublin), although due to ill health he did not take his degree. He spent several winters travelling in Mediterranean countries with his mother and a tutor. He married at twenty-five and spent some further time travelling abroad, before returning to take up public duties and church matters in his own county. He had an interest in egg-collecting and ornithology from boyhood, and he spent his summers at Ardmore (Waterford) and became an excellent cliff climber. He amassed an important egg collection, which he later donated to the Natural History Museum (Dublin), before concentrating on the field study of birds. He was an expert explorer of caves, especially in his native Waterford, but also further afield, and his discoveries of bones of prehistoric birds and mammals are well known. He also discovered bones of the Great Auk (*Pinguinus impennis*) in sand dunes in Waterford.

Ussher was one of the most outstanding Irish naturalists of his time, and he made a major contribution to many aspects of Irish ornithology. His place among those who made an important contribution to the ornithology of Cork is assured alone by his articles on the 'Birds of County Cork', published in the *Journal of the Cork Historical and Archaeological Society*. These sixteen articles were written 'specially for this journal' by Ussher, presumably at the invitation of the editors.[35] Ussher was conscious that he might not have the fullest knowledge of the birds of the county, given that he did not live there. Nevertheless, few others anywhere at that time could better his knowledge. However, his list is not without problems, and it has the appearance of being hurried. He overlooked several published records (e.g. the Velvet Scoter of 1850 and the Dotterel of 1844), and he got the dates mixed up in many more, but all these were corrected for *Birds of Ireland*, and that publication has the hallmark of the precision for which Ussher was renowned. Nevertheless, his list is the most useful to that date on the birds of Cork and several new species were added to the Cork List. The journal articles were later republished as 'A Catalogue of the Birds Observed in this County' in the version of Charles Smith's book edited by Day and Copinger.[36] Although stated to be re-written by Ussher, there is no new information in the latter list, and errors and omissions were not corrected. Ussher died in 1913 after a brief illness and was buried in the family vault at Whitechurch (Waterford), near where he lived.[37;38]

Clive Desmond Hutchinson (1949–98)

Clive Desmond Hutchinson was born at Cork in 1949, but tragically died before he was fifty, arguably with his most productive years still ahead. He had an interest in ornithology from an early age and had his first publications as a teenager. He was educated at Midleton College (Cork) and Trinity College (Dublin), and worked in accountancy in the city of his birth. Like many of his predecessors, Hutchinson worked in ornithology wholly as an amateur. From the mid-1960s he immersed himself in the migration work at Cape Clear Bird Observatory, but as well as his work at the observatory he held administrative roles with the IWC (now BWI).

Hutchinson was most of all a field man; he loved survey work which brought him into contact with his subject. He organised the *Wetlands Enquiry*, the results soon appearing in book form. This was the first real effort at a quantitative assessment of the birds of Irish wetlands, work which is ongoing today as the Irish Wetland Bird Survey (I-WeBS), and which paved the way for designation of important wetland sites under the European Union Birds and Habitats Directives. He was a prolific author, and he did not delay in publishing the results of his work; he was impatient with those who procrastinated. He founded the journal *Irish Birds* in 1977 and edited it for eight years. His finest publication was *Birds in Ireland*, a worthy successor to Thompson, Ussher and Warren, and Kennedy, ensuring Hutchinson's position as 'a major figure in Irish ornithology'.[39] Such was his commitment to ornithology that in the year before his untimely death he was preparing data for a revision of *The Natural History of Cape Clear Island*, and he made a public presentation of this work at a conference at UCC in November 1997. He played a key role over three decades in the study and conservation of birds in Ireland.

It was typical of Hutchinson that while living in Dublin (1967–75) he would immerse himself in the ornithology of that area (Dublin and Wicklow), and produce the first Irish regional avifauna for many years.[40] His papers on the birds of the North Bull Island with Joe Keys and John Rochford are good examples of his work ethic. His finest early paper was on the status of the Little Gull, and this was followed by another in collaboration with Brian Neath. Of his return to Cork, John Rochford wrote, 'I suspect that a *Birds of Cork* cannot be too far off'.[41] This did not happen, probably because he was too busy with greater things. Such was his generosity of spirit, we are sure he would approve of this book.

Clive Hutchinson was personally known to the authors, and an event is recalled in spring 1997 one month after he had invasive surgery which illustrates his dedication to fieldwork. Clive, along with his wife Rachel, John O'Halloran and one of the authors (PS), spent several hours on the bank of the River Blackwater watching Little Egrets displaying in trees and discussing the possibility of breeding, discussion which later continued in a local hostelry. During the following weeks it was possible to prove breeding at this site, the first for Ireland, a discovery that pleased Clive immensely. His contribution to ornithology lives on in the pages of his written works.[42;43;44]

Chapter 4

Research and Monitoring

River Blackwater

Introduction

As indicated in Chapter 3, the level of knowledge of bird population changes was almost non-existent before the 1960s. This does not mean the ornithologists of the era were remiss in their work; all that can be said is that the concept of monitoring and surveying for conservation was not then appreciated and it would take some time to evolve. It must be said that in those early days there were no career ornithologists working for universities, government agencies or conservation bodies.

Systematic research and monitoring have become the norm since then. The rest of this chapter briefly describes the types of surveys that have been undertaken, and those that are now carried out annually. It is these monitoring tools that now serve as early warning systems for change, and which enable ornithologists to state with confidence which species are increasing or decreasing. Targeted research can then be undertaken to determine the cause of observed changes. Chapter 7 should be consulted for the results of surveys on a species-by-species basis.

Breeding and wintering bird atlas surveys

Surveys of breeding birds using the atlas method have been taking place in Ireland and Britain since the late 1960s. This method involves the compilation of a list of species breeding in each 10 km square of the national grid. The first such atlas took place in Ireland over the five years 1968–72, and despite concerns that coverage would be incomplete, it was a resounding success. Observers visited examples of all habitats within each square to obtain the maximum coverage, and therefore to list the maximum number of species. The range of each species was then plotted on a map to illustrate its distribution.

Two subsequent breeding atlases have been carried out at twenty-year intervals. The two most recent atlases introduced new methods which enabled researchers to plot density, calculate total numbers for each species, and examine habitat associations at an increasingly fine scale. It is now possible to examine population change over time by comparing the results of all three atlases across a forty-year period (Fig. 4.1).

Two atlas surveys of wintering birds have been carried out, the first over the three winters of 1981/2–1983/4 and the second over the four winters

Fig. 4.1 Blackcap; has increased recently as a breeding species and as a winter visitor, often seen at bird feeders in gardens.

of 2007/8–2010/11. Rather like the later breeding atlases, timed visits were made to each 10 km square and all individuals of each species were counted. This enabled the production of maps which showed distribution, abundance and population change, in addition to examining the effects of factors such as severe winter weather.[1]

Monitoring seabird populations

Breeding seabirds have been monitored in a series of surveys since the first in 1969–70. This survey set out to count the number of breeding seabirds at all coastal colonies in Ireland, and it was largely successful at doing so. Subsequent surveys, including inland breeding seabirds, have been carried out at intervals and the results published in books, reports and papers.[2]

Localised surveys have been carried out at many sites in County Cork, for example at Cape Clear Island and other Roaringwater Bay islands,

Fig. 4.2 Common Tern; periodic surveys monitor tern populations, and annual fieldwork takes place at a colony in Cork Harbour where conservation measures are ongoing.

as well as at individual colonies, such as Old Head of Kinsale. Targeted survey work has also taken place on species such as Fulmar, Gannet, Great Cormorant, Kittiwake, Black Guillemot, and tern species (Fig. 4.2).

Waterbird surveys

Only minimal surveying of the birds of wetlands had taken place anywhere in Ireland before the 1970s. Although some work had been done on certain species, especially swans and geese, there was an almost complete lack of knowledge of which species occurred at different wetlands. Numbers of each species were known in only the sketchiest detail, and knowledge of seasonality of occurrence of most species was incomplete. The 1960s was also a time of great concern about many wetlands as freshwater sites were threatened with drainage while coastal wetlands (especially estuaries) were under threat from the placement of landfill sites (dumps), from reclamation

Fig. 4.3 European Golden Plover; a large flock at Pilmore Strand with Capel Island in the background, this species is monitored annually at many estuaries.

for industry and other uses and from the perceived negative effects of the spread of cord-grass.

The first serious attempt at a co-ordinated and quantitative assessment of the birds of Irish wetlands began in 1971/2 when the *Wetlands Enquiry* was established, initially covering only estuaries, but from 1972/3 onwards covering both estuarine and freshwater habitats. A considerable quantity of data was collected over four winters from 1971/2 to 1974/5. There then followed a period of relative inactivity, although some surveyors continued counting, until a follow-up known as the *Winter Wetlands Survey* was carried out over three winters from 1984/5 to 1986/7 (Fig. 4.3).

While some surveyors continued their monitoring projects at a local level after these surveys ended, it became obvious that a more permanent monitoring scheme was needed for wetlands and their birds, especially in the light of accelerating changes ranging from the establishment of shellfish industries, growth in renewable energy projects and growth in recreational activities, to the spectre of continuing climate change. Therefore, from the

Fig. 4.4 Red Knot and Black-tailed Godwit; both species are monitored annually at many estuaries.

winter of 1994/5, I-WeBS began operating on an ongoing basis. At the time of writing this chapter (September 2021) the twenty-eighth season of this survey was beginning. Additionally, there was a need to provide a strong scientific basis for site selection and designation of SPA under European Union directives and for reporting on the long-term monitoring of these wetland sites.[3] This survey is now the principal tool used to achieve this. It has also provided new data on population estimates and trends, and on the responses of waterbirds to global warming.[4;5] Additionally, many studies of bird populations or of individual species have been carried out at most of the important coastal wetland sites (Fig. 4.4).

Birds of non-estuarine coastal habitats

Many waterbirds, including ducks, waders, gulls and some other species, often occur away from the estuarine wetlands and coastal lagoons which are covered by I-WeBS (see above). This means that, for some species, an important proportion of their total numbers may not be recorded during I-WeBS counts. Oystercatcher, Common Ringed Plover, Sanderling and

Turnstone may occur in significant numbers at sandy beaches, with Purple Sandpiper populations often occurring at shingle and rocky coastlines, while offshore waters may hold important numbers of Red-throated Diver, Great Cormorant, Common Scoter, and gull species. Three surveys have been carried out on the Irish coastline, and each has proved an important addition to knowledge of the distribution of coastal waterbirds. The most recent survey was in the winter of 2015/16.[6]

Countryside Bird Survey

The Countryside Bird Survey (CBS) was initiated in 1998, and it has been annual since, although there was no survey work in 2001 (due to foot-and-mouth disease) or 2020 (due to Covid-19). This survey aims to monitor population change in common and widespread breeding species on terrestrial habitats, largely on farmland, woodland and open areas, including uplands (Figs 4.5 and 4.6). Although this is a relatively new monitoring

Fig. 4.5 Farmland; much farmland is intensively managed with herbicides and insecticides widely used but is an important habitat for several species.

Fig. 4.6 Farmland; smaller fields with well-developed hedgerows are extremely important habitats for insects, small mammals and wild plants, all of which provide food for a wide range of bird species.

Fig. 4.7 Yellowhammer; formerly widespread on farmland, has now disappeared from much of west and north Cork, but is still widespread on east and south Cork farmland.

tool, it has produced useful results on population estimates and trends of a suite of species which heretofore were poorly monitored.[7] It shows that Yellowhammers are still numerous in south and east Cork, although absent in west and north Cork, Meadow Pipits are scarce or absent on improved pasture and are mostly found in west Cork, and that the European Stonechat has expanded inland from the coast (Fig. 4.7). However, this survey began too late to monitor the effects of the significant landscape changes and intensification which took place on farmland following Ireland's entry to the European Economic Community (later European Union) in 1973.[8]

Garden Bird Survey

Gardens are increasingly recognised as important areas for birds, for two reasons. Gardens are an increasing habitat as urbanisation envelops former farmland, and gardens provide a refuge for at least some of the species

Fig. 4.8 Goldfinch; has increased in recent years and is often seen foraging in weedy areas and at garden bird feeders in winter.

affected by agricultural change in the wider countryside (Fig. 4.8). Birds are also increasingly provisioned with food in gardens as citizen scientists enjoy watching and monitoring them. Results from the Garden Bird Survey are published periodically.[9]

Migration studies

Early observations on bird migration in Ireland involved the recording of individuals killed at light-stations around the coast. Richard Barrington asked lighthouse-keepers to collect a leg and a wing of dead birds at their stations, and these were later identified by ornithologists.[10] This recording system was at its peak in the last two decades of the 1800s. It operated for some time in the 1900s, but at a much lower level after 1915. It could not operate today since all light-stations are run automatically.

The ringing of birds with numbered metal bands developed from the early 1900s, and this became one of the main ways in which migration was studied for the next 100 years.[11] This method worked extremely well for large species and for species associated with humans, such as waterbirds which are shot. Return rates were very low for smaller species as these birds were rarely found when they died. Other bird-ringers frequently captured birds with rings, and this increased the return rate for many small species. However, as the decades moved on, new techniques which improved the return rate were developed, such as coloured neck and leg bands for large species of waterbirds and coloured wing tags for birds of prey. High-quality telescopes and digital photography has also been used to read various types of rings.[12]

Further new techniques have been developed in the last few decades, such as satellite telemetry, with geolocation and global positioning system tools increasingly used which provide a continuous record of the movement of birds carrying such markers, compared to traditional bird-ringing which only gave a start and end point of movement.[13]

In the last few years there has been an increased interest in sound recording of birds. This has been facilitated by the availability of relatively cheap digital recorders and increasing online resources, including a library of bird calls (Cornell Laboratory of Ornithology), which allows the uploading and identification of calls. The recording of nocturnal migration of birds by call is known as 'nocmig', and several observers in Cork use the system. Some surprises have been revealed, such as the regular movement of Little Grebe, Water Rail and waders over Cork city, a spring overland passage of terns, and the expected April–May northwards migration of Common Whimbrel. The first county record of Semipalmated Plover has also been detected using this system. There is no doubt but that it has a useful future as an aid to migration studies, but it is still in the early stage of development.

However, the joy of birdwatching for many involves going into the countryside, sometimes to remote places, such as islands, and recording migrating birds visually, rather than sitting at a screen watching a moving dot representing a bird on migration carrying a satellite device. There is room for all types of study, and the concept of systematic recording of migrants by observation has long been enshrined in Cork ornithology, especially since the establishment of Cape Clear Bird Observatory.

Other surveys

A series of surveys, or reviews of status, of individual species and groups of species at different seasons have been carried out since the 1970s. These have included Whooper Swan, Bewick's Swan, Gannet, Great Cormorant, Little Egret, Hen Harrier, Common Buzzard, Peregrine Falcon, Corncrake, Kittiwake, Little Gull, Mediterranean Gull, tern species, Common Kingfisher, Chough and Common Reed Warbler (Fig. 4.9).

Fig. 4.9 Bewick's Swan; this species has disappeared from Cork (and Ireland) resulting from climate warming as birds remain closer in winter to their Arctic breeding grounds.

Fig. 4.10 Hen Harrier; has declined as a breeding species, but has remained relatively stable in recent years, although threatened by developments in its upland breeding areas.

Research at University College Cork

A wide range of species have been studied for higher degrees at University College Cork over the last forty years. Species studied have included Mute Swan, Hen Harrier, Corncrake, Lesser Black-backed Gull, Barn Owl, Chough, Dipper, European Robin, European Stonechat, Song Thrush, crow species, tit species, farmland birds and various seabirds. Long-term research is ongoing on Hen Harrier and on bird communities of semi-natural and planted forests (Fig. 4.10). Researchers there have published a long list of scientific papers, and relevant ones have been consulted in the preparation of Chapter 7.

Chapter 5

Recent Changes of Status

Musheramore

Breeding species

This chapter briefly reviews changes in status of breeding species at the level of new arrivals and extinctions, rather than at the level of changes in population size, although this is referred to where relevant. Species lost as breeders before 1800, such as Capercaillie, Red Kite and Common Crane, are not dealt with and for a species-by-species account the reader is referred to Chapter 7.

There were more extinctions than new breeders in the 1800s (Table 5.1). Several of the extinctions involved the large birds of prey (four species) along with the Great Bittern, European Golden Plover and three passerines. A major cause of the loss of the birds of prey was persecution, while a combination of drainage of wetlands and persecution caused the loss of the Great Bittern. Perhaps a county so far south as Cork could never be expected to have anything other than a small European Golden Plover population, and the Irish Garden Warbler population and range has always appeared to be restricted. Human activities may also be implicated in the demise of the Jay and Woodlark.

The Irish breeding range of the Tufted Duck was known to be increasing in the 1800s, as was the case with the Woodcock and Siskin, and the Mistle Thrush had only reached Ireland at the beginning of that century. The Common Crossbill is irruptive and breeding in earlier decades cannot be ruled out. Gannet numbers were at a low level in the 1800s in Britain and Ireland, largely due to killing for food, and the colonisation of Bull Rock was probably from the nearby Little Skellig (Kerry). However, the ornithological record for the 1800s is fragmentary and needs to be viewed with caution, mainly because of the paucity of reliable observers to chronicle changes.

It was indicated earlier (Chapter 3) that the first half of the 1900s was a period of stagnation in Irish ornithology, despite the presence of a small number of excellent observers. The Shoveler, Red-breasted Merganser and Fulmar bred for the first time, as did the Stock Dove, all species that were increasing across their respective ranges for some years. The Jay returned as plantations matured and as the craze for cabinet specimens waned. However, two gamebirds (Grey Partridge and Quail) effectively disappeared as breeders. These gamebirds are open habitat species found mainly on farmland, and perhaps their demise was a taster of what was to come in that habitat.

New breeders 1801–1900	Returned breeders 1801–1900	Ceased breeders 1801–1900
Tufted Duck		Great Bittern
Gannet		White-tailed Eagle
Woodcock		Marsh Harrier
Mistle Thrush		Common Buzzard
Siskin		Golden Eagle
Common Crossbill		European Golden Plover
		Jay
		Woodlark
		Garden Warbler

New breeders 1901–58	Returned breeders 1901–58	Ceased breeders 1901–58
Shoveler	Jay	Grey Partridge
Red-breasted Merganser		Quail
Fulmar		
Stock Dove		

New breeders 1959–2018	Returned breeders 1959–2018	Ceased breeders 1959–2018
Gadwall	White-tailed Eagle	Corncrake
Common Pochard	Common Buzzard	Common Curlew
Little Egret		Little Tern
Great Crested Grebe		Roseate Tern
Common Gull		Turtle Dove
Little Tern		Nightjar
Roseate Tern		Ring Ouzel
Collared Dove		Tree Sparrow
Turtle Dove		Twite
Blackcap		Corn Bunting
Common Reed Warbler		
Tree Sparrow		

Table 5.1 New breeders, returned breeders, and those that ceased to breed during different eras since 1801. To qualify, the species had to breed regularly for several years in the relevant era. Introduced species, sporadic breeders and one-off breeders have been excluded. Note that the return of the White-tailed Eagle was the result of re-introduction.

Fig. 5.1 Corncrake; formerly a common bird of hay meadows, has disappeared as a breeding species mainly due to intensification of farming practices, especially silage harvesting.

The sixty-year period 1959–2018 has seen significant changes in species composition, range and population size on a scale not previously seen. This period coincides with an increasing number of experienced ornithologists who diligently documented these changes.

Ireland began to emerge from the economic stagnation prevailing since the foundation of the state around the beginning of the 1960s. Probably the most significant development economically was Ireland becoming a member of the European Economic Community (later European Union) in 1973. The most immediate change was in agriculture, where small fields became larger, monocultures prevailed, and rural housing improved immeasurably as spending power grew. Changes were rapid, leading to declines in many well-known species such as Yellowhammer, and the complete extermination of the Corncrake, Twite and Corn Bunting (Fig. 5.1). It must be said that declines were taking place in some of these species since the 1940s (Chapter 7), as even then the mechanisation of agriculture was having an effect. Of course, declines in farmland birds were not confined to Ireland; they occurred right across Europe.[1;2]

Fig. 5.2 Gadwall; a scarce duck found mainly at east Cork lakes, decreased following habitat change at Ballycotton in the 1990s, exacerbated more recently due to climate warming.

Changes have also taken place among non-farmland birds for various reasons, some not so obvious. Little and Roseate Terns have flirted with breeding on the Cork coast but have not remained for long anywhere. Today, the headquarters of these species are on the east coast of Ireland at continuously wardened sites without which they would be very scarce indeed. There seems to be no reason why Tree Sparrows have disappeared, but the history of the species shows that they fluctuate markedly, and Cork being at the edge of their range in Ireland does not help. Factors outside Ireland also influence the stability of species here, and the migrant Turtle Dove has declined significantly in Britain and Europe at least partly due to conditions in the African wintering area and due to hunting pressure in Europe. Other migrant species have declined in Ireland, especially the Spotted Flycatcher, although a reduced breeding population hangs on.

Waterbirds such as Gadwall, Common Pochard and Great Crested Grebe colonised Cork during this period, but numbers remain small (Fig. 5.2). Waders breeding on wet grassland declined (Northern Lapwing and

Fig. 5.3 Skylark; has disappeared from farming landscapes as a breeder but occurs in stubble in winter where spring-sown, but not autumn-sown, cereals are grown.

Common Snipe) as this habitat reduced in size or was eliminated through intensification, and these species may follow the Common Curlew into oblivion. Skylarks have also gone from pastures as breeders and are now confined to marginal uplands and coastal heaths and dunes where small numbers continue. Meadow Pipits do not generally breed on lowland farmland either, but both occur commonly in winter, the lark in stubble and the pipit in pasture. However, the switch to autumn-sown cereals has removed a rich winter food source from the Skylark (Fig. 5.3). Pheasants now appear to be incapable of breeding successfully on farmland, and winter shooting depends almost entirely on released birds, at least in some areas. Agri-environmental schemes have a local positive effect, but these are not applied on a sufficiently wide scale to stem the overall declines.

Uplands have largely been reclaimed, forested or used for wind energy projects (or in combination). The Common Curlew and Ring Ouzel have both disappeared from these habitats, and the Nightjar occurs sporadically

Fig. 5.4 Little Egret and Great Egret (and two Black-headed Gulls); the former has bred in Cork since 1998 following expansion of its European range, the latter may breed in the future as it continues to increase in Europe.

at best. The Hen Harrier holds on in some areas, mainly in second rotation forestry, but the adjoining feeding areas are much reduced. It is a distinct possibility that the Red Grouse will cease to breed in Cork in future decades, without management to improve the habitat.

A diverse range of species have bred in Cork for the first time in the last few decades. First of these was the Collared Dove, here since the 1960s, and doing well about farmyards and grain stores. The Common Reed Warbler began to haunt reedbeds from the early 1980s and appears to have consolidated its position. It is unique in that it is one of the few migrant colonists of recent decades. The Little Egret has also settled as a breeder, and it may be followed by its relatives the Cattle and Great Egret, both of whom are occurring recently in increasing numbers (Fig. 5.4). These species may have benefited from a combination of a more benign winter climate and an increasing population after the cessation of exploitation for their plumes. The increase in breeding among the Blackcap has been spectacular;

every woodland and patch of woody scrub now has breeding birds, yet it was completely absent except as a passage migrant in the early 1960s.

The White-tailed Eagle has returned to breed again in Cork and elsewhere. This came about not by natural colonisation but by the hard work of dedicated ornithologists over the last twenty years who nurtured the species through a re-introduction programme often beset with expected (and some unexpected) difficulties. On the other hand, the arrival of the Common Buzzard is entirely natural, and the species is now widespread.

The examples of success described in this chapter show what can be achieved through conservation action in bringing back and maintaining populations of vulnerable species (terns and birds of prey), but it will be much more difficult to maintain populations of others such as Common Curlew, and migrants such as Corncrake and Nightjar. Conservation efforts need to extend to habitat improvement and restoration at a landscape level; it does work at local and site-based levels.

Wintering species

It is only since about 1970 that there has been an understanding of population sizes of most species of waterbirds. Ireland traditionally was a winter refuge from severe weather in Europe for many species, and numbers were high for most. However, there have been significant declines in most species since the early 2000s, by over 50 per cent in some cases.[3] The species most affected negatively from the Cork point of view include Bewick's Swan, Common Shelduck, Common Wigeon, Mallard, Shoveler, Common Pochard, Tufted Duck, Red-breasted Merganser, Common Coot, Oystercatcher, European Golden Plover, Grey Plover, Northern Lapwing, Dunlin and Common Curlew. However, increases have taken place in Whooper Swan, Brent Goose, Little Egret and Sanderling.

Pressures and threats to waterbird populations were examined at a national level, but many of these threats also apply to Cork. One of the most important threats to waterbird numbers was shown to be the effects of climate change. Many of the species breeding in Fennoscandia and the Siberian Arctic which formerly wintered in high numbers in Ireland now stay in ice-free waters in northern Europe much closer to their breeding grounds. The Bewick's Swan has ceased to visit Cork, where formerly there was an important flock at Ballycotton and on the River Blackwater callows.

Other threats identified were sea-level rise, which is associated with climate change, land-based and water-based recreational disturbance, aquaculture, urbanisation and water pollution.

Summary

It could be said that the fortunes of birds in Ireland, and in Cork, are in a state of flux at present, with several species extinct as breeders and others decreasing rapidly, especially in the winter season. In contrast, others are increasing, and new species are colonising or may do so in the future. Fifty-four (26 per cent) of Ireland's regularly occurring species (including breeders and winterers) are now on the 'red' list of conservation concern, with upland and farmland birds having the highest proportions of red-listed species (85 per cent combined).[4]

These ups and downs are the result of a combination of changes in the environment and climate warming, which has been shown to be largely driven by human activity.[5] Therefore, the direction taken in future decades by policy-makers will determine how populations fare through the rest of this century. Climate warming is having an effect well beyond the jurisdiction of Ireland and is impacting on many species that visit Ireland either to breed or to spend the winter.

The ranges of species in Europe are changing as the climate warms. It has been said that 'the distribution of native species has consistently and significantly moved northwards since the last atlas' (i.e. since the 1990s), suggesting that 'climate change is a major factor driving bird distribution shifts in Europe'.[6] Even within Ireland, it has been shown that some species are extending their range into the north-west where they were previously scarce or absent, with a corresponding thinning of the population in the south-east (see especially Willow Warbler, Chapter 7). When bird populations decrease, the change is often felt more at the edge of the range than in central parts, and Ireland is firmly at the western edge of the range of many species which are centred on the continent.

National policies in agriculture, forestry, fisheries (including aquaculture) and other areas have had unintended negative consequences; this being so, there is no reason why policies in the future cannot be designed to have minimal negative impact. The data is clear, and it is there in abundance, although in mitigation it must be said that it was not there before about 1970. We know most of the factors likely to have negative impacts on species, so new policies should be designed and implemented so they have minimal effect on vulnerable species. Maintaining viable populations during the rest of this century will not be an easy task, and it will depend largely on how governments and people respond to climate change and other environmental issues. The use of fossil fuels will have to decrease, which will lead to increased power generation from renewable resources on land and at sea, leading to more fragmentation of habitats and the inevitable risks of direct kills and avoidance by birds (Fig. 5.5). Stabilisation of declining species will need a worldwide response as many are migrants wintering in Africa; we need to be as concerned about what is happening south of the Sahara Desert as we are about what is happening in our own mountain ranges and river valleys (Fig. 5.6).

Fig. 5.5 Wind turbines; wind energy expansion in uplands may threaten populations of Red Grouse and Hen Harrier due to habitat fragmentation.

Fig. 5.6 Sedge Warbler; a common trans-Saharan migrant breeding mainly in wetland vegetation, some migrant species have declined due to habitat change in their African winter range.

Chapter 6

Introduction to Systematic List

Barley Cove and Lissagriffin Lake

English names

The account of each species begins with its English name. The most appropriate names for use across the English-speaking world have been debated, but there is divergence among ornithologists and journal editors. The problem has been exacerbated by the significant number of taxa that have been raised to specific level due to 'splits' following DNA analysis of taxonomic relationships. We have broadly followed the names used in Ireland in recent decades, but with some changes.[1] For example, there are several 'paired' species (e.g. Redshank and Spotted Redshank) on the Cork List that can give rise to confusion. In all such cases we have expanded one of the names by adding a modifier, in this case Redshank becoming 'Common' Redshank. However, we have reverted to using the shorter name within the species' texts, but not in the heading (e.g. Curlew, rather than Common Curlew), in the interests of readability. We include subspecies (and some colour morphs) within the accounts of the respective species, in line with normal practice among recording authorities.

Irish names

The Irish name of the species is also given, these having been taken directly from the main published source, but where species have been added to the Cork List since 1998, new names have been coined, as appropriate.[2]

Sequence, taxonomy, scientific names, species and subspecies

There have been considerable changes in the sequence in which species are listed and in taxonomy in recent years. The main reason for these changes is the extensive use of DNA in establishing relationships between and within species and families.[3] These changes are usually incorporated into annual reports of the Irish Rare Birds Committee. The rate of change creates instability and often leads to radically altered sequences within some families between successive reports. Stability is desirable but probably not attainable, at least in the medium term, such is the rate of research in this field. Compromise is necessary and desirable. We here follow the familiar sequence of Parkin and Knox.[4] However, we follow the taxonomy, scientific names, species and subspecies of Hobbs, with minor variations in both where appropriate.[5] A county avifauna, such as this one, is not a statement on taxonomy, rather it is a statement of the past and present status of each taxon (species and subspecies) occurring within the boundary of the study area. We believe that once each taxon is recognisable to readers and to science, this approach is appropriate until a more stable taxonomic situation is reached at some time in the future.

Readers desiring to know more about early and local names (English, Irish and scientific) are referred to the work of Anderson.[6]

Categories

All species on the Cork List are assigned a category, in keeping with standard practice. The appropriate category appears after the Irish name

Fig. 6.1 Short-eared Owl; a scarce winter visitor to coastal and inland marshes, feeding on small to medium waders and small mammals in coastal districts.

Category A	Species that have been recorded in an apparently natural state in County Cork at least once since 1 January 1950.
Category B	Species that have been recorded in an apparently natural state in County Cork at least once up to 31 December 1949 but have not been recorded subsequently.
Category C1	Species that, although originally introduced by man, have established feral breeding populations in County Cork which apparently maintain themselves without necessary recourse to further introduction.
Category C2	Species that have occurred in County Cork but are considered to have originated from established naturalised populations outside Ireland, or from elsewhere within Ireland, but outside County Cork.
Category D1	Species that would otherwise appear in Categories A or B, except that there is a reasonable doubt that they have ever occurred in County Cork in a natural state.
Category D2	Species that have arrived in County Cork on board ship, or by other human assisted means.
Category D3	Species that have only been recorded in County Cork as tideline corpses.
Category D4	Species that would otherwise appear in Category C1 except that their feral populations in County Cork may or may not be self-supporting.
Category E	Species that have been recorded in County Cork only as introductions, transportees, or as escapes from captivity.

Table 6.1 List of categories to which birds on the Cork List are assigned.

of the species. The main Cork List is comprised only of species assigned to categories A, B and C.[7] There is also a supplementary list which contains species (or records) assigned only to categories D and E (see Appendix 1). The categories, with the wording slightly modified to make them appropriate to County Cork, are shown in Table 6.1.

Status

Each species account is preceded by a table, which gives a brief outline of its status in the world, in Ireland and in (County) Cork, respectively, followed by its conservation status.

World status is condensed mainly from the *Handbook of the Birds of the World*, from *The Birds of the Western Palearctic*, and from Parkin and Knox.[8;9;10] These brief statements indicate where the species concerned breed, usually given at the level of continents, regions or (in the case of some seabirds) oceans, and are intended only as a guide to where else in the world the birds on the Cork List occur. Many species are migratory, with those breeding in northerly regions moving south into our latitude for the winter, and those breeding around our latitude moving still further south to southern Europe and Africa, while others commute through Ireland to their wintering and breeding grounds further south and north, respectively (Fig. 6.2). For more precise detail on the range and status of any species throughout the world, the listed references should be consulted.

Ireland status is based on recent standard works and other sources.[11;12;13] For rare species the number of individuals that have occurred is based on figures published by the Irish Rare Birds Committee in its most recent annual report where that species is listed, terminating in 2018.

Cork status summarises the data given in the main body of the text for each species. The number of individuals given for rare species is based on criteria similar to that for the Ireland status, but with slight variations where appropriate.

Conservation status is based on the most recent assessment of 'birds of conservation concern' in Ireland, supplemented by recent European works.[14;15;16] The conservation status of vagrants from other continents has been assessed using the *Handbook of the Birds of the World*.[17]

Fig. 6.2 Curlew Sandpiper and Buff-breasted Sandpiper; the former migrates through Ireland in variable numbers in autumn from its Siberian breeding range, the latter occurs as an autumn vagrant from North America.

Recording area

The recording area is the entire land surface of County Cork, and all offshore islands and rocks as well as the sea area extending out to 30 km from the nearest point on the mainland, island, or rock, whichever is appropriate.[18] It follows that for inclusion in the main Cork List each species must be recorded either on or over land, or at sea out to the 30 km limit, but not beyond.

Place names used have been taken from the Discovery Series of maps produced by the Ordnance Survey of Ireland (sheets 72, 73, 74, 79, 80, 81, 84, 85, 86, 87, 88 and 89). We have deviated from the map names only in a small number of cases where local usage dictated otherwise. All places mentioned (with variants) are shown in Appendix 5.

Species accounts

The account of each species has been written in a standard way appropriate to the status of that species in County Cork. For all species where the number of occurrences (records) falls between one and ten (regardless of the number of individuals) the entire record is quoted, without further comment in most cases. A graph is provided showing the monthly distribution of individuals of some rare and scarce species recorded in the sixty-year period 1959–2018 (1959 being the year of establishment of Cape Clear Bird Observatory after which the true status of many species in the county was clarified for the first time). The account of each species follows a broad chronological sequence, with the breeding status usually given first followed by the winter status and generally ending with migration information. In a small number of cases there are minor deviations from this sequence.

In a book such as this which aims to cover the history of every species in the county it would be impossible to include every snippet of published or unpublished data available. Therefore, only the information most likely to be useful in a county-wide context has been used. A vast amount of data has been accumulated in recent decades on a wide range of species, especially on wetland birds, seabirds and countryside birds. Much of these data have been summarised in books, reports and papers, and we refer to the most important of these rather than re-analyse the data.

Similarly, a considerable amount of data has been accumulated on bird migration by observations at headlands and on islands since the 1960s and 1970s. Much of this information is unpublished, and we hope that it will appear in the future. Bird-ringers, and others, also possess large datasets rather few of which have been published.[19]

Many species exhibit varying levels of albinism or other types of plumage and soft part aberrations. However, these variants, and hybrids, have generally not been considered for inclusion here, but we do include information on well-marked colour morphs, such as 'Polish' Swan and 'Blue' Fulmar. Studies dealing mostly or entirely with behavioural aspects are not included either, although we realise that almost all aspects of the ecology of birds are behavioural in nature. We also generally exclude studies on biometrics, genetics and parasites, except in a small number of cases where these have an important bearing on the species concerned in County Cork.

Our over-riding aim is to provide the reader with a picture of the past and present status of each species on the Cork List in as broad a sense as possible, and to provide those interested with the means, through key references, for further research.

Records of certain rare birds which were not specifically identified are included, but excluded from the statistics on the number of Category A, B and C species given at the end of this chapter. One species has been included after the cut-off point of 2018, a Two-barred Warbler recorded in 2019, but excluded from county statistics. Six species from 2020 and 2021 which were still pending acceptance by the Irish Rare Birds Committee at the time of going to press (February 2022) have been added at the end of Chapter 7 and are also excluded from county statistics.

Sources of data and verification of records

The sources of data used in this book almost all relate to previously published material. This book brings these diverse sources together and blends and interprets them into a single piece of (hopefully) cohesive writing on each species. We have generally ignored sources such as newsletters, not because they might be unreliable, but because such sources may comprise early thoughts on certain subjects rather than fully analysed data. While some 'grey' literature has been used, it too has been mostly ignored. Internet sources have been checked and used when it has been possible to confirm the reliability of data. Information in the form of personal communications from colleagues has been used where such information is new and where it is unlikely to be published elsewhere. The authors took a sceptical view when examining sources of data in the literature, and elsewhere, and included information only when it was evidently reliable. Hopefully, this position has helped to make the final product as unbiased and as reliable as possible.

One of the greatest sources of unverified data is in the form of reports of rare and scarce species (including records of rare breeding birds). It is important to set out the standards against which such records have been included as verified, and to set out those where some published records have been considered unverified, and therefore excluded. To do this, it is necessary to give a brief overview of the system of bird recording in Ireland generally.

The recording system for birds in Ireland has evolved over the years from one where almost every rare bird was verified only after it was obtained, usually shot, to one where almost all are identified in the field by observers using high-quality optical aids and cameras. The change from shooting to observing came about around 1950 with the establishment of coastal bird observatories, a natural follow-on from the migration work of Richard Barrington in the late 1800s. The change was assisted by the publication of field guides and identification papers, and by the desire of observers to trap and ring migrant birds. The establishment of an annual publication, the *Irish Bird Report*, in 1953 was a milestone, and it encouraged the concept of a written description of every rare bird before it could be published in the report as verified. An expert panel (Irish Records Panel) was established in 1971 to assess records of rare and unusual occurrences of birds in Ireland. The system set up in 1971 still operates today, although the panel changed its name in 1985 to the Irish Rare Birds Committee. Verification of a record of a rare bird is confirmed once it is included in an annual report of the committee, now published in the journal *Irish Birds*. It follows that only records of rare species accepted and published in the way described have been considered for inclusion in this book. Records of rare birds dated before 1953 are taken from the standard works quoted and listed in the bibliography.

Rare breeding birds have been treated in a similar way, and the standard source of information since the establishment of the Irish Rare Breeding Birds Panel has been its annual report published in *Irish Birds*.[20] The results of research are usually published in scientific journals, reports and books, and all such material relating to birds in County Cork has been examined, assessed, and used in this book where appropriate.

Records of scarce species have been taken mainly from published sources as above, with additional data from the *Cork Bird Report*, *Cape Clear Bird Observatory Report*, *Sherkin Island Bird Report* and from irishbirding.com, the latter courtesy of Joe Doolan. Graphs showing monthly distribution of records of individuals and tables showing the number of individuals recorded per decade have been provided within the accounts of many scarce species for the period 1959–2018.

County statistics

The statistics in Table 6.2 show a total of 427 species recorded in County Cork to the end of 2018; the first occurrence of ninety-four (22 per cent) of these was also the first Irish record of the species.

Species and category	**No.**
Species in Category A	411
Species in Category B	14
Species in Category C	2
Total for Cork	**427**

Table 6.2 Statistics for the Cork List up to 31 December 2018.

Chapter 7

Systematic List of the Birds of County Cork

Mudflat, with Brent Geese

Mute Swan

Cygnus olor | Eala bhalbh | A, C1

(monotypic)

World	Europe and Asia, introduced elsewhere
Ireland	Widespread breeding resident
Cork	Widespread breeding resident
Conservation	Amber-listed

Whether or not the Mute Swan was formerly native to Ireland and Britain has been debated. Fossil Mute Swan material from the late Pleistocene or Holocene and bone material from later periods occurs at British sites. In Ireland, fossil material probably from Mute Swan was found at Castlepook, suggesting it was almost certainly indigenous to Ireland, but no Mute Swan bones were discovered at Lagore (Meath) (AD 750–950), despite the presence of Whooper and Bewick's Swan bones.[1] If the Castlepook material truly belongs to Mute Swan, then this species may have been native to Ireland, but its occupation was not continuous if information from the 1700s and 1800s is correct. Other opinion suggests it was a post-Norman introduction to Ireland.[2] There is some evidence from population genetics, including Cork Lough birds, that the Irish Mute Swan stock originated in Britain, with a limited number of individuals involved.[3]

In the mid-1700s the Mute Swan was said to be frequently met with near gentlemen's seats, and on their ponds and reservoirs in Cork. However, it is likely that it was not then a bird of the wider countryside. Castlemartyr, where the River Kiltha had recently been altered by Lord Shannon to form a canal and a small lake, was the only site mentioned (poetically) by Smith.[4] A century later a female bred with a male Black Swan at Castlemartyr; presumably Mute Swans had not wandered far outside the grounds of the great houses by the 1840s.

Mute Swans were breeding on lakes and rivers in many parts of Ireland by the early 1880s, having escaped from private grounds. An 1873 photograph shows a brood of cygnets at Cork Lough, while boating and shooting was prohibited in 1881 to prevent frightening and injuring the swans.[5] In the 1890s it was resident in a semi-domesticated state on some rivers and lakes in the county. It is evident there was a major expansion of range and numbers outside the private waters of the great houses in the latter half of the 1800s. It was breeding freely by 1900 and several favoured localities were mentioned where flocks were found, for example Rostellan (on tidal waters) and Ballycotton (50–60 formerly on the bog), and eleven were noted on the enclosed inlet at Rostellan in July 1907.[6]

The three recent breeding atlas surveys revealed that the Mute Swan was spread across the county, including at the western peninsulas, but was generally absent from higher ground in the west and north which lack suitable breeding waters. Density was highest in lowland areas along the Rivers Blackwater, Lee, Bandon and Argideen.

A breeding density of just under two pairs per 10 km square was recorded in 1978.[7] However, a density of seven pairs per 10 km square occurred in east Cork during a study on breeding biology.[8] In peak years ten pairs bred at Ballycotton, with up to eight pairs on a lagoon of 33 ha. These did not breed colonially, each pair nesting in reedbeds near the edge of the lake, but out of view of its neighbour. Most small to medium standing water-bodies appear to hold one to three pairs, but in 2007 ten pairs laid eggs at Lough Aderry. Seven nests were in a confined area of about 500 m^2, and each was visible to all others in a true colonial situation. Mute Swans also breed at some islands, for example Sherkin Island, but only eight records of vagrants occurred at CCBO during 1959–69.

Winter distribution does not differ from the breeding season. The most complete census of the winter population to date recorded 667 swans at fifty-five sites in January 1991.[9] This was believed to underestimate the true population by about fifty birds. Most sites (43) held one to nine birds, and ten sites held 10 to 100. Over 100 birds were present at two sites, 105 at the Gearagh and 144 at Cork Lough.

Large flocks occur at favoured sites, but there has been significant change in flock distribution over the last fifty years, presumably mainly related to the suitability of these sites for feeding. Kilkerran Lake formerly held a moulting flock of over 300 and was known locally as 'Swan Lake', but numbers declined after 1968. Following flooding of the River Lee valley in the 1950s this site gained in importance and moulting flocks in late summer reached 300 in the 1970s, 241 in July 1988 and 250 in August 1993, and in winter a mean of 131 between 1996 and 2000. The River Lee around Cork city held up to 270 until Irish Distillers left the site now occupied by University College Cork. Cork Lough has held a flock for many years with peaks of 232 in October 1985 and 266 in September 1988.

The main change in east Cork has been the demise of the Ballycotton flock (peak 185 in August) with the drainage of the lagoon after 1990. This affected the much smaller moulting flocks at Ballymacoda and Youghal Harbour, which disappeared, but led to an increase at Lough Aderry. Numbers at Lough Aderry declined again recently after the public were advised to cease feeding with bread.

Lead, the heavy metal, is a serious environmental pollutant and a threat to grazing birds such as Mute Swan. It originates mainly from angling and shooting. Studies have been carried out on the flock at Cork Lough, and blood lead levels compared with sites in rural areas. High lead levels can kill swans directly or make them more prone to collisions with objects such as electricity cables.[10] Lead is more prevalent in urban areas (Cork Lough) than rural areas, but levels have reduced over time.[11]

Movements of 660 Mute Swans marked at Cork Lough with yellow plastic leg rings during 1984–94 have been reported.[12] A total of 432 birds was recorded at least once after ringing, and 205 (47 per cent) were ever seen again only at the place of ringing, suggesting sedentary behaviour. Most (77 per cent) of the remaining 227 moved between 16 and 48 km and eight moved more than 80 km. Most birds moved to sites in east Cork, and to the Gearagh. Swans which travelled more than 80 km were recovered at Skibbereen (1), Lissagriffin (2), Limerick (1), Wexford (3), Wicklow (1) and Staffordshire (England) (1). The longest movement within Ireland was 220 km to Wicklow, while the one that went to Staffordshire moved 460 km. A cygnet ringed in Warwickshire (England) in August 1975 was recovered at Rosscarbery in September 1976, and movements from Dublin and Wicklow to sites in east Cork and Cork city have also occurred.

Polish Swan: There is one record of a Polish Swan. One obtained at Kinsale in January 1881.[13] There was severe weather that month and this bird may have migrated from Europe, where this form is relatively common.

Bewick's Swan

Cygnus columbianus | Eala Bewick | A

(subspecies: *bewickii; columbianus*)

World	Arctic regions of North America, also north-east Europe and Asia
Ireland	Scarce winter visitor, recent decrease
Cork	Recent decrease, now rare winter visitor
Conservation	SPEC 3; Red-listed

The Bewick's Swan (*C.c. bewickii*) was not identified as a separate species from the Whooper Swan until 1830.[14] Therefore, references to 'wild swans' in the 1700s and 1800s are not always possible to identify to species (see Whooper Swan). The first occurrence of a Bewick's Swan appears to be one obtained at the River Bandon early in 1879; another was obtained (location unknown) early in 1881. Its Cork status did not change up to the 1950s when it was described as rare or absent and only one record was cited, a late bird seen near Bandon on 20 May 1949.

There were no other published accounts of the Bewick's Swan until 1964 when eight were seen at Ballincollig (Tanner's Pond, since filled in) in February. Through the rest of the 1960s small flocks (maximum 9) occurred regularly during November to February at sites ranging from Lissagriffin to Ballycotton, but they did not stay longer than a few days at any site. Most occurrences were at or near the coast.

The early 1970s brought about a major change in status in Cork. By then the River Blackwater callows between Fermoy and Lismore (Cork/Waterford border) regularly held up to 100 birds (peak of 117 in 1980), while Ballindangan Lough, near Mitchelstown, also held a population (numbers not stated). The presence of birds at the latter site depended on flooding. When dry, the birds moved to the River Blackwater, and birds from the Blackwater also visited the River Bride (north) near Conna. The only other site then known to hold a wintering population (numbers not stated) was Lissagriffin Lake.[15]

A population of Bewick's Swans developed at Ballycotton from the winter of 1969/70 and peaked at 194 in January 1991. A rapid decline set in after the winter of 1991/2 due to habitat change (see below for a wider decline in the population), and no birds have wintered there since 1996/7.

These foraged by day on agricultural land in the Cloyne and Ballymaloe areas (6 km away) but returned to roost at Ballycotton at night. During the 1980s a temporary food source (unharvested cereals) was available at Ballymacoda (8 km away) and up to 107 birds utilised it for several winters. These birds also returned to Ballycotton at night, but occasionally roosted on the estuary at Ballymacoda. When the Ballycotton flock was in decline in the early 1990s, some birds visited Ballybutler Lake (10 km away) for short periods. Other sites in the east of the county were often visited by small numbers of Bewick's Swans, but they rarely stayed very long, and it is likely all were part of the Ballycotton flock.

A flock averaging twelve birds occurred at Kilcolman during 1971–90, the maximum count being fifty-seven birds. However, this flock declined soon afterwards, and a maximum of only three was recorded after 1994/5.[16]

Winter atlas surveys showed the restricted distribution of the Bewick's Swan in the 1980s. The importance of Ballycotton and Ballymacoda was evident, as was that of Lissagriffin Lake and the River Blackwater callows. Elsewhere, just a few birds occurred at Kinsale, Ballincollig and Kilcolman. Twenty years later the Bewick's Swan occurred in only two 10 km squares (Ballymacoda and Dunmanway), both involving very small numbers which were not present every winter.

There has been a significant decline in the number of Bewick's Swans visiting Ireland in recent years. The most recent national census recorded only twelve birds in January 2020, with none in Cork.[17] It was known that Bewick's Swans formerly visited Ireland in greater numbers in cold winters, and the succession of mild winters in recent decades has resulted in birds not migrating so far south and west as Ireland from their Siberian breeding grounds (the phenomenon known as short-stopping).[18] The decline of the Cork population is part of this general trend. Although habitat change was taking place at Ballycotton at the same time as the national decline, if the habitat change had not taken place, then the Bewick's Swan population would have disappeared anyway, as has been the case elsewhere without habitat change.

Whistling Swan *C.c. columbianus*: There are two records of this subspecies, both adults. One at Ballycotton on 24 February 1979. One at Ballycotton on 3–18 February 1983, Ballymacoda on 18 February 1983, and Ballycotton on 15 January–27 February 1985.

Whooper Swan

Whooper Swan

Cygnus cygnus | Eala ghlórach | A

(monotypic)

World	Iceland, northern Europe, and northern Asia
Ireland	Very rare breeder, winter visitor
Cork	Locally scarce winter visitor, has increased
Conservation	Amber-listed

Some statements in the 1700s and 1800s regarding 'wild swans' must be viewed against the confusion then prevalent concerning identification of this species and Bewick's Swan. The first definite records of Whooper Swan appear to be of three birds obtained at Cork Harbour in January 1879. Nine Whooper Swans were seen at the River Ilen, near Skibbereen, in the hard frost of January 1881, and others were obtained at the same time, probably around Kinsale. The two migratory swans occurred in Ireland in the proportion of one Whooper to twenty-five Bewick's in the late 1800s, a strong indicator of the rarity of the former, and they chiefly occurred in hard winters.[19]

Whooper Swans increased in Ireland in the 1930s and 1940s, but it was not until about 1950 that they became more numerous than the Bewick's Swan.[20] Whooper Swans first appeared near Crookhaven (Lissagriffin Lake) in 1938, and by the early 1950s they were regular visitors, while they occurred near Glengarriff from 1944 and were particularly numerous there in the winter of 1946/7, although there is no further mention in the literature of them frequenting this area. In the 1950s, small numbers (up to 7) visited other sites in the south-west (Corran Lake, Shepperton Lake, Skibbereen, Lough Allua). Whooper Swans occurred for the first time at Ballynacarriga Lake and at the River Bandon, near Dunmanway, with records of thirty-eight at the former in March 1955 and fifteen at the latter in November 1957. The increase and spread continued in south-west Cork and by 1960 they were considered common. During this time there was only a single occurrence recorded for east Cork, one near Midleton in March 1959.

Whooper Swans first appeared at Kilcolman in the late 1940s, and they have been occurring ever since. Daily counts there from 1971 to 1990 revealed an increase of about 6 per cent per annum.[21] Maximum numbers increased from about twenty birds in the early 1970s to 120 in the late 1980s. Arrival dates in autumn varied from 1 October to 15 November, and the mean duration of stay each winter was 174 days (range from 138 days in 1971/2 to 203 days in 1981/2). The duration of stay increased significantly over the years. In years when the arrival date was early, departure was also early, and when arrival was late, departure was late. In all years except two, the largest numbers occurred in spring.

The River Blackwater callows between Fermoy and Lismore (Cork/Waterford border) was regularly holding about fifty birds by the early 1970s.[22] This flock has increased over the years to become the most important in the county. Numbers also increased at Ballycotton in the 1960s, and later peaked at seventy-six birds.

The distribution of Whooper Swans since the 1980s has included flocks at Kilcolman and the River Awbeg floodplains, around Glanworth on the River Funshion, the Rivers Blackwater and Bride (north), the River Lee lakes and reservoirs, River Bandon, and River Argideen floodplains. The main change has been the loss of birds from Ballycotton and other east Cork sites, due mainly to the unsuitability of Ballycotton as a safe night roost since about 1990. Small numbers also occur at Lissagriffin Lake and around Ballydehob and Skibbereen. Whooper Swans may visit freshwater sites across the county, especially during the autumn arrival period, but most stay only a short time before eventually finding their way to one of the main wintering areas.

Whooper Swans have been surveyed at five-year intervals in recent decades and the results show an increasing trend, with the highest count to date in the most recent survey (Table 7.1), mirroring the increase in the Icelandic breeding population.[23]

Irish and most British wintering Whooper Swans breed in Iceland. Soon after arrival there is some movement between sites, and there have been several sightings in Cork of colour-ringed birds that were recorded earlier and later at other sites within the same winter and between winters in Wales, northern England and Scotland.[24]

Survey month/year	Total birds	Total flocks
January 1986	185	7
January 1991	341	9
January 1995	219	9
January 2000	190	6
January 2005	320	4
January 2010	215	10
January 2015	244	9
January 2020	488	9

Table 7.1 County Cork totals for wintering Whooper Swan, 1986–2020.

Taiga Bean Goose

Anser fabalis | Síolghé Taiga | A

(subspecies: *fabalis*)

World	Subarctic taiga of northern Europe and northern Asia
Ireland	May have been commoner in 1800s, now rare winter visitor
Cork	Perhaps was commoner in 1800s, 33 individuals (3 records) since 1900
Conservation	Green-listed

There is evidence that a bean goose species occurred more commonly in Ireland in the 1800s, at least during severe winters. However, the practice of applying different names to the same species in different parts of the country must have caused confusion and misidentification. The former bean goose has recently been split into two species, Taiga Bean Goose and Tundra Bean Goose.[25]

On geographical grounds, the Taiga Bean Goose should occur more frequently in Ireland than the Tundra Bean Goose, though this has not been the case in recent years. However, assuming it was commoner in the past (but we cannot be certain), then the Taiga Bean Goose was occasionally plentiful in Cork in winter in the early 1800s, although said to be rare about Youghal. It was said to be very abundant about 1880, while at the end of the 1800s it was reported that a few were occasionally shot in west Cork.

There have been three records (33 individuals) of the Taiga Bean Goose since 1900, with most (32) occurring as part of an influx to Ireland in winter 1981/2. Two at Ballymacoda on 25 October 1981, joined by three others on 21 December 1981–January 1982. Up to twenty-seven at Ballintotis on 5–28 January 1982; some of these, possibly all, also visited Ballycotton and Ballymaloe. One at Ballymacoda on 28 April–16 May 1984. The thirty-three geese reported here were large, long-billed birds and fitted the criteria for identification as Taiga Bean Goose.

Tundra Bean Goose

Anser serrirostris | Síolghé Tundra | A

(subspecies: *rossicus*)

World	Arctic tundra of northern Europe and northern Asia
Ireland	37 individuals
Cork	2 individuals
Conservation	Green-listed

Adult at Kilcolman on 3–5 March 2006. One at the Gearagh on 27–9 April 2013.

Unidentified bean goose species

One bean goose, either Taiga Bean or Tundra Bean (*Anser fabalis/serrirostris*), was recorded at Dursey Island on 29 October 2015.

Pink-footed Goose

Anser brachyrhynchus | Gé ghobghearr | A

(monotypic)

World	Greenland, Iceland, and Svalbard
Ireland	Scarce winter visitor
Cork	107 individuals
Conservation	Green-listed

There were no records of Pink-footed Geese in Cork until November 1974. Since then, this goose has occurred with increasing frequency, and there have been records on the coast from Dursey Island to Ballymacoda. Sixty per cent of birds have occurred inland, mostly in the River Lee valley. Other inland records have been in the River Bandon valley, Cork Lough, east Cork and Kilcolman. Most birds occur during October to January (83 per cent), but there are records of apparently new birds in all months except

August. Most birds have been seen on one or a few days, suggesting onward movement to winter elsewhere, but a few have made stays of one or two months and occasionally up to four months. All but two records are of one to five birds, the exceptions being seven at Cork Harbour in October 1985 and twenty-nine in the River Lee valley in November and December 1992.

Pink-footed Geese wintering in Ireland are of Icelandic origin, and this population has increased in recent decades. The increase in Cork records probably reflects this population-level status change. However, there is a probability that some of the recent River Lee valley records originate from several birds reared at Farran Wood in the 1980s by the National Parks and Wildlife Service (NPWS). Since it is impossible to distinguish between wild and possible feral birds, all records from this area have been included in the table so that future population changes can be monitored.

A Pink-footed Goose at Ballycotton from 30 September to 7 October 1977 – believed to be of feral origin and which associated with feral Canada Geese and accompanied them to Cork Lough in October and remained to 29 May 1978 – has been excluded from the table. This bird was present until 1987, visiting Ballycotton in summer and autumn and returning to Cork Lough in winter.

1959–68	1969–78	1979–88	1989–98	1999–08	2009–18
0	3	15	33	12	44

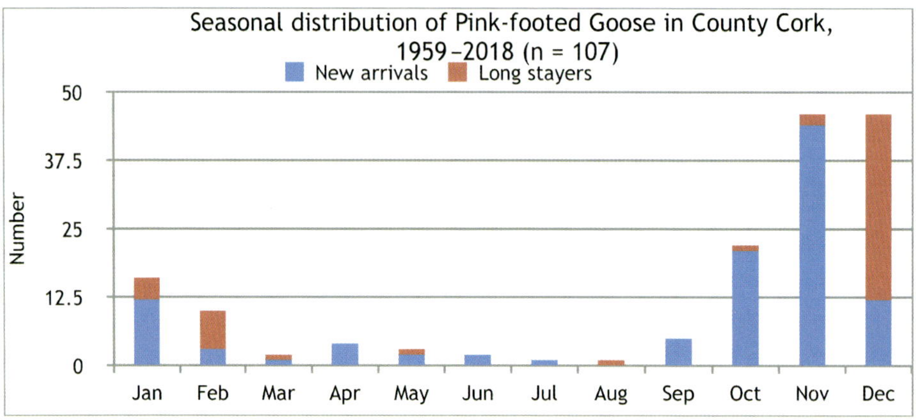

Fig. 7.1 Pink-footed Goose.

Greater White-fronted Goose

Anser albifrons | Gé bhánéadanach | A

(subspecies: *flavirostris; albifrons*)

World	Arctic regions of North America, also Greenland, north-east Europe, and Asia
Ireland	Winter visitor
Cork	Recent decline, scarce autumn passage migrant
Conservation	Amber-listed

The Greenland White-fronted Goose (*A.a. flavirostris*) was considered rare in the 1800s but had been obtained at Kilkerran Lake and near Youghal. It was sometimes common in severe weather and many birds were shot in fields during the frost of January 1881. It was regarded as a regular winter visitor to certain localities by 1900, but no details are available.

Very small numbers irregularly visited the Fermoy area in the early 1900s, and at the same time up to fifty visited Kilcolman. Between thirty and fifty geese occurred at Kilcolman in the 1950s, but numbers declined due to shooting pressure. Between thirty-five and forty were present in 1969, but by 1971 there were fewer than twenty despite legal protection of the habitat, and later in the 1970s only occasional (0–10) birds occurred.[26;27] An against-the-trend peak of forty-five was present in spring 1972.

Numbers at Kilcolman increased from 1979/80 and peaked at forty-two in 1986/7 but declined again to fifteen in 1990/1 and six in 1996/7. Through the 2000s the species was almost a rarity (maximum 3) and was absent in some years, and the first record since 2004 was on 31 May 2011.

In winters 1987/8 and 1989/90 peaks of eight and seventeen to thirty-eight, respectively, wintered at Ballycotton, possibly birds displaced from Kilcolman. Water levels were very low at Kilcolman in 1989/90, and flocks of fifteen and seven were seen overflying the site.

Elsewhere in the county this goose is rare and irregular, although small parties can appear anywhere in late autumn and winter at coastal sites, and occasionally inland. There has been a late autumn movement of small numbers passing east over coastal east Cork since at least the 1970s, and if they stop, they usually stay only a short time. Small parties may occasionally stay for longer, or even overwinter. A flock of thirty-five at Ballywilliam in November 1990 exceeded what might normally be expected.

A typical winter situation of a few birds occurring at sites in east Cork and at Kilcolman is well illustrated in the winter atlas for 1981–4. The complete lack of records during 2007–11 reflects the general decline of this goose in Ireland. There is an unseasonal record of a flock of eleven geese arriving exhausted on the sea off Ballycotton on 15 July 1969.

European White-fronted Goose *A.a. albifrons*: There are six records (13 individuals) of this subspecies. One at Ballymacoda in December 1980–May 1981. One at Cloyne and Ballymaloe in January–February 1982. Seven at Kilcolman in December 1997–March 1998. Two at Ballymacoda in December 2011. One at the Gearagh in November 2016–April 2017. One at Kilcolman in April 2017.

Greylag Goose

Anser anser | Gé ghlas | A, C1

(subspecies: *anser*)

World	Iceland, Europe, and Asia
Ireland	Scarce naturalised resident breeder, winter visitor
Cork	Scarce naturalised resident breeder, scarce winter visitor
Conservation	Amber-listed

The Greylag Goose was the only 'grey goose' species mentioned for Cork in the 1700s. It was said to be common in winter, frequenting uncultivated districts, and 'wild geese' visited a small lake at Dower during the great frost of 1739. A Greylag was obtained at Old Head of Kinsale in December 1878 and several others were shot there in the severe winter of 1880/1, and fifteen stayed some time near Cobh in January 1881.

It seems certain that the Greylag was scarcer in Ireland in the mid-1800s than either bean goose or Greater White-fronted Goose. During the early decades of the 1900s a wintering flock developed at Ballymacoda, which numbered fewer than fifty in the late 1940s and early 1950s.[28] The species appeared to be unknown elsewhere, apart from seventeen flying near Blarney on 28 August. The Ballymacoda flock continued until the mid-1960s, but in decreased numbers, and no regular wintering has taken place there since.

Published records indicate the presence of a small but regular wintering flock (up to 11 birds) at the River Lee reservoirs in the 1960s and early 1970s. A flock of thirty-two was seen at Castlemartyr in November 1968, but this site appears not to have been a regular haunt.

Since the 1970s wild Greylags have been rare and irregular in Cork, and the records in the recent winter atlases are typical, although some probably refer to naturalised birds. Only nine wild birds were recorded (in Cork Harbour) during a survey in March 1986.

Naturalised Greylag Goose: A large flock of 'wild geese', probably Greylags, were kept near Cobh in 1811, where they bred and were used for food. There are growing populations of naturalised Greylags at several sites in Cork. These populations originated from three sources. Those at the River Lee reservoirs are from a wildfowl collection established by NPWS at Farran Wood from the 1970s onwards. Some geese from Farran Wood were transported to Doneraile (managed by NPWS), and escapes from here established a breeding population at Kilcolman from the mid-1980s onwards. Greylags (and hybrids) have been kept at Cork Lough since at least the 1970s. Fota Wildlife Park was established in the early 1980s, and a population of Greylags has been there since. This collection is the main source of records of Greylags in and around Cork Harbour in recent decades.

A survey in 1994 gave a county total of 115 (92 adults and 23 goslings at six sites).[29] Numbers have increased since then and most birds are at Kilcolman (peak 100), the Gearagh (peak 250, September 2020) and Great Island (peak 35). Breeding continues at Kilcolman, Doneraile, River Lee system, and at least occasionally at sites in and around Cork Harbour. There are increasing feral breeding populations elsewhere in Ireland, in Britain and in Europe.

Barnacle Goose

Branta leucopsis | Gé ghiúrainn | A

(monotypic)

World	Greenland, Svalbard, and Novaya Zemlya
Ireland	Rare feral resident breeder, winter visitor
Cork	Rare in 1800s, 238 individuals since 1959
Conservation	Amber-listed

The evidence suggests the Barnacle Goose was scarce or rare on the coast of Cork in the 1800s. It was not recorded again until 1964.

Barnacle Geese have occurred at Dursey Island and CCBO more frequently than at all other sites during 1959–2018, these sites hosting 80 per cent of all birds. Elsewhere, a few records have come from sites along the coast from the Beara peninsula to Pilmore Strand. There is, however, a feral population of about five birds at Cork Harbour. All occurrences there since 1981 are almost certainly from Fota Wildlife Park, and these are excluded from the table. Small numbers are increasingly recorded inland at sites along the Rivers Lee and Bandon, and at Kilcolman. There is no evidence one way or the other regarding the origin of these birds, and in the absence of same they are included in the table. Nevertheless, the fact some occur late in the season (March to July) may indicate feral origin, although there is one record of two apparently wild birds at CCBO in June; it may be possible to determine their origin with a longer run of data.

Seasonal distribution is dominated by October and November records (71 and 16 per cent, respectively), and 6 per cent have occurred in March to July. The birds recorded at Dursey Island and CCBO are usually seen in flight on migration, predominantly in October. Those recorded at other coastal sites are mostly seen on one or a few days. Those that remain longest, sometimes for months, are often seen on the River Lee system. Most occur in flocks of from one to five birds, but at Dursey Island and CCBO flocks of forty-four, twenty-four, nineteen (twice) and eleven have been seen.

1959–68	1969–78	1979–88	1989–98	1999–08	2009–18
1	5	21	91	84	36

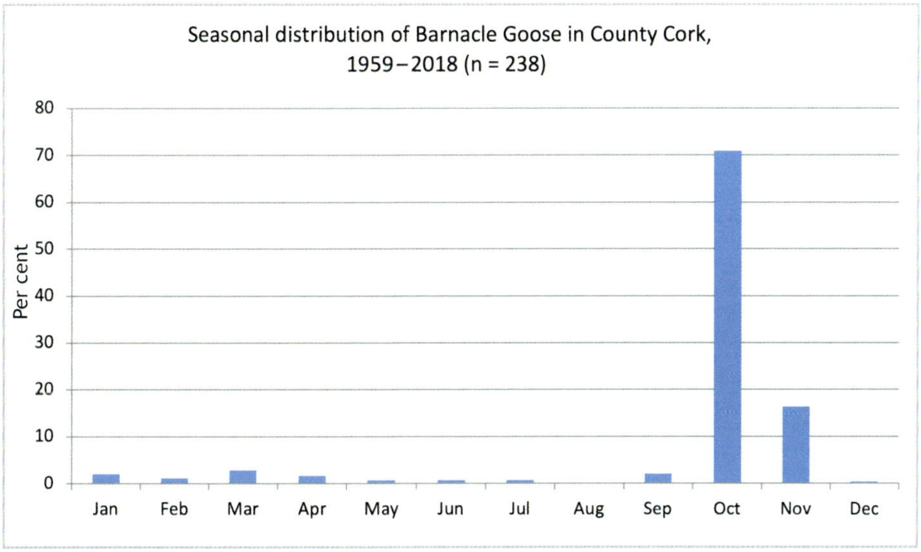

Fig. 7.2 Barnacle Goose.

Brent Goose

Branta bernicla | Cadhan | A

(subspecies: *hrota; bernicla; nigricans*)

World	Arctic North America, Greenland, Svalbard, Franz Josef Land, and Siberia
Ireland	Winter visitor
Cork	Winter visitor, recent increase
Conservation	Amber-listed

The Pale-bellied Brent Goose (*B.b. hrota*) was known as a Cork bird in the 1700s, although often then referred to, incorrectly, as the Barnacle Goose. The Brent was said to be more frequent and occurred in greater numbers than the Barnacle in the mid-1800s, and Cork Harbour and Youghal Harbour were among the chief haunts known at the time. During severe weather in January 1867 considerable numbers were shot near Youghal, and several were shot there also in winter 1878/9. Brent Geese were never seen at Bantry Bay, and they were said to be rare on the south coast between

Tralee and Wexford. The above comments must be viewed in the context of the large numbers present in Kerry and Wexford. The Brent was the most common goose on tidal waters in Cork by the late 1800s; it was said to be absent as a rule from Bantry Bay and occurred at Cork Harbour and Youghal Harbour in hard winters.

Its status in Cork in the early to mid-1900s is poorly documented, and it was variously described as rare and a straggler. No mention was made of it in the late 1940s and early 1950s.[30] However, it was apparently present in small flocks at a few places.[31] About thirty birds were known to winter regularly at Ballymacoda in the 1940s, but they declined and disappeared in the 1950s.

The Brent occurred at least occasionally from the 1940s onwards, and twenty-five were at Roaringwater Bay in September 1948, while others had been seen and obtained at Glandore and Rosscarbery. About thirty birds were also at Courtmacsherry Bay in September 1948, and abnormal numbers for the locality were passing north-west near Skibbereen in October 1949. Small flocks occurred fairly frequently in the 1960s and 1970s (e.g. 14 at Courtmacsherry Bay in September 1962; 16 at Ballycotton in February 1964; 11 at Ballymacoda in November 1964).

There has been a considerable increase in Cork since about 1980, and birds now winter regularly at several sites. The first winter atlas in the early 1980s showed the beginnings of the return of the species to sites in east Cork. The first wintering of birds at Ballymacoda was in 1986/7, and they have progressively settled at sites moving in a broadly westerly direction to Ballycotton, Cork Harbour, Courtmacsherry, Clonakilty and Roaringwater Bay. Numbers at Ballymacoda peak in spring before departure for the breeding grounds, when up to 700 have been counted. Numbers at other sites are smaller, and the peak count at Cork Harbour is 152. One out-of-season bird was at Ballymacoda from July to September 1978.

Brent Geese occasionally occur inland, e.g. one obtained at Buttevant in September 1903, one obtained at the Gearagh in October 1972, one intermittently at Cork Lough from November 1993 to April 1994 and December 1994, and two at Dooniskey in February 1998.

Dark-bellied Brent Goose *B.b. bernicla*: There are seven records (8 individuals) of this subspecies. One at Cork Harbour in March 1986. One at Ballycotton in January 1991. One at Ballycotton in October 1997. Two at Ballycotton in March–April 1999. One at Ballycotton in March

2001. One at Rosscarbery in December 2004. One at Dunboy Castle in January–February 2006. Seven birds at Courtmacsherry Bay (December 1978–February 1979) and two at Ballymacoda (January 1979) described as belonging to one of the dark-bellied subspecies cannot be accepted as either *B.b. bernicla* or *B.b. nigricans*.

Black Brant *B.b. nigricans*: There is one record of this subspecies. Adult at Cork Harbour on 6 December 2009–5 April 2010.

Ruddy Shelduck

Tadorna ferruginea | Seil-lacha rua | B

(monotypic)

World	Eastern Europe, Asia, and north-west Africa
Ireland	75 individuals
Cork	14 individuals (5 records)
Conservation	Not globally threatened

One obtained off Clonakilty in January 1871. Six (one female obtained) at Kinsale on 26 June 1886. Three (one male obtained) at Banteer on 16 July 1886. Three (one female obtained) at Cork Harbour on 4 August 1892. One obtained at Whitegate Bay on 30 January 1946.

Common Shelduck

Common Shelduck

Tadorna tadorna | Seil-lacha | A

(monotypic)

World	Europe and Asia
Ireland	Resident breeder, winter visitor
Cork	Local resident, winter visitor
Conservation	Amber-listed

The Shelduck was known in the 1700s, and breeding was reported near Midleton in the 1840s, with another breeding site near Youghal. It bred at inlets of Cork Harbour in the late 1800s, was noted as breeding at Youghal and Rostellan in July 1907 and was reported to be increasing in Cork about 1900.

Although Shelducks and their broods are occasionally seen on the sea, having bred on islands (such as Capel or Ballycotton) or mainland heathland backed by farmland, most favour a sheltered estuarine habitat. Breeding atlas surveys reflect this preference where they are scarce or absent from the large bays of the west which have little if any such habitat, although they breed sparingly on low islands in Roaringwater Bay.

In the first survey breeding was confined to the coast, and they occurred continuously from Crookhaven to Youghal, there being little change in distribution during 1968–2011. The breeding population has been estimated at 100–200 pairs. Single pairs bred inland at Charleville in 1978, 1987 and 1995, at Mallow lagoons in 1994 and 1995, and at Ovens in 1995. The main change has been a recent move into the lower reaches of the River Lee.

Winter distribution mirrors the breeding season, with a few inland occurrences on the River Lee and elsewhere. Complete counts of the county in January 1992, 1993, 1994 and 1995 gave totals of 2,877, 3,174, 3,532 and 2,514, respectively. Cork Harbour is the most important site for Shelducks. Totals of 2,415 were counted in March 1968, 2,683 in January 1979 and 3,765 in December 1981. However, numbers there have declined significantly in the last decade to an all-time low of 694 birds in 2019/20. Elsewhere, flocks are much smaller, such as at Clonakilty, Courtmacsherry, Bandon estuary, Ballycotton and Ballymacoda where numbers rarely exceed 200 birds. Numbers at some of these sites, such as Ballymacoda, also increased through the 1970s and 1980s, but have declined again in recent decades.

The Shelduck performs a moult migration to Heligoland Bight in north-west Germany, and ring recoveries show that at least some Irish birds go there. There is one record of a bird ringed there while in moult and recovered in Cork in winter. Pairs which have bred depart for the moulting grounds in July, leaving the young in a crèche led by one or two pairs of adults. Adults arrive back on the wintering grounds from October. Young also leave the breeding area after reaching the flying stage, and few are present in October at Ballymacoda. One was killed striking the light at Fastnet Rock in October 1920.

Common Wigeon

Mareca penelope | Rualacha | A
(monotypic)

World	Iceland, northern Europe, and northern Asia
Ireland	Very rare breeder, common winter visitor
Cork	Common winter visitor
Conservation	Amber-listed

Although the Wigeon has occasionally been seen in the breeding season at coastal and inland sites, there are no records of breeding for Cork.

The Wigeon occurred in bays and harbours, and inland at Longueville duck decoy lake in winter during the 1800s. Recent winter atlas surveys showed that Wigeon are widespread at coastal and lowland inland sites. On the coast, Wigeon are commonest at estuaries and the population thins out or disappears in the deeper bays of the west, such as Bantry, Dunmanus and Roaringwater. They are common on the River Lee system of lakes and reservoirs, some parts of the River Blackwater valley and in Cork Harbour, and at several sites in east Cork.

Counts across the whole county in January 1992, 1993, 1994 and 1995 gave totals of 9,772, 6,818, 9,605 and 7,414, respectively. The first birds arrive for the winter in late August. Numbers build up to a peak in December and January, and the main exodus takes place in March. The most important sites are the Gearagh (2,000), Cork Harbour (2,000), Kilcolman (1,000), Ballymacoda (1,000) and Courtmacsherry (1,000). Smaller numbers occur at several other sites, such as Clonakilty, Ballycotton and Charleville lagoons. It should be noted that numbers have declined at all sites, for example only two counts at Cork Harbour exceeded 1,600 in the 2010s.

Wigeon wintering in Ireland (and Cork) originate in Scandinavia, Siberia and Iceland. There are recoveries in Cork of birds ringed in mainland Scotland, Shetland, Netherlands and Sweden.

American Wigeon

Mareca americana | Rualacha Mheiriceánach | A

(monotypic)

World	North America
Ireland	Rare winter visitor
Cork	39 individuals
Conservation	Not globally threatened

The American Wigeon has occurred regularly in Cork, but only since 1985. Like some other Nearctic ducks, this species may stay for many months, and return to the same site in successive winters, which makes it difficult to calculate total numbers. New birds have been recorded between September and May, with most from October to December. At least thirty-nine different birds have occurred, with about twenty remaining for periods ranging from one to six months; one other remained for more than seven months (8 April–22 November), the only one to stay throughout a summer season. On the other hand, some new birds (11) have been recorded on only one day, including at sites where they are unlikely to have been overlooked. Many birds apparently returning in successive winters showed a similar occurrence pattern to new birds in terms of length of stay, both long and short. The same birds have, it seems, been present at nine sites during two to six successive winters, with Ballintubbrid, 1986/7–1991/2 (absent in 1987/8), having the longest successive run. However, Ballycotton holds a similar record and may possibly exceed Ballintubbrid by two winters, but the presence of three birds in the early years makes identification of individuals uncertain.

American Wigeons have occurred at coastal sites from Adrigole to Ballymacoda, and at inland sites at the River Lee reservoirs (5), Charleville lagoons (3) and Kilcolman (2). Most have been recorded at sites within Cork Harbour (12), Ballycotton (4), Reenydonagan Lake (3), Clonakilty Bay (3), Timoleague (2) and Ballymacoda (2), with singles at Adrigole, Lissagriffin and Rosscarbery. Most records involved single birds, but three occurred together twice (Ballycotton and Rostellan) with two together on four occasions. Of the thirty-nine new birds, the sex of thirty-eight was established on first arrival, thirty-five males and three females. Of those aged, seventeen were adult and seven were in their first year. Assuming there is an equal probability of males and females, and adults and young occurring as vagrants, then females and young birds are likely to have been overlooked because of their more cryptic plumage and similarity to Common Wigeon.

1959–68	1969–78	1979–88	1989–98	1999–08	2009–18
0	0	6	18	7	8

Earliest and latest occurrence in winter (all records): 26 September–12 May.

Gadwall

Mareca strepera | Gadual | A

(monotypic)

World	North America, Iceland, Europe, and Asia
Ireland	Rare resident breeder, uncommon winter visitor
Cork	Rare breeder, scarce and declining winter visitor
Conservation	Amber-listed

The earliest record in Cork appears to be one obtained in December 1849 on the River Lee (location unknown). One was obtained near Blarney in January 1850, while others were obtained at Bishopstown (winter 1878/9), Ballycotton and near Upton (both January 1881), the latter out of a flock of six. The Gadwall, although not a common duck in Ireland, had been obtained in all parts of the county and had visited the Longueville duck decoy lake every year, specimens having been obtained there in four winters between 1893 and 1897, three being taken in January 1894 alone.

Three pairs were at Garryvoe until 17 April 1946, but there was no proof of breeding. The first breeding record was in 1977 when a pair bred at Lough Rhue. Breeding was not repeated, and the lake was drained in the early 1980s. Breeding took place at Ballycotton in 1981 (5 pairs), 1983 (1 pair), 1987 (5 pairs) and 1988 (5 pairs). A pair bred at Calf Island East in 1990, 1991, 1992, 1996 and 2005 (and probably also in 1997, 1999 and

2004), and the species also bred at the Calf Islands in 2017. Breeding was suspected at Ovens gravel pit in 1993. There was one record of breeding in the Glandore area during 2008–11, and several cases of probable breeding elsewhere in west Cork and east Cork, some of which may relate to lingering winter visitors.

Annual in winter in small numbers since before 1950, the Gadwall increased in the 1950s and 1960s, especially in east Cork. Numbers at Ballycotton peaked at 270, only to decrease to zero after 1990 with the draining of the lagoon. Ballycotton birds often visited other lakes in east Cork, especially Lough Aderry (peak of 256, December 1987), but here the numbers steadily declined in succeeding years, and currently numbers are about fifty.

Cork Harbour has a small population, mainly at Rostellan and Whitegate, but numbers have declined to single figures in recent years. There is also a regular flock at Kilcolman, which has an average winter population of fewer than ten. The River Lee system is also visited by small numbers (peak of 20), and Blarney Lake had fourteen in December 2006. Elsewhere, the Gadwall is very scarce, although flocks of fewer than ten can turn up anywhere on the coast or on lakes near the coast.

Ring recoveries suggest that Gadwall wintering in Ireland originate in Iceland, Britain and northern Europe, but there are no recoveries relating to Cork.

Common Teal

Anas crecca | Praslacha | A

(subspecies: *crecca*)

World	Iceland, Europe, and Asia
Ireland	Uncommon resident breeder, common winter visitor
Cork	Scarce breeder, common winter visitor
Conservation	Amber-listed

The Teal was probably a widespread breeding species before the drainage of wetlands. It was described as resident and breeding in small numbers in the late 1800s, but the only breeding location mentioned was Annagh Bog (Inishannon) in 1849.

One pair bred at CCBO in 1960, and breeding was suspected at Sheep's Head in 1967. During 1968–72 breeding was proved (or considered probable) in only five 10 km squares (Beara peninsula, south-east of Bantry, Kilcolman, Minane Bridge and Cork Harbour). There was no proof of breeding anywhere during 1988–91, and birds were seen at only a few scattered sites. One pair bred at Calf Island East in 1992. Records of Teal were more widespread in 2008–11 than in previous breeding season surveys, especially along the course of the River Lee, and most related to cases of probable breeding, while there was one case of confirmed breeding at Kilcolman.

Teal are widespread in winter, inland as well as on the coast. Counts of the whole county in January 1992, 1993, 1994 and 1995 gave totals of 3,710, 4,789, 6,498 and 3,757, respectively, although only a few sites held flocks of over 150. The earliest birds arrive in late July, and numbers slowly increase to a peak in December. Some Teal are already beginning to disperse by January, and few remain by April.

The most important sites are Cork Harbour (2,000), Ballymacoda (1,000), Ballycotton (1,000), Inishcarra reservoir and the Gearagh (1,000), Kilcolman (500) and Charleville lagoons (900), with smaller numbers at Clonakilty, Courtmacsherry and Bandon estuary. Note that declines have taken place at many sites in recent years, including at Cork Harbour.

Most Teal wintering in Ireland (and in Cork) originate in Scandinavia and the Baltic states, many of which are ringed when on migration in the Netherlands. Smaller numbers come from Iceland and Britain, and there may be considerable onward movement to France and Iberia during cold spells.

Green-winged Teal

Anas carolinensis | Praslacha ghlaseiteach | A

(monotypic)

World	North America
Ireland	Rare visitor, mainly in winter
Cork	97 individuals
Conservation	Not globally threatened

Green-winged Teals have occurred on the coast from Lissagriffin to Ballymacoda, with 53 per cent seen at Cork Harbour, Ballycotton and Ballymacoda combined. Clonakilty Bay is the most frequented west Cork site at 10 per cent. Twenty-four per cent have occurred inland, some at lakes within 10 km of the coast, but most at the River Lee lakes and reservoirs, and at Kilcolman and Charleville. Most have occurred from November to February, but there are records for October through to May. While many birds have been seen on a single day, many also make extended stays, some over a full winter period. The occurrence pattern is of a winter visitor with most first seen in November and December. The indications are that the same individuals occur in successive years at the same sites, especially in east Cork, Cork Harbour and on the River Lee, suggesting that a small population is permanently resident in Europe. Only males have been recorded, females being identical to Common Teal. Four occurrences have involved two birds, at Cork Harbour, Ballycotton, Ballymacoda and Kilcolman, all other records involving single individuals.

1959–68	1969–78	1979–88	1989–98	1999–08	2009–18
2	12	12	27	27	17

Earliest and latest occurrence in winter (all records): 9 October–22 May.

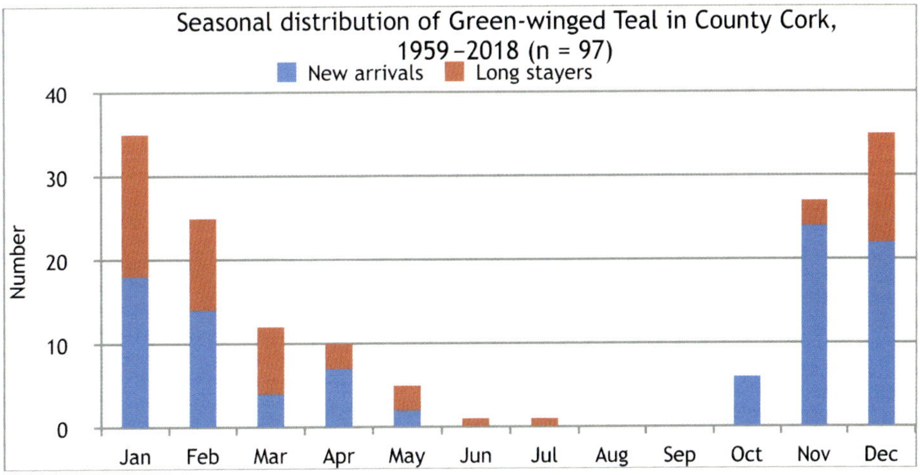

Fig. 7.3 Green-winged Teal.

Mallard

Anas platyrhynchos | Lacha fhiáin | A, C1
(subspecies: *platyrhynchos*)

World	North America, Europe, Asia, and north Africa, introduced elsewhere
Ireland	Common resident breeder, winter visitor
Cork	Widespread breeding resident and winter visitor
Conservation	Amber-listed

The Mallard is common and widespread as a breeding species and was present throughout the county during all three recent breeding atlas surveys. Mallard breed in all types of wetland habitats, including rivers, lakes, ponds and marshes, but also often on hills far from water, on small islands without freshwater (Capel Island), and occasionally in unusual locations such as suburban and urban gardens. Most Mallard breed among vegetation on the ground, although nests have been found at elevated sites. A nest 9.15 m up in a Scots pine was found at Castlemartyr in the 1800s, and a nest at a similar elevation was seen at the same location in the 1980s. Breeding density data are scarce, and 120 pairs were estimated at Ballycotton (33 ha) in 1968. The number of pairs breeding at Kilcolman (53 ha) varies from a few to fifty, and up to twenty-five pairs bred at Ballymacoda in the 1980s.

Mallards are equally widely distributed outside the breeding season. Large numbers gather in post-breeding flocks from August onwards, but these disperse after the start of the shooting season in September. Many hundreds frequented the decoy lake at Longueville in winter, and up to 300 were taken annually. Flocks of 200 plus are widespread and the largest numbers occur at the Inishcarra reservoir and the Gearagh (1,750), Kilcolman (1,500) and Cork Harbour (formerly 1,500, but 400 recently). Smaller and more variable numbers occur at other sites, such as Ballymacoda and Ballycotton.

Numbers have frequently been supplemented with birds released by gun clubs. A total of 797 ducklings was released during 1982–4. Releases take place in some locations for shooting (e.g. Rostellan), and in others to populate protected areas (e.g. Lough Aderry) and urban waterways (e.g. Cork Lough and Atlantic Pond). Some releases include hybrids or birds of domestic origin. Many such birds remain tame but often breed and at least some of their offspring presumably naturalise into the wild population.

Some British hand-reared Mallard occur in Cork, while many wild birds migrate into Ireland (and Cork) in winter from Britain, with smaller numbers coming from Europe.[32] One ringed as a duckling in Latvia in June 1985 was recovered near Coachford in January 1987.

American Black Duck

Anas rubripes | Lacha chosrua | A

(monotypic)

World	North America
Ireland	22 individuals
Cork	7 individuals (6 records)
Conservation	Not globally threatened

Male at Ballycotton on 26–8 January 1993, and Lough Aderry on 30 January–15 March 1993. Male at Kinsale Marsh on 13 March 1994. Male at Lissagriffin on 3 September–mid-October 1994, and 19 August–12 October 1995. Male at Kilcolman on 4 March–7 May 2006. Two males at Mizen Head and Crookhaven on 7 February–9 March 2013. Male at Baltimore on 11–29 February 2016. All seven individuals were males; perhaps some females are overlooked as dark or melanistic female Mallard.

Pintail

Anas acuta | Biorearrach | A

(subspecies: *acuta*)

World	North America, Iceland, Europe, Asia, and north Africa
Ireland	Very rare breeder, winter visitor
Cork	Scarce winter visitor
Conservation	SPEC 3; Amber-listed

The Pintail was variously described as occasionally occurring in Cork in considerable numbers in winter, and as occurring annually, but sparingly, in the 1800s, without mention of locations. It occurred annually in small numbers in winter at Longueville duck decoy, where thirty to forty males were once seen (presumably females also occurred).

During the first half of the 1900s the Pintail was considered of irregular occurrence, except in Cork Harbour. There were also stray reports from near Fermoy, and a pair once stayed to 20 April at Lough Rhue. It regularly occurred in flocks of fifty to seventy at Cork Harbour (possibly totalling 100 birds) during the 1950s and 1960s. Up to eighty-five also occurred at Courtmacsherry in the 1950s and 1960s. It was stated in 1969 that Pintails could be met with widely in small numbers in west Cork, and the same applied at Ballycotton and Ballymacoda in east Cork.

During winter atlas surveys the population was concentrated at coastal sites from Roaringwater Bay to Ballymacoda, with some inland at the Gearagh and Kilcolman. There are no breeding records for Cork, although a few birds may linger late at the end of winter.

There were high counts in Cork Harbour of ninety-three in 1995/6 and sixty in January 2002. Various parts of Cork Harbour have been favoured, such as Whitegate, Rostellan and Great Island Channel, and movement between sites probably takes place. Seventy-one were on flooded fields at Cloyne in January 1996. Small numbers occur at Ballymacoda in most years, with a peak of thirty-three. A flock of forty flew east off Dursey Island on 15 October 1993. Ten were at the Gearagh on 26 November 2004. Numbers have declined at all sites, and most counts at Cork Harbour in the 2010s were less than thirty, with a peak of fifty-one in 2018/19.

A first-year male ringed at Ottenby (Sweden) in August 1976 was recovered at Clonakilty in January 1977.

Garganey

Spatula querquedula | Praslacha shamhraidh | A

(monotypic)

World	Europe and Asia
Ireland	Very rare summer visiting breeder, scarce passage migrant
Cork	143 individuals
Conservation	SPEC 3; Amber-listed

There were four records involving six individuals in the 1800s, one before 1845, one about 1863, three in March 1878 and one in March 1880. The latter four were all at Cork Harbour. Three singles were recorded between 1900 and 1958 (Lissagriffin, April 1946; Whitegate, December 1947; Cobh, November 1957). The Cobh bird had been ringed as full-grown in the Netherlands in August 1955.

The Garganey has occurred frequently, but not annually, during 1959–2018, most regularly over the last thirty years when a mean of thirty per decade have been recorded. Most occur in two periods, in April and May, and in August and September. They have occurred at many coastal sites from Dursey Island to Youghal, most frequently at Ballycotton and Cork Harbour. The Garganey has also occurred inland at nine sites, mostly at lakes within 10 km of the coast, but others have been recorded at the Gearagh, Mallow, Charleville, and especially at Kilcolman where most inland records have occurred, although there has been no evidence of breeding.

About 75 per cent of birds have been seen on one to five days, but some make longer stays, one for sixty-six days (mean stay of 4.7 days, n = 99). Most birds occur singly, but five and six have occurred together twice each. Males have been recorded about three times more frequently than females (62 males, 23 females), probably due to the cryptic plumage of the latter.

1959–68	1969–78	1979–88	1989–98	1999–08	2009–18
9	22	12	32	29	30

Earliest and latest occurrence in spring and autumn (1959–2018): 10 January–24 June and 7 August–6 November.

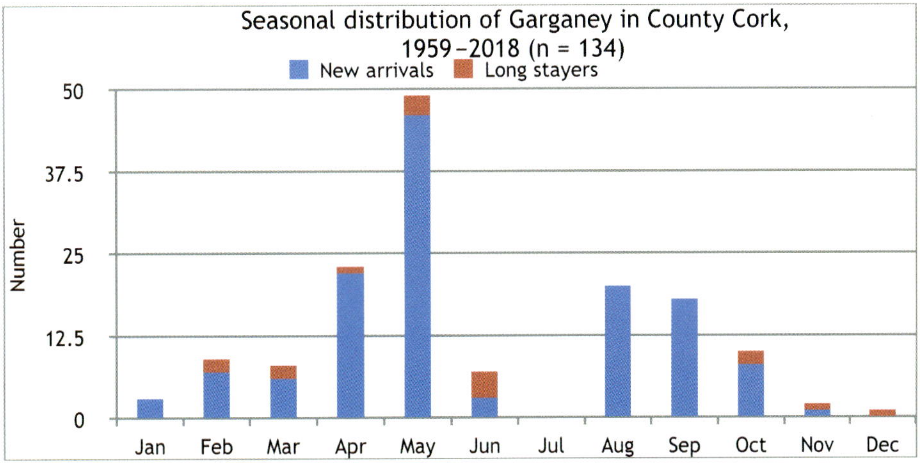

Fig. 7.4 Garganey.

Blue-winged Teal

Blue-winged Teal

Spatula discors | Praslacha ghormeiteach | A

(monotypic)

World	North America
Ireland	Rare winter visitor
Cork	18 individuals
Conservation	Not globally threatened

Three Blue-winged Teals were obtained (all shot) before 1959; at Ballycotton in September 1910, Garryvoe in December 1950, and Corran Lake in November 1955.

A further fifteen individuals occurred during 1959–2018. There was one in 1975, followed by two clusters of records, one involving nine during 1985–99, and one involving five during 2010–14. Most birds (11) were first recorded in September to December, but there were also records in February and May (2 each). Twelve occurred at scattered coastal districts from

Crookhaven to Ballymacoda, and three occurred inland at Ballyhonock Lake, the Gearagh and Charleville lagoons. Three sites have had more than one individual: Clonakilty (2), Ballycotton (2) and Ballymacoda (3). One record refers to more than one individual, a pair at Ballymacoda in February 1995. Blue-winged Teals are generally present for one to three days (10), but four have remained for longer periods, one for seventy-three days (mean stay = 9.1 days). Of those that have been aged and sexed, three were adult, two were first-year, seven were male and six were female. There is no evidence of a return to the same site in successive winters, as is the case with some Nearctic ducks.

1959–68	1969–78	1979–88	1989–98	1999–08	2009–18
0	1	3	5	1	5

Earliest and latest occurrence in winter (1959–2018): 19 September–26 February (7–27 May).

Shoveler

Spatula clypeata | Spadalach | A

(monotypic)

World	North America, Iceland, Europe, and Asia
Ireland	Scarce resident breeder, winter visitor
Cork	Rare breeder, scarce winter visitor
Conservation	Red-listed

The Shoveler was said to occur frequently in Cork in the 1800s, particularly in immature plumage. Individuals were obtained at Cork Harbour in January and April 1846, with one or two offered for sale at the Cork market each winter, birds being often shot around Cork and Bandon. It occurred as a winter visitor in small numbers at Longueville duck decoy lake where four were taken in winter 1845/6 and nine seen on 25 January 1891.

There are very few records of breeding in Cork. Ussher and Warren indicated that it had bred but cited only two cases of birds seen in May and

June, at Kilcolman (8 birds) and Longueville (1 bird), respectively. Neither of these can be accepted as a breeding record. Therefore, the statement that it bred occasionally in the late 1800s must be discounted.[33] It was described as a rare breeder in the first half of the 1900s, but no date of first breeding or the sites frequented was given.

Shovelers were known to breed occasionally at Kilcolman before 1978 but had not done so for some time. One to four pairs bred there from then until 1985, when four pairs produced thirty-six young. Two pairs bred in 1993 (12 or 13 young), two pairs summered in 1994, and two pairs bred in 1998 (1 with 9 young), and breeding took place during 2006–10. One pair bred at Lough Aderry in 1979. One pair bred at Ballycotton in 1981 and summering birds were recorded there in nine other years during 1968–89. Summering birds were recorded at Kilcolman and Charleville lagoons in 1968–72, while in 1988–91 there were records of a few birds at east and north Cork sites, but without evidence of breeding. There were probable breeding records at six 10 km squares and proved breeding at one (Kilcolman, 2006) during 2008–11.

The main concentrations during winter atlas surveys were at the River Lee lakes and reservoirs, at Cork Harbour and at several sites in east Cork. Important concentrations also occurred at Kilcolman, Charleville and at the River Funshion valley around Glanworth. Shovelers were absent from coastal areas, except as vagrants, west of Cork Harbour.

The Shoveler has declined at all sites in the last decade. Ballycotton, which held up to 200 in the 1980s, has been deserted due to habitat change. Several sites within Cork Harbour, including Cork Lough, held a total of about 100 birds in the 1980s and 1990s; Cork Harbour numbers rarely exceeded thirty birds in the 2010s. Kilcolman usually holds up to 100, with maximum counts of 300–400 in February 1974. Charleville lagoons has had counts of up to 205 in December 1993, but usually fewer. Up to fifty occur at Carrigadrohid, but numbers at Lough Aderry now rarely exceed ten. Shoveler numbers at Ballymacoda exceptionally reach sixty-five but are usually much lower.

There are two recoveries of Shoveler. A full-grown bird ringed in the Netherlands in September 1957 was recovered at Buttevant in January 1958, and one ringed as a duckling in Latvia in June 1980 was recovered at Pilmore Strand in November 1980.

Red-crested Pochard

Netta rufina | Póiseard cíordhearg | A

(monotypic)

World	Europe and Asia
Ireland	70 individuals
Cork	5 individuals
Conservation	Not globally threatened

Male obtained at Reenydonagan Lake on 29 December 1927. Adult male obtained at Youghal on 30 January 1977. Female at Cork Lough in early November 1990. One at Curraghalicky Lake on 21–2 July 1995. Female at Rostellan on 27–30 October 1996. The Youghal bird was ringed as a first-year male on 17 July 1975 at Guadalquivir Delta, Seville (Spain). There is an increasing feral population in Britain, largely confined to the south and east and originating from escapes and releases from waterfowl collections.

Common Pochard

Aythya ferina | Póiseard | A

(monotypic)

World	Iceland, Europe, and Asia
Ireland	Rare resident breeder, common winter visitor
Cork	Rare breeder, common winter visitor
Conservation	SPEC 1; Red-listed

The Pochard was a winter visitor to inland and tidal waters in the 1800s, and it was described as rare at Youghal, although not rare at Cork Harbour.

The Pochard bred for the first time in 1978 with one pair at Ballycotton. One to three pairs continued to breed there to 1988 or 1989, after which the lagoon was drained. Two pairs bred at Kilcolman in 1995, and one pair bred in 1998, 2008 and 2010.

The first winter atlas showed the Pochard had four main centres of population in Cork. These were at the River Lee lakes and reservoirs, Cork

Harbour along with lakes in east Cork, Kilcolman and Charleville in the north, and lakes in the Clonakilty area. While the range did not change significantly during the second winter atlas, there were more losses than gains from 10 km squares.

Pochards are predominantly winter visitors arriving in September and peaking in December. Few remain after the end of March, although small numbers were often present in summer in the late 1960s and early 1970s. Kilkerran Lake held up to 400 birds in the 1950s, but numbers declined in the 1970s and 1980s and hardly any visited the site by the early 1990s. The main population centre shifted to the Lee reservoirs, particularly Carrigadrohid. A maximum count of 550 occurred at the Lee reservoirs in November 1977, but numbers vary greatly and there has been a decrease in recent years. Ballyhea gravel pit held up to 260 in the late 1990s. Other important locations then were Charleville lagoons (peak of 156 in November 1986), Castlemore gravel pit (up to 100), Rostellan (usually fewer than 100), Kilcolman (115 in January 1994), Ovens gravel pit (fewer than 50), and Castlenalact Lake (fewer than 50). Many other sites regularly held smaller numbers.

Numbers have declined throughout the county in recent years and the above sites now hold many fewer birds, for example none now occur in Cork Harbour. Reenydonagan Lake formerly held up to eighty birds, but few were present by the late 1990s. Ballycotton is no longer used since drainage, and sites that held small numbers in east Cork in the 1970s, such as Lough Aderry, now usually hold none.

Redhead

Aythya americana | Póiseard ceannrua | A

(monotypic)

World	North America
Ireland	1 individual
Cork	1 individual
Conservation	Not globally threatened

Male at CCBO on 12–15 July 2003.

Ring-necked Duck

Aythya collaris | Lacha mhuinceach | A

(monotypic)

World	North America
Ireland	Rare winter visitor
Cork	77 individuals
Conservation	Not globally threatened

Occurrences of this Nearctic duck are characterised by its propensity to remain for several months, to move about between nearby waters, to occur in small groups, and to re-appear at the same site in successive winters. Occurrences in successive winters are here treated as different for clarity of expression, even when it is likely the same individuals are involved. Most birds (33) have occurred on the River Lee system of interconnected lakes and reservoirs. Twenty-one have occurred at east Cork lakes, inclusive of Cork Harbour, Cork Lough, Blarney and Garryhesta quarry. Nineteen have occurred at a series of small nearby west Cork lakes, with three in north Cork (Mallow, Kilcolman and Charleville) and one at CCBO. The spread of sites visited extends from west Cork to Youghal, and north to Charleville.

Only one has occurred outside the October to April period, in June at Charleville lagoons. While some birds have been seen on only one or a few days, many others have remained for several months or throughout the winter period. Single individuals are the norm, but two (not necessarily pairs) have been seen on several occasions, and three were at Kilkerran Lake in October to December 1979, while four and six were at the River Lee system in November 1977 to January 1978 and February to April 1990, respectively, and eight visited Gallanes Lake in December 2018. Only two birds have remained into early May, Gallanes Lake in 1974 and the Gearagh in 1990. Birds seen for the first time late in the winter may be individuals that had arrived earlier but went undetected perhaps at other sites. There has been a tendency for birds to move about between sites, especially in west Cork, the River Lee system and the Cork Harbour area, and this behaviour could lead to some going undetected for some time at sites less frequented by birdwatchers. The sex ratio of sixty-six individuals has been forty-seven males and nineteen females.

1959–68	1969–78	1979–88	1989–98	1999–08	2009–18
0	9	12	14	15	27

Earliest and latest occurrence in autumn and spring (all records): 3 October–14 April.

Ferruginous Duck

Aythya nyroca | Póiseard súilbhán | A

(monotypic)

World	Europe and Asia
Ireland	35 individuals
Cork	3 individuals
Conservation	Not globally threatened

Adult male at Kilcolman on 20 January–14 February 1991; subsequently seen every winter to 8 October 1999, mainly present at Kilcolman but also visited Ballyhea gravel pit and Charleville lagoons; earliest date was 8 October and latest 1 April. Adult male at Castlenalact Lake on 11 December 1994–2 March 1995, and 1 November 1995–6 (also visited the Gearagh on 29 December 1994). Adult male at Kilcolman on 1 May 2001. The 1991 Kilcolman bird was believed to be the same as one seen at Lough Gur (Limerick) on 17 February 1991. Note the number of sites visited by this individual, and the fact it returned to the same area over ten successive winters.

Tufted Duck

Tufted Duck

Aythya fuligula | Lacha bhadánach | A

(monotypic)

World	Iceland, Europe, and Asia
Ireland	Resident breeder, common winter visitor
Cork	Scarce breeder, declining winter visitor
Conservation	SPEC 3; Amber-listed

The Tufted Duck was described as common in winter and five specimens were known from around Youghal in the 1800s, and it also occurred at Cork Harbour. The earliest evidence of breeding is from Harold Barry who told Ussher that the nest had been found at Kilcolman, where Ussher himself saw birds in May 1899. The Tufted Duck was probably a scarce breeding species at the end of the 1800s.

During the first breeding atlas distribution was mainly centred on the River Lee reservoirs and lakes. Other populations were in the north (Kilcolman area), Shreelane Lakes, Kilkerran Lake, near Bantry, near Blarney and Lough Aderry. The total population was then estimated at 100 pairs. Thirteen pairs bred at Inishcarra in 1977. There was a major range contraction in 1988–91 with breeding recorded at only two sites, the Gearagh and Charleville lagoons. However, the range expanded again in 2008–11 with breeding mostly confined to the River Lee system and north Cork. Some new 10 km squares were occupied near the coast in west Cork, but without evidence of breeding, although none occurred at Lough Aderry then, or since.

Tufted Ducks mainly occur in freshwater habitats and were present at three main centres during the first winter atlas survey. These were at the River Lee lakes and reservoirs, Cork Harbour and east Cork, and lakes in the Clonakilty area; other lakes held very small numbers. In 2007–11 the main populations were at the River Lee system and north and west Cork lakes, with rather few in east Cork.

In the early 1950s up to 670 Tufted Ducks wintered at Kilkerran Lake, but numbers declined significantly thereafter, and 100–250 occurred there in the 1970s. The decline at this site may be related to the creation of the River Lee reservoirs during the 1950s, the centre of population shifting to these new habitats (see Common Pochard). In the late 1990s the River Lee reservoirs regularly held 400–550 birds. Most other lakes hold up to fifty and sometimes over 100 birds.

Tufted Ducks have declined at all sites in the last decade. Numbers at Cork Harbour have fallen from a peak of 123 in 1978–81 to a mean of twenty-five in the 2010s. None now usually winter at east Cork lakes.

Tufted Ducks wintering in Ireland originate in northern Europe and Iceland, and one from Scotland has been found in Cork.

Greater Scaup

Aythya marila | Lacha iascán | A

(subspecies: *marila*)

World	North America, Iceland, northern Europe, and northern Asia
Ireland	Has bred, winter visitor
Cork	Scarce winter visitor
Conservation	SPEC 3; Red-listed

The Scaup was a common winter visitor to tidal waters in the 1800s but was rare in the Youghal area. It was less common in the south than elsewhere in Ireland, although it had been obtained at an unnamed inland lake in

Cork. A notable exception to the scarcity on the south coast was the regular presence of a small flock at Bantry Bay during the first half of the 1900s where only females occurred.

Cork Harbour held a regular flock in the late 1960s with a peak of twenty-six in the winters of 1965/6 and 1966/7.[34] They were apparently scarce or absent in the early 1970s, but numbers increased in the mid-1970s and early 1980s with a peak of ninety-one in February 1979. They have declined again and occurred only in low single figures throughout the 1990s, 2000s and 2010s.

A few typically arrive in late September but the main arrival is in late October and November. They remain to March, with a few to April or May. A pair at Lough Beg in June 1982 and individuals at Reenydonagan Lake in August 1978 and Ballycotton in August 1979 are the only unseasonal records.

The Scaup is scarce and irregular away from Cork Harbour, and recent winter atlas surveys show it as occurring at several sites along the coast and inland. Inland records were scarce until the mid-1980s, but since then have become increasingly frequent and one to four have occurred at several sites: Bateman's Lake, Castlenalact Lake, Cork Lough, Kilcrea quarry, Castlemore gravel pit, Mallow lagoons, Lough Aderry, River Lee lakes and reservoirs (repeatedly), Ballyhea gravel pit, Charleville lagoons (repeatedly) and Kilcolman (repeatedly).

Most Scaup wintering in Ireland are from Iceland, but some are known to come from Fennoscandia and Siberia. There are three recoveries of Icelandic birds in west Cork.

Lesser Scaup

Aythya affinis | Mionlacha iascán | A

(monotypic)

World	North America
Ireland	42 individuals
Cork	2 individuals
Conservation	Not globally threatened

First-year male at Dooniskey on 4–25 January 2004. Female at the Gearagh on 10 December 2015.

Common Eider

Somateria mollissima | Éadar | A

(subspecies: *mollissima*)

World	Northern North America, Europe, Asia, also Greenland and Iceland
Ireland	Resident breeder
Cork	At least 138 individuals
Conservation	SPEC 1; Red-listed

The earliest records of Eider were of birds obtained at Cork Harbour in December 1878 and winter 1887/8. Another was obtained at Old Head of Kinsale on 6 November 1903.

Eiders have been recorded during 1959–2018 on the coast from Dursey Island to Youghal. Most have occurred at Cork Harbour, followed by Ballycotton, Dunmanus Bay, CCBO, Galley Head and Roaringwater Bay. Most occur from November to January, but there are records of new birds in all months except June and August. While some birds have been seen on one or a few days, many have made protracted stays over the course of a winter. One first-year male seen at Ballycotton in late March 1982 remained in the area until early January 1984. Most records are of one to four birds. Flocks of three and four have been recorded at CCBO in July 1974 and July 1982, respectively. However, eight flew west past Galley Head in November 2002, thirteen were at Dunmanus Bay in January 1969, and fifteen were at Cork Harbour in January 2006. About fifteen, including a flock of ten, were seen 19 km west of Dursey Island on 5 September 1986, at least some of which were in moult and flightless. Where age and sex were documented, a majority were immature or female, and adult males were rarely recorded. Two birds at Cork Harbour in September 1994 may have escaped from Fota Wildlife Park.

There has been one ring recovery of an Eider from abroad. A male ringed as a duckling in the Netherlands in July 1983 was drowned in a fish net at Ballymacoda Bay in December 1983. This recovery raises the possibility that some Eiders seen on the south Irish coast in winter may originate from the continent rather than the northern Irish population.

1959–68	1969–78	1979–88	1989–98	1999–08	2009–18
3	27	38	35	21	11

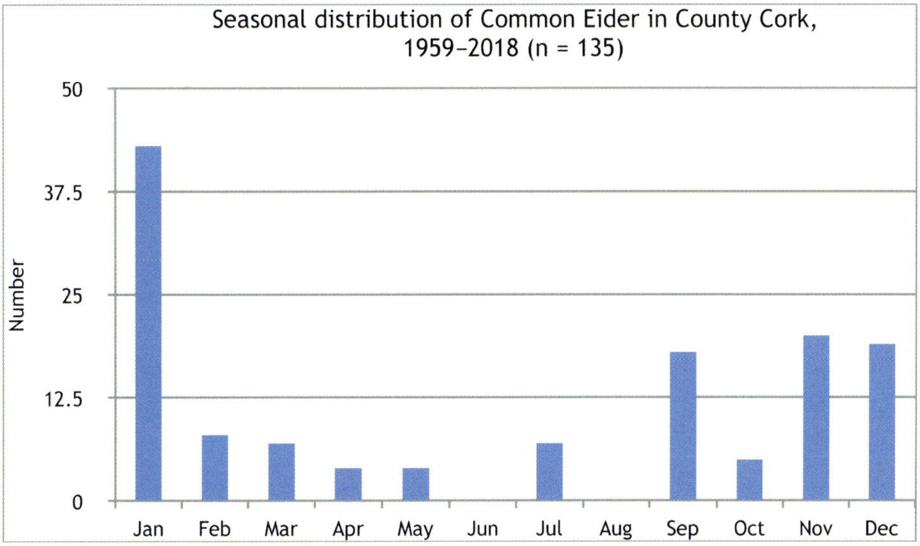

Fig. 7.5 Common Eider.

King Eider

Somateria spectabilis | Éadar taibhseach | A

(monotypic)

World	Arctic regions of North America, also Greenland, north-east Europe and Asia
Ireland	36 individuals
Cork	3 individuals
Conservation	Not globally threatened

Male at Baltimore on 29 January–24 February 1959. Male at Cahermore on 19 January–9 March 2014. First-year female at Toormore on 10 March–17 April 2018.

Long-tailed Duck

Clangula hyemalis | Lacha earrfhada | A

(monotypic)

World	Northern North America, Europe, Asia, also Greenland and Iceland
Ireland	Uncommon winter visitor
Cork	Scarce winter visitor
Conservation	SPEC 1; Red-listed

The earliest record is that of Payne-Gallwey, who stated that three had been obtained at Cork Harbour, including one collected by himself in January 1878, while another record from Castletownshend in the late 1800s was mentioned without a date. A pair was reported at Glengarriff for a week in June 1933, and seven were at Cork Harbour in December 1943.

Winter atlas surveys have shown small numbers at a few west Cork sites such as Bantry Bay and extending from Roaringwater Bay to Clonakilty. There were also records at Cork Harbour and Cork Lough. Some winters were better than others for this species during 1959–2018 and those of 1967/8, 1969/70, 1975/6 and 1991/2 stand out. Many birds in most winters occur in the Aghada area of Cork Harbour, although records come from all parts of the coast from Dursey Island to Youghal, with certain sheltered bays such as Roaringwater, Glandore, Clonakilty, Courtmacsherry, Cork Harbour and Ballycotton being favoured. A few have occurred inland at Cork Lough, Ovens, the Gearagh, Carrigadrohid and Charleville.

Long-tailed Ducks occur mainly from October to April, but there are records in this period for all months except June. Many birds remain at the same site for long periods, although some may be absent for spells during any winter.[35] A peak in March is more likely to be an observer effect than a spring passage. It is likely that some of the same individuals occur again in successive years at the same sites. Most records are of one or two individuals, but up to fourteen have occurred together at Cork Harbour. Most birds are either immature or female, adult males being rarely seen.

1959–68	1969–78	1979–88	1989–98	1999–08	2009–18
34	49	71	66	17	22

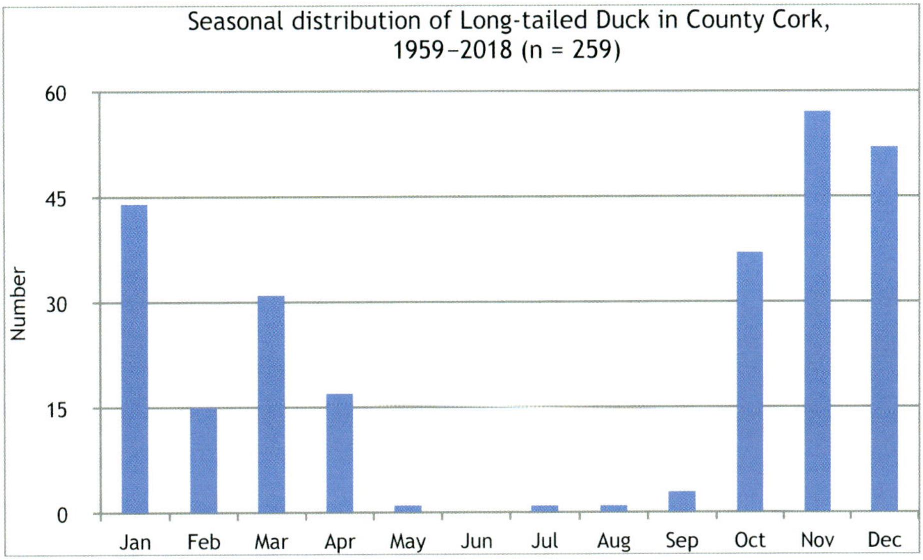

Fig. 7.6 Long-tailed Duck.

Common Scoter

Melanitta nigra | Scótar | A

(monotypic)

World	Iceland, northern Europe, and northern Asia
Ireland	Rare breeder, recent decrease, winter visitor, autumn passage migrant
Cork	Passage migrant and scarce winter visitor
Conservation	Red-listed

This scoter was described in the 1840s as common in winter on the coast and at Cork Harbour, although regarded as rare about Youghal. However, Payne-Gallwey, who did so much punt-shooting at Cork Harbour, made no mention of it there, or even in the county, and said it was rare in the south of Ireland.[36] It was considered a very rare visitor to Bantry Bay in the early 1900s, and four there in February 1934 was considered noteworthy. Three males were off Cork Harbour in August 1948, and one was obtained on that coast in winter 1947/8.

Clonakilty Bay held a regular winter flock averaging about thirty birds, but this declined and disappeared by the late 1960s as the species declined nationally.[37] However, birds returned in a few winters since, notably in 1975/6 with a peak of 124 in December. In general, winter flocks are now absent from the Cork coast but there are irregular records of small numbers, mostly at Cork Harbour, but also Inch Bay and Ballycotton where ten to twenty may occur (40 in Cork Harbour in January 2006 was exceptional by recent standards). Recent winter atlas surveys probably exaggerate the numbers on the coast since they form a composite picture of three or four winters.

Westerly passage has been noted offshore throughout the year. Spring passage is often very light and rarely exceeds about sixty birds at any site. The species is most numerous from mid-June to October. Movements are most noticeable off CCBO where peak counts often exceed 100 per day. Similar passage movements are observed at Dursey Island, Galley Head, Old Head of Kinsale, Power Head and Ballycotton in spring and autumn. Forty birds were in Ballycotton Bay on 3 July 1955.

A male was seen at Lough Allua on 18 June 1988 and a male was at the Gearagh on 11 August 2006, but there was no indication of breeding.

Surf Scoter

Melanitta perspicillata | Scótar toinne | A

(monotypic)

World	North America
Ireland	Rare winter visitor
Cork	29 individuals
Conservation	Not globally threatened

There was one record of a Surf Scoter before 1959, an immature male obtained at Crookhaven Bay on 5 November 1888.

This sea duck was not recorded again until 1969, and new birds have occurred in seventeen of the fifty years during 1969–2018. Birds have occurred at many sites from Toormore Bay to Ballycotton Bay, but most have been recorded at Cork Harbour (5), Clonakilty Bay (4), Courtmacsherry

Bay (3), Garrettstown (3) and Ballycotton Bay (3). One or two birds have occurred at eight other sites. Many sites favoured by this species are large and open sandy bays with a southerly or easterly aspect, while fewer sites tend to be enclosed, like Cork Harbour. Nearly all occurrences are of single individuals, but there is one record of four together at Clonakilty Bay and another of two together at Cork Harbour. Only one Surf Scoter appears to have returned to the same site in successive years, an adult female at Broad Strand being present in three successive winters (1989/90–1991/2). Males at Garrettstown from January to April 2012 and in September 2014 may have been the same but are treated here as different. Where ages and sexes have been published, fourteen were first-years and six adults, while eleven were male and six were female. Although eleven birds were recorded on a single day, most made longer stays, some remaining for many months. Most birds were first recorded in October and November, and several remained through to the following April, and it is possible that those birds seen for the first time from December onwards had arrived earlier but remained undetected.

1959–68	1969–78	1979–88	1989–98	1999–08	2009–18
0	5	3	5	3	12

Earliest and latest occurrence in winter (1959–2018): 27 September–14 February (15 April) (16 June).

Velvet Scoter

Melanitta fusca | Sceadach | A

(monotypic)

World	North America, northern Europe, and Asia
Ireland	Scarce winter visitor
Cork	69 individuals
Conservation	SPEC 1; Red-listed

There were five records (8 individuals) of Velvet Scoters in Cork before 1959. These were at Youghal (March 1850), Castletownshend (3 from

December 1889 to February 1890) and Glengarriff (October 1898, 2 in February–March 1925, and January 1930).

Velvet Scoters do not occur every year, but they have been recorded on all parts of the coast from Bantry Bay to Youghal. Cork Harbour, the coast from Rosscarbery to Old Head of Kinsale and the large west Cork bays are the most frequently visited sites. They occur between September and March, with one record in August and two in April. Many records are of birds seen on only one day, but a few have stayed for longer periods and some have wintered, especially at Cork Harbour, Clonakilty Bay and Courtmacsherry Bay. Some birds have been recorded flying past headlands and islands (10 individuals) in September, October, January and April. Most records refer to single birds, but flocks of two, three, four and five have occurred. Few have been aged or sexed, but males and females have been identified an equal number of times (9 each).

1959–68	1969–78	1979–88	1989–98	1999–08	2009–18
5	11	3	26	4	12

Earliest and latest occurrence in winter (1959–2018): 21 August–5 April.

Bufflehead

Bucephala albeola | Órshúileach mór-cheann | A

(monotypic)

World	North America
Ireland	3 individuals
Cork	2 individuals
Conservation	Not globally threatened

Female at the Gearagh on 18 January–8 March 1998. Adult female at Ballynacarriga Lake on 26 November–15 December 2018.

Common Goldeneye

Bucephala clangula | Órshúileach | A
(subspecies: *clangula*)

World	North America, Europe, and Asia
Ireland	Has bred, winter visitor
Cork	Scarce winter visitor
Conservation	Red-listed

The Goldeneye may have declined in numbers before 1845, by which time it was not common in Cork Harbour. In the late 1800s it was a winter visitor to coastal and inland waters. A change in status apparently took place in the first half of the 1900s when it was reported as a common winter visitor to parts of Cork Harbour, and it became regular in small numbers at Glengarriff where no adult males were seen.

Recent winter atlas surveys showed the main population of Goldeneye was at Cork Harbour and on the River Lee lakes and reservoirs west to Lough Allua. The only other significant population was at Roaringwater Bay and around Skibbereen. Elsewhere there were just one or two birds at Bantry Bay, Clonakilty Bay, Ballycotton, Lough Aderry and Kilcolman.

Goldeneye usually do not appear before early November, reach maximum numbers in December and January, and have mainly departed by late March, although a pair was at Castletownbere on 25 May 1988, a male was at Carrigadrohid on 22 July 1995, and a male has summered at Rostellan. A regular wintering flock of sixty to ninety (maximum 177, January 1980) occurred in the Great Island Channel of Cork Harbour, although few are there now possibly due to shellfish farming disturbance. Up to sixty occur at the Gearagh, about thirty at Inishcarra reservoir and about twenty at Lough Allua, up to thirty at Lough Hyne and up to twenty on the River Ilen between Skibbereen and Baltimore. The Great Island Channel birds may have transferred to the Lough Mahon and Douglas estuary area where about fifty were present during the 2000s, although numbers dropped to about ten in Cork Harbour during the 2010s.

Hooded Merganser

Lophodytes cucullatus | Síolta chochaill | B

(monotypic)

World	North America
Ireland	6 individuals
Cork	2 individuals (1 record)
Conservation	Not globally threatened

Male and female obtained at East Ferry in December 1878. These birds were kept under observation by Ralph Payne-Gallwey for some time at a creek (probably Saleen) in Cork Harbour where they were in company with Red-breasted Mergansers during the severe frost of December 1878. They were too wild to approach at first, although he did eventually shoot them both, and he preserved one, an adult male.

Smew

Mergellus albellus | Síolta gheal | A

(monotypic)

World	Northern Europe and northern Asia
Ireland	Rare winter visitor
Cork	19 individuals
Conservation	SPEC 3; Amber-listed

The only records for Cork up to 1958 were a female obtained near Enniskean in March 1895, and a male seen near Cobh in January and December 1950, both records possibly relating to the same individual.

Sixteen individuals occurred during 1959–2018 at Kilkerran Lake, Clonakilty, and Cork Harbour on the coast, and inland at Castlenalact Lake, River Lee reservoirs and Kilcolman. Fourteen occurred between October and February, with singles in May and July. Most have been recorded on a single day, while others have remained for up to twenty days, with three for up to five months over the winter. Two have occurred together twice, all others have been seen singly. Of fourteen where the sex was established, four were male and ten were female.

1959–68	1969–78	1979–88	1989–98	1999–08	2009–18
1	1	2	6	3	3

Earliest and latest occurrence in winter (1959–2018): 24 October–15 February.

Red-breasted Merganser

Mergus serrator | Síolta rua | A
(monotypic)

World	North America, Greenland, Iceland, northern Europe, and Asia
Ireland	Uncommon breeder, winter visitor
Cork	Scarce breeder, decreasing winter visitor
Conservation	SPEC 3; Amber-listed

The Red-breasted Merganser was described as frequent in winter, although it was considered rare at Cork Harbour until some years before 1850. Thereafter, flocks were observed regularly, and they were more numerous than usual in the winter of 1849/50. In that winter between 100 and 200 birds were seen at Great Island Channel on 11 January, and in the severe winter of 1878/9 reports ranged from 100 to even 500–600 birds in the outer harbour east of the Man-of-War Roads. Bantry Bay was also visited regularly in winter.

Red-breasted Mergansers began to breed in the county in (or before) 1920. There is no information on how the population developed. A pair with seven young was seen at Barley Cove in September 1966. Atlas fieldwork in 1968–72 revealed breeding at Bantry, Dunmanus and Roaringwater Bays. There followed a decline and breeding was proved at only one 10 km square at Bantry Bay in 1988–91. However, it was breeding at several islands in Roaringwater Bay in 2004–5, and a pair bred at Ballydehob in 2006. There was no case of proved breeding during 2008–11, although it was considered probable in three 10 km squares at Bantry Bay.

Winter atlas surveys showed the Red-breasted Merganser was thinly spread along the coast from Bantry Bay to Youghal Harbour in sheltered waters and was absent from exposed coasts. The main populations were at Cork Harbour and Roaringwater Bay, but flocks of up to thirty were at sites such as Rosscarbery, Clonakilty, Courtmacsherry, Ballycotton and Ballymacoda.

About 400 birds, moving in small parties, were reported to fly out of Cork Harbour each evening during winter to spend the night on the sea off Roche's Point, before returning to the harbour in the morning (1967). Cork Harbour continued to host numbers ranging about 140–304 during the 1970s and 1980s. However, numbers have declined and in the 2010s counts ranged from fifty to eighty-nine. Numbers have declined at all wintering sites and birds have disappeared from Ballycotton and Ballymacoda.

Goosander

Mergus merganser | Síolta mhór | A

(subspecies: *merganser*)

World	North America, Iceland, Europe, and Asia
Ireland	Very rare breeder, scarce winter visitor, recent increase
Cork	Probably more regular in 1800s than later, 109 individuals since 1959
Conservation	Amber-listed

The Goosander was described in the 1800s as rare in winter, adults and immatures having been recorded. It sometimes occurred at Cork Harbour, but not every winter, one being obtained there in 1849/50. It was seen almost annually at Glengarriff, which it frequented commonly in the winter of 1848/9, but not the next winter, which was a severe one. Goosanders were more plentiful than previously known and were obtained in every part of Ireland in the severe frost of January 1881 (Cork not specifically mentioned). Twenty-four were known to have been obtained in Munster to 1900, but no county breakdown was given. Apart from the locations mentioned, Goosanders were recorded in the late 1800s at Timoleague, Longueville and the River Blackwater.

Goosanders have occurred regularly only since 2004 during the 1959–2018 period. A majority have been recorded at the River Lee system of lakes and reservoirs, where it now occurs annually. A few individuals have occurred at a suite of coastal sites from Bantry Bay to Cork Harbour, with one at Kilcolman in January 2013. There are records from October to March, with one in August, but most occur between December and February. Before 2004 most birds were present on a single day, with one staying for five weeks. Since 2004 most records are of birds staying for several months through the winter period. Most records are of one or two birds, but flocks of six and seven have occurred. Where the sex has been recorded, twelve were males and twenty were 'redheads', a category which includes juveniles.

1959–68	1969–78	1979–88	1989–98	1999–08	2009–18
2	2	2	4	40	59

Earliest and latest occurrence in winter (1959–2018): (27 August) 4 October–19 March.

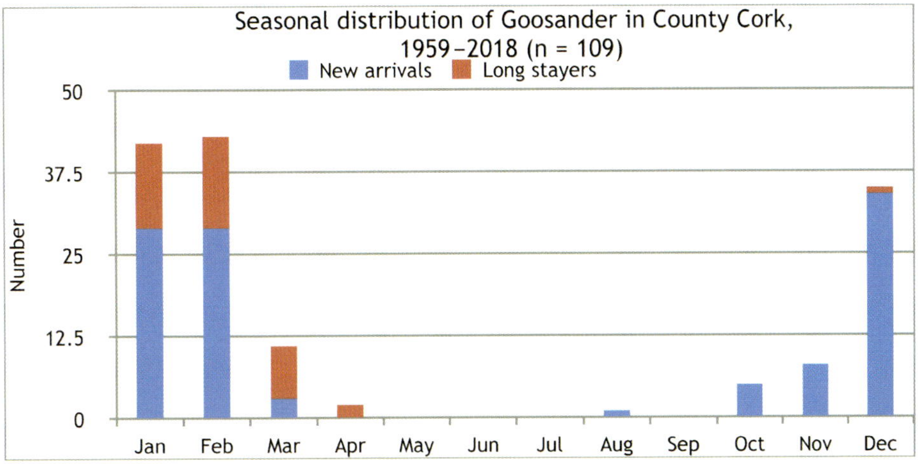

Fig. 7.7 Goosander.

Ruddy Duck

Oxyura jamaicensis | Lacha rua | C2

(subspecies: *jamaicensis*)

World	North and South America, introduced Britain
Ireland	Rare resident breeder, uncommon winter visitor, decreasing
Cork	53 individuals
Conservation	Not globally threatened

The Ruddy Duck is a recent Irish colonist, having established a breeding population in Northern Ireland.[38] Breeding has not been confirmed for Cork. The first occurrence in Cork was on 7 January 1979 when two birds were at Ballycotton. A series of records have occurred at several east Cork sites (Ballycotton, Lough Aderry, Ballybutler Lake, Rostellan, Great Island and Cork Lough), and birds have been known to move between them. Others have occurred in north Cork at Charleville lagoons and Kilcolman, again with known movement between sites. Similarly, a series have occurred around the River Lee reservoirs, especially the Gearagh, and others at Courtmacsherry Bay, Castlenalact Lake and Mallow lagoons.

Many birds have made lengthy stays, some for several months, while others have been known (or suspected) to return to the same site, or a nearby one, in several successive years with periods of absence in between. This makes calculation of total numbers difficult, and the number calculated (53 individuals) is almost certainly higher than the true value, but it serves to illustrate the occurrence pattern. Ruddy Ducks have occurred in all months, with peaks in winter (January), spring (May and June) and autumn (October). Only at one site (Charleville lagoons) has there been a consistent pattern, presumably involving birds returning in successive years. The pattern of low numbers in late summer and autumn suggests that birds become secretive during moult, only to emerge again later. The January peak is suggestive of wandering birds, perhaps responding to cold weather spells.

Most of the records are for inland lakes and lagoons. Only one bird has occurred in a truly marine environment, at Courtmacsherry Bay for two days in November 1991. Those at Ballycotton and Rostellan occurred in brackish lagoons. Most records are of single birds, but two and three, and once five, have occurred together. Males (20) and females (17) have been recorded almost equally.

Declining numbers in Cork since the early 2000s reflects the reduction in the presumed source in Britain due to an eradication programme there.

1959–68	1969–78	1979–88	1989–98	1999–08	2009–18
0	0	10	26	15	2

Red Grouse

Lagopus lagopus | Cearc fhraoigh | A
(subspecies: *scotica*)

World	Northern regions of North America, Europe, and Asia
Ireland	Resident breeder, recent decrease
Cork	Small and declining breeding population
Conservation	SPEC 3; Red-listed

The Red Grouse was a common resident on the highest mountains in the 1700s and 1800s. Although it remained common in in some places, a decrease was reported in certain localities in the late 1800s. Richard Ussher was informed by G.H. Kinahan about 1900 that he had seen large flocks of grouse consisting of several packs on mountains in Munster (Cork not mentioned).

Hardly any information on distribution is available for most of the 1900s, although it is believed there was a marked decrease in Ireland after 1920. During the breeding seasons of 1968–72 Red Grouse were present at the Mullaghareirk, Ballyhoura, Derrynasaggart, Boggeragh and Nagles Mountain ranges, and at moorland about Bantry. A decline had taken place throughout the range by 1988–91, and birds had apparently disappeared from the Nagles Mountains. There was a further contraction of range by 2008–11, although some birds were present at the Mullagh-areirk, Derrynasaggart and Boggeragh Mountains, and at moorland between Bantry and Dunmanway. Overall, there has been a considerable

Red Grouse

decline in range and numbers since the late 1960s. The Red Grouse is sedentary, and winter range almost mirrors that of the breeding season.

A national survey of Red Grouse in 2006–8 partly coincided with the third breeding atlas project.[39] Examination of maps produced for both surveys indicates that Red Grouse were breeding only at the Shehy, Derrynasaggart, Boggeragh and Mullaghareirk ranges. No breeding birds were recorded at either the Ballyhoura or Nagles ranges. However, the second winter atlas showed a slightly wider distribution with records from the Ballyhoura range and from the Hungry Hill area of the Beara peninsula as well as from those ranges mentioned above. Clearly, there has been a considerable decline since the early 1970s, and any differences between breeding and winter season distributions may be explained by the low-density populations now present at many sites, which makes them easy to miss.

County Cork falls into two of the regional divisions used by surveyors, 'south-west' and 'east and south'. The decrease in occupied range at 'south-west' was 47 per cent and at 'east and south' was 58 per cent over the period 1968–2008. Changes in habitat availability and quality were considered the primary causes of decline.[40] Changes to habitat are caused by several factors such as vegetation burning leading to loss of heather and development of a grassy sward, an increase in grazing pressure from hill farming of sheep, conifer afforestation, mechanical peat extraction, and habitat fragmentation from wind farm developments.

Red Grouse are present in very low numbers on the Ballyhoura range and persist despite high levels of habitat fragmentation due to afforestation and wind energy developments, and high levels of recreational use (quad biking, motor-cycle scrambling, mountain biking, hiking).[41]

The Irish Red Grouse was formerly considered a distinct subspecies (*hibernica*) due to its paler plumage (compared to the British form). Birds and eggs from Britain have frequently been introduced to Ireland (Cork not mentioned), but no change towards the darker British form has apparently taken place. However, no genetic difference was found between Irish and British populations using mitochondrial DNA, and it was concluded there was little molecular support for *hibernica*.[42]

Capercaillie

Tetrao urogallus | Capall coille | B

(subspecies: *urogallus*)

World	Europe and Asia
Ireland	Former breeder, extinct as Irish since 1700s
Cork	Extinct
Conservation	Not globally threatened

The Capercaillie apparently abounded in the woods of Ireland in the twelfth century. It was said to be rarely found in Ireland by 1750 since the woods had been destroyed.[43] Smith does not explicitly mention its presence in Cork in 1750, and a quote from John Scouler that 'it remained in the County of Cork until as late as 1750' sounds like it originated from Smith's work, albeit as a misinterpretation.[44] However, it is unlikely Smith would have included it in his list of Cork birds if he did not have reasonable cause. With specific reference to County Cork, when mentioning the use of Irish names by several authors, including Don Philip O'Sullivan Beare, a connection with the mid-Cork barony of Muskerry was suggested.[45]

An attempted reintroduction about 1842 of three pairs of Swedish birds at Glengarriff failed within about nine months when they all died.

There has been much debate on the question of whether the Capercaillie was ever a native Irish species.[46;47] However, there is no doubt about its former presence in many parts of Ireland, and it is generally accepted that it became extinct during the 1700s. The discovery of Capercaillie bones near Coleraine (Londonderry) in 1982 showed that this gamebird was part of the native Irish avifauna, at least in prehistoric times. Since then, bones have been found at sites in Dublin, Waterford and Wexford.[48]

Grey Partridge

Perdix perdix | Patraisc | B

(subspecies: *perdix*)

World	Europe and Asia, introduced elsewhere
Ireland	Rare resident breeder, recent decrease, recent introductions
Cork	Former breeder, now extinct as a wild bird
Conservation	SPEC 2; Red-listed

The Grey Partridge was a very common resident in the 1700s and early 1800s. It became scarce in Ireland from the late 1830s, possibly due to a series of wet and cold summers, but an increase took place in the late 1840s. Grey Partridges were said to be extremely scarce in the neighbourhood of Cork city until 1847, but they were found in greater numbers thereafter. Two or three coveys, and in well-preserved grounds five coveys, could be seen in a day in Cork. George Jackson described the Grey Partridge as abundant at Glengarriff, but in 1846 nearly all disappeared. Young were reported to have hatched, but the coveys diminished in size in August. He also said coveys were often found at the Beara peninsula far distant from cultivated land. It was possible to shoot forty birds in a day in the Mallow area about 1866, but by the 1920s the Grey Partridge had disappeared from the south of Ireland.[49] It was said to be resident throughout the county in the late 1800s, but all correspondents of Ussher agreed it was scarce, or had recently become scarce, one remarking that 'Sunday sportsmen' were responsible for its near extermination.

Numbers remained low in the country up to 1930, when shooting of partridges was banned under the Game Protection Act, 1930. A short and limited shooting season was allowed in 1932 and 1933 following the release of birds. The position continued to improve in the country due to an increase in tillage, but in Cork in 1943 and 1944 no open season was allowed, due presumably to low numbers.[50] Numbers in Cork obviously did not recover, and there appears to be no authenticated record of wild birds breeding in the county since the late 1940s.

Attempts to re-establish viable populations have been made many times in recent decades by the release of birds by gun clubs in many parts of the county. During 1982–4 about 300 birds were released. A project began in 2009 at Castletownroche with wild birds translocated from Boora (Offaly) (the Offaly population had earlier been augmented by Estonian stock). Birds from this project bred in the wild for a while, but this soon ceased. Another project using birds mostly originating from game farms was established at Rathcormack. Unfortunately, the evidence to date suggests minimal success, and viable breeding populations have not yet been established.

Quail

Coturnix coturnix | Gearg | A

(subspecies: *coturnix*)

World	Europe, Asia, and Africa
Ireland	Very rare summer visiting breeder, passage migrant
Cork	Very rare summer visitor
Conservation	SPEC 3; Red-listed

The Quail was very common in its season in the 1700s. This statement implies that its migratory habits were known. Around the 1840s it was said to be a summer visitor to the county but was not regarded as common; it was distributed over cultivated districts in Ireland, and some remained throughout the winter. Robert Ball knew of nests exposed to view in early summer during the mowing of grass in the south of Ireland, presumably about Youghal, while George Jackson indicated it may have been breeding as far west as Glengarriff in the 1840s. With reference to the Mallow area, it was said to have not been seen for many years before 1926, but that some sixty years before (i.e. about 1866) two or three were seen in winter and breeding took place.[51] The Quail was widely regarded as extinct, or nearly so, in Ireland after 1880, although formerly believed to be common throughout the year. One reason posited for the decline was a reversion to pasture and moorland from tilled land after the Great Famine.[52]

There were influxes to Ireland in 1892, 1893 and 1896, but they were very scarce in 1894 and 1895. During these so-called Quail years birds occurred in many districts and were generally distributed across the county; in 1896 during an unusually warm and dry period they were more numerous than elsewhere in the Midleton district. It was concluded that drought conditions in Europe in spring led to many birds appearing in Ireland. In Wexford, Moffat recorded rainfall for the years 1877–96 and found that the lowest rainfall in March, April and May was in 1887, 1892, 1893 and 1896, the latter three being good Quail years.[53] Repeated releases of hand-reared Quail took place in different parts of Ireland in the late 1800s, without mention of Cork, but the birds soon disappeared in every case.

In the 1900s numbers fluctuated in the country, with a tendency to be very scarce in the early years, but with a trend towards increase from about 1940 onwards, and it was noticed in Cork (among other counties) in the period 1935–47, although breeding was not proved. However, the situation does not appear to have improved as the 1900s progressed. In the 1968–72 atlas the Quail was found in only two contiguous 10 km squares in Cork, both on the east side of Cork Harbour, and breeding may have taken place. In the 1988–91 and 2008–11 atlases only three 10 km squares had records, all in coastal areas and mostly in the east of the county, but breeding was considered unlikely to have occurred.

Only sixty individuals were recorded in Cork in the sixty years 1959–2018, including those discussed above in the breeding atlases. Many have been spring migrants (with a few in autumn) occurring at sites such as CCBO and Dursey Island, and one spring migrant was in a small suburban garden near Cork Lough. Others have occurred widely at headlands from Mizen Head to Old Head of Kinsale and Ballycotton. Twenty-six of the sixty individuals have occurred in three years, 1989 (13), 1990 (9) and 1992 (4). In these years singing males were heard in cereal and pasture crops, especially around Oysterhaven, Ballycotton, Ballymacoda, Killeagh and Midleton. Only two have been recorded any distance from the coast, at

Kilcrea and Ballinagree, but this may reflect the distribution of competent observers.

Migrant birds at coastal sites are typically recorded on a single day, but those that occur in potential breeding habitats can usually be heard singing over a period of days or weeks. Most records refer to a single bird, but there have been occurrences of three (Oysterhaven), four (Roche's Point) and five (Ballycotton).

The Quail is a trans-Saharan migrant, and the only migratory gamebird in Europe. It winters in the Sahelian zone where conditions may have affected its survival, and it is a popular huntable species in Mediterranean countries. In 1945, a year in which Quail were plentiful in Ireland, individuals were killed at Fastnet Rock (3 September) and Bull Rock (10 October) lighthouses.

1959–68	1969–78	1979–88	1989–98	1999–08	2009–18
3	10	3	30	9	5

Earliest and latest occurrence in spring and autumn (1959–2018): 21 April–8 November.

Pheasant

Phasianus colchicus | Piasún | C1

(subspecies: *colchicus; torquatus*)

World	Asia, introduced elsewhere
Ireland	Very common resident breeder
Cork	Widespread breeding resident
Conservation	Not globally threatened

The Pheasant was believed absent from Ireland in the 1100s and 1300s, but by the end of the 1500s it was numerous. In the period 1599–1603 large numbers of Pheasants were often served up at feasts.[54] It has been described as a post-Norman introduction to Ireland, and the earliest mention of it dates to 1589, but the date of introduction to Cork is unknown.[55] The earliest Irish archaeological records date to the Anglo-Norman period.[56] The Pheasant was said to be very rare in the 1700s, most woods having been felled. This implies it was commoner at an earlier time. It was also mentioned that it was commoner in Kerry in the 1700s than in the more cultivated county of Cork.

During 1968–72 the Pheasant was very widespread across the county and appeared to be absent only from the Beara and Sheep's Head peninsulas. Several new 10 km squares were occupied in 1988–91, especially in coastal districts, and by 2008–11 the western peninsulas were fully populated. Density was highest east of Cork Harbour. The Pheasant was also widespread in winter, but the population was most dense in the east and north and least dense in the west, especially at the peninsulas. At Sherkin Island, a population of up to twenty-one females was present at least during the 1960s–90s, and breeding was regularly recorded. At nearby CCBO there were no official records during the early years of the bird observatory, but residents often reported the presence of single birds which were presumed to be strays from Sherkin Island. From 1981 onwards records of Pheasants at CCBO became regular, and it later transpired that a resident had introduced six birds. The population still exists, and breeding takes place. Pheasants have also been seen at Capel Island.

The Pheasant is an important species to the sport shooting community, and thousands are reared and released every year to augment the wild population, for example 8,760 birds were released in Cork in 1983. The Pheasant population is generally self-sustaining at low population levels, but many gun clubs release varying numbers each year on a put-and-take basis, particularly in areas where the habitat is incapable of supporting birds at high density. Driven shoots for Pheasants exist entirely on a put-and-take basis because of the large numbers required for management purposes.

While the Pheasant is the most important species of gamebird to sport shooters over most of the county, mainly because of severe declines in native populations of other gamebirds, such as Quail, Grey Partridge and Red Grouse, very little information has been published on any aspect of its ecology in Ireland, and none since the 1980s, and it was suggested that habitat management practices would be more cost-efficient than rearing and releasing.[57] A study at Ballymacoda in April and May 1981 showed that over an area of mixed farmland a minimum of 6.7 territorial male birds per 100 ha was present, with a minimum mean harem size of 1.6 females per male.[58] The figures of density quoted were believed to be above average for east Cork at that time, and higher than at present due to agricultural intensification.

The old brown Pheasant (*colchicus*) was apparently the type first introduced to Ireland, but by 1900 there were few places where this was found uncrossed with the ring-necked type (*torquatus*). No appraisal of the subspecies found in Ireland has been carried out. However, two are commonly seen (*colchicus* and *torquatus*), although due to captive breeding and releasing for sport shooting, intergrades between these (and probably other subspecies) are frequent, and many melanistic forms occur.

Pheasants often respond to loud noises by crowing and wing-flapping, even at night. Thunder will cause some agitation and crowing, and the passage of the supersonic aircraft *Concorde* (discontinued since 2003) off the Cork coast could be predicted by the crowing of Pheasants several seconds before the sonic boom was heard, Pheasants being able to sense the air vibrations before the human ear could hear the slower-moving sound. Introduced Pheasants are largely sedentary, rarely moving more than a few hundred metres. However, exceptional circumstances may cause birds to turn up in unusual places, like the male on a quay wall in Cork city in January 1963 during a cold spell, and the female rescued at sea by Courtmacsherry lifeboat between Wood Point and Coolmain Point in 1984.

Red-throated Diver

Gavia stellata | Lóma rua | A

(monotypic)

World	Northern North America, Europe, and Asia, also Greenland and Iceland
Ireland	Very rare breeder, winter visitor, passage migrant
Cork	Passage migrant and winter visitor to the coast
Conservation	SPEC 3; Amber-listed

Red-throated Divers were described in the 1800s as common in immature plumage in winter; they arrived in September and stayed to late April and were known to occur also at Cork Harbour and Glengarriff Harbour. The status and distribution of the species in the late 1900s and early 2000s was very similar to that of the 1800s, with birds generally arriving in September, departing in April, and occurring along the entire coast. They are somewhat less widely distributed, being scarcer in the west and almost absent from Bantry Bay and some other western bays.

Returning individuals occasionally occur as early as August, but most do not appear before November. Most shallow and sandy bays generally hold a few, and loose flocks of ten or more are regularly seen. Counts exceeding fifty birds are relatively numerous: 153 at Ballybranagan Strand in March 1990, 122 at Broadstrand Bay in December 1991, 116 at Barley Cove in January 1986, 100 at Inch Bay in March 1998, 100 at Galley Head in March 2009, and eighty at Rosscarbery in December 1994. Small numbers occur at Youghal Harbour, and at Cork Harbour where fourteen seen on 1 January 1977 is the peak count. The only inland record concerns one at Kilcolman after south-east gales on 22 February 1994. Most have departed by the end of April, but some occasionally remain into late May or early June. There is a single July record, involving one (ringed as a nestling in Orkney four years previously) found dead in a fishing net in Youghal Bay on 1 July 1988.

Ring recoveries indicate that at least part of the Irish wintering population has its origin in Scotland, although an Icelandic and Greenlandic origin cannot be ruled out for some. Several ring recoveries in Cork are of birds born in Orkney and Shetland.

Black-throated Diver

Gavia arctica | Lóma Artach | A

(subspecies: *arctica*)

World	Northern regions of Europe and Asia
Ireland	Scarce winter visitor
Cork	218 individuals
Conservation	SPEC 3; Amber-listed

Although Black-throated Divers were reported from Glengarriff in January 1928 and January 1930, the accounts lack detail and leave room for doubt about the identification.

The Black-throated Diver has occurred frequently since the first verifiable record in 1962, and annually in recent decades. Birds occur along the coast from Dursey Island to Ballymacoda Bay, but most have been in west Cork. Most have occurred at six areas: Bantry Bay, Roaringwater Bay, Rosscarbery to Clonakilty Bay, Cork Harbour, Ballycotton Bay, and Knockadoon Head and Ballymacoda Bay. Two occurred at freshwater lakes adjacent to the coast: Cork Lough and Reenydonagan Lake. While many are seen on only one day, some have remained for months, as at Bantry Bay, Galley Head area and Cork Harbour, and some probably return to the same location in successive winters (Bantry Bay and Cork Harbour). This together with the monthly occurrence pattern suggests a small wintering population, with most records from November to April. July is the only month without a record of a new bird. Most records refer to one or two birds, but flocks of three, four, five, six and seven have occurred, and eight have been seen between Galley Head and Rosscarbery.

1959–68	1969–78	1979–88	1989–98	1999–08	2009–18
5	20	38	22	51	82

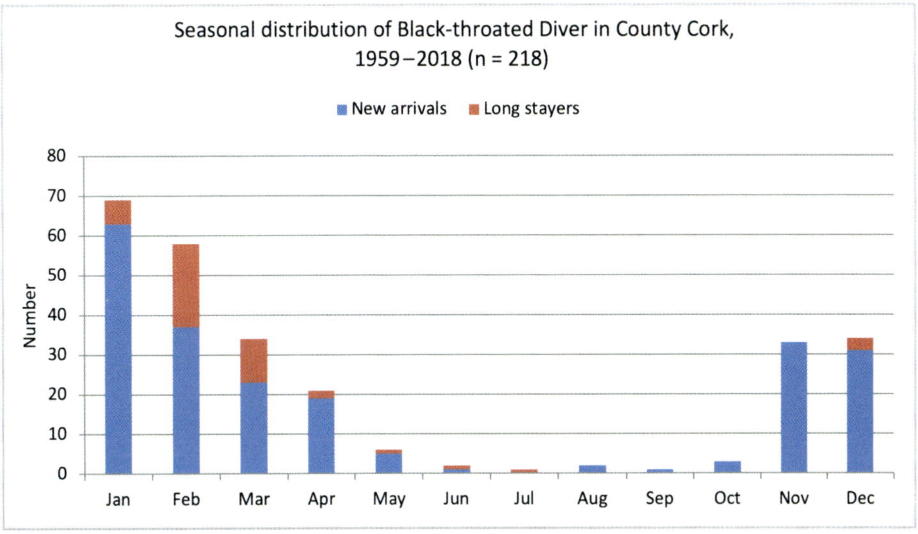

Fig. 7.8 Black-throated Diver.

Pacific Diver

Gavia pacifica | Lóma Áiseach | A

(monotypic)

World	North-east Siberia and northernmost North America
Ireland	4 individuals
Cork	1 individual
Conservation	Not globally threatened

Adult at Crookhaven on 18 January–18 April 2018, 7 October 2018–28 March 2019, and 19 October–22 November 2019.

Great Northern Diver

Gavia immer | Lóma mór | A

(monotypic)

World	North America, Greenland, Iceland, and Bear Island
Ireland	Winter visitor
Cork	Passage migrant and winter visitor to the coast
Conservation	SPEC 3; Amber-listed

The Great Northern Diver was described in the 1800s as common in immature plumage in winter on the coast. It occurred at Cork Harbour and Bantry Bay, and sometimes remained to April. It was also reported from Fastnet Rock and Spit Bank. One still in summer plumage was obtained at Kinsale on 12 December 1912. Little change has apparently occurred in the status and distribution of the species since the 1800s.

This is the most widely distributed diver on the Cork coast. Some wintering birds appear as early as August, but most arrive during September and October and the wintering population is usually present by November. By this time the species can be encountered either singly or in parties of up to ten anywhere along the coast and in estuaries, including Cork Harbour, and occasionally Youghal Harbour. Flocks rarely exceed fifty birds, although fifty-five were at Garrettstown in February 1974, fifty at Bantry Bay in

Great Northern Diver

February 2012, forty-four at Dunmanus Bay in January 1971, forty-four at Courtmacsherry Bay in February 1995, and thirty at CCBO in January 1992. They tend to favour deeper waters than other divers, and highest densities occur at Bantry Bay, between Glandore and Seven Heads and along the coast east of Cork Harbour.

Most have departed by April, although a few regularly linger until mid-May and in some years a distinct westerly passage has been noted at CCBO. This westerly movement is rarely obvious from mainland observation points, but a total of 108 passed Galley Head and Toe Head between 9 and 21 May 1986, thirty-four passed north-west at Dursey Island between 8 and 19 May 1995, and 128 birds mostly in summer plumage passed Dursey Island between 2 May and 5 June 2000. June records are scarce but regular, and birds are mostly in summer plumage. There are at least five records of singles in July: CCBO 1981, Garrettstown 1990, Ringaskiddy 2004, Galley Head 2005 and 2011.

There are a few inland records, including one at Ovens gravel pit in winter 1977/8, and one at Castlenalact Lake on 5 December 1992. There have also been a series of records involving single birds at various locations in the River Lee reservoirs, the Gearagh and Lough Allua in January 1994, January 1998, February 1998, December 1999, January 2000, December 2008 and December 2017.

White-billed Diver

Gavia adamsii | Lóma gobgheal | A

(monotypic)

World	Arctic regions of North America, also northern Scandinavia and Siberia
Ireland	24 individuals
Cork	3 individuals
Conservation	Not globally threatened

One at Lough Hyne on 3 February 1974. Probable first-year at CCBO on 31 May 2004. Adult at CCBO on 27 October–2 November 2006.

Black-browed Albatross

Thalassarche melanophris | Albatras dú-mhalach | A

(subspecies: undetermined)

World	Southern Ocean
Ireland	13 individuals
Cork	6 individuals (5 records)
Conservation	Not globally threatened

Adult at CCBO on 24 September 1963. Two adults at Old Head of Kinsale on 10 September 1976. Adult at CCBO on 2 May 1995. Adult or sub-adult 3 km south-west of CCBO on 9 September 1995. Adult at CCBO on 22 August 1996.

Unidentified albatross species

Four albatrosses, probably Black-browed (*Thalassarche melanophris*), have been recorded. One at CCBO on 2 September 1967. One 8 km south of CCBO on 8 July 1988. One at Dursey Island on 8 October 1991. One at Mizen Head on 21 October 2008.

Fulmar

Fulmarus glacialis | Fulmaire | A

(subspecies: *glacialis*)

World	Arctic and boreal zones of North Atlantic and North Pacific Oceans
Ireland	Common resident breeder
Cork	Widespread breeding resident on coast
Conservation	SPEC 3; Amber-listed

The earliest records of Fulmars in Cork are of single birds obtained at Inchydoney in 1832, and at Castlefreke in October 1845. There were no more records for Cork in the 1800s, but Fulmars were regularly encountered some distance off the south-west Irish coast in the early 1900s. Fulmars increased and spread into many new areas of the North Atlantic Ocean over at least the last 200 years.[59] In Ireland, colonisation began in the early 1900s, and the first breeding birds reached the Cork coast in the 1930s. The colony at Great Skellig (Kerry) acted as a point of dispersal for the south coast of Ireland. The spread of breeding birds along the coast was, therefore, from west to east as far as Cork Harbour, but the coast east of Cork Harbour was colonised from the east by birds that had earlier colonised the Waterford and Wexford coasts. Colonisation of Cork followed the pattern of first settling at the prominent headlands, and later the intermediate coasts.

Year(s)	Number of AOSs
1939	57
1944	57
1949	171
1954	204
1959	215
1969–70	540
1985–88	2,059
1998–02	1,569

Table 7.2 County Cork totals for breeding Fulmar, 1939–2002.

The first birds to show breeding intent in Cork were prospecting cliffs at Mizen Head in 1935 and in the following year (1936) six nest sites were occupied, but breeding was not proved for the first time until 1938 when there were fifteen to twenty-five nest sites.

The Irish Fulmar population has been surveyed at intervals since first breeding. Table 7.2 contains figures for the whole county from national censuses. The counting unit used is the Apparently Occupied Site (AOS), and the figures given include subsequent revisions. The figures show a considerable increase in numbers. The national trend shows a stabilising of the increase, and a decline at some sites.[60] Fulmars have a long period of immaturity and some birds occupy sites for many years before they breed, thus the number of AOS includes some non-breeders, failed breeders and those that breed successfully.

The recent history of individual colonies (or sections of coast) has been documented. There were 101 AOS at CCBO in 1960, and this number increased to 716 in 1996, but declined to 525 in 1998. Fulmars bred at Sherkin Island for the first time in 1969 (2 pairs) and increased to thirty-six AOS in 1990 but declined to nineteen in 1994. Elsewhere in Roaringwater Bay, numbers increased at Goat Island from fifteen to seventy-nine AOS between 1982/3 and 1994. The number of AOS increased from 69 in 1969 to 164 in 1985 on the coastline between Cork Harbour and Youghal Harbour.

In 1982 and 1984 single eggs were laid on a flat concrete roof at Youghal, but no young fledged. Fulmars have prospected other roofs at Youghal, and breeding may have taken place out of view. Two young fledged from a cliff 3.5 km from the open sea at Youghal in 1984, but this site has been abandoned. In Cobh, St Colman's cathedral was prospected from 1976, but none have bred. Fulmars have been recorded using old nests of Raven as breeding sites at Knockadoon Head and Ballycotton. Highest concentrations were at Dursey Island (275 AOS) and between Firkeel and Cahermore (253 AOS) in a 1985/6 survey.

Distribution of Fulmars in winter is slightly more restricted than in the breeding season. The distribution of Fulmars off the Cork coast was studied in the 1990s and 2000s and the highest inshore densities were in the west, especially off the Beara peninsula, generally corresponding with the areas of highest breeding density.[61;62]

There is a large offshore passage of birds in spring and autumn, and the situation at CCBO serves to illustrate the point. Peak passage periods are early April and early August, when movements may exceed 800 birds per hour. Most passage is in a westerly direction, often in south-west winds. The numbers passing CCBO showed a marked increase in the years 1970–80 compared with 1959–69, mirroring the increase in the breeding population. Spring passage reached 2,223 birds per hour in April 1991, while autumn passage that year peaked at 1,500 birds per hour in August. Significant numbers also pass CCBO in winter, and an hourly rate of 4,912 was recorded in November 1982 and 2,756 in December 1982.[63] Birds are almost completely absent for several weeks between late September and early November. Passage movements have been recorded at almost all prominent headlands along the coast of Cork.

Blue Fulmar: A 'blue' form of the Fulmar occurs among some Arctic populations. Records of 141 individuals have been published for Cork during 1961–2005. This form has occurred in every month, peaking at 23 per cent in August and 44 per cent in November and December. Most have occurred at CCBO (93 per cent), but there are records on the coast from Dursey Island to Redbarn Strand.

Unidentified gadfly petrel species

Seventy gadfly petrels, either Desertas, Fea's or Zino's Petrel (*Pterodroma deserta/ feae/ madeira*), have been recorded off the Cork coast since the first in 1974. These birds have been recorded at the following sites: Dursey Island (1), Mizen Head (10), CCBO (14), Baltimore (2), Toe Head (2), Galley Head (32), Dunowen Head (1), Seven Heads (2), Old Head of Kinsale (4) and Ballycotton (2). Three birds seen from boats in inshore waters off Galley Head in 2014 and 2015, but within the 30 km recording distance from land, have been included in the total for Galley Head. Most birds have been recorded as single individuals passing headlands on autumn passage migration, August (33) and September (20) being the months of most frequent occurrence, and with a strong peak in late August. There are only three records before 26 July and only two records after 5 October. It is probable that the increase in records in the last three decades is at least partly the result of a greater awareness of the presence of these species in Irish waters, and the increasing skills of identification among observers using high-quality optical equipment. All three species of gadfly petrel are rare or endangered in their breeding habitats on the Atlantic islands off north-west Africa, and specific identification away from these areas presents an almost impossible task to seabird migration watchers.

1959–68	1969–78	1979–88	1989–98	1999–08	2009–18
0	1	0	14	14	41

Earliest and latest occurrence in spring and autumn (all records): 2 July–15 October.

Bulwer's Petrel

Bulweria bulwerii | Peadairín Bulwer | A
(monotypic)

World	Atlantic and Pacific Oceans
Ireland	1 individual
Cork	1 individual
Conservation	Not globally threatened

One at Galley Head on 1 August 2013.

Cory's Shearwater

Calonectris borealis | Cánóg Cory | A
(monotypic)

World	Azores, Canary, Madeira, and Berlengas Islands
Ireland	Scarce autumn passage migrant
Cork	Passage migrant
Conservation	SPEC 2; Amber-listed

The Cory's Shearwater was first recorded in 1962, and it has occurred in every year since except 1966, 1976, 1986 and 1988. Numbers vary significantly between years, and totals have exceeded 500 in twelve years, and 10,000 in two years, most moving west. The reasons for such variation have not been satisfactorily explained.[64] The majority have been seen off CCBO (97 per cent), but it is probably under-recorded at mainland watch points. A total of 272 was recorded at Old Head of Kinsale between 5 and 7 October 1971.

Occasional birds occur in spring, thirty-seven in April and nine in May, mostly at CCBO. The main passage usually takes place from early July to early October, although a raft of 374 was present off CCBO on 16 June 1968. Although numbers can peak at any time during this period, most typically occur in late July or August. An unprecedented passage took place in 1980. The total for that year was 14,365 birds, and a peak of 10,900 was recorded at CCBO on 16 August. Very high numbers also occurred in 1999 with a total of 10,609 birds, and a peak of 1,929 was recorded at CCBO on 10 October.

1959–68	1969–78	1979–88	1989–98	1999–08	2009–18
830	380	15,686	2,434	14,950	7,246

Great Shearwater

Ardenna gravis | Cánóg mhór | A

(monotypic)

World	South Atlantic Ocean
Ireland	Uncommon autumn passage migrant
Cork	Passage migrant
Conservation	Green-listed

Great Shearwaters have been known off the Cork coast since the early 1850s, the first dated record being two obtained off Youghal in December 1854. Others were seen off the coast of west Cork in August 1890, August 1892 and September 1899. In September 1901, a flock of 200–300 was seen between Mizen Head and Clear Island.[65;66]

Great Shearwaters are summer and autumn visitors to the North Atlantic. In autumn 1955, before the scale of autumn passage was fully appreciated, abundant remains were found washed ashore at Bantry Bay.[67] These were believed to have been captured and eaten by Spanish fishermen, inedible portions discarded overboard as the boats sailed up the bay to Bantry.

This species is now known as a regular migrant between July and November, with the heaviest passage during August and September. There are very few occurrences outside this period with fewer than forty birds having been recorded in April to June. Numbers vary greatly between years, the reasons for which are not fully understood.[68;69] Annual totals have exceeded 1,000 birds in ten years since 1959, with 11,248 in 1965, 14,374

Great Shearwater

in 1973 and 12,454 in 1997. Most birds are recorded at CCBO with much smaller numbers seen from other watch points.

1959–68	1969–78	1979–88	1989–98	1999–08	2009–18
12,435	17,899	4,130	23,676	13,209	8,882

Sooty Shearwater

Ardenna griseus | Cánóg dhorcha | A

(monotypic)

World	Southern South America, Australian and New Zealand seas
Ireland	Autumn passage migrant
Cork	Passage migrant
Conservation	Green-listed

Two Sooty Shearwaters were seen 5 km off Cork Harbour on 24 August 1849. In September 1899, at least nineteen were seen at various points off the coast between Bull Rock and Old Head of Kinsale, and at least ten were seen between Mizen Head and CCBO in September 1901.[70]

The Sooty Shearwater occurs mainly as an autumn visitor and is the most numerous of the scarcer shearwaters off the Cork coast. In many years over 1,000 are reported at CCBO, although numbers vary greatly between years. In 1970, 1980 and 1990 about 6,000 were recorded at CCBO, including 1,735 on 20 August 1970, and 1,486 on 15 September 1980. High numbers are regularly seen from Dursey Island, including 1,073 in two hours on 6 November 1978 and 1,672 on 17 September 2011. Smaller numbers (often several hundred) occur off Mizen Head, Galley Head and Old Head of Kinsale.

Sooty Shearwaters have occurred in every month except February and March. Small numbers (single figures) occur in April, May and June, but regular passage takes place from July to October, peaking in August and September. Large numbers may occur in calm anticyclonic conditions which can induce flocks (or rafts) to gather offshore. Under such conditions a raft of approximately 6,000 was present off Mizen Head on 22 August 1970.

Manx Shearwater

Puffinus puffinus | Cánóg dhubh | A

(monotypic)

World	North Atlantic Ocean
Ireland	Common breeder, passage migrant
Cork	No proved breeding, common passage migrant
Conservation	Amber-listed

There are confusing statements in the literature regarding the breeding of Manx Shearwaters in Cork. Birds were often heard calling over land at night at CCBO, but evidence of breeding was lacking in the 1960s, although could not be entirely discounted. Several birds were heard in burrows at CCBO in 1983, and three corpses of fledglings were found in 1989, which could have been brought ashore by a foraging Peregrine Falcon. There remains a suspicion of breeding there, and at other sites including Dursey Island,

Mizen Head and Old Head of Kinsale, but there is no satisfactorily proved breeding record.

The Manx Shearwater occurs commonly offshore in spring, summer and autumn, but is rare in winter. It is the commonest species involved in offshore passage at CCBO from May to August. In spring and autumn, the majority pass west, but there is an easterly dawn movement and a larger westerly one at dusk between late June and early August. The largest movements of the year peak in early July. Passage frequently exceeds 10,000 birds per hour, and on occasions has been as great as 30,000 per hour. Only fifty-two birds were recorded at CCBO between December and February during 1959–2018.

Large numbers pass predominantly westwards offshore at other islands and headlands also, e.g. peaks of 12,000 per hour at Dursey Island, 13,000 at Old Head of Kinsale, 12,000 per hour at Sherkin Island, 8,000 at Seven Heads and 7,000 per hour at Galley Head. Surveys from ships off south-west Ireland in 1990–3 (March to October) showed greatest numbers were present in July and August, with feeding observed only in July.[71] An estimated total of at least 5,000 birds was observed along a transect 7–10 km offshore between Seven Heads and Ballycotton in April 1980.[72]

From the end of July onwards rafts may be found in calm weather in Bantry Bay.[73] At least 1,000 birds were seen feeding and rafting around Hornet Rock between Bere Island and the mainland on 24 August 2006.

Irish and British Manx Shearwaters are trans-equatorial migrants that winter off the Atlantic coast of South America. Few birds have been ringed in Cork, but there have been many recoveries along the Cork coast in spring, summer and autumn of birds ringed as nestlings elsewhere at Irish and British colonies, especially at those in south Wales. It has recently been shown that birds from Wales forage off the Cork coast as far west as Old Head of Kinsale during the breeding season.[74]

Single individuals have been seen inland at the Gearagh on 29 May and 8 September 2017. Birds occasionally enter Cork Harbour, but rarely penetrate very far, but two were at Cobh on 6 June 1977 and one at Cork city quays on 30 August 1994.

Balearic Shearwater

Puffinus mauretanicus | Cánóg Bhailéarach | A

(monotypic)

World	Balearic Islands in Mediterranean Sea
Ireland	Scarce autumn passage migrant
Cork	Passage migrant
Conservation	SPEC 1; Red-listed

The Balearic Shearwater has been recorded regularly since the first in 1961, and has occurred every year to 2018, apart from 1970 and 1987. A large majority occurred at CCBO, but small numbers have been recorded at many sites from Dursey Island to Ballycotton. Most occurred from late July to October, but small numbers have been seen in all other months. At least seventeen have been recorded in the months of December and January. In most years fewer than forty individuals are recorded, but over 100 have been seen in three years. Numbers have increased since 1990; during 1959–89 the peak year was 1982 when thirty-two occurred, while during 1990–2018 the peak exceeded forty in fifteen years. Most records involve sightings of from one to five or six individuals, most of which were moving in a westerly direction.

1959–68	1969–78	1979–88	1989–98	1999–08	2009–18
63	93	122	568	358	540

Barolo Shearwater

Puffinus baroli | Cánóg bheag | A

(monotypic)

World	Islands in eastern North Atlantic Ocean
Ireland	26 individuals
Cork	6 individuals (5 records)
Conservation	Not assessed

One obtained off Bull Rock on 6 May 1853. One at CCBO on 24–6 September 1978, joined by a second bird on 25–6 September 1978. One at CCBO on 14 August 1992. One at CCBO on 26 August 1993. One at Toe Head on 12 August 2015. The Bull Rock bird was taken alive after coming aboard a sloop and was brought ashore at Valentia Harbour (Kerry).

Wilson's Storm Petrel

Oceanites oceanicus | Guairdeall Wilson | A

(subspecies: undetermined)

World	Southern Ocean
Ireland	Rare passage migrant, mainly in July and August
Cork	309 individuals
Conservation	Not globally threatened

Wilson's Storm Petrels occur between late May and early October, but mainly in the period from late July to late August. Most birds have been seen during boat trips in inshore waters of up to 30 km from the nearest land within the official recording area. Small numbers have been seen from headlands and islands, such as Mizen Head, CCBO and Galley Head. The sea area in which most have been recorded extends from Mizen Head to Galley Head; only one has been seen at sea off Seven Heads, two from land at Old Head of Kinsale and one from land at Power Head, the latter being the only record from east Cork waters. Most sightings have been of single birds or of small groups not exceeding ten, but there are records of sixteen, forty, fifty-six and fifty-nine birds seen on single boat trips, although not all recorded birds may have been seen together.

This species was regarded as very rare in Irish waters until recently, and it only became apparent that many were present in offshore waters off the south-west once observers sailed out and searched for them.[75] In more recent times they have been increasingly seen in inshore waters (inside 30 km) and small numbers have been identifiable from headlands. The decadal occurrence pattern likely reflects the realisation that this species was present in numbers mainly out of view from land, rather than any recent change in status. The very high numbers in the most recent decade results largely from regular boat trips into the waters between Mizen Head and Galley Head each autumn between 2014 and 2018.

1959–68	1969–78	1979–88	1989–98	1999–08	2009–18
0	0	1	10	12	286

Earliest and latest occurrence in spring and autumn (all records): 24 May–10 October.

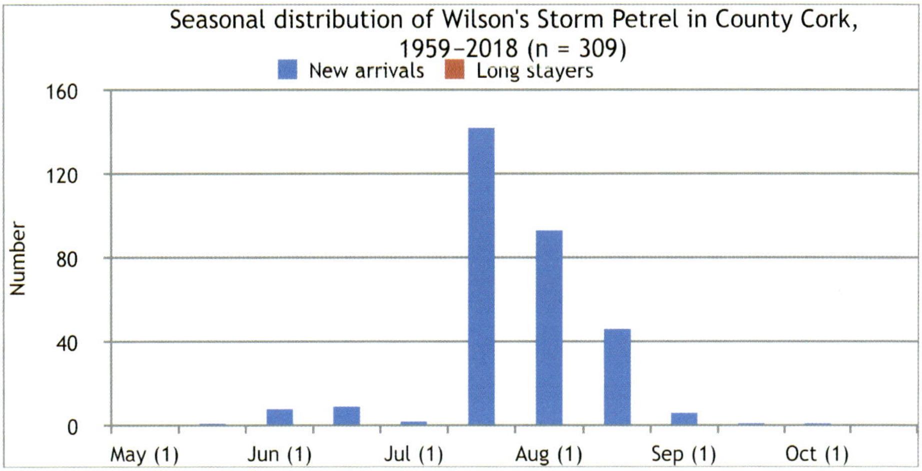

Fig. 7.9 Wilson's Storm Petrel.

European Storm Petrel

Hydrobates pelagicus | Guairdeall | A

(subspecies: *pelagicus*)

World	North-east Atlantic Ocean and Mediterranean Sea
Ireland	Common breeder, passage migrant
Cork	Very local breeder, common passage migrant
Conservation	Amber-listed

The breeding status of the Storm Petrel in Cork in the 1800s is uncertain, although a reference by a resident of CCBO suggests it may have been breeding there.[76] There was no evidence of breeding in the late 1800s,

European Storm Petrel

although the suitability of Bull and Cow Rocks was mentioned. In fine weather birds were frequently seen off CCBO and Mizen Head, mostly in late summer and autumn. They were sometimes seen at Bantry Bay after storms, and occasionally blown inland, and a large rush took place at Fastnet Rock and Bull Rock lighthouses in July 1889.

Storm Petrels were nesting abundantly at CCBO in August 1937.[77] Thousands have recently been ringed at CCBO, and although breeding has not been proved, it was suspected in 1965 when birds were reported ashore, and in 1983 when a bird was calling in a burrow. It is probable that a colony is present, but it is unlikely large numbers are involved.

There was no further evidence of breeding until the discovery of a colony of about 150 pairs at Bull Rock in 1955.[78] The colony was estimated at 2,000–5,000 pairs in 1969. Breeding may also take place at Cow and Calf Rocks.[79] A pair of Storm Petrels bred at Fastnet Rock in 1972, the first occasion any bird species is known to have nested there. One pair bred there in 1981 and two pairs during at least the five-year period 1988–92.

Spring passage is usually very light, but 590 were recorded at CCBO on 8 June 1983. A westerly passage takes place chiefly between July and September at CCBO. A peak occurs in August and passage almost invariably takes place during south-west winds, often in drizzle or fog. Numbers often exceed 1,000 birds per hour, and over 2,000 on occasions. On 13 August 1981, 7,500 birds were passing per hour. Passage can also be seen off other headlands and islands, but the pattern does not differ significantly from that at CCBO. Given suitable weather conditions other sites may record large numbers, 8,000 at Dursey Island on 2 September 1997 being the largest.

Following severe weather in September 1983 large numbers occurred at Ballycotton Bay peaking at 1,500 on 3 September, some remaining to October. Storm Petrels are rarely seen on the Cork coast following departure from the breeding colonies until return the following spring, although strays are occasionally seen on the coast or are blown inland during November and December storms. There are no records for January or February.

Many Storm Petrels have been ringed on the Cork coast at sites where they do not breed. Largest numbers have been ringed at CCBO, Toe Head, Seven Heads, Old Head of Kinsale and Power Head, and smaller numbers at other sites, such as Knockadoon Head. The catching technique (using recordings of calls) selects for a population class of wandering pre-breeders. This class consists of birds of two years and older that have not yet bred but may also include failed breeders.[80;81]

Irish and British breeders do not normally return north in their first year but do so typically when two or three years old. By this stage they are in the wanderer class and they come close inshore at night to prospect for nest sites and to visit colonies before breeding for the first time at four or five years old.[82] Recent evidence suggests they may also come inshore to feed on benthic organisms that migrate to the surface during nocturnal high tides.[83]

The results of ringing show that there is considerable movement of birds on the coast during the late summer and autumn, and there have been many exchanges between Cork and Scandinavia, France, Spain, Portugal and West Africa, as well as between many sites elsewhere in Ireland, Scotland, Wales, England, Isle of Man and Channel Islands. Although Storm Petrels winter around the southern tip of Africa, there appears to be no recovery south of Senegal of a bird ringed in Cork. Some of these movements can be very rapid, with no more than a few days separating ringing and recapture and can be in a northerly direction within the same autumn, when the general direction of movement is south, emphasising the wandering propensity of this population class.

Leach's Storm Petrel

Oceanodroma leucorhoa | Guairdeall gabhlach | A

(subspecies: *leucorhoa*)

World	North Atlantic and North Pacific Oceans
Ireland	Scarce breeder, scarce autumn passage migrant
Cork	Scarce passage migrant, subject to 'wrecks' in severe weather
Conservation	SPEC 1; Red-listed

The first dated record was one obtained about 15 km inland on a mountain in September 1818. Three others (undated) were obtained before 1894, one of them near Youghal. Three were obtained at light-stations (Old Head of Kinsale, November 1895; Fastnet Rock, May 1896; Castletownbere, October 1897). Another was obtained at Roancarrigmore lighthouse in September 1946, and two were found dead at Roaringwater Bay in December 1948.

Large numbers of exhausted birds may be blown ashore during storms, and any such event is termed a 'wreck'. A major 'wreck' took place in west Cork in late October 1952 during and following a severe south-west storm.[84] Large numbers, including fifty at Snave, were seen at Bantry Bay and many were found dead. Large numbers were also seen at Dunmanus Bay, and 200–300 at Roaringwater Bay where small numbers were found dead. At least ten others were blown inland to Inishcarra, Rathcormack, Mallow, Buttevant and Doneraile.

Leach's Petrels occurred sparingly during 1959–2018, mostly between July and December, with single records in January, February and June. Most occurred as passage migrants off CCBO and Galley Head, with smaller numbers at Dursey Island, Mizen Head and elsewhere. Records of migrants usually involve single birds, and twelve at CCBO in July 1963 was exceptional. One was trapped at CCBO in July 1983 after being attracted to a tape lure for European Storm Petrels.

As indicated above, this species is prone to exhaustion during severe weather, and at least twenty-nine (possibly up to 48) were recorded at Cork Harbour in early December 2006. Others affected by severe weather involved two in November 1977 (Ballymacoda and Little Island), two in December 1989 (Garrettstown and Charleville), one in October 1994

(Inishannon), one in January 1996 (Ballycotton) and one in November 1998 (the Gearagh). One was also at the Gearagh on 12 April 2020. One captured on board a boat off Great Skellig (Kerry) in September 1986 and released at Clonakilty Bay has been excluded from the decadal totals.

1959–68	1969–78	1979–88	1989–98	1999–08	2009–18
20	7	3	14	41	13

Gannet

Morus bassanus | Gainéad | A

(monotypic)

World	North Atlantic Ocean
Ireland	Common breeder, scarcer in winter
Cork	Local breeder, one colony
Conservation	Amber-listed

The Bull Rock is the only breeding place of the Gannet in Cork, although it was stated incorrectly that breeding took place at Fastnet Rock and Stags Rock off Toe Head in some early literature. The colony at Bull Rock was first recorded as having nests in 1856, although birds were seen there three years earlier. Numbers increased quickly to about 1,000 pairs by 1884. A lighthouse was completed in 1888. Blasting operations during construction caused considerable disturbance, and a reduction in breeding numbers. After the building works ceased numbers increased slowly through the 1900s to about 500 pairs by 1955 (excepting some high counts at variance with the trend).[85;86] Numbers increased to 1,500 pairs in 1969–70 and to 1,815 in 1995, before rocketing to 3,694 in 2004 and to 6,388 in 2014.[87]

Gannet populations in Ireland and Britain increased and established many new colonies during the 1900s. Food supply is not currently a limiting factor, but recent European Union fisheries policy changes on fish discarding may reduce supply and availability in the future.[88] It has been suggested the

Gannet

increase should not be looked on as a population explosion or an extension of range, but as a recovery from intensive persecution of past centuries.[89]

The Gannet is present offshore throughout the year and is sometimes seen in considerable numbers in winter. Winter distribution is probably related to food availability with flocks congregating where there are shoaling fish; 3,250 were feeding on sprats at Youghal Bay on 14 December 1990. However, density in all seasons is greatest off the western peninsulas.

A predominantly westward movement offshore in spring and autumn has been noted by many authors. This movement is more extensive from August to October, and up to 4,000 birds per hour have been seen passing CCBO. Spring passage is normally light, with typically up to 300 birds per hour at CCBO between March and May. Significant numbers also pass other prominences, such as Dursey Island and Old Head of Kinsale, particularly in autumn. Offshore rafts and fishing flocks have occasionally reached 2,000 individuals at CCBO.

Gannets are sometimes blown inland, especially during autumn and winter storms. There are many cases of such storm-blown individuals found in different parts of the county (adults and juveniles), and one was seen flying down the River Lee at Cork city on 13 April 1963. Gannets regularly enter Cork Harbour but usually do not go upriver beyond Cobh.

Gannets in their first winter move south to the coasts of southern Europe and West Africa. Nestlings ringed at Bull Rock have been found off Senegal in the following winter. As birds become older, they return to natal waters and many adults remain near the breeding colonies in winter. Gannets from British, Channel Islands and other Irish colonies have been recovered in Cork, frequently drowning after entanglement in fishing nets, and one ringed as a nestling at Andoy (Norway) in July 1980 was found at Whitegate in December 1987. Although Icelandic Gannets have been recovered in Ireland, none have been found in Cork.

Great Cormorant

Phalacrocorax carbo | Broigheall | A

(subspecies: *carbo; sinensis*)

World	Canada, Greenland, Iceland, Europe, Asia, Africa, Australia, and New Zealand
Ireland	Resident breeder
Cork	Common resident and breeder
Conservation	Amber-listed

The Cormorant (*P.c. carbo*) was a common resident in the 1800s. Several breeding sites were named: Cow Rock, CCBO, near Glandore Harbour, Black Rock (there are at least 11 sites so named), Old Head of Kinsale, Reanies Bay (over 20 nests in 1848) and various sites between Reanies Bay and Sovereign Islands. It was common at Cork Harbour, Youghal and Kinsale in July 1907.[90]

Cormorants formerly bred at inland sites. More than eighty nests were on Scots pines beside a lake at Castlemartyr about 1833. Nesting was also noted on cliffs over the River Blackwater below Mallow, and above Fermoy. The latter may refer to (or include) Cregg where they were said to breed in trees.[91] The nearest cliff above Fermoy is Ballyhooly, while there are several other cliffs between there and Mallow. Cliffs also occur below Fermoy at Ileclash, but there is no record of breeding at these. Breeding on the River Blackwater had ceased many years before 1954. No inland colony has been documented since.

Year(s)	Number of AONs
1969–70	38
1985–88	477
1998–02	366

Table 7.3 County Cork totals for breeding Great Cormorant, 1969–2002.

Censuses of the breeding population have been carried out since the late 1960s. Table 7.3 contains figures for the whole county from national censuses. The counting unit used is the Apparently Occupied Nest (AON), and the figures given include subsequent revisions. The figures show a considerable increase in numbers, and this trend continues.[92] Records show that three east Cork colonies increased from 11 nests in 1979 to 109 nests in 1998, and a colony at Sovereign Islands increased from 35 nests in 1985 to 147 nests in 1998. Twenty-eight nests were on the remote Cow Rock in 2002.

Man has persecuted Cormorants for a long time, mainly because of alleged damage to fish stocks. Birds were shot, breeding colonies raided, and fishery managers paid a bounty for birds destroyed. This has made assessment of population change difficult, and their habit of frequently shifting breeding sites has added to the problem. There is no doubt that Cormorant breeding populations in Cork were depressed in the past due to persecution. However, legal protection was afforded the species by the Wildlife Act, 1976. Persecution was reduced and populations responded by increasing at existing colonies and establishing new ones. An increase in winter food supply through an increase in coarse fish, especially perch and roach, and the removal of predatory fish such as pike at some lakes may also be factors.[93]

Cormorants are present at all parts of the coast at all times of the year, and they also visit inland waters at all times of the year, but mainly in autumn and winter when birds can be found on lakes and rivers, big and small, especially the River Blackwater. All estuarine habitats hold some Cormorants in winter, but numbers are greatest at Cork Harbour where roost surveys have shown numbers ranging from 442 to 599 during 2013–19.[94] The highest Cork Harbour count on record is of 991 birds in December 1995. Winter numbers in Cork Harbour have mirrored the increase in the breeding population. Exceptional numbers may occur in estuaries during stormy weather, as at Ballymacoda in January 1988 when 517 were counted. Coastal and estuarine roosts of 100–200 birds are not unusual. The largest regularly used inland roost involves up to fifty birds at Carrigadrohid.

Nestlings have been ringed at colonies at Capel Island, Ballycotton Islands and Sovereign Islands. Recoveries from these sites show a dispersal inland through the river systems and along the south Irish coast to adjoining counties, but with some going as far south and east as the Bay of Biscay. Birds from Wexford, the northern half of Ireland and from Wales and north-west Britain also occur in Cork, inland as well as on the coast, in the non-breeding season.

Continental Cormorant *P.c. sinensis*: There are three records of this subspecies. One at Baltimore in January 2012. One at Lough Aderry in February 2012. One at Lough Aderry in November 2015.

Shag

Phalacrocorax aristotelis | Seaga | A

(subspecies: *aristotelis*)

World	Iceland, Europe, and north Africa, Mediterranean and Black Seas
Ireland	Resident breeder
Cork	Common resident and breeder
Conservation	SPEC 2; Amber-listed

The Shag was noted as not numerous in the 1800s but was breeding on cliffs between Reanies Bay and Sovereign Islands (single pairs at several locations), including one or two pairs at the latter site. It was also breeding at Old Head of Kinsale.

Censuses of the breeding population have been carried out since the late 1960s. Table 7.4 contains figures for the whole county from national censuses. The counting unit used is the Apparently Occupied Nest (AON), and the figures given include subsequent revisions. The figures show an increasing trend nationally.[95]

Year(s)	Number of AONs
1969–70	215
1985–88	419
1998–02	221

Table 7.4 County Cork totals for breeding Shag, 1969–2002.

Shags were continuously distributed along the coast from Dursey Island to Capel Island in the breeding seasons of 1968–72 and 1988–91.

A small breeding population in east Cork had disappeared by the late 1990s and has not returned.

Although the Shag is a bird of wilder coasts than the Great Cormorant, breeding does not take place at Bull and Cow Rocks, but they may breed at Calf Rock, where eleven birds were seen in 1969.[96] Shags are common at CCBO, and up to 139 pairs (1963) have been recorded breeding, although they declined to thirty-two pairs (1990). Numbers also declined at Roaringwater Bay (exclusive of CCBO) from 120 pairs in 1982/3 to seventy pairs in 1994.

Recent winter atlas surveys show a strictly coastal distribution, densest concentration in the west, with numbers diminishing eastwards. Highest counts are usually recorded at CCBO in August and September when a peak of 381 occurred. About 250–300 birds, mostly in a tight flock, were present off Barley Cove in October 1994. Numbers decline in mid-winter and increase again in spring, with a peak in east Cork of 136 at Ballybranagan Strand in March 1989. Shags are scarce in Cork Harbour and are rare inland (Gearagh in August 2019 and August 2020), although birds have been seen at Patrick's Bridge (Cork city) in March 1977 and December 1978.

The Shag is dispersive rather than migratory, with young birds moving some distance in their first autumn and winter. A few birds from Scotland and the south-west English coast reach Cork, but most recoveries of ringed birds are of nestlings from Irish colonies, especially from Great Saltee Island (Wexford), but some also come from the north-west (Sligo) and north-east (Down). Recoveries of Great Saltee Island birds may be over-represented as few birds are ringed at other south coast colonies in Ireland.

Unidentified frigatebird species

One frigatebird, probably Magnificent (*Fregata magnificens*), was recorded at CCBO (probable male) on 24 August 1973.

Great Bittern

Botaurus stellaris | Bonnán | A

(subspecies: *stellaris*)

World	Europe, Asia, and north and South Africa
Ireland	Former breeder, now rare winter and spring visitor
Cork	Former resident, extinct as breeding species, vagrant since 1900
Conservation	SPEC 3; Amber-listed

The Bittern was breeding in Cork in the 1700s. It was said to be resident in the mid-1800s, but rare of late years. Booming birds were heard at Youghal many years before 1850 (probably by Robert Ball when 13–14 years old). A man living at Rathcoursey told Ussher that his father, when young, heard Bitterns on a bog that had long since been drained, well before the 1840s. This bog was probably in the valley at the source of the River Womanagh north-east of Rathcoursey. It was said to be formerly common along the River Blackwater and bred there before the river was embanked some forty years before, although this was probably a reference to the Waterford sector around the confluence of the River Bride (north), where Ussher was informed breeding took place many years before 1856. The Bittern was probably a reasonably widespread species in wet and marshy habitats in Cork into the early 1800s, but it was extinct as a breeding species in Ireland by about 1840.

After breeding ceased some birds continued to turn up, particularly in winter, but it was a scarce species by 1850, due principally to drainage of bogs and marshes and incessant shooting. One was obtained at Glengarriff in autumn 1838, one at Youghal in winter 1840/1, and another near Cork city in the same winter. The Bittern was described as not very rare in Ireland in severe winters, especially in the south and west.[97] Some winters were marked by the occurrence of unusual numbers. No fewer than thirteen were shot in various parts of the county in winter 1874/5, twelve in about one month leading up to 9 January 1875. One was shot at Ballyvergan Marsh on 15 December 1875, and five were shot in Cork in the winter of 1878/9, one being caught alive at Douglas. One was shot at Old Head of Kinsale in February 1883 and another at Ballyvergan Marsh a few days before 19 December 1885. Another was seen in County Cork on 13 January 1900, one of the few encountered by shooters that was laudably not shot. Inclusive of those quoted by Thompson, thirty-seven records of birds in Cork in winter were known since breeding ceased, the most for any Irish county.

There have been six records in Cork during 1901–2018, all single birds: Ballycotton in 1901 and 1903, Ardfield on 23 January 1963, Ballyvergan Marsh on 19 March 2006, Ballyhonock Lake on 19 March 2012 an Kilcolman on 31 March 2014.

American Bittern

Botaurus lentiginosus | Bonnán Meiriceánach | A
(monotypic)

World	North America
Ireland	23 individuals
Cork	4 individuals
Conservation	Not globally threatened

First-year obtained at Myross Wood, Glandore in early October 1875. Adult obtained at Annagh Bog (Inishannon) on 25 November 1875. One obtained at Youghal–Cork city in December 1875. First-year at Castlefreke on 25 November–14 December 2015. It is remarkable that three individuals should occur in Cork in the same year (1875), with no other occurrence for the next 140 years. The Castlefreke bird is unique in that it is the only one to occur in Ireland and survive the experience, all others having either been found dying or having been killed in one way or another.

Unidentified bittern species

One bittern, either Great or American (*Botaurus stellaris/ lentiginosus*), was recorded at Kilcolman on 28 December 2002.

Little Bittern

Ixobrychus minutus | Bonnán beag | A

(subspecies: *minutus*)

World	Europe, Asia, Africa, and Australia
Ireland	56 individuals
Cork	22 individuals
Conservation	Not globally threatened

Fourteen birds occurred singly before 1959, all but one at or near the coast. Eleven of these were in the 1800s: about Cork Harbour and city (4) and Youghal (3), with others at Schull, near the River Bandon, Rathcormack (the only truly inland record) and an unknown location. Between 1901 and 1958 another three were recorded: at Fastnet Rock, Galley Head and Clonakilty Bay. A month of occurrence was recorded for seven of the fourteen birds, March (2), April (2), and May, June and November. Another, from Youghal, was probably found dead at the end of May as it was exhibited before the Dublin Natural History Society on 3 June 1864, while one was obtained at 'Woodside' (Carrigrohane) in summer. All fourteen birds except one (Clonakilty, 1929) were either obtained (mostly shot) or found dead or dying, including one killed by a blow from a stick and one taken on board a steamer; one from Fastnet Rock (1953) was killed when it struck the light. Of four aged birds, two were adult and two were immature, while three were male and one female.

Eight individuals (7 records) occurred during 1959–2018. Two at CCBO on 1 May 1966 (1 male was exhausted and died on 2 May, the other was caught by a cat). Exhausted adult male at Myrtleville on 21 April 1978, released at Fountainstown on 22 April. Male at CCBO on 15 and 16 May 1979. One at Ballycotton on 31 May 1981. Adult male at Inchigaggin Marsh on 4 May 1989. Exhausted female at Ballycotton on 16 March 1990, died on 17 March. Adult male at Sherkin Island on 20 and 21 April 1992.

Across both time periods (pre- and post-1959), eight birds were either found dead, dying, or obtained in unusual circumstances. All occurred in spring, suggesting severe exhaustion after a long flight following overshooting of their European breeding range. A female captured by hand at Ballycotton in 1990 died overnight and weighed only 85 g (average weight of healthy females is 146 g). All birds that apparently survived remained for only one or two days.

Earliest and latest occurrence in spring and autumn (all records): 16 March–6 June and 8 November.

Night Heron

Nycticorax nycticorax | Corr oíche | A

(subspecies: *nycticorax*)

World	North and South America, Europe, Asia, and Africa
Ireland	84 individuals
Cork	39 individuals
Conservation	Not globally threatened

Eight Night Herons were obtained singly in Cork, most of them shot, before 1959, including seven in the 1800s. One of the eight has been recorded without date, but a few years before 1845, while another occurred in winter. The other six birds occurred in March, May (4) and October. Six occurred in coastal districts between Castlefreke and Youghal, while two were at inland locations at Doneraile and near Fermoy. Those that have been aged and sexed were three adults, three immatures, and four males.

The records (31 individuals) in the 1959–2018 period show a strong peak in May, with only five occurring in autumn (2 in July and August, and 3 in October and November), and one in winter (January). The only months without a record are September and December. Most records refer to singles, but there have been four occurrences involving two and three birds, Carrigrohane and Clonakilty (2 birds each) and Ballymaloe and CCBO (3 birds each). Most records have been spaced out along the coast from Castletownbere to Youghal, but some have penetrated short distances inland to Abisdealy Lake (Lissard), Skibbereen, Carrigrohane, Ballymaloe and Lough Aderry, with one much further inland near Fermoy.

Slightly more birds were recorded at and near the coast east of Kinsale (17) than west of Kinsale (13). Somewhat surprisingly, given the general and widespread nature of the occurrences, the coastal wetlands between Rosscarbery and Kinsale have produced only two birds, both at Clonakilty.

There has been an increase in occurrence of this species over the sixty years of this analysis, but the increase is modest if the fourth decade is excluded, when fourteen birds were recorded, including all four multiple sightings. More than half (17) were observed on only one day, while eight birds made protracted stays of between ten and twenty-nine days (mean stay = 6.1 days).

There was a general tendency for birds observed in poor feeding habitats (such as islands) to remain for a very short time, while those observed in richer feeding habitats (such as Clonakilty, Ballycotton and Youghal) remained for longer periods. Of those aged, thirteen were adult and sixteen were immature birds. None were sexed in this period, compared with the 1800s, reflecting the fact that no birds were shot. Only one Night Heron was found in a moribund state, in marked contrast with the Little Bittern, several of which have died following arrival. Why the Night Heron should suffer lower mortality following spring overshooting of their breeding range is unknown, especially since both species are long-distance trans-Saharan migrants with a broadly similar diet.

1959–68	1969–78	1979–88	1989–98	1999–08	2009–18
1	2	3	14	5	6

Earliest and latest occurrence in spring and autumn (1959–2018): 18 February–22 June and 6 July–29 November.

Green Heron

Butorides virescens | Corr glas | A

(monotypic)

World	North and Central America
Ireland	1 individual
Cork	1 individual
Conservation	Not globally threatened

First-year at Schull on 11–13 October 2005. This bird, recognisable on plumage details, was seen at Anglesey (Wales) from 30 October to 20 November 2005.

Squacco Heron

Ardeola ralloides | Corr scréachach | A

(monotypic)

World	Europe, Asia, and Africa
Ireland	23 individuals
Cork	7 individuals
Conservation	Not globally threatened

Adult obtained at Killeagh on 26 May 1849. One obtained at Ballycotton in summer 1850. One obtained at Castlemartyr on 26 October 1860. Female obtained at Kilkerran Lake on 15 July 1877. Female obtained at Skibbereen on 13 May 1913. Adult at Ballycotton on 13–17 June 1996. One at Ballycotton on 22 May 2015.

Cattle Egret

Bubulcus ibis | Éigrit eallaigh | A

(subspecies: *ibis*)

World	North and South America, Europe, Asia, Africa, Australia, and New Zealand
Ireland	Rare visitor, recent increase
Cork	214 individuals
Conservation	Not globally threatened

There were three records of single Cattle Egrets in Cork up to 2005, all at coastal west Cork sites between Rosscarbery and Clonakilty (March 1976, January 1999, April 2005). There followed two significant influxes during the winters of 2007/8 and 2008/9 involving at least 108 and 38 birds, respectively. A further three individuals were recorded in October 2009, May 2012 and March 2016. Another influx (27 birds) took place in 2016/17, followed by three birds in November and December 2017 and three more in April 2018. Finally, another influx took place in late 2018 (29 birds).

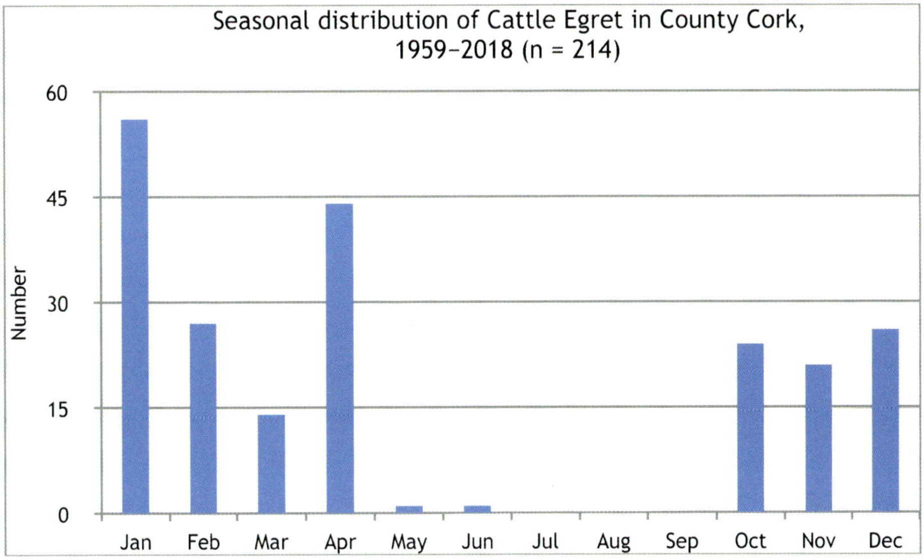

Fig. 7.10 Cattle Egret.

Although there were many records of single birds, most observations involved flocks of between two and thirteen. Some birds have occurred at almost every site along or near the coast from Ardgroom to Youghal, including at CCBO and Dursey Island, although the only truly inland records were at the Gearagh and Enniskean, as well as a flock in the Rossmore and Ballynacarriga Lake areas of west Cork. During each of the influxes, birds were widely spread from west to east, especially in 2007/8 and 2008/9. Each influx began in the October to December period, with new birds appearing over the following few months, before petering out. It is possible that at least some of the apparent later arrivals had arrived earlier but remained undetected at sites not visited by observers. Flocks and individuals were also suspected of moving between several nearby sites, and numbers at some sites varied over their stay, which made it difficult to accurately calculate the total number. In addition, some birds recorded in 2008/9 were believed to be the same individuals as those involved in the 2007/8 influx. Some movement across the border with Waterford was also recorded in the area around the estuary of the River Tourig.

There was a general tendency for larger groups to stay longer at one place and for single birds or small groups to stay for only one or a few days at one place, although there were exceptions to both rules. The tentative suggestion is that single birds and small groups moved about until they found a larger group to which they attached themselves.

1959–68	1969–78	1979–88	1989–98	1999–08	2009–18
0	1	0	0	111	102

Earliest and latest occurrence in winter (all records): 9 October–2 June.

Little Egret

Egretta garzetta | Éigrit bheag | A

(subspecies: *garzetta*)

World	Europe, Asia, Africa, Australia, and New Zealand
Ireland	Scarce or uncommon breeder, recent increase
Cork	Resident scarce, but increasing, breeder
Conservation	Green-listed

The first record for Cork (and Ireland) was of one present near Skibbereen from 26 May to August 1940. The next record was of five birds, again near Skibbereen, in May 1957, followed by a single bird at the same place in May 1958 which moved about between several west Cork sites. There were regular occurrences during 1960–89 which increased in frequency over time. During this period the Little Egret was mainly a spring overshooting migrant (70 per cent) with most records in May. Some birds remained for several months, and one for more than a year. Numbers increased from 1990 onwards, and the status changed from one of spring migrant to autumn migrant and winter visitor.[98] Little Egrets are present in all months since 1990, often in small flocks feeding or roosting on estuaries by day, or tree-roosting by night, with counts of twenty to sixty being not unusual in most years at favoured resorts.

The Little Egret bred for the first time in Cork (at Ballymacoda) in 1998, having bred for the first time in Ireland the previous year at the River Blackwater (Waterford). Expansion of the breeding population continued in the following years with new colonies established at Cork Harbour.[99]

Little Egret

Over the next ten years (to 2011) new colonies were established from Ballymacoda Bay to Roaringwater Bay in fourteen 10 km squares. One colony is in an urban habitat at Cork city. Most sites are close to rich tidal river and estuarine habitats where birds mainly feed. The limited amount of feeding ecology data suggests that shrimps are an important food item.[100;101] Only one breeding site has been inland and upriver of tidal influence, on the River Blackwater near Clondulane.

Little Egrets are frequently seen tens of kilometres inland up the main Cork rivers (Blackwater, Lee, Bandon), as well as on many of the smaller rivers and on lakes during the breeding and winter seasons, and individuals are also often seen on rocky coastal habitats far from the nearest breeding site. Its winter range is more extensive than that of the breeding season, with all but the upland areas occupied where, presumably, few feeding areas exist.

One ringed as a nestling at Oranmore (Galway) in May 2009 was seen at Rosscarbery in September 2017 and in July 2019.

Great Egret

Ardea alba | Éigrit mhór | A

(subspecies: *alba*)

World	North and South America, Europe, Asia, Africa, Australia, and New Zealand
Ireland	Rare visitor
Cork	22 individuals
Conservation	Not globally threatened

The Great Egret was first recorded in Cork at CCBO on 26 October 1997, this individual later moving to the mainland at Croagh Bay where it stayed until 3 November. The history of the next occurrence is interesting: one colour ringed as a nestling on 28 May 2000 at Loire-Atlantique (France) was seen at Barley Cove on 23 August 2000. The same bird was then reported in Wicklow in October 2000. This bird was next seen at Careysville, near Fermoy, on 1–2 November 2000, later at Midleton, Lee Fields and Ballincollig up to 11 November. Another bird seen at Blackrock on 30 April 2005 was reported to have been subsequently seen in Wicklow in May and in Wexford in June of that year. There is a small peak of occurrences in April and May, and in September and October, but only three months are without records of new birds. While many birds have remained for only one day, several have made protracted stays (as above), the longest being one seen on and off at Cork Harbour from 28 July 2008 to 6 May 2011. The Great Egret has occurred at various coastal or near coastal sites from Ballylicky to Midleton, with inland records at Careysville (as stated) and the River Lee system. All records have involved single birds.

1959–68	1969–78	1979–88	1989–98	1999–08	2009–18
0	0	0	1	6	15

Grey Heron

Ardea cinerea | Corr réisc | A

(subspecies: *cinerea*)

World	Europe, Asia south to Java, and Africa
Ireland	Resident breeder
Cork	Common resident
Conservation	Green-listed

Nesting sites (heronries) of the Grey Heron were well known in the 1700s, particularly near the coast in the barony of Imokilly, including one at Clonpriest. A list of sixty-six heronries, excluding those with fewer than four nests, was published by Ussher and Warren.[102] These had a combined minimum total of 512 nests. Only five heronries had twenty or more nests (40 at Myross Wood, and 20 each at Creagh, Drishane, Doneraile and Shippool). It was said the break-up of large estates and extensive tree felling during the Second World War made heronries fluctuate and become unstable.[103]

The usual nesting site is in tall trees, but low trees and bushes are also used. Nesting on sea cliffs at Old Head of Kinsale and Coolim Cliffs was reported in the 1700s. Four or five nests were on sea cliffs at Bere Island in the late 1800s, and this site was still occupied in 1953. Nests are sometimes built on the ground in marshy areas. One pair nested on the ground at Ballymacoda in 1974, and one pair nested on the ground at Ballycotton in 1980.

Nesting Grey Herons were spread across the county during all three breeding atlases, but there were gaps in distribution on high ground, especially in the north-west, and away from the coast and major rivers. There were concentrations along the Rivers Blackwater and Lee, and around Cork Harbour.

Heronries were censused in east Cork in 1979 and 1980 (47 nests, 9 colonies, largest 11 nests), and around Cork city (44 nests, 9 colonies, largest 15 nests). Based on these figures, the total population for the county was estimated at a conservative 200 pairs (now considered probably a serious underestimate). A colony at Clonakilty declined from thirty-six to twenty pairs between 1975 and 1978 when a nearby refuse site closed, the herons having previously fed on the remains of chickens dumped there.

The winter distribution is almost identical to the breeding season. Counts across the whole county in January 1992, 1993, 1994 and 1995 gave totals of 148, 650, 533 and 128, respectively (some are clearly underestimates). Communal ground roosts are a feature of the autumn and early winter, particularly around estuaries. Winter counts of fifteen to thirty occur at Ballycotton, Lough Beg, Cuskinny and Kinsale Marsh. Up to forty were regularly counted at Clonakilty Bay in 1971, with maximum counts of seventy-five and 100. Cork Harbour is the most important site in the county with a peak count of 130 in 1995/6. Numbers there have remained high with a peak count in the 2010s of 122.

Post-breeding dispersal in autumn is indicated by observations at CCBO, where up to six may be present from August to October. A nestling ringed at Shanagarry in April 1982 was found dead in Wexford in November 1982. One ringed as a nestling in southern Norway in June 1949 was found at Castletownbere in January 1950, and Scottish ringed birds have occasionally been recovered in Cork in winter. One was at Fastnet Rock in December, although it is uncertain if this was a migrating bird, as some visit remote rocks to feed.

Purple Heron

Ardea purpurea | Corr chorcra | A

(subspecies: *purpurea*)

World	Europe, Asia south to Timor, and Africa
Ireland	29 individuals
Cork	14 individuals
Conservation	Not globally threatened

The Purple Heron is mainly an overshooting spring migrant, and there are only two autumn occurrences. It has been recorded at ten coastal sites ranging from Lissagriffin to Knockadoon Head. Only four sites have had more than one bird: CCBO, Clonakilty, Garrettstown and Lough Beg, each with two. Half of the birds have been seen on only one day, but others have remained from two to six days, with two remaining seventeen and nineteen days, respectively (mean stay = 4.5 days); the latter two were seen at Garrettstown, the only record involving more than one bird. Ages have been published in five cases, four adults and one first-year bird. Although a very rare bird in Ireland, the Cork records have remained fairly constant over the last few decades.

1959–68	1969–78	1979–88	1989–98	1999–08	2009–18
1	1	3	4	2	3

Earliest and latest occurrence in spring and autumn (all records): 19 March–24 June and 26 August–1 October.

White Stork

Ciconia ciconia | Storc bán | A

(subspecies: *ciconia*)

World	Europe, Asia, and north and South Africa
Ireland	35 individuals
Cork	7 individuals (5 records)
Conservation	Not globally threatened

Three (1 obtained) at Fermoy in late May 1846. One obtained at Hop Island on 7 August 1866. One at Pilmore Strand on 1 May 1978. One at Clonakilty on 22–5 April 1995. One at Cloyne on 22–3 April 2002. What is presumed to be the Cloyne bird was seen at Tallow (Waterford) on 28 April 2002.

Glossy Ibis

Plegadis falcinellus | Íbis niamhrach | A

(monotypic)

World	Southern North America, West Indies, Europe, Asia, Africa, and Australia
Ireland	Rare visitor
Cork	At least 132 individuals
Conservation	Not globally threatened

Fifteen individuals occurred in the 1800s, but no record was accurately dated. These involved two (many years before 1850, at an unknown location), eight (a long time before 1882, at the River Blackwater) and five (October or November in the 1870s, at Ballymacoda). Four singles occurred in the first quarter of the 1900s: Buttevant (late 1903), Conna (1 October 1903), Clonakilty (10 September 1906) and Annagh Bog (Churchtown) (21 February 1924).

There were no further records until one was seen at Lough Beg on 7 March 1981. Four other singles occurred between then and 2008. The status of the species changed dramatically from 2011 onwards as there was an influx of twenty individuals in October. There followed influxes of fifteen in January 2012, four in December 2013, nine across several months from January 2014, seventeen between October and December 2015 followed by a further twenty in January 2016 (some of which remained for a few months), thirteen between October and December 2017, eight between January and March 2018, ending with two in September 2018.

These birds occurred along the coast, or very close to it, from Bantry to Youghal, and only four were inland, at Carrigrohane, Blarney, Kilmichael and the Gearagh. Twenty-seven birds occurred west of Skibbereen, fourteen in the Clonakilty area, twenty-one in the Courtmacsherry area, six between Garrettstown and Cork city, and forty-one at several east Cork sites (excluding birds which moved between sites). The accompanying figure, which excludes birds that remained across more than one month, shows that most occurred between October and March, but few new birds occurred in November and February. Some birds were seen on only one or a few dates, but many were present for weeks or months, especially during the four-month period from October 2015 to January 2016. The longest stay has been of one from 7 March 1981 to 8 May 1983. Most sightings relate to one or two birds, but up to six together have occurred, and there were two flocks of seventeen, at Courtmacsherry and at Midleton. Some of the larger flocks apparently broke into smaller ones, and some moved about between sites, especially in the winter of 2015/16.

1959–68	1969–78	1979–88	1989–98	1999–08	2009–18
0	0	1	2	2	108

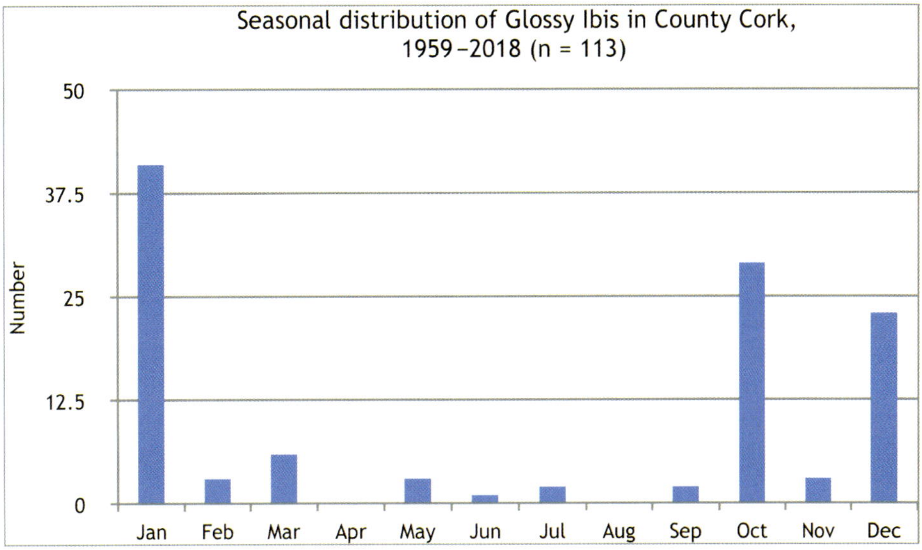

Fig. 7.11 Glossy Ibis.

Spoonbill

Platalea leucorodia | Leitheadach | A

(subspecies: *leucorodia*)

World	Europe, Asia, and Africa
Ireland	Rare visitor
Cork	55 individuals
Conservation	Green-listed

Eighteen individual Spoonbills occurred in the 1800s, and a further seven during 1901–58. Apart from three where the location was unknown, all occurred on the coast between Lissagriffin and Youghal. Seven (including 3 together) occurred at Youghal, four together near Inishannon and two together at Timoleague. Other records were of single individuals. All, except four, of the eighteen recorded in the 1800s were shot, therefore few survived long enough to exhibit the long-staying and nomadic trait of the species known today. Dates of occurrence, where stated, were March, April, June, and August to December.

A total of thirty individuals occurred during 1959–2018. Most occurred in May to July and in September to November. New birds occurred in all months except January. The spread of records along the coast extended from Lissagriffin to Ballymacoda, with inland records from the Gearagh and Blarney. More individuals occurred at Ballycotton (9) than at other sites, with Cork Harbour (6) and Kinsale Marsh (5) next in importance. In this period occurrences were characterised by long-staying individuals or small groups, and while many birds remained for only one or a few days, others made protracted stays, some for several months, or years.

Two birds were at Ballycotton and Ballymacoda in 1961/2 and 1962/3, and a single bird was present in every year to May 1974; although often absent during summer, the evidence suggests there was continuous presence during the thirteen-year period, first by two and then by one individual. Three birds also remained at Ballycotton and Ballymacoda from July to October 1977. A bird, originally seen at Clonakilty in September 2003, remained to June 2005, and visited several sites between Lissagriffin and Kinsale Marsh during its stay. Five birds (the largest number ever seen together in Cork) occurred at Kinsale Marsh in October 2009. This group split and individuals visited Courtmacsherry Bay, Clonakilty, Rosscarbery and Baltimore, and one or two were present to March 2011. Another individual originally seen at Rosscarbery visited several sites east to Kinsale Marsh over five days in October 2015. Two immatures at Kilkerran Lake in September 1979 remained for one day. Of the fifty-five individuals, six were aged as adult and twenty-two as immature. None of those recorded since 1901 have been shot, contrasting with the 1800s.

One of the five birds at Kinsale Marsh in October 2009 had been ringed as a nestling in the Netherlands on 8 June 2009, while one seen at Lough Beg on 22 June 2007 was identified (from photographs) on the Wexford coast during 27 June–12 August 2007.

1959–68	1969–78	1979–88	1989–98	1999–08	2009–18
3	6	5	1	6	9

Pied-billed Grebe

Podilymbus podiceps | Foitheach gob-alabhreac | A

(subspecies: undetermined)

World	North and South America
Ireland	12 individuals
Cork	3 individuals
Conservation	Not globally threatened

One at Rostellan on 1 February–23 March 1997, 10 October 1997–5 April 1998, and 29 June 1998–20 April 1999. One at Lough Aderry on 24 October–8 November 2000. One at Little Island and Rosslague on 11 December 2010–3 February 2011. The Rostellan bird was present in three successive winters, but there was no indication of where it was during its periods of absence.

Little Grebe

Tachybaptus ruficollis | Spágaire tonn | A

(subspecies: *ruficollis*)

World	Europe, Asia south to New Guinea, and Africa
Ireland	Resident breeder
Cork	Common resident
Conservation	Green-listed

The Little Grebe was a common resident in the 1700s and 1800s, and parties visited rivers and estuaries in winter. It was noted as numerous at the enclosed inlet at Rostellan in July 1907.

Little Grebes were discontinuously spread across the county during the three breeding atlas surveys. The main change over forty years was the apparent loss of birds from the Beara peninsula during 1968–72 and 1988–91, from which they are still absent. During each survey there were different 10 km squares with suitable habitat from which none was recorded, suggesting that birds were missed rather than absent, while new 10 km squares were also occupied in each period. The main areas from which birds were apparently truly absent were high ground and those with few or no suitable waters. Breeding density was highest around Cork Harbour and west to Glandore. In some areas, small irrigation ponds (Great Island) and slow-flowing rivers (River Womanagh) are used as breeding habitats, as are artificially created ponds on golf courses. Food provisioning of chicks has been studied at one such site near Midleton.[104] Occasional pairs are easy to overlook at some small sites. Formerly a winter visitor to CCBO, it now breeds annually.

Winter distribution is slightly more widespread as some birds partly withdraw from inland sites and congregate at sheltered brackish and saline lagoons and estuaries, especially in Cork Harbour. Perhaps those that leave are juveniles, with adults remaining to ensure their territories are not taken by wandering birds.

Numbers are highest in winter but counts rarely exceed fifty birds. Counts across the whole of Cork Harbour are usually between 50 and 90 birds but have exceeded 100 in three recent winters (peak of 134 in 2019/20). Most of the Cork Harbour birds occur at Rostellan. Up to forty-five were present at Bantry Bay in the winter of 1995/6.

There are no ring recoveries to indicate whether migrants from Britain or Europe reach Ireland. Two Little Grebes have been obtained in December at Berehaven lighthouse (1898), and Old Head of Kinsale lighthouse (1910). Up to four birds arrived at CCBO between July and September in the 1960s (they did not then breed there) and remained to early March. Such movements may be little other than local in nature.

Great Crested Grebe

Podiceps cristatus | Foitheach mór | A

(subspecies: *cristatus*)

World	Europe, Asia, Africa, Australia, and New Zealand
Ireland	Resident breeder
Cork	Scarce breeder, common winter visitor, mainly to the east of the county
Conservation	Amber-listed

The Great Crested Grebe was evidently much scarcer in the 1800s than it is today. It was known at Cork Harbour where five were seen in January and February 1849, and one was shot at Ballyvergan (no date). It was common in winter at Cork Harbour by the early 1950s, but no indication of numbers was given.

During the breeding seasons of 1963–6 between three and six grebes were recorded on the River Lee reservoirs and Lough Allua, but breeding was not proved. Two or three pairs bred at Lough Allua and two pairs were present at the Gearagh in 1967, the first known breeding in Cork. The pairs at the Gearagh were not successful as there was a sudden drop in water levels and the site was deserted. Two pairs again bred at Lough Allua in 1968. Breeding was taking place at both sites in 1970 and 1975. Breeding continued during 1988–91 and 2008–11, and subsequently, although breeding may not take place at each site every year (5 pairs bred at the Gearagh in 2021). The River Lee between Inishcarra and the Gearagh was flooded in the 1950s by the building of a dam for electricity generation and breeding by the Great Crested Grebe followed development of the reservoirs, but also coincided with an increase in wintering numbers at Cork Harbour.

Great Crested Grebe

In east Cork, single pairs bred at Ballycotton during 1968–70, and although one or two birds occurred there occasionally up to 1982, no further breeding took place. The habitat at Ballycotton became unsuitable for breeding after 1990/1. East Cork was re-colonised in 1984 when a pair bred at Ballybutler Lake (breeding continues most years). Pairs also bred at Ballyhonock Lake (successfully) and Lough Aderry (unsuccessfully) in recent years, but have not become established at either site, probably due to disturbance by fishermen. One pair bred (successfully) at Rostellan Lake in 1991, and breeding has recently occurred at Blarney Lake.

Breeding has taken place in north Cork at Ballyhea gravel pit and Charleville lagoons since the 1990s. Two pairs bred at Kilcolman in 1994, and a single pair in 1995.

Winter atlas surveys showed the distribution of Great Crested Grebes was almost entirely confined to coastal east Cork, with a few in west Cork and at the River Lee reservoirs. The increase in winter numbers evident in Cork Harbour in the early 1950s continued during the 1960s. A total of sixty-nine was counted in November 1967 and a minimum of seventy-five in December 1969. From the mid-1970s onwards numbers exceeded 100 birds in most winters, for example 106 in December 1976, 146 in December

1977, 153 in November and December 1978 and 169 in January 1981. Regular counts in the early 1990s produced totals varying from 233 to 365. Numbers declined in the 2010s and only one recent count has exceeded 200 birds. Grebes roost communally in flocks on the water in Cork Harbour, and this behaviour and the impact of boat traffic has been studied.[105]

Numbers are small at sites away from Cork Harbour and usually do not exceed about ten birds, so a count of forty-eight at Ballymacoda in December 1990 was exceptional.

Red-necked Grebe

Podiceps grisegena | Foitheach píbrua | A

(subspecies: *grisegena*)

World	North America, Europe, and Asia
Ireland	Rare winter visitor
Cork	25 individuals
Conservation	Not globally threatened

There were two records of this grebe in the 1800s, both referring to single birds obtained at Bantry Bay, in December 1842 and December 1850. Two others occurred between 1901 and 1958, at Crookhaven in March 1940, and at Glandore Harbour in April 1952.

Twenty-one Red-necked Grebes occurred in coastal waters mainly between November and February during 1959–2018, although there may be some duplication due to returning birds. They have occurred at most major bays and harbours from Dunmanus Bay to Cork Harbour. More have occurred at Cork Harbour (13) than any other site, all since 1966. All records, except one of three, refer to single birds. Most sightings were of birds seen on only one day, but a few stayed for longer periods, one for 121 days from November to March 1998.

In addition to the above, a grebe obtained at East Ferry in December 1903 was so badly damaged that certain identification was not possible.

1959–68	1969–78	1979–88	1989–98	1999–08	2009–18
3	1	1	5	5	6

Earliest and latest occurrence in winter (1959–2018): 19 August–2 April.

Slavonian Grebe

Podiceps auritus | Foitheach cluasach | A

(subspecies: *auritus*)

World	North America, Iceland, Europe, and Asia
Ireland	Scarce winter visitor
Cork	At least 204 individuals
Conservation	SPEC 1; Red-listed

There was one record of Slavonian Grebe in the 1800s, one obtained at Cork Harbour in early 1879. In the first half of the 1900s one was at Glengarriff from December 1941 to February 1942, and another was at Garryvoe in January 1947.

The species occurred almost annually during 1959–2018, and some winter at several sites, especially Roaringwater Bay and Cork Harbour. Most occur between October and March. Slavonian Grebes occurred at almost all sites between Bantry Bay and Youghal, and occasionally inland at Gallanes Lake, Castlenalact Lake, Cork Lough, the Gearagh and Kilcolman. While many records are of birds seen on only one day, some stay for several months over the winter, and there is no doubt that some of the same birds

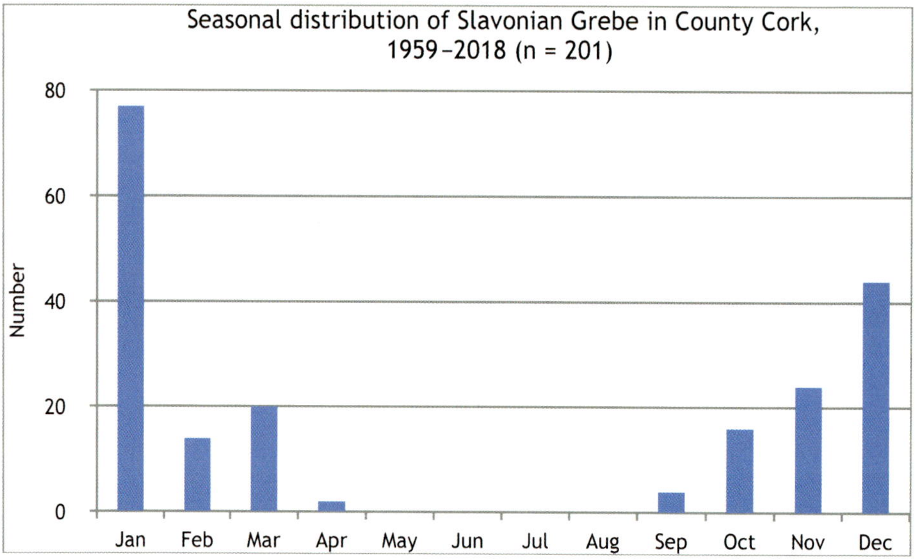

Fig. 7.12 Slavonian Grebe.

occur at the same sites in successive years. Most records are of one or two, but counts of three to five birds are not unusual, and ten were at Roaringwater Bay in January 1980.

1959–68	1969–78	1979–88	1989–98	1999–08	2009–18
11	27	49	54	33	27

Earliest and latest occurrence in winter (1959–2018): 5 September–13 April.

Black-necked Grebe

Podiceps nigricollis | Foitheach píbdhubh | A

(subspecies: *nigricollis*)

World	North America, Europe, Asia, and Africa
Ireland	Former breeder, now scarce winter visitor
Cork	82 individuals
Conservation	Red-listed

There were at least four records of this grebe in the 1800s. The first was obtained in the barony of Muskerry in 1847, with others obtained at Cork Harbour in December 1878 and in the winter of 1878/9. A fourth record was mentioned, but no details were given, while Ralph Payne-Gallwey informed Ussher that he had obtained two or three others at Cork Harbour at different times, also without details. There were no records of this grebe for Cork in the first half of the 1900s.

Seventy-eight Black-necked Grebes were recorded during 1959–2018, although this number may be inflated due to the same birds returning in successive winters. All but two have occurred between September and February. Records were sporadic during the 1960s and 1970s, and none occurred between 1981 and 1991, but this species has been more regular since then. There have been records at many coastal sites from Bantry Bay to Youghal, but only one inland on the River Lee system. Most have occurred at Cork Harbour, where from one to five have regularly wintered since the 1990s. Bantry Bay, Roaringwater Bay and Clonakilty Bay have also had multiple records. Most records are of singles, but two to five have also been seen together.

1959–68	1969–78	1979–88	1989–98	1999–08	2009–18
1	7	3	26	18	23

Earliest and latest occurrence in winter (1959–2018): 20 August–4 March.

Honey Buzzard

Pernis apivorus | Clamhán riabhach | A

(monotypic)

World	Europe and Asia
Ireland	35 individuals
Cork	2 individuals
Conservation	Not globally threatened

One at CCBO on 23 August 1972. First-year at CCBO on 25–6 September 2000.

Black Kite

Milvus migrans | Cúr dubh | A

(subspecies: *migrans*)

World	Europe, Asia, Africa, and Australia
Ireland	23 individuals
Cork	7 individuals
Conservation	Not globally threatened

One at Garryvoe and Ballycotton on 20 April–13 May 1980. One at Midleton on 20 April 1991. One at Lissagriffin on 13–15 August 2009. One at Cobh on 7 April 2010. One at Toe Head on 9 October 2010. Adult at Galley Head on 1 May 2012. First-year north of Midleton on 23 April 2017.

Red Kite

Milvus milvus | Cúr rua | A, C1, C2

(subspecies: *milvus*)

World	Europe, south-west Asia, north-west Africa, and Cape Verde Islands
Ireland	Former breeder, now rare visitor, recently reintroduced
Cork	Former resident; 28 individuals since 1968
Conservation	SPEC 1; Red-listed

There has been considerable debate regarding the former status of the Red Kite in Ireland. Authors have been divided, and those opposed to the view that it was once widespread often point to the confusion regarding the names applied (in Irish and English) to the larger birds of prey. In the past the same name was often applied to several different species and several different names to the one species.

The Red Kite was a scavenger in many English and European towns and cities in the Middle Ages, and there is evidence that it may have been present at Cork city in 1621.[106] Archaeological evidence of Red Kites has been found in the adjoining counties of Limerick and Tipperary dating from the thirteenth to the seventeenth century, but it is unlikely that it survived as a breeding species anywhere in Ireland into the 1800s. Evidence has been produced suggesting that kites were present in Antrim, probably throughout the year, until the second half of the 1700s.[107]

A convincing account of the Red Kite in Cork was given in 1750 when it was said to be so common that a description was not needed, and that it remained throughout the year. However, no mention was made of the Hen Harrier, and it is this species with which the greatest degree of confusion over names has occurred. The Hen Harrier, Marsh Harrier, and other species, were often referred to as kites in different parts of Ireland. However, Charles Smith's view was supported by Joshua Harvey who said there was little doubt that the Red Kite was a resident in the county a century before.

Scepticism that kites existed in Ireland as late as the 1700s (as implied by Smith) was expressed by some authors, but others conceded that when the country was more wooded and less populated by humans it would have been better suited to them. There are only two records for Cork given for the 1800s, both in 1827. One of these was seen near Blarney, and the other at Ballincollig. Thompson knew of no other occurrences anywhere in the south of Ireland.

There were no further records of the Red Kite in Cork until 1968 when one was seen at Ballycotton. There followed two further records in the 1970s, both at CCBO, in October and November, and no more until the 1990s, since when occurrences of the species have become somewhat more regular.

There have been numerous reintroduction programmes of this species in England and Scotland in recent decades, and on the east coast of Ireland (Wicklow 2007, and Down 2008). Wing-tagged birds from the Scottish programmes have been recorded in Cork, one at CCBO in October 1997, and a pair near Cloyne between January and March 1998. A bird bearing wing tags was seen at Lisgoold in November 1991; its provenance was not established, but it is presumed to have been of British origin. At least sixteen individuals have occurred in Cork in the ten years 2009–18. These have included several wing-tagged birds of Irish origin.

Since the 1990s Red Kites have occurred in many parts of Cork, mostly on or near the coast from Dursey Island to Ladysbridge, but inland as well. Five have occurred at CCBO and five in the Clonakilty area, including at least three different individuals in December 2009 and January 2010. Four have occurred in the Cobh area, including three together in June 2011. Inland, records have come from Fermoy, near the Gearagh, Millstreet and Derrylahan near the Kerry border in north-west Cork. Apart from the Cloyne and Cobh occurrences already mentioned, all records have involved single individuals. Most birds have been seen on one day, the exception being the pair at Cloyne which were present for two months. Little in the way of a consistent seasonal pattern of occurrence has developed, but most have been in the months from September to January. Frequency of occurrence is likely to change considerably in the future as the re-introduced Irish populations on the east coast grow and expand into adjoining counties.

1959–68	1969–78	1979–88	1989–98	1999–08	2009–18
1	2	0	8	1	16

White-tailed Eagle

Haliaeetus albicilla | Iolar mara | B, C2

(monotypic)

World	Greenland, Iceland, Europe, and Asia
Ireland	Former breeder, now very rare visitor, recently reintroduced
Cork	Former breeder, recent breeding by reintroduced stock
Conservation	Red-listed

The potential historical range of the White-tailed Eagle in Ireland (and Britain) was examined using documentary and placename evidence.[108] The authors showed that between the years 500 and 1800 AD the range of the species in Cork mainly encompassed the western peninsulas, with little apparent change over the 1,300-year period. This eagle was said to be common on sea cliffs in the 1700s. Lord Kinsale had one that had been obtained near Old Head of Kinsale, and eyries were known at either side of Courtmacsherry Bay (Old Head of Kinsale and Coolim Cliffs). A century later it was still described as a common resident; it had several eyries in the county, and was occasionally met with about Youghal.

The White-tailed Eagle is predominantly a lowland species of coastal and inland wetlands across much of its European range and is largely tree-nesting, and the valley of the River Blackwater between Lismore and Youghal Bridge (Waterford) is today considered prime breeding habitat. However, it is unknown if the historical Youghal reference was to breeding birds or perhaps to winter visitors. In 1854 or 1855 an eyrie with young was seen at Crow Island on the south-west extremity of the Beara peninsula, while Sheep's Head and Bere Island were also known as breeding sites. The Beara peninsula was described as a great haunt of this eagle prior to 1855. It is not known exactly when breeding ceased in Cork, but from about 1854 onwards there was systematic persecution by poisoning, made easy by its partiality to feeding on carrion. It had ceased to breed by 1894 and was then known only as a rare straggler, presumably from the adjoining county of Kerry.

There are no confirmed records of wild White-tailed Eagles in Cork between the late 1800s and the present time. However, five eagles seen at CCBO on 1 September 1965 were almost certainly this species.

White-tailed Eagle

A captive breeding programme began in 1991 based at Fota, with the aim of re-establishing a wild breeding population in Ireland. However, this programme was not successful and ceased soon after. A reintroduction programme involving juvenile Norwegian birds began in Kerry in 2007.[109] Between 2007 and 2013, released birds were seen in various parts of Cork (CCBO, Ballycotton, River Lee valley, lower River Blackwater valley, Mullaghanish and other locations). A pair took up residence in the Glengarriff area in 2013, nesting annually from 2014 to 2019. In 2016, they fledged their only chick to date. In 2018, a breeding attempt was made at a second site in County Cork, but the chick died before fledging. In addition to these two sites, two pairs breed on the Kerry side of the Beara peninsula and have part of their home ranges on the Cork side.

In recent years, released birds and immature Irish-bred birds have been seen in many places in the county, especially along the south coast (including Ballycotton and Capel Islands), the mountains along the Cork/Kerry border, Beara peninsula and the upper River Lee valley, including Gougane Barra. Within these locations, some unpaired adults and immature birds have been resident for extended periods. Based on current rates of expansion it is predicted that, in addition to the Beara peninsula, the other western peninsulas (Sheep's Head and Mizen Head) and the River Lee valley are likely to hold breeding pairs by 2030.

Griffon Vulture

Gyps fulvus | Bultúr gríofa | B
(subspecies: *fulvus*)

World	Southern Europe, southern Asia, and north Africa
Ireland	1 individual
Cork	1 individual
Conservation	Not globally threatened

Immature obtained at Cork Harbour in spring 1843. This bird was captured alive and kept for several months, until it died, at the home of Lord Shannon at Castlemartyr.

Marsh Harrier

Circus aeruginosus | Cromán móna | A
(subspecies: *aeruginosus*)

World	Europe, Asia, and north-west Africa
Ireland	Former breeder, now passage migrant and summer visitor
Cork	Has bred, 129 individuals since 1959
Conservation	Amber-listed

The Marsh Harrier was a resident in the 1800s and breeding sites were mentioned for Blarney and Inchigeelagh (among others). It was common around Youghal, and Robert Ball was brought young birds from the neighbourhood, while William Corbet recorded it breeding at Ballycotton up to about 1870. It was occasionally seen west of Bandon in winter, and one wintered at Castlemartyr in 1879/80 and fed on Moorhens.

Marsh Harriers became extinct as breeding birds in Ireland about 1917 due to extensive drainage of marshes and human persecution, such as shooting. Since extinction as a breeding species, the Marsh Harrier has occurred in Cork as a passage migrant and summer visitor.

Marsh Harriers have occurred in every month and in most years during 1959–2018, with the majority in April to June and smaller numbers in

September and October. They have occurred at coastal sites from Dursey Island to Youghal, with most records at CCBO, Ballycotton and Ballyvergan Marsh, and smaller numbers at Clonakilty, Garrettstown and Cork Harbour. Marsh Harriers have occurred inland most regularly at Kilcolman, but also at Lough Allua and Lisgoold.

Most birds were seen on just one day, but some remained for up to ten days, and exceptionally for longer. Three birds remained for up to forty-five days during winter months. A further five birds remained for extended periods in summer months at potential breeding sites, the longest being ninety-two days. One seen carrying nest material remained only two days. Two birds were seen together three times, all other records refer to singles. Of those where the sex was recorded, females predominated by almost four to one, and of sixteen males only two were adult.

Numbers have increased in Cork over the decades, probably associated with an increase in the British population which breeds there mostly on the east coast. More birds are now wintering in Britain than previously, and the few Cork winter occurrences may be reflecting this trend.

1959–68	1969–78	1979–88	1989–98	1999–08	2009–18
4	6	17	34	41	27

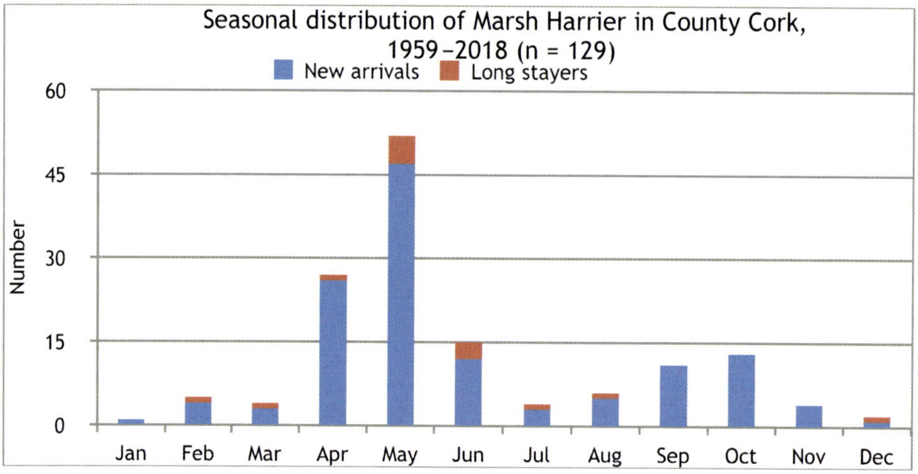

Fig. 7.13 Marsh Harrier.

Hen Harrier

Circus cyaneus | Cromán na gcearc | A

(monotypic)

World	Europe and Asia
Ireland	Scarce resident breeder, recent decrease
Cork	Scarce resident breeder
Conservation	SPEC 3; Amber-listed

The Hen Harrier was a common resident in the mid-1800s and was occasionally recorded around Youghal. Fifty years later it was decreasing and had almost disappeared as a breeding species, at least from some areas, but the mountains south of the Mallow to Killarney railway line (Boggeragh and Derrynasaggart ranges) remained a stronghold. The apparent inability of the Irish sportsman to resist shooting Hen Harriers was noted, although there may have been several factors involved in the decline.[110]

It was widely quoted that the Hen Harrier became extinct as a breeding species in Ireland in the early 1900s. Later evidence suggests this was not so, but it did decline significantly. It is unknown if it ceased to breed in Cork at any stage. The following account is based mainly on the work of O'Flynn and Nagle.[111; 112] Breeding increased (or resumed) from 1956 onwards as it adapted to nesting in new conifer plantations. Liam O'Flynn recalled seeing up to seven birds at one time from vantage points in the Nagles Mountains and on the Cork/Waterford border near Tallow (Waterford) during the late 1960s and early 1970s. Tony Nagle estimated the Cork population in the early 1970s at forty to forty-eight pairs, close to Ken Preston's estimate of fifty pairs.

The recovery lasted until the late 1970s, when another decline began, probably largely related to the maturation of conifer forests. By the mid-1980s Ken Preston estimated the population at twelve to fifteen pairs. The population has recovered from this low level, mainly due to adapting to nest in second rotation forestry plantations.[113] Populations in heavily forested areas, such as the Nagles and Ballyhoura Mountains, are subject to cyclical fluctuations as habitat availability is dependent on tree-harvesting and re-planting regimes. Widespread land use change in previously open habitats, primarily commercial afforestation, agricultural intensification

Hen Harrier

and wind energy developments, may have a negative impact on the species.[114] The population was estimated by Ken Preston at fifteen to twenty pairs in 1991. National surveys in 1998–2000 recorded at least twenty-seven pairs in Cork. Subsequent surveys have shown the population has remained relatively stable, although declines in some areas have been reported.[115] Breeding pairs are largely confined to the north and north-west, with a small population in the east in the Kilworth area (1–2 pairs), and no breeding has been recorded east of Watergrasshill since 2005, although Hen Harriers still occur widely during winter at coastal sites in east Cork.

There is little historical information about the winter range of the Hen Harrier in Cork. Winter atlas surveys showed it was sparsely distributed at inland sites across the county, the main difference from the breeding season being that coastal sites were occupied as well. Hen Harriers appear in coastal districts in September, and they occur almost annually in small numbers at CCBO and Dursey Island in autumn. Some birds stay on or near the breeding grounds in winter, as at the Nagles, Ballyhoura and Mullaghareirk Mountains. Thirty communal roosts at wetland coastal habitats and at upland inland habitats have been identified. Studies at these winter roosts, which usually hold one to five birds, have been carried out since the 1980s.[116]

Hen Harrier utilisation of a limestone fen at Kilcolman was studied during 1994–2005.[117] Birds used the fen, which is close to the Ballyhoura Mountains breeding area, in all months, but use was most intensive between August and February.

Winter diet has been studied at east Cork sites.[118] Many small mammal species are taken as prey, but Common Snipe and small passerine species were taken most frequently.

There are several winter recoveries in Ireland (and in Cork) of nestlings ringed in Scotland. Nestlings ringed in Cork disperse locally during winter and use mainly coastal roosts across southern Ireland as far east as Wexford.

Pallid Harrier

Circus macrourus | Cromán bánlíoch | A

(monotypic)

World	Europe and Asia
Ireland	8 individuals
Cork	4 individuals
Conservation	Not globally threatened

First-year male at Ballyvergan Marsh on 22–3 April 2011. First-year at Power Head and Ballycotton on 29 October–16 November 2011. First-year female at Power Head on 7–29 November 2011. First-year at Barry's Head on 7–30 October 2017.

Montagu's Harrier

Circus pygargus | Cromán liath | A

(monotypic)

World	Europe, Asia, and north Africa
Ireland	Has bred, 88 individuals, mainly in spring and autumn
Cork	Has bred, 11 individuals (10 records)
Conservation	Not globally threatened

Male and female at undisclosed breeding location in 1957. Immature at CCBO on 1 November 1969. Immature at CCBO on 19 August 1973. Ringtail at CCBO on 24 October 1981. Adult male at Ballymacoda on 28 April 1984. Male at Dursey Island on 10 May 2000. First-year male at Dursey Island on 13 May 2006. First-year male at Old Head of Kinsale on 7 May 2011. Adult male at Mitchelstown on 14 May 2012. Adult male at CCBO on 29 April 2017.

There is no information regarding success, or otherwise, of the breeding pair in 1957. The Old Head of Kinsale bird had been wing-tagged in France in 2010. An adult male Montagu's Harrier was seen between Youghal (Cork) and Tallow (Waterford) on 22 July 1948. This bird was in (or very close to) breeding habitat, but it was most likely on the Waterford side of the county bounds and cannot be claimed with certainty as a Cork bird.

Goshawk

Accipiter gentilis | Spioróg mhór | A

(subspecies: *gentilis*; *atricapillus*)

World	North America, Europe, Asia, and north-west Africa
Ireland	Very rare breeder, recent increase
Cork	Former resident, 22 individuals since 1955
Conservation	Amber-listed

There has been much debate about the former status of the Goshawk (*A.g. gentilis*) in Ireland. This arose mainly from the many different names applied to the species, which has been confused mostly with the Peregrine Falcon, and the use of falconry terms has compounded this. However, that it occurred in Ireland in the past when the country was more wooded is proven by various means and Cork was one of several counties in which it was breeding around the beginning of the Early Modern period (1400s and 1500s).[119] It probably died out as a breeding species during the 1600s and 1700s with the widespread clearance of woodlands. The Goshawk was listed as a Cork species in the mid-1700s, but no comment on its status was made, which may imply it was then a well-known bird, or had been in the recent past. However, some authors in the 1800s were unable to accept the Goshawk as a Cork species with any degree of certainty, or that it had ever occurred in Ireland. There were no Cork records from the mid-1850s until one was seen near Rosscarbery on 29 April 1955.

Twenty-one individuals were recorded in Cork during 1959–2018, there being none in that period until 1973. Most occurred in autumn, with a few in spring and two in winter. All occurred singly, except in one case when a pair was seen together. The majority have been seen on a single day, but a few have remained for up to five weeks. Most birds have been seen on or near the coast, while five have been seen at inland sites. On the coast, they have occurred at Dursey Island, Crookhaven, CCBO, Galley Head, Inchydoney, Cork Harbour and Ballycotton, while inland, they have occurred at Dungourney, near Mallow, Kilcolman and the Gearagh. Males and females, adults and immatures have been recorded, and one in September was aged as a juvenile male. The Irish Raptor Study Group suggested that Goshawks may now be breeding in several Irish counties, and there are persistent rumours that breeding has been occurring in Cork.[120] There is some evidence that birds have been seen at several inland potential breeding locations in the last ten years, and three 10 km squares in north Cork had records during the breeding atlas of 2008–11, although there has been no firm evidence of breeding. Outside of the 1959–2018 study period, one record of an adult has been published for April 2019, but details were withheld, lending some support to the suggestion that some birds may be breeding in Cork.

American Goshawk *A.g. atricapillus*: There is one record of this sub-species, an individual at CCBO on 5 October 1974. The Irish Rare Birds Committee intends to review all records of American Goshawk.

1959–68	1969–78	1979–88	1989–98	1999–08	2009–18
0	6	3	6	4	2

Earliest and latest occurrence in spring and autumn (1959–2018): 6 April–25 May and 26 August–15 November.

Sparrowhawk

Accipiter nisus | Spioróg | A

(subspecies: *nisus*)

World	Europe, Asia, and north Africa
Ireland	Resident breeder
Cork	Widespread resident and breeder
Conservation	Green-listed

The Sparrowhawk was resident and common but did not frequent cliffs either on the coast or mountains in the 1800s, and many were destroyed by game preservers because of alleged killing of game and small birds.

During the 1950s and 1960s it is known that Sparrowhawk populations, along with other birds of prey, notably Peregrine Falcon, declined in Britain and elsewhere.[121] The decline was caused by use of organochlorine pesticides as seed dressing in agriculture. Declines also took place in Ireland, although perhaps a little later than in Britain. Since there was no monitoring scheme in Ireland capable of documenting change in a widespread species, the extent of decline here must remain conjecture, but the evidence suggests it was not as marked as in Britain.[122] Evidence for a decline in Cork was provided by the Cork Ornithological Society, which stated in 1965 that reports from different areas indicated the species, although scarce, was not as near extinction as was feared a few years before. It would be surprising if no decline took place, especially in the arable agricultural landscape of the south, east and north. Records of individual birds published by the society during the 1960s probably reflect the areas in which members were active rather than being a true reflection of distribution at that time. Factors other than chemicals used in agriculture can also have a local influence on distribution, as at Castlemary where Sparrowhawks were said to have disappeared in 1964 following tree felling.

The 1968–72 breeding atlas showed the Sparrowhawk was widely spread across the county from the western peninsulas to Youghal and north to the border with Limerick. However, there were large areas where no birds were recorded, especially in the west and north, and it is possible that the true distribution was under-recorded, as the adjoining counties of Kerry and Limerick showed fewer gaps, but this could also reflect a population still in recovery. By 1988–91 several of the vacant areas in the west had been filled, but although birds were recorded in new places in the north, no new 10 km square there had proved breeding. However, the indications overall suggested an increase. This continued into 2008–11, particularly in the west and north, giving a continuous occupancy across the county. An estimate of about 1,350 pairs for the county was given in 2006, with highest densities believed to be in wooded lowland areas.[123]

The first winter atlas showed the Sparrowhawk was widely distributed at inland sites, but with gaps at the western peninsulas and in a coastal band from Glandore to Cork Harbour. There were gains at many of these areas in the second winter atlas, but they were generally thinly spread at the western peninsulas, perhaps reflecting the barrenness of these areas at that season.

The Sparrowhawk feeds almost entirely on small birds. One was seen capturing and killing a Hoopoe at Shanagarry in April 1948, and another was seen chasing a Common Kingfisher at Oysterhaven in August 1964, but the outcome was not recorded. On many occasions during fieldwork on Dippers in east Cork Sparrowhawks have been seen flying in hunting mode beneath the tree canopy along the river corridor. Although a Dipper has never been seen to be taken, the latter species always reacted to the presence of the hawk. A female Sparrowhawk was observed at a roost of soprano pipistrelle bats near Macroom in July 2007 where it tried, unsuccessfully, to capture emerging bats.[124]

Irish breeding Sparrowhawks appear to be largely resident. Recoveries of nestlings ringed in Ireland and Britain reveal random dispersal from the natal territory with a median movement for males of 6 km, and for females of 13 km. There is little evidence from ringing to suggest an arrival of British or continental migrants into Ireland at any season. However, at CCBO (and Dursey Island) there is a strong passage between August and October and a smaller passage between March and May, and one was obtained at Bull Rock in October 1912. It should be noted that the heavy autumn bias of observer coverage may have distorted the relative importance of the two peaks. There is a similar spring and autumn passage at Sherkin Island. More ringing of Sparrowhawks at CCBO during the passage periods could be revealing.

Common Buzzard

Buteo buteo | Clamhán | A

(subspecies: *buteo*)

World	Europe and Asia, and Azores, Madeira, Canary and Cape Verde Islands
Ireland	Resident breeder, recent increase
Cork	Resident breeder, recent increase
Conservation	Green-listed

The Buzzard was included among the birds of Cork in the 1700s without comment, perhaps indicating it was well known then. It was not common a century later apart from locally, for example about Youghal where it was probably still breeding in the early 1800s. Whatever the situation, a rapid decline set in and by 1865 it was breeding only in the north coast counties of Ireland. By the 1890s it was described as uncommon and only an occasional visitor to Cork. Only two records can be cited between 1850 and 1900 (2 near Kilbrittain before September 1886 and 1 west of Bandon in November 1891), and one during the first half of the 1900s (1 near Fermoy in February 1931).

The recent range expansion of the Buzzard has been one of the most spectacular ornithological events of the last fifty years in Cork. Not one bird was recorded in the county in the breeding atlas surveys of 1968–72 and 1988–91. However, by 2008–11, Buzzards were present and breeding over much of the east and north. The winter range in 2007–11 reflected the breeding season distribution, but with a slightly wider spread in parts of west Cork, where breeding birds were absent.

The spread of the Buzzard from north to south in Ireland has been documented.[125] There were no records of the Buzzard in Cork in the 1950s, but in the 1960s, 1970s and 1980s there were four, three and four records of individuals, respectively. These ranged from Barley Cove to Ballymacoda, and inland to Mallow and Kilcolman. There was an increase to at least sixteen records of individuals in the 1990s, with an equally wide range across the county at coastal and inland sites. This was the decade in which the vanguard of the invasion that was to come arrived. The increase in the number of records continued in the following years, and by 2001 it was estimated that at least four pairs were present throughout the year.

Breeding was proved for the first time in 2006 when two pairs bred (Cloyne and Belgooly), although may have begun as early as 2001. A minimum of seventy-eight pairs were recorded in 2011–12, involving a total of 209 individuals.[126] The increase has continued, although Buzzards are still absent or rare in parts of the west of the county.

Densest populations occur in lowland areas where tillage forms a high proportion of land use, particularly in east and south Cork and north of the River Blackwater. Small woods and hedgerow trees are used as nesting sites. Rabbit, brown rat, Woodpigeon, Magpie, Rook and Jackdaw formed 84 per cent of diet based on prey weight (n = 238).[127] In winter, Buzzards feed on earthworms and probably on other invertebrates in ploughed ground and pastures. Buzzards also feed on carrion, and this makes them susceptible to poison baits laid in animal carcases. Poisoning and woodland clearance are believed to be reasons for the decline and disappearance of the Buzzard from all but the north-east of Ireland in the 1800s. Persecution of birds of prey still occurs and twenty-three Buzzards were poisoned by Carbofuran in the Bandon to Timoleague area of west Cork in 2020.[128]

Buzzards are mainly sedentary, although young disperse locally. A nestling from Ballymacoda travelled to Tramore (Waterford), and a nestling from Baldonnel (Dublin) travelled to Ardgroom. A south-west movement was recorded in September and October 2018, with up to fifty-five seen at Eyeries and similar numbers at Mizen Head, and up to twenty-five in a day at CCBO.

Rough-legged Buzzard

Buteo lagopus | Clamhán lópach | A

(subspecies: *lagopus*)

World	Northern regions of North America, Europe, and Asia
Ireland	48 individuals
Cork	3 individuals
Conservation	Not globally threatened

Male obtained at Mitchelstown on 18 November 1906. One at CCBO on 13 October 1963. One at CCBO on 16 October 1980.

Greater Spotted Eagle

Clanga clanga | Iolar breac | B

(monotypic)

World	Europe and Asia
Ireland	2 individuals
Cork	2 individuals (1 record)
Conservation	Not globally threatened

Two (1 immature obtained) at Castlemartyr–Claycastle in January 1845, having been present in the locality for several weeks; the second bird was not preserved.

Golden Eagle

Aquila chrysaetos | Iolar firéan | B

(subspecies: *chrysaetos*)

World	North America, Europe, Asia, and north Africa
Ireland	Former breeder, now very rare visitor, recent reintroduction attempt
Cork	Extinct as breeding species since late 1800s, no recent records
Conservation	Red-listed

The potential historical range of the Golden Eagle in Ireland was examined using placename and documentary evidence.[129] These authors showed that in the years 500 and 1800 AD the range in Cork probably encompassed the western peninsulas and several of the inland mountain ranges, with little apparent change over the 1,300-year period. The Golden Eagle was a resident in Cork until the latter part of the 1800s. Breeding sites were known in the Caha Mountains on the Beara peninsula at Glen Lough, Barley Lake and Hungry Hill, each being within a few kilometres of the border with Kerry.[130] Other eyries were across the border in Kerry, and eagles from these probably ranged into Cork. The Beara peninsula was described as a great haunt of this eagle prior to 1855. It was believed that Golden Eagles also bred at Gougane Barra. Although it is not known exactly when breeding ceased in the county, it appears to have been extinct by the early 1890s, and thereafter there were only isolated occurrences.

It is unlikely that Golden Eagles bred on coastal cliffs in Cork. Although William Thompson said he never knew of an eyrie on marine cliffs anywhere in Ireland, recent research suggests that the Golden Eagle may have replaced the White-tailed Eagle on marine cliffs at Achill Island (Mayo) after the latter was exterminated in the late 1800s.[131] However, this does not seem to have occurred in Cork or Kerry, suggesting that the Golden Eagle may have been exterminated in these counties earlier, or perhaps around the same time as the White-tailed Eagle.

In 1894 both the old eagles at Lough Inchiquin (Kerry) on the Beara peninsula were caught in traps at their nest, which contained young. These were probably the last breeding Golden Eagles on this peninsula. It became extinct as a breeding bird throughout most of its Irish range between 1850 and 1900 resulting from incessant shooting, trapping and poisoning by gamekeepers and shepherds, as well as by collectors taking the eggs and young of those that remained.

There has been no confirmed record of the Golden Eagle in Cork since the late 1800s, although an eagle seen at CCBO on 6 September 1967 was considered almost certainly an immature of this species.

A reintroduction programme to Donegal (from Scotland) has been underway since 2001 and some birds have dispersed across the north-west and Northern Ireland. One bird moved as far south as Clare in 2011, but none have, to date, been recorded in Cork.

Osprey

Pandion haliaetus | Coirneach | A

(subspecies: *haliaetus*)

World	North America, Caribbean islands, Europe, Asia, north Africa, Australia
Ireland	Former breeder, rare summer visitor
Cork	103 individuals since 1848
Conservation	Not globally threatened

Nine individuals were recorded during 1848–93 (including three in October 1881). Dated records occurred in May (1), July (1), September (1) and October (4). The next did not occur until November 1948, at Fermoy.

Ninety-three individuals were recorded during 1959–2018, evenly distributed at coastal sites from Dursey Island to Youghal, with some inland at Inchigeelagh, the Gearagh and Mallow. Ospreys are mainly autumn passage migrants, with a smaller passage in spring. All have occurred singly. Only a few have been aged, two adults and six immatures. Most have been recorded on only one day, but a small number have stayed for up to eighteen days, and one for seventy days.

The increase in Cork is associated with an increase of the species in Britain, particularly Scotland, and recent breeding in Wales. A nestling ringed in Scotland in July 2014 was seen at Clonakilty two months later when on its southward migration.

The Osprey is believed to have been an Irish species, presumably breeding at least to the twelfth century when Giraldus Cambrensis visited Ireland.[132] It is not known when it became extinct, other than it had disappeared before the 1700s.

1959–68	1969–78	1979–88	1989–98	1999–08	2009–18
2	3	4	17	26	41

Earliest and latest occurrence in spring and autumn (1959–2018): 22 April–3 November.

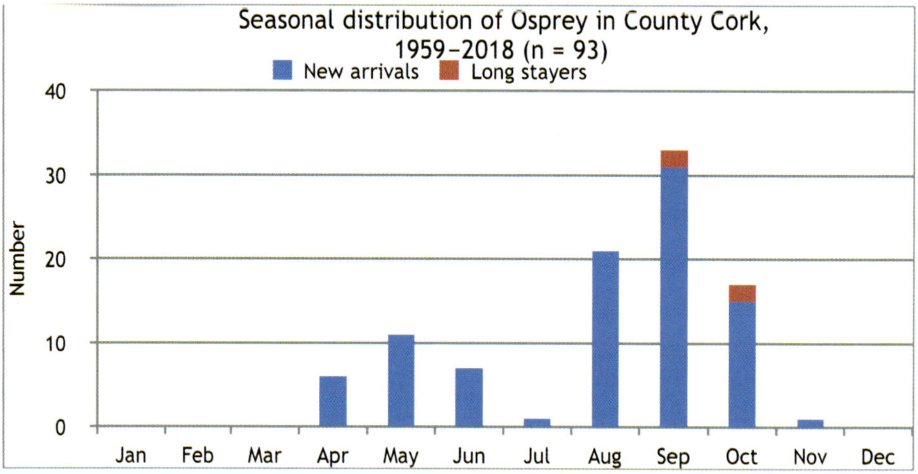

Fig. 7.14 Osprey.

Common Kestrel

Falco tinnunculus | Pocaire gaoithe | A

(subspecies: *tinnunculus*)

World	Europe, Asia, and Africa
Ireland	Resident breeder and winter visitor
Cork	Resident breeder and winter visitor
Conservation	SPEC 3; Red-listed

The Kestrel bred at Old Head of Kinsale and Coolim Cliffs and at other parts of the coast in the 1700s. It was resident and common in the late 1800s, breeding on sea cliffs and ruins inland, as well as using old nests of crows. Ussher said it was frequently shot mercilessly, being mistaken for the Sparrowhawk. The breeding sites described hold true today, but inland quarries can be added to the site list.

The Kestrel was the most widespread bird of prey breeding in Cork during 1968–72. It was found throughout the county, including coastal cliffs, and there appeared to be only one major gap in the range, in the mid-west. The indication in 1988–91 was of a general decrease with many vacant 10 km squares appearing, chiefly in the centre and east of the county. The greatest density was in the western peninsulas and upland areas with fewer in the south and east, and between the two surveys there was an overall decrease in east Cork. There are few 10 km squares with several pairs of Kestrels, in contrast to the Sparrowhawk. However, some of the earlier losses had been made good again by 2008–11, and the Kestrel was once more widespread across the county. A county total of 596 pairs was calculated in 2006.[133]

Distribution during the 1981–4 winter atlas was widespread but sparse. However, they were virtually absent from the western peninsulas and from some upland 10 km squares, but these gaps were filled in 2008–11 when the range encompassed the whole county. However, there has been a recent decline in numbers in autumn and winter, at least across the intensive farming areas of east Cork. From the 1970s through to at least 2010, several Kestrels could be encountered hovering over stubble fields, rough ground, coastal heaths and dunes during a drive across east Cork, especially from September onwards, but by 2018 they were rarely encountered.

The diet of this species has been studied in east Cork farmland habitats. Although they take small birds, common frog and common lizard, rodents and shrews form the most important part of their diet. Their adaptability is shown by the way in which the switch to a new prey species takes place, as demonstrated by high numbers of greater white-toothed shrew taken when that species was present.[134]

Little is known of the movements of Irish-bred Kestrels. However, it is likely that birds breeding in the more exposed upland and coastal sites move to lower ground. This appears to be the case in Cork, where in the lowland east of the county Kestrels are more widespread during the autumn and winter than they are at other times. At least some of these are birds that have arrived from abroad. There have been autumn and winter recoveries in Cork of birds ringed in Wicklow, central and northern England, Scotland, Belgium and Finland. There is evidence of a regular passage migration from August to early November at CCBO, with up to nine birds recorded in a day, and a much lighter passage from April to mid-May. One was obtained at Bull Rock on 27 December 1902 and one at Fastnet Rock on 20 October 1911, and birds have also been obtained at Old Head of Kinsale lighthouse in October and November.

Red-footed Falcon

Falco vespertinus | Fabhcún cosdearg | A

(monotypic)

World	Europe and Asia
Ireland	35 individuals
Cork	7 individuals
Conservation	Not globally threatened

Adult male at CCBO on 31 May 1991. Second-year male at CCBO on 4 May 1994. First-year male at Ballymacoda on 15 April 1995. Adult female at Churchtown South on 9–17 May 1997. Female at CCBO on 27 March 2012. Adult male at Owenahincha on 24 May 2012. Male at Barryroe–Dunworly on 25 May 2017.

In addition, an adult female was captured aboard a fishing boat in an exhausted state 32 km south of Fastnet Rock on 3 May 1994. It was taken into care and released on 17 May 1994 at Ballyvergan Marsh. However, the finding location lies outside the official recording area, therefore this record has not been included in the total number of individuals for Cork.

Merlin

Falco columbarius | Meirliún | A

(subspecies: *aesalon*; *subaesalon*; *columbarius*)

World	North America, Iceland, Europe, and Asia
Ireland	Scarce resident breeder, uncommon winter visitor and passage migrant
Cork	Very scarce breeder, scarce winter visitor to lowland and coastal areas
Conservation	Amber-listed

The Merlin (*F.c. aesalon*) was resident in the mid-1800s, and young birds were frequently brought to Robert Ball from the vicinity of Youghal. Towards the end of the 1800s it occurred in small numbers, frequented mountains where a few pairs bred on the ground, and was sometimes seen in lowlands in winter.

The Merlin was proved to breed in only two 10 km squares during 1968–72. One of these was at the Beara peninsula on the Cork/Kerry border (and may refer to Kerry), while the other was in the Riverstown area of east Cork. This was an unusual breeding area, but the record was verified by Liam O'Flynn. There was also a probable breeding record at the Ballyhoura Mountains. The few scattered sightings elsewhere may include late spring migrants at coastal sites. The 1988–91 survey revealed mixed fortunes for the Merlin, with some losses and some gains. There were then small populations in the Derrynasaggart and Boggeragh Mountain ranges and in the area around Bantry. There was a suggestion of a few also holding on in the Ballyhoura Mountains, but breeding was not proved. There was little change in 2008–11 with small breeding populations in the Mullaghareirk, Derrynasaggart and Boggeragh Mountains.

It is very likely that the breeding distribution in all three atlas surveys underestimated the range of the species due to its elusiveness on the breeding grounds. It is possible that Merlins have been affected by afforestation, intensification of hill farming, and fragmentation of habitat. In the early 1990s the known breeding population never exceeded three pairs. In 1990 there were pairs at Hungry Hill, Millstreet and Nadd. In 1991, there were three pairs in the Ballyhoura Mountains and two pairs in 1992 with a further pair in east Cork. The first tree-nesting in Cork by this species was in 1990. Analysis of past and present data on breeding records in 2006 indicated a county population of ten to twenty pairs.[135]

The winter atlas survey of 1981–4 showed that the Merlin was found in only a handful of 10 km squares across the county from Dursey Island to Youghal and from the coast to the border with Limerick. More than half of the records were from, or near, coastal areas, but there were records also from breeding areas in the Ballyhoura Mountains. There were more records from east of Cork Harbour than elsewhere, but this may be an observer effect. The 2008–11 winter situation showed the main concentration of records to be from coastal areas of south and east Cork, but there were also records from west Cork and from within the breeding areas already mentioned.

The Merlin is best known to most birdwatchers as an autumn and winter visitor to lowland and coastal districts, especially in east Cork, where it hunts over marshes, beaches and estuaries in its pursuit of small waders, larks and finches. An unusual potential prey species was a Great Skua being chased by a Merlin off Cobh in January 1998.

There is a light spring passage in March and April at CCBO, but the autumn passage is much heavier and more noticeable and extends from August to November. Most birds occurring at this season are females, adult males being rarely seen. This suggests that males may stay at or near their breeding territories in winter. Birds are usually seen at coastal districts from early October to late March, but there have been occurrences as late as 30 May (Ballycotton in 1964).

Icelandic Merlin *F.c. subaesalon*: The Icelandic subspecies occurs regularly in Ireland in winter and there are several ring recoveries, although none appear to have been found in Cork. However, there is little doubt that it occurs in Cork.

Taiga Merlin *F.c. columbarius*: There is one record of this subspecies. First-year at CCBO on 29 September 2000.

Hobby

Falco subbuteo | Fabhcún coille | A

(subspecies: *subbuteo*)

World	Europe, Asia, and north Africa
Ireland	Rare spring and autumn visitor; has recently bred
Cork	122 individuals
Conservation	Green-listed

There is no convincing evidence the Hobby bred on the Cork coast in the 1700s, or earlier. Clearly, there was confusion regarding names of birds of prey in that era.[136]

There are five authenticated records of the Hobby in Cork for the period before 1959. The first was obtained at Carrigrohane in summer about 1822. Single birds were seen in a wooded area at Glengarriff in the summers of 1925, 1926 and 1927, but it was clearly stated that breeding did not take place. The next record was not until 1952 when one was seen at Gokane House near Skibbereen on 18 April. Reports of the Hobby in the vicinity of Glengarriff during summer between 1890 and 1917 are too vague to be accepted, although note the later proved occurrences in that region.

The Hobby has occurred in Cork in thirty-five of the sixty years during 1959–2018, and it has been annual since 2004. It occurs regularly in spring and autumn with peaks in May and October, respectively. The numbers occurring in June and July blend together, giving the impression of a summer visitor as well as a passage migrant. There is one record each for January and November.

The Hobby occurs in Cork at a wide range of sites on the coast ranging from Dursey Island to Youghal, with many extending inland for about 10 km. Six birds were recorded at four sites more than 10 km inland, including three together at Millstreet, the others being at Bottlehill, the Gearagh and Kilcolman. Most occurred at the Mizen peninsula (19), CCBO, Cork Harbour catchment and Ballycotton Bay (13 each). Many sightings refer to birds close to wooded habitats especially, but not exclusively, when seen away from the coast, as at Glengarriff, Leap, Connonagh, Kilbrittain, Belgooly, Minane Bridge, Glanmire, Midleton, Killeagh and Bottlehill. However, there are no confirmed breeding records for Cork.

Most records of the Hobby are of single birds, but two occurred together thrice, and three together once; three of these records involved birds in seasons and at (or near) habitats that could be considered suitable for breeding. Nearly all records (103) involved birds seen on only one day, and only one bird remained for more than ten days (mean stay = 1.6 days). The bird that made the longest stay (27 days) was the only one seen in winter, in the wooded River Bandon valley at Inishannon.

There has been a reasonably even spread of ages where this has been recorded (25 adult birds and 32 first-year birds). A recent increase has occurred, particularly since 2005, with 2010, 2011, 2012 and 2013 being especially good years. The high numbers may be related to the success of the British population where there has been a 295 per cent increase in range over the forty years since the first breeding atlas in 1968–72.

The Hobby migrates to tropical Africa for the winter from its breeding grounds in Britain and Europe, and the winter record in Cork is unusual, particularly for a predominantly insect-feeding falcon.

1959–68	1969–78	1979–88	1989–98	1999–08	2009–18
3	2	6	13	19	74

Earliest and latest occurrence in spring and autumn (1959–2018): 9 April–30 June and 1 July–7 November (25 January).

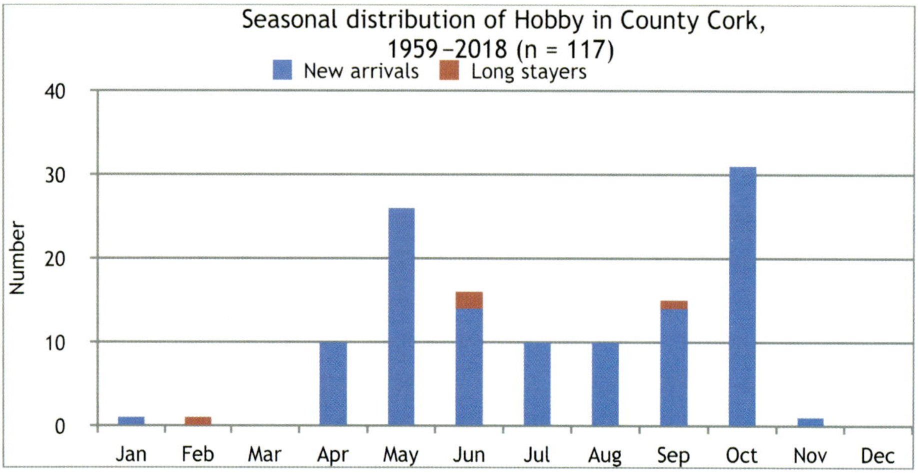

Fig. 7.15 Hobby.

Gyr Falcon

Falco rusticolus | Fabhcún mór | A

(monotypic)

World	Northern regions of North America, Greenland, Iceland, Europe, and Asia
Ireland	Rare visitor, mainly between October and April
Cork	8 individuals
Conservation	Not globally threatened

Adult female obtained at Blackrock on 23 November 1883. One at Fastnet Rock on 5 December 1888. Immature female obtained at Mizen Head in March 1905. One at Glengarriff in June 1925. One at Glengarriff in December 1926. Immature at Castletownbere on 19 October 1931. Immature male obtained at Midleton on 6 March 1970. One at Lisgoold on 13 April 2009.

This falcon was formerly considered to have several subspecies, although it is now regarded as monotypic with three morphs. The white morph is most abundant in the extreme north. All Cork records refer to the white morph. The Blackrock individual was sent alive to England in the spring of 1884.

Peregrine Falcon

Falco peregrinus | Fabhcún gorm | A

(subspecies: *peregrinus*)

World	North and South America, Greenland, Europe, Asia, Africa, and Australia
Ireland	Scarce resident breeder, recent increase
Cork	Resident breeder, recent increase
Conservation	Green-listed

The Peregrine bred at Old Head of Kinsale and Coolim Cliffs, and at other parts of the coast, in the 1700s. Ussher said there was a chain of eyries (he knew of 11) on the coast and islands of Cork in the late 1800s, one on each

Peregrine Falcon

precipitous island, but there was no mention of inland breeding although it was sometimes seen inland in winter, and one caused commotion among wildfowl at Longueville duck decoy.

Sixteen to twenty eyries were known in Cork from the 1930s to the mid-1950s.[137] A serious decline in the Irish breeding population took place from the late 1950s or early 1960s. By 1967/8 the population was drastically reduced, and it was considered by Ken Preston that there were only two or three pairs in Cork.[138] The population in the Republic of Ireland in the 1970s was believed to be well below its pre-1956 level, with single figures only breeding in Cork. The cause of the population crash in Ireland was the same as that in Britain: the widespread use of organochlorine insecticides in agriculture, notably dieldrin, aldrin, heptachlor and DDT. These chemicals resulted in increased adult mortality and caused sub-lethal effects such as eggshell thinning, leading to breeding failure.[139]

Following a ban on use of the contaminants, a significant recovery occurred. There were sixteen pairs in 1982 (13 coastal, 3 inland), and Ken Preston estimated twenty-seven to twenty-nine pairs in 1985.[140] However, they were noted as still scarce or absent west of Baltimore in 1986, although mountain ranges were occupied.[141]

Surveys of the national population in 1981 and 1991 did not include Cork, but a sample of ten coastal Cork eyries had a mean spacing density of 4.5 km. David Norriss stated that in 1991 the Peregrine had recovered nearly all its lost ground throughout Ireland, but that on the Cork coast there was ongoing occupation of a small number of still vacant territories, while inland the mountains of west Cork were poorly known. An estimate of the Cork population in 1992 put the number at twenty-eight pairs.[142;143;144]

A survey in 2002 did include Cork, but a county breakdown was not given.[145] However, figures for the 2002–6 period indicate a population of sixty pairs in the county, with thirty-six at coastal cliffs, eight at inland crags, fourteen at quarries, and one each at an urban area and on a ruined building.[146]

Traditionally, the Peregrine nested on coastal cliffs and natural cliffs inland, usually in mountains and uplands, but also in river valleys, and since the recovery of the population from the declines of the 1950s and 1960s, it has taken to nesting in quarries, both active and disused. In the 2002 survey, the numbers nesting on natural cliffs remained stable, but there was an increase in quarry nesting, with twelve occupied sites in Cork. Birds were also nesting on ruined buildings in 2002, and a pair has bred on a cliff in an urban area for many years.[147;148]

Distribution is somewhat more widespread in autumn and winter as individuals move to bird-rich estuaries and other coastal wetlands, as well as to river valleys and lakes, but high ground is largely avoided.

One was killed striking a window at Fastnet Rock lighthouse in September 1898, and one was found dead at Cloyne in January 1985 having been ringed as a nestling in Cumbria (England) in June 1980.

Water Rail

Rallus aquaticus | Rálóg uisce | A

(subspecies: *aquaticus*)

World	Iceland, Europe, Asia, and north Africa
Ireland	Resident breeder, winter visitor
Cork	Resident, probable winter visitor
Conservation	Green-listed

Historical records show that the Water Rail was resident and common in marshy ground, and Robert Ball found nests at Roxborough (near Midleton), and birds were often trapped in snares set for Common Snipe at Youghal by Bent Ball (Robert's brother) when a boy in the early 1800s.

The Water Rail was recorded in only eleven 10 km squares during the breeding seasons of 1968–72. These ranged from coastal areas in the west and east to inland areas in mid- and north Cork. However, there were large areas with no records, many of which had suitable habitat. This is a skulking species easy to overlook, and birds were probably present in many more squares, perhaps at low density. However, more 10 km squares had records of Water Rails in 2008–11 than in previous surveys. Most records were from sites within 20–30 km of the coast, but they were absent from the Beara and Sheep's Head peninsulas, where suitable habitat is limited. Breeding also takes place at CCBO and Sherkin Islands.

It is likely the distribution of the Water Rail was underestimated during the 1981–4 winter atlas, as it was during the breeding atlases. Most were recorded in and around Cork Harbour, and at sites east of Cork Harbour. Apart from scattered records, the main populations elsewhere were around Bantry, Skibbereen and Kilcolman. Winter distribution in 2007–11 was more extensive than previously and almost mirrored the breeding season, but with more occupied squares near the coast. This may reflect migrant birds occupying smaller sites in winter on the coast which are unsuitable as breeding habitats. High ground is avoided at all seasons due to lack of suitable habitats.

Highest densities are in coastal reedbeds at east Cork sites, although there is little information on breeding numbers. About ten pairs were breeding at Ballymacoda in the 1980s, and although probably an underestimate at the time, it is likely there are fewer than ten today. Habitat change is believed to have reduced the number breeding at Ballycotton. Water Rails are present in summer and winter at Ballyvergan Marsh and Rostellan.

The Water Rail is most obvious in autumn, and counts of nine at Lissagriffin between 27 August and 11 October 1989, eight at Ballycotton on 29 August 1981 and five at CCBO on 9 October 1983 are high by Irish standards. Because of the elusive nature of the species the real numbers present in reedbed habitats may be greater. It is likely that some Water Rails also occur as winter visitors from Europe and Iceland. Many birds have been obtained at lighthouses from Bull Rock to Ballycotton during

September to November, especially at Fastnet Rock. Following a rush of birds at Fastnet Rock on 28–30 September 1913, no fewer than eleven dead Water Rails were found. One ringed as a nestling in Poland was found in Cork in February 2014.

Spotted Crake

Porzana porzana | Gearr breac | A

(monotypic)

World	Europe and Asia
Ireland	Has bred, rare visitor, mainly in autumn, 54 individuals since 1950
Cork	At least 22 individuals
Conservation	Amber-listed

It has been remarked that this species may have been more common in the 1800s than was generally supposed, having been obtained by Adam Parker on two or three occasions. Harvey said that Parker and his brother, when shooting, had met with it frequently over several years. Ussher agreed with Harvey and said it had been obtained occasionally in different parts of the county. Nine occurrences for Cork, more than for any other Irish county, were known up to 1900. The records for Ireland mainly related to autumn, indicating a passage migration with little evidence of wintering. The possibility of breeding was difficult to assess since most records were of birds shot during the autumn and winter duck-shooting season, shooters being less active during the summer months. Only eleven occurrences for Cork were known up to the early 1950s.

Unpublished notes by Ussher provide locations and dates of seven birds, only some of which have previously been published.[149] For the record, these seven birds, all obtained specimens, are as follows: one at Claycastle in October 1843, two at Claycastle in 1847, one at 'Longan' on 25 March 1850 (this location is unknown to us, and it may be the surname of the person who obtained the bird), one at Kanturk probably in winter 1878/9 and preserved at Longueville, one at Cork Lough in September 1893, and one at Fastnet Rock just before 20 August 1895.

A further two Spotted Crakes were obtained between 1900 and 1958, one at Buttevant in January 1904, and one at Fastnet Rock on 28 August 1949.

Eleven individuals (10 records) have been recorded since 1959. One at CCBO on 9–15 October 1959. One at CCBO on 10 September 1967. Adult at CCBO on 8–9 October 1975. One at Ballycotton on 22 September 1982. Two adults at Ballycotton on 25–31 August 1983. One at Kilcolman on 2–3 June 1988. One at Lough Beg on 26 August–10 September 1990. First-year at Inchydoney on 24 August 2002. One found injured at Owenahincha on 28 October 2004, died 29 October. One in breeding habitat at west Cork on 4–6 May 2014.

Breeding by this species is a possibility, and most post-1959 records have been from habitats that could be considered suitable. The Spotted Crake is not difficult to locate by its call, but the discovery of a nest would be much more difficult. However, most of the post-1959 records are for short stays, often only a day or two, therefore it is considered unlikely that any of these involved breeding. It is much more difficult to assess the possibility of breeding in the 1800s and the records are inconclusive, but suitable habitat for breeding was more extensive then than at the present time.

Baillon's Crake

Zapornia pusilla | Gearr Baillon | B

(subspecies: *intermedia*)

World	Europe, Asia, Africa, Australia, and New Zealand
Ireland	3 individuals
Cork	1 individual
Conservation	Not globally threatened

One obtained at Claycastle on 30 October 1845.

Corncrake

Crex crex | Traonach | A

(monotypic)

World	Europe and Asia
Ireland	Scarce summer visiting breeder, recent decrease
Cork	Former breeder, rare visitor
Conservation	SPEC 2; Red-listed

Corncrakes were common in summer during the 1700s and 1800s, breeding on islands such as Dursey, and sometimes remaining into the winter. Corncrakes usually arrived in April and May (occasionally in late March) and departed between August and early October. A decrease in Ireland began in the early 1900s, became more general before 1920, and more so by 1939. The decrease was very marked in Cork and several other counties. Although declining in Cork in 1939, local fluctuations took place, and an increase was reported around Buttevant in 1941 and 1942.[150] Corncrakes then often bred in fields and waste ground close to Cork city, and Freda Shorten remembers them at Centre Park Road in the 1930s.

Despite reported declines, 420 calling Corncrakes were recorded in the area between Adrigole, Glengarriff, Bantry and Kealkill in May and June 1942.[151] This is the only measure of their abundance before the 1950s. There was a general decrease in the 1950s and 1960s with a considerable thinning of the population, and birds disappeared from many areas.

There were large gaps in the breeding range in 1968–72, with most birds at the western peninsulas and islands and on the borders with Kerry and Limerick. Elsewhere, the main population extended from Kinsale to the northern suburbs of Cork city. Gaps in the south, mid, east and north were considered recent, Corncrakes being present at many locations to 1966/7, but not thereafter. For example, 1966 was the last year birds were heard calling at one site near Ballymacoda. None was recorded at Bantry and Durrus after 1969 or at Ballydehob after 1976.

Censuses at CCBO in 1967 and 1969 revealed at least twenty pairs (calling males were present in all but a single hayfield in 1967), but only two were heard in 1973 although the same number of hayfields were available.[152] None was recorded for the first time at CCBO in 1976.

A survey of the Irish population in 1978 showed the decline was continuing, and only twenty-one birds were recorded in Cork (possibly representing 21 pairs). Most of these were at Mizen Head and Sheep's Head, with a handful of records from elsewhere, mostly involving migrants.[153] Survey work from 1978 to 1985 showed further decline in the Schull area (1978, 7 birds; 1979, 4 birds; 1981, 2 birds), with none in 1980 and 1982–5.[154] Another national survey took place in 1988, with only six calling birds recorded in Cork.[155] None was recorded in 1993.[156]

The 1988–91 atlas revealed the extent of decline since 1968–72. Birds were recorded in only six 10 km squares scattered widely along (or near) the coast with two in the north, indicating that breeding was sporadic at best. One Corncrake was present in the breeding season in 1998, at the Beara peninsula, and one passage migrant was heard early in the season near Schull.[157] Only one coastal record was obtained during the breeding seasons in 2008–11. Breeding was not proved at any of these sites, and it is probable that the records referred to transient migrants or unmated males. The Corncrake has occurred since 1990 only as a vagrant, mostly in autumn, more rarely in spring and summer, and there is no evidence that breeding has taken place anywhere in Cork since the 1980s.

There are several records of Corncrakes in November and December. One was obtained on 6 November 1848, and another near Old Head of Kinsale on 20 December 1884. Another was recorded at Carrignavar on 19 December 1911. More recently, a few early winter records have come to light: Roche's Point on 1 November 1982, near Killeagh on 6 November 1983, and Carrigtwohill on 1 November 1984, the latter two shot and verified.

The long-term decline of the Corncrake has been attributed to the effects on breeding success of changes in land use, especially the loss of tall vegetation, and the introduction of mowing machines and intensification of hay and silage production.

Moorhen

Gallinula chloropus | Cearc uisce | A

(subspecies: *chloropus*)

World	North and South America, Europe, Asia south to Indonesia, and Africa
Ireland	Common resident breeder
Cork	Common resident breeder
Conservation	Green-listed

The Moorhen was a common resident in the 1800s, and it was numerous on the enclosed inlet at Rostellan in July 1907. The 1968–72 survey showed it was the most widespread breeding waterbird in Cork and was distributed throughout the county, with breeding taking place on several islands, such as CCBO and Sherkin. The 1988–91 survey showed a serious decline in the west, including an almost complete withdrawal from the peninsulas and from around Roaringwater Bay, and the loss of birds from upland areas in the Boggeragh Mountains. Although there were some further losses by 2008–11, the situation had not changed significantly since the previous survey.

Winter atlas surveys in 1981–4 and 2007–11 showed the range as quite widespread, mirroring the breeding atlases. Density was generally low over most of the county, but an area of higher density extended from east of Cork Harbour along part of the River Lee system to near Macroom.

Most Moorhen nests are found in low vegetation in water, or on or near the ground beside water. A pair nested at Lough Aderry in an old Rook nest on an alder about 4 m above the ground. Moorhen habitats range from the edge of large lakes to small water-filled ditches. Estimates of up to thirty-three pairs were given for Ballycotton in the 1960s, 1970s and 1980s, although habitat change is believed to have halved that number. About forty pairs bred at Ballymacoda in the 1980s, but this number is now down to not more than five pairs due to small-scale agricultural drainage works. Up to sixty may be present at Kilcolman, although there is no breeding estimate for this site, and Fota Wildlife Park and Cork Lough have populations which benefit from food provided to the waterfowl collections.

Moorhens form loose flocks in winter, although there have been few counts. Highest numbers recorded have been fifty-six at Kilcolman on 1 September 1995, fifty at Mallow lagoons on 3 November 1991, forty-eight at Ballyhonack Lake in October 1995, forty at Kilcolman on 18 July 1989, up to thirty-four at Cork Lough in winter 1992/3, thirty at Cuskinny on 20 September 1989, and thirty at Kilcolman in April 1995. A winter flock of about thirty at Castlemartyr has declined by half. A peak count of 115 was obtained at Cork Harbour in 1996/7, small numbers being present at many sites.

While the distribution of the species across the county has remained widespread, it is likely there has been a significant thinning of the population. Several small wetlands which held populations in east Cork have disappeared through drainage, such as Lough Rhue and Cloyne. The possible role of predation by the introduced mink is unclear, but it is likely to be another negative factor alongside habitat loss.

The extent of migration involving Ireland is uncertain, although there have been a few Scottish and continental ring recoveries. The Moorhen was reported from Fastnet Rock during a rush of birds in October 1914.

Common Coot

Fulica atra | Cearc cheannann | A

(subspecies: *atra*)

World	Europe, Asia, north Africa, Australia, and New Zealand
Ireland	Resident breeder, winter visitor
Cork	Common resident and breeder, winter visitor
Conservation	SPEC 3; Amber-listed

The Coot was said to be resident in the 1800s, occurring on lakes and sluggish rivers, which in hard winters it left for the coast. It was noted as numerous on the enclosed inlet at Rostellan in July 1907.

The 1968–72 breeding atlas showed the Coot had a more restricted breeding range than the Moorhen because it requires larger water-bodies which are in short supply in some areas. However, Coots will sometimes

Common Coot

nest on small ponds, although never in drains, ditches or reedbeds, all of which are used by Moorhens. The main breeding population was centred on the water-bodies in the River Lee system, but lakes such as Reenydonagan, Lissagriffin and Curraghalicky also had breeding birds, as had smaller water-bodies such as Bogaghard Ponds, and urban settings such as Cork Lough. The 1988–91 survey recorded an almost complete absence from the west and south, with density remaining highest on the River Lee system. There was little change by 2008–11, the main feature being an almost complete absence west of Rosscarbery. There are few data on breeding densities, but thirty-two pairs were at Ballycotton in 1968.

Although the Coot is more widespread during the winter season, its preference for larger water-bodies remains. During winter atlas surveys the main population was around Cork Harbour, extending to east Cork, and west along the River Lee system to Lough Allua. It was extremely scarce, and mostly absent in the west, and only a few sites in the north had any significant numbers.

Numbers increase from September with the arrival of wintering birds from continental Europe. Peak numbers occur from October to January with the largest concentrations at the Gearagh (up to 600), Lough Aderry (350 in October 1983), Ballybutler Lake (324 in January 1985), Inishcarra reservoir (mean of 312 in mid-1980s) and Gallanes Lake (100 in December 1984). Declines have continued during the last two decades, and the species is now very scarce in east Cork at all seasons. Ballycotton formerly held up to 500 (800 in December 1984), but this flock has disappeared due to habitat change.

Coots wintering in Ireland are presumed to come from northern Europe, but data from ring recoveries are scanty. One ringed in Lincolnshire (England) in August 1971 was recovered at Kinsale in September 1971.

American Coot

Fulica americana | Cearc cheannann Mheiriceánach | A

(subspecies: undetermined)

World	North and Central America, and north-west South America
Ireland	4 individuals
Cork	1 individual
Conservation	Not globally threatened

Possible first-year male (on call) at Ballycotton on 7 February–4 April 1981.

Common Crane

Grus grus | Grús | A

(monotypic)

World	Europe and Asia
Ireland	Former breeder, now rare winter visitor, but has recently bred again
Cork	Former resident, 87 individuals since 1739
Conservation	Not globally threatened

The Crane was resident in Ireland many centuries ago, and it is believed it was extinct as a breeding species by the fourteenth century.[158] Giraldus Cambrensis noted an abhorrence of Crane flesh among the Irish, but archaeological evidence indicates that Bronze Age people exploited and ate Cranes at Ballycotton, suggesting they occurred, and probably bred, in the extensive marshes then present in the area.[159;160] It probably occurred in the alluvial woodland of the Gearagh and its associated marshes; it would have ceased to breed well before the 1700s, although it may have continued as a visitor.[161]

A Crane was obtained at Cork Harbour in the severe frost of 1739. At least eight Cranes visited the Kinsale area in November 1851. Three or four were first seen near Kinsale and three were obtained, including one at Annagh Bog (Inishannon) on 17 November. Later in the same month five more were seen east of Kinsale and one was obtained.

There were no further records until 1972. Between that year and 2010 a total of twelve Cranes occurred. There followed a major influx in 2011 involving an unprecedented total of sixty-three birds. In the years 2012–18 there has been a return to its former rarity with only three birds being recorded.

The 2011 influx began with a single bird at Long Strand in late October, followed by another single bird at CCBO in early November. Four significant flocks then occurred between 8 and 13 November, ten at the Beara peninsula, fifteen at Castletownroche, fourteen at Ballincollig (also seen at Power Head) and nineteen at Midleton. Two birds were seen at Wellington Bridge on the western edge of Cork city on 14 November, and finally one was seen at Kinsale on 13 December. Most birds were recorded on one or two days, but the Ballincollig and Power Head birds remained for seven days, the Midleton birds for thirteen days, while the single Kinsale bird remained for seventeen days. There was no evidence to suggest that the Cork birds dispersed to sites elsewhere in the country.

The fifteen birds recorded during the period 1972–2018 (but excluding 2011) occurred at eleven sites across the county from Dursey Island to Ballymacoda, and from coastal Croagh Bay and Kinsale to inland Castletownroche and Kilcolman, with no site hosting more than two birds. Eight of these were present on only one or two days, but others remained for up to two months (e.g. Churchtown South and Ballymaloe, 1987/8) and eighty days (e.g. Kilcolman, 2009/10).

Vagrant Cranes sometimes tend to be rather mobile and have often been seen only in flight, and some individuals or groups have been presumed, or are known, to have occurred at two or more sites widely separated from each other. A bird at Kilcolman in late August 1976 was believed to be the same as one seen near Mallow a week or two later, while two at Dursey Island in early December 1978 were believed to be the same as two seen at Union Hall and at Cork Harbour during the following month. It is possible that the Kilcolman bird of August 1976 was the same as one seen at Ballymacoda in September 1976, while three single birds recorded in west Cork in late October 2008 may refer to a single individual, but in the absence of distinguishing marks it is impossible to know.

Cranes have been breeding in eastern England since the early 1980s and have been increasing slowly, so it is possible that birds from this population may occur in Ireland. However, with an English wintering population of about fifty birds in 2010, these could not have accounted for the 2011 influx; it is much more likely that they were of continental European origin.

1959–68	1969–78	1979–88	1989–98	1999–08	2009–18
0	6	2	0	3	67

Earliest and latest occurrence in winter (1959–2018): 31 August–27 December (7 February).

Sandhill Crane

Antigone canadensis | Grús Ceanadach | B

(subspecies: *canadensis*)

World	North America and eastern Siberia
Ireland	1 individual
Cork	1 individual
Conservation	Not globally threatened

One obtained at Castlefreke on 14 September 1905 had been present for about three days.

Little Bustard

Tetrax tetrax | Bustard beag | B

(monotypic)

World	Europe, Asia, and Morocco
Ireland	10 individuals
Cork	2 individuals
Conservation	Near threatened

Probable first-year male obtained at Ballycotton on 24 December 1860. One obtained at Ballymacoda on 14 November 1883.

Great Bustard

Otis tarda | Bustard mór | B

(subspecies: *tarda*)

World	Europe, Asia, and Morocco
Ireland	3 individuals
Cork	1 individual
Conservation	Vulnerable

Female obtained at Castletownbere on 9 December 1925.

Oystercatcher

Haematopus ostralegus | Roilleach | A

(subspecies: *ostralegus*)

World	Iceland, Europe, and Asia
Ireland	Resident breeder, common winter visitor, autumn passage migrant
Cork	Resident, common winter visitor, passage migrant
Conservation	SPEC 1; Red-listed

The Oystercatcher was resident, bred numerously in west Cork, and occurred in considerable flocks in winter in the 1800s. During the three atlas surveys (1968–2011), it was breeding on the mainland coast and on islands from Roaringwater Bay westwards to the border with Kerry, the most easterly point being near Glandore in 2008–11. Eleven pairs bred at CCBO in 1963 and 1967, twelve pairs in 1983, five pairs in 1990, but there was evidence of breeding at only one site in 1995. Five or six pairs bred at Dursey Island in 2000. A pair bred on the flat roof of a building at Marina Industrial Park in Cork city in 1994 and 1995 (probably also in 1993). A territory-holding pair was at Spike Island in 2018, and a single bird was displaying at Ballyandreen in 2019.

Small flocks of non-breeders remain on estuaries and beaches during May and June and are joined by young birds and breeders in July and August. Autumn passage peaks at most sites in September and October. Many birds come from outside Ireland, and there have been recoveries at this time and in winter of Oystercatchers ringed in summer in Wales, northern England, mainland Scotland, Orkney, Shetland and Iceland.

Cork Harbour is the most important site for Oystercatchers in Cork. Tivoli, a reclaimed area in the north-west corner of Lough Mahon, was formerly the main roosting location, with a maximum of 3,000 in September 1971. As this site became industrialised, nearby Dunkettle and other sites were occupied. From the late 1970s to the early 1990s Cork Harbour regularly recorded mean peak counts of over 2,000 birds. Numbers decline slowly after late October, not rising significantly again until April and May when a modest spring passage occurs.

Ring recoveries indicate the destinations of some of these birds, with movements from Cork to Donegal, northern England, mainland Scotland, Orkney and Faeroes. Clonakilty, Courtmacsherry, Ballycotton and Ballymacoda hold lower numbers, usually about 200–600 birds. Numbers have declined at all sites recently, and the peak at Cork Harbour has exceeded 2,000 birds on only one occasion since 2011.

Oystercatchers are distributed along the coast in winter. Small numbers often extend their diurnal feeding range inland for several kilometres, especially in the lower reaches of the River Lee where about 100 feed in the Lee Fields, but there are no real inland populations. However, chance events occasionally bring birds inland; a flock of about thirty was seen in a field near Killavullen on 3 March 2018 immediately following heavy snowfall over the county. One was present at Charleville lagoons on 11 August 1985.

Black-winged Stilt

Himantopus himantopus | Scodalach dubheiteach | A

(subspecies: *himantopus*)

World	North and South America, Europe, Asia, Africa, Australia, and New Zealand
Ireland	57 individuals
Cork	26 individuals
Conservation	Not globally threatened

There were three records (4 individuals) of Black-winged Stilts in Cork before 1959: Youghal (winter 1823 or 1824), Clonakilty (April to June 1942) and Timoleague (2, September and October 1949).

Since 1959 the Black-winged Stilt has occurred mainly as a spring overshoot between March and May, with one record (2 birds) in October. They have occurred at several sites along the coast from CCBO to Ballycotton, with one 8 km inland at Bateman's Lake. A few occurrences have been of single birds, but most were of small flocks, six on one occasion. There was a remarkable influx of at least eleven individuals (possibly more) in March and April 1990. These occurred at five sites between Ballycotton and Bateman's Lake, and the peak count (11) was recorded on 1 April. Several records relate to a stay of just one day, but many have remained for several days and some for up to two weeks, although because of movement of birds between sites in the 1990 influx it is difficult to calculate the length of stay of many birds. Four adults have been recorded, as well as four males and two females, a pair (male and female) being recorded on two occasions. A single bird at Clonakilty on 30–1 May 2014 had been seen previously at Finnamore Lake (Offaly) on 28 May 2014.

1959–68	1969–78	1979–88	1989–98	1999–08	2009–18
0	0	2	16	2	2

Earliest and latest occurrence in spring and autumn (1959–2018): 17 March–30 May and 15 October.

Avocet

Recurvirostra avosetta | Abhóiséad | A

(monotypic)

World	Europe, Asia, and Africa
Ireland	Has bred, rare winter visitor
Cork	49 individuals
Conservation	Not globally threatened

There were eight records (11 individuals) of Avocets in Cork before 1900, all except one having been obtained. Six individuals occurred at Cork Harbour, two at Youghal and one at Castletownbere, a further two have no precise location data but may be from east Cork. Of the seven individuals dated to month, all occurred between November and February, five occurred as singles, while two occurred together on three occasions.

There were no further records of the Avocet in Cork until the winter of 1955/6 where for the next ten winters to 1964/5 between one and eight birds occurred in the Great Island Channel of Cork Harbour, usually between December and February, and once to 3 March (peak counts in each of the ten winters = 2, 3, 6, 2, 2, 5, 5, 8, 1, 1). The Irish Rare Birds Committee regards these occurrences as involving a minimum of twelve new birds, but there is no way of knowing. Many more may have been involved, although wintering waterbirds (among other species) are site faithful across seasons, so it is likely that at least some were birds returning to the same site.

Excepting those wintering birds in Cork Harbour (described above), twenty-six individuals have occurred in the six decades 1959–2018. These records indicate arrivals from November to January, with a small peak in May possibly indicating a passage movement. One was seen flying west past CCBO on 23 April 1963. Others have occurred at Clonakilty Bay (4), Courtmacsherry Bay (4), Cork Harbour (8), Ballycotton (6) and Ballymacoda (2), with one bird inland at the Gearagh in October 2013. Eighteen records have been of single birds, while two and three together have occurred once and twice, respectively. Fourteen of the twenty-six birds occurring in this period remained for between one and eight days (mean stay = 2.4 days), with twelve birds remaining for longer periods, effectively overwintering, especially at Ballycotton (1, December 1964–March 1965;

1, December 1992–March 1993) and Cork Harbour (up to 3, November 1968–February 1969). The numbers occurring in Cork have declined over the sixty years of this analysis, although it has continued to increase in numbers and range in Britain with 1,600 breeding pairs and a wintering population of 7,500 birds by 2010, albeit mainly on or near the east coast of England.

1959–68	1969–78	1979–88	1989–98	1999–08	2009–18
6	6	2	8	2	2

Earliest and latest occurrence in autumn and spring (1959–2018): (6 August) 2 October–12 May.

Stone Curlew

Burhinus oedicnemus | Crotach cloch | A
(subspecies: *oedicnemus*)

World	Europe, Asia, and north Africa
Ireland	28 individuals
Cork	6 individuals
Conservation	Not globally threatened

Adult female obtained at Castletownshend on 24 February 1913. One at CCBO on 10 May 1988. One at CCBO on 8–9 May 1996. One at Dursey Island on 26 April 1999. One at Sherkin Island on 20–1 March 2010. One at Bantry on 3 June 2013.

Little Ringed Plover

Charadrius dubius | Feadóigín chladaigh | A

(subspecies: *curonicus*)

World	Europe, Asia south to New Guinea, and north Africa
Ireland	Has bred, rare spring and autumn visitor
Cork	57 individuals
Conservation	Amber-listed

Little Ringed Plovers have occurred in spring (24) and autumn (29), and very occasionally in winter (4). They have mostly occurred at coastal sites from Dursey Island to Ballymacoda Bay (37), although they are increasingly recorded at inland sites (20). Only six have been recorded west of Kinsale: four at Clonakilty Bay and singles at Kinsale and Dursey Island. There is a record from Cork Harbour involving four birds, while Ballycotton Bay (20) and Ballymacoda Bay (5) provide the remainder from coastal sites. They have occurred at four inland sites: the Gearagh (11), Dooniskey (3), Mallow lagoons (7) and Kilcolman (1).

They have been suspected of breeding, but this has not been proved beyond doubt at any site. An adult female and two juveniles were present at Mallow lagoons for three days in early August 2006, but evidence of breeding fell short of proof, contrary to reports.[162] A pair was at Mallow lagoons in 2007, and a female was present for eleven days in June 2008. Five and two juveniles were at the Gearagh and Dooniskey, respectively, on the same day in early September 2011. One adult was seen at Dooniskey on one day in April 2013, while at the Gearagh in 2018 one juvenile arrived in early July, followed by four other juveniles later in July. Four adults (2 males and 2 females) visited Harper's Island (Cork Harbour) between 5 April and 27 May 2018; display behaviour was observed, breeding was suspected off site, but no nest was found. One was seen at the same site on 27 March and 30 June 2019.

Most birds turn up as single individuals, but flocks of two (4), three (1), four (2) and five (1) have also been seen. The age profile, where this has been published, involves fifteen adult birds and twenty-eight first-year birds. Thirty-one birds have remained for one to four days, while twenty-four have been present for longer periods of up to fifty-three days, with one wintering bird remaining for 117 days (mean stay = 11.9 days). There has been a significant increase in occurrence of the Little Ringed Plover in Cork since the first was recorded in 1968, and this increase has been mirrored in Ireland as a whole. The increase in Ireland should be viewed against the 42 per cent increase in range recorded for the British breeding population over a twenty-year period since 1988–91; they first bred in Britain in 1938 and have increased steadily over succeeding decades, with most of the population nesting at inland sites, such as gravel and sand pits, other static water-bodies and river shingle.

Almost all west Palearctic breeding Little Ringed Plovers, including the British population, winter in Africa south of the Sahara Desert. However, one was at Mallow lagoons from late December 2003 to mid-February 2004 and one was at Ballycotton from late December 2004 to late April 2005. One was at Ballymacoda from mid-December 2005 to early January 2006, and possibly the same bird was recorded at Ballycotton later in January and remained to mid-March 2006.

Although only fifty-seven individuals have occurred in Cork, the records show several trends: an increasing tendency towards inland occurrences at sites with breeding potential, an arrival of adults usually early in the season, and of juveniles usually late in the season. This suggests the possibility of dispersal from small breeding populations, at least in some years, from sites yet to be discovered. Finally, the occurrence of birds in winter may reflect the increasing mildness of Irish winters and the beginnings of a tendency for this species to remain closer to its breeding grounds (see Little Egret, Hobby and Lesser Black-backed Gull).

1959–68	1969–78	1979–88	1989–98	1999–08	2009–18
1	0	3	4	17	32

Earliest and latest occurrence in spring and autumn (all records): 19 March–20 June and 4 July–18 October.

Common Ringed Plover

Charadrius hiaticula | Feadóg chladaigh | A

(subspecies: *hiaticula; tundrae*)

World	North-east Canada, Greenland, Iceland, northern Europe, and Asia
Ireland	Resident breeder, winter visitor, autumn passage migrant
Cork	Breeder, winter visitor and passage migrant
Conservation	Amber-listed

The Ringed Plover (*C.h. hiaticula*) bred on the coast, and numbers increased in winter in the 1700s and 1800s. Recent atlas surveys have shown a patchy coastal breeding distribution, but it is possible that in the first two surveys the range was underestimated. Most breeding pairs are found either in the far west or the far east of the county, with a general absence from many sites between Cork Harbour and Roaringwater Bay. One or two pairs breed inland at the Gearagh.

Ringed Plovers were more numerous as breeders in the past. There were fifty pairs in 1964 and nineteen pairs in 1968 at Ballycotton, but fewer than five pairs in recent decades. Up to ten pairs bred at Ballymacoda Bay in the 1970s and 1980s, but breeding is now sporadic at best. Human beach recreational use is at least partly responsible for the decline. Nesting has been recorded in cereal and sugar beet fields adjoining the beaches at Ballymacoda, Garryvoe and Ballybranagan. Nesting also takes place at industrial yards and car parks in Cork Harbour, and at rock and gravel quarries close to Cork Harbour.

The first migrant Ringed Plovers of the autumn arrive in late July. At Ballymacoda, numbers usually peak in September and decrease thereafter, while at Cork Harbour, peak numbers often do not occur until later. The main concentrations occur at Rosscarbery, Clonakilty, Cork Harbour, Ballycotton and Ballymacoda with numbers ranging from about 100 to 400 birds. There was an exceptional estimate of over 1,000 at Harbour View on 8 November 1967. Up to fifty have occurred inland at the Gearagh in September. The winter range is wider than during the breeding season and is entirely coastal.

Although there is no conclusive evidence of spring migration, it is possible that very small numbers pass through in April and May. Birds of

Common Ringed Plover

the North Scandinavian subspecies (*C.h. tundrae*) have been suspected at Ballymacoda and Ballycotton at this time.

One ringed as a nestling at Heligoland (Germany) on 13 May 1956 was recovered at Courtmacsherry on 7 October 1956, and one ringed as a nestling at Nijkerk (Netherlands) on 13 June 1964 was found dead at Old Head of Kinsale on 7 February 1965. There have also been movements in both ways between Cork, Wales and England.

Killdeer

Charadrius vociferus | Feadóg ghlórach | A

(subspecies: *vociferus*)

World	North America, Caribbean islands, and north-west South America
Ireland	22 individuals
Cork	9 individuals
Conservation	Not globally threatened

One obtained at Crookhaven on 30 November 1938. One at Crookhaven on 11 December 1938, found dead on 19 March 1939. One at Ballycotton on 7–31 March 1979. One at Ballymacoda on 15 February–24 March 1984. Adult at Ballycotton on 7 January–2 March 1991. One at Ballycotton on 8–22 October 1995. First-year at CCBO on 18 October 1996. One at Saleen on 3 February 2003, and at Ballycotton on 9 February 2003. One at Sherkin Island on 10 November 2011. The Sherkin Island bird was not seen, but its distinctive call was heard at night as it flew off, recalling the Ballymacoda bird; although it remained for more than a month, it too was frequently heard calling as it flew at night.

Kentish Plover

Charadrius alexandrinus | Feadóigín chosdubh | A

(subspecies: *alexandrinus*)

World	North and South America, Europe, Asia, and north Africa
Ireland	16 individuals
Cork	3 individuals
Conservation	Not globally threatened

One at Ballycotton on 23 April 1970. First-year at Ballycotton on 22–7 September 1980. One at Ballycotton on 24 April 1984.

Lesser Sand Plover

Charadrius mongolus | Feadóigín ghainimh | A

(subspecies: *mongolus*)

World	Central and east Asia
Ireland	1 individual
Cork	1 individual
Conservation	Not globally threatened

Adult at Ballymacoda on 27–8 July 2013.

Dotterel

Charadrius morinellus | Amadán móinteach | A

(monotypic)

World	Europe and Asia
Ireland	Has bred, rare passage migrant
Cork	34 individuals
Conservation	Not globally threatened

There was one record of a Dotterel in Cork before 1959, a female obtained inland just west of Cork city in September 1844. This remains the only inland occurrence for Cork.

The next record was not until 1966, and thirty-three birds have occurred up to 2018. Most have been in autumn (30), mainly September and October, with three birds in spring. All have been west of Cork Harbour, apart from two in east Cork (2013 and 2015). Most birds have been recorded at CCBO (15) and at Dursey Island and the Beara peninsula combined (9), while other sites have produced a maximum of two: Lissagriffin (2), Galley Head (1), Old Head of Kinsale (2), Robert's Cove (2) and Ballycotton (2). Most records refer to single birds except in three cases when two were seen together (CCBO twice, and Robert's Cove). Nineteen have been aged, seventeen as first-year birds and two as adult birds. Most birds have been present for one to three days (29), the maximum stay being eight days (mean stay = 1.8 days).

1959–68	1969–78	1979–88	1989–98	1999–08	2009–18
1	4	6	6	5	11

Earliest and latest occurrence in spring and autumn (1959–2018): 7 April–5 May and 17 August–22 October.

American Golden Plover

Pluvialis dominica | Feadóg bhuí Mheiriceánach | A

(monotypic)

World	Northern regions of North America
Ireland	Rare visitor, mainly in autumn
Cork	74 individuals
Conservation	Not globally threatened

The bird formerly known as the 'lesser golden plover' was split into two species in the 1980s, thus establishing the American and Pacific Golden Plover.[163;164] Before the split, observers made little effort to establish the subspecies, understandable bearing in mind the poor optical and near total lack of photographic equipment available, and the belief that all 'lesser golden plovers' occurring in Ireland were of the North American subspecies *dominica*. However, after the split, the 'lesser golden plover' was looked upon in a different light, especially following identification of the first Pacific Golden Plover in Wexford in 1986. This led to a partial review of 'lesser golden plover' records (documentation was unavailable for some records), and as far as Cork was concerned, one of the older records was upgraded to American Golden Plover (Ballycotton, 10–18 September 1971), becoming the first record for Cork. National statistics show that 4.7 per cent of identified birds were Pacific Golden Plovers (17 of 359), while comparable statistics for Cork show that 3.9 per cent of identified birds were Pacific Golden Plovers (3 of 77). Ninety 'lesser golden plovers' occurred in Cork during 1959–2018, the first dating to 1966; seventy-four of these have been identified as American Golden Plovers, three as Pacific Golden Plovers with thirteen remaining unidentified.

American Golden Plovers occur in Cork mainly in September and October (64), but there are records of new birds in November, December, May, June and August. The three May and June records were in 2011 and 2012, and have occurred since the recent increase in occurrences of this species in Ireland, and therefore may have been birds that had arrived in a previous autumn. American Golden Plovers have been recorded at many coastal sites from Dursey Island to Ballymacoda Bay, with four at the Gearagh, the only significantly inland records. Most have been recorded at

Ballycotton Bay (24, including one 5 km inland at Ballymaloe), Rosscarbery (10), the Mizen peninsula (7), Clonakilty Bay (7) and Ballymacoda Bay (5), while smaller numbers have occurred at eleven other sites, including three sites within Cork Harbour. Two birds have been seen together on four occasions (Clonakilty Bay 1996; CCBO 2009; Ballycotton 2011 and 2016), while all other records have involved single birds. Sixty-one birds have been aged, nineteen adult and forty-two first-year birds. Most birds have been recorded on one or two days (38), but others have remained for longer periods, including four for more than twenty days (28, 34, 37 and 44 days) (mean stay = 5.8 days).

Thirteen unidentified birds are included in parentheses in the decadal table. The American Golden Plover has occurred annually in Cork since 2005. Numbers have varied between years, but there have been few in some years (2005, 2006, 2007, 2014, 2018) and many in others (2010, 2011, 2012, 2015, 2016), which may suggest high populations in some years or greater vagrancy in others, perhaps due to increased storminess during migration.

1959–68	1969–78	1979–88	1989–98	1999–08	2009–18
0 (3)	1 (1)	3 (7)	11 (2)	15 (0)	44 (0)

Earliest and latest occurrence in spring and autumn (all records): 17 May–1 June and 24 August–10 November (2 December).

Unidentified 'lesser golden plover' species

Thirteen unidentified 'lesser golden plovers' have been recorded. All but one of these were probably American Golden Plovers, but one showed some characters of Pacific Golden Plover (Ballycotton, 22–3 July 1983). Their seasonal pattern of occurrence matches those that have been identified, with twelve seen between August and October, and one in July as noted. The geographical spread of unidentified birds is more restricted, with nine at Ballycotton Bay, three at the Mizen peninsula and one at Cork Harbour; two at Ballycotton in 1979 was the only sighting involving more than one bird. The mean stay of these thirteen birds was six days, while of those aged, six were adult and two were first-year birds.

Pacific Golden Plover

Pluvialis fulva | Feadóg bhuí Áiseach | A

(monotypic)

World	Siberia from Yamal Peninsula to Chukotskiy Peninsula, and western Alaska
Ireland	17 individuals
Cork	3 individuals
Conservation	Not globally threatened

Adult at Kinsale Marsh on 5–8 October 1991. Adult at Inchydoney on 19 October 1991. Adult at Ballycotton on 8 August 1993.

European Golden Plover

Pluvialis apricaria | Feadóg bhuí | A

(monotypic)

World	Greenland, Iceland, and northern regions of Europe and Asia
Ireland	Scarce breeder, common winter visitor, passage migrant
Cork	Former breeder, common winter visitor
Conservation	Red-listed

The Golden Plover was described as breeding on mountains in the 1800s, but no locations were given. Correspondents had no evidence of breeding in the 1940s. It was said in the early 1970s that breeding areas had been evacuated for thirty to sixty years, suggesting it may have continued to the 1910–40 period. While there have been records from upland areas during recent breeding atlas surveys, there were no records of breeding.

Winter atlas surveys showed the main populations centred around Cork Harbour and east Cork, especially Ballycotton and Ballymacoda. Other significant populations occurred at Roaringwater Bay and between Rosscarbery and Courtmacsherry, with inland populations around the Gearagh and valleys of the Rivers Awbeg and Funshion. Small flocks occurred at several other sites, particularly in fields near the River Blackwater. In general, flocks are very mobile and coastal birds fly inland to feed in pastures. The Golden Plover is subject to considerable movements during cold weather. During such conditions on 8 December 1882 flocks were going south-west all day at Dursey Island, while at lighthouses on the Cork coast during the following week many birds continued to fly south.

Numbers of between a few hundred and 3,000 occur at Rosscarbery, Clonakilty Bay, Courtmacsherry Bay, Ballycotton, Kilcolman, Charleville lagoons and the River Lee reservoirs. Particularly high numbers were present in the winter of 1990/1 with 5,800 at Clonakilty in November and 5,300 at Courtmacsherry in December. Counts of 10,000 or more were normal at Ballymacoda during the 1970s and 1980s, with a maximum of 15,000. Cork Harbour has also recorded a peak of 15,000, and throughout the 1970s and 1980s over 5,000 were present in winter. However, numbers are now well below 10,000 at Ballymacoda, while in the 2010s most counts at Cork Harbour were below 3,000 birds. Complete counts of the county gave January totals of 15,925 in 1992, 18,439 in 1993, and 18,159 in 1994.

Departure is rapid at the end of February and in March, and small flocks often occur on migration in upland areas (e.g. 25 in summer plumage at Mushera Mountain on 17 April 1992, and 100 at Mullaghareirk Mountains on 24 April 2004).

Most wintering Golden Plovers are probably referable to the southern form (*apricaria*) which breeds from mid-Scandinavia southwards. A flock of 100 of the northern form (*altifrons*) which breeds predominantly in Iceland and northern Scandinavia was noted at Ballymacoda in April 1965, and others of this form have been seen there in spring. The only ring recoveries of this species are of birds from Iceland, and therefore presumably referable to *altifrons*; one ringed in July 1929 was recovered at Carrigaloe in January 1930, one ringed in June 1947 was recovered in Cork in January 1948, and one ringed in October 1955 was recovered at Banteer in January 1956.

Grey Plover

Pluvialis squatarola | Feadóg ghlas | A

(monotypic)

World	Arctic regions of North America, north-east Europe and Asia
Ireland	Winter visitor, passage migrant
Cork	Winter visitor, passage migrant
Conservation	Red-listed

The Grey Plover was an uncommon winter visitor, well known to Robert Ball at Youghal where it was found sparingly in the 1800s and was regularly occurring in winter in many parts of Cork Harbour by about 1950.

Winter atlas surveys showed the main populations were at Cork Harbour, Ballycotton and Ballymacoda, with smaller numbers at estuaries between Clonakilty and Kinsale, and in Roaringwater and Bantry Bays.

Some Grey Plovers are present in Cork throughout the year, although most occur in winter and at migration times in autumn and spring. Complete counts for the county in January 1993, 1994 and 1995 gave totals of 469, 558 and 667, respectively. Ballymacoda is the main wintering site and peaks of 514 and 565 were recorded in February 1995 and February 1984, respectively, but numbers have declined since then. The winter peak at Cork Harbour was usually around 100 in the 1970s and 1980s, and a maximum of 175 was recorded, but numbers did not exceed fifty birds in the 2010s. Mean peak numbers at Ballycotton during 1970–2005 were about 100 with a peak of 174 in February. Numbers here have also declined recently. Smaller numbers occur at Courtmacsherry and Clonakilty Bays.

Inland, four were seen at Charleville lagoons on 1 December 1985, and one or two have occurred at the Gearagh in spring, autumn and winter.

Northern Lapwing

Northern Lapwing

Vanellus vanellus | Pilibín | A

(monotypic)

World	Europe and Asia
Ireland	Resident breeder, recent decrease, common winter visitor
Cork	Resident breeder, recent decrease in both seasons
Conservation	SPEC 1; Red-listed

The Lapwing was a resident, breeding commonly on rush-covered boglands in the 1800s. It was noted at Fastnet Rock on 5 March 1887 when an extensive movement took place on the south coast. During severe frost and snow Lapwings have often been noticed moving towards the south-west of Cork. A similar westerly movement took place in severe weather in January and February 1963 when up to 1,400 were present at CCBO, and where 368 corpses were found.

Its breeding range in 1968–72 was patchy with large vacant areas, and it was mainly confined to three localities: (a) from west of Dunmanway eastwards to Macroom and the River Lee reservoirs, and north to the Boggeragh Mountains, (b) wet pastures around the River Awbeg near Buttevant and Doneraile, and north to the Ballyhoura Mountains, and (c) from Carrignavar and Ballyhooly eastwards to the Waterford border, including coastal marshes at Ballycotton and Ballymacoda.

The 1988–91 breeding survey showed a contraction of range. It was by then nearly absent from the county, and proof of breeding was obtained in only two 10 km squares (the Gearagh and Ballyhoura Mountains area). In 2008–11 proved breeding was reported at four 10 km squares (Harper's Island, Minane Bridge, Cecilstown and Kilcolman) and probable breeding at one (Mallow). Twenty-nine pairs nested at Ballycotton in 1964 but only nine in 1968, of which only one produced young. No birds have nested in recent years at this site.

The 1981–4 winter atlas showed the main coastal populations were at Clonakilty and Courtmacsherry Bays, and at Cork Harbour, Ballycotton and Ballymacoda. Significant inland populations occurred along parts of the Rivers Awbeg, Funshion, Blackwater and Lee. Many other inland 10 km squares held small populations, and the only parts of the county with few or none were the western peninsulas and upland areas in the north-west. There was little change in distribution along the coast in 2007–11, but there was a significant thinning of the inland population.

Non-breeding Lapwings arrive in mid-June at coastal sites, although numbers remain low through most of the autumn. Flocks exceeding 500 are unknown before October. Thereafter, numbers build up rapidly to peak in February, this varying between years depending on weather conditions. In late February and early March there is typically a sudden decrease as most depart for the breeding areas.

Many sites regularly hold over 1,000 birds and sometimes up to 3,000 in winter (e.g. Clonakilty, Courtmacsherry, Bandon estuary, Ballycotton, the Gearagh, and Charleville lagoons). The main concentrations are at Cork Harbour and Ballymacoda. These sites regularly hold more than 5,000, and sometimes as many as 10,000 birds. Counts in December 1985 were exceptional: 26,413 at Cork Harbour, 15,350 at Ballymacoda and 5,500 at Ballycotton. These figures are best appreciated in terms of the then mid-winter Irish population of 200,000 Lapwings.[165] However, numbers have plummeted to between 1,000 and 2,000 at Cork Harbour during the 2010s, and the decline at Ballymacoda has been on a similar scale. Total counts of the county in January 1992, 1993, 1994 and 1995 recorded 16,472, 22,707, 22,253 and 34,216, respectively.

Most recoveries of ringed Lapwings between December and February are from Scotland and northern England, but many birds ringed as nestlings also come from the Netherlands, Norway, Sweden and Poland.

Red Knot

Calidris canutus | Cnota | A

(subspecies: *islandica*)

World	Arctic regions of North America, also Greenland and Asia
Ireland	Common winter visitor, autumn passage migrant
Cork	Winter visitor and passage migrant, chiefly in autumn
Conservation	SPEC 1; Red-listed

The Knot was known at Cork Harbour from a few specimens obtained in different years between August and January in the early 1800s. However, its true status then is difficult to judge because of conflicting statements. In the early 1900s it was reported to winter at Cork Harbour (and in neighbouring estuaries), where flocks of hundreds had been seen. It was absent from Glandore, but one in summer plumage was seen at Rosscarbery in August 1946.

In the 1960s and early 1970s, Cork Harbour held about 3,000 birds. The most favoured sites were Douglas estuary and the nearby shores of Lough Mahon, and Tivoli. However, numbers declined to less than 100 by the mid-1970s. Numbers remained low during the 1980s, and peak counts have varied from less than 100 to 230 in the 2010s. Poor weather conditions in Arctic breeding grounds apparently led to high adult mortality and caused the decline.[166]

There are no data for Ballymacoda before 1970. A particularly high count of 1,000 was obtained in February 1971, but regular counts since then have never reached this high point. Although numbers peaked at about 350 in the mid-1990s, the current winter peak ranges around 100–150 birds. A flock of about 200 occurred at Courtmacsherry Bay in the 1960s, but had disappeared by the early 1970s; fewer than 100 were present in the late 1990s, and 390 in January 2006. The only other regular wintering site is Clonakilty Bay, where up to 500 occurred in the early 1970s, but the highest recent count has been 190 in February 1994. Small numbers occur at Rosscarbery, Ballycotton (110 in 2005/6) and Youghal.

Singles were obtained at Old Head of Kinsale lighthouse in September 1902 and December 1909, and a few have occurred recently during migration times at CCBO and Dursey Island.

One was obtained inland near Doneraile after a storm in December in the late 1800s. One was at Charleville lagoons on 12 January 1997, and another at the Gearagh on 4 December 2005. In recent years, low numbers have occurred at the Gearagh in most autumns, with one in May.

A Knot ringed as a nestling at Stavanger (Norway) on 29 May 1957 was found dead at CCBO on 19 February 1963, and one ringed at Cork Harbour in January 2017 was captured in the Netherlands in September 2018.

Sanderling

Calidris alba | Luathrán | A

(monotypic)

World	Arctic regions of North America, also Greenland, Svalbard and Asia
Ireland	Winter visitor, passage migrant
Cork	Winter visitor, passage migrant
Conservation	Green-listed

Robert Ball considered the Sanderling rare at Youghal, with small flocks of three or four occurring on Claycastle beach in winter in the 1800s. It remained a winter visitor in limited numbers into the 1900s.

Recent winter atlas surveys showed the very restricted distribution of this species, which is largely confined to the beaches of east Cork, and to a few sites in west Cork. Ballymacoda and Ballycotton Bays hold most of the population, with between 100 and 200 at each site in winter. The first birds of autumn begin to appear in July, and autumn numbers peak in August and September (300 at Ballycotton in August) before levelling off for the winter. Numbers peak again in spring (April and May). Most Sanderling have departed for the breeding grounds by mid-May, and June records are extremely rare.

Away from east Cork, small numbers occur at sites such as Lissagriffin, Sherkin Island, Rosscarbery, Long Strand, Inchydoney, Broadstrand Bay and Garrettstown, often in autumn. It rarely occurs at CCBO or Cork Harbour, mainly because of a lack of suitable sandy habitat. One was obtained at Fastnet Rock lighthouse on 4 September 1913.

One was inland at the Gearagh on 19 May 2012 and two were there on 3 August 2018.

Semipalmated Sandpiper

Calidris pusilla | Gobadáinín mionbhosach | A

(monotypic)

World	North America
Ireland	Rare visitor, mainly in autumn
Cork	53 individuals
Conservation	Not globally threatened

Semipalmated Sandpipers occur between July and November, mainly in September (30), and two have occurred in May. Apart from one at the Gearagh, the only inland record, the remainder have occurred at coastal sites from Lissagriffin to Pilmore Strand. Most have been recorded at Ballycotton (30), and other significant sites are the Lissagriffin area (5), Clonakilty Bay (7) and Ballymacoda Bay (5). Most records relate to single individuals, but there are five cases of two, and two cases of three together. Of those that have been aged, thirty-three have been first-year birds and fourteen have been adult birds. Adults occur earlier than first-years in May, July and September. The earliest first-year bird has not been recorded until 21 August, and they occur through September, October and November. Typically, Semipalmated Sandpipers remain between one and five days, but nine birds have remained for more than ten days (mean stay = 7.6 days). One bird remained at Clonakilty Bay for seventy-one days from 28 September to 7 December 2017.

In the 1960s and 1970s identification criteria for Semipalmated Sandpiper were poorly known outside the United States, and vagrant birds presented a difficult identification task. There has been an increase in records since the 1980s, but with stable numbers in the last two decades. The increase is likely to be the result of increased skills and understanding of identification criteria among Irish observers, rather than a real increase in occurrences. Birds recorded in Ireland are likely to originate from the east Canadian breeding population, which migrates south over the west Atlantic in autumn making it vulnerable to displacement by eastward-moving depressions.

1959–68	1969–78	1979–88	1989–98	1999–08	2009–18
1	1	8	4	19	20

Earliest and latest occurrence in spring and autumn (1959–2018): 12 May–31 May and 4 July–6 November.

Western Sandpiper

Calidris mauri | Gobadáinín iartharach | A

(monotypic)

World	Chukotskiy Peninsula and Alaska
Ireland	5 individuals
Cork	1 individual
Conservation	Not globally threatened

First-year at Ballydehob on 1–8 September 1999.

Red-necked Stint

Calidris ruficollis | Gobadáinín píbrua | A
(monotypic)

World	Arctic regions of Siberia and Alaska
Ireland	5 individuals
Cork	2 individuals
Conservation	Not globally threatened

Adult at Ballycotton on 2–5 July 1998. Adult at Ballycotton on 31 July–1 August 2002.

Little Stint

Calidris minuta | Gobadáinín beag | A
(monotypic)

World	Arctic regions of Europe and Asia
Ireland	Uncommon autumn passage migrant
Cork	Scarce passage migrant
Conservation	Green-listed

There were two records of Little Stint before 1959, one in September 1932 at Crookhaven and one in September 1957 at Rostellan. The next record involved six at Clonakilty in October 1964, and it has occurred annually on autumn passage since 1966.

Autumn passage begins in mid-August (the earliest was at Ballycotton on 18 July 2002), and peaks in September and October with a few lingering into November. A few coastal sites produce most of the records, such as Ballycotton and Clonakilty. Numbers vary considerably between years; fewer than twenty birds are usually recorded and counts of over fifty are rare. Highest counts have been at Ballycotton with ninety on 22 September 1996 and eighty on 21 September 1993. Highest counts elsewhere have been of seventy at Clonakilty and Rosscarbery (combined) on 21 September 1996 and fifty at Clonakilty on 9 October 1993. Little Stints are regularly seen elsewhere at sites from Lissagriffin eastwards to Youghal, but usually in numbers of fewer than five birds. It has also occurred at CCBO.

The species is rare on spring passage, with records of one to three birds between March and June, most being concentrated in the 1960s, with very few in recent decades.

The first recorded wintering by Little Stints was in 1975/6 when one or two were present at Tivoli from 1 November to 25 February. One was at the same place from 17 December 1976 to 1 January 1977, and up to three were at Ballycotton in 1978/9. There have been a few subsequent winter records, the most significant being of five birds at Courtmacsherry Bay in 1993/4.

Inland records are rare, although the species has occurred at Charleville lagoons on several occasions in August, September, November and December (up to 4), Kilcolman in August and October (singles), the Gearagh in July, August, September and November (singles) and Mallow lagoons in August (1).

Temminck's Stint

Calidris temminckii | Gobadáinín Temminck | A

(monotypic)

World	Arctic regions of Europe and Asia
Ireland	44 individuals
Cork	3 individuals
Conservation	Not globally threatened

First-year at Ballycotton on 15–19 September 1981. Adult at Ballycotton on 3 August 1987. One at the Gearagh on 25 May 2016.

Long-toed Stint

Calidris subminuta | Gobadáinín ladharfhada | A

(monotypic)

World	Central and eastern Asia
Ireland	1 individual
Cork	1 individual
Conservation	Not globally threatened

One at Ballycotton on 15–16 June 1996.

Least Sandpiper

Calidris minutilla | Gobadáinín bídeach | A

(monotypic)

World	Northern regions of North America
Ireland	15 individuals
Cork	5 individuals
Conservation	Not globally threatened

One at Clonakilty on 13 September 1966. Adult at Inchydoney on 9–11 September 1967. Adult at Ballycotton on 9–15 August 1984. Adult at Ballycotton on 7–12 August 1988. Adult at Kinsale Marsh on 29–30 August 1993.

White-rumped Sandpiper

Calidris fuscicollis | Gobadán bánphrompach | A

(monotypic)

World	Arctic regions of North America
Ireland	Rare autumn visitor
Cork	77 individuals
Conservation	Not globally threatened

White-rumped Sandpipers occur in coastal districts between July and November, with most in September and October. They have been recorded at least once at almost every suitable habitat on the coast from Dursey Island to Youghal, although more have been recorded at Ballycotton (35) and Lissagriffin (10) than at other sites; Clonakilty Bay and Ballymacoda Bay have each hosted six birds. Unlikely sites have been CCBO (4) and Dursey Island (1), where little typical sandpiper habitat occurs. There are also two inland records, both at the Gearagh, in September 2012 and October 2015. Most records refer to single birds, although there have been five occurrences of two and one of three. White-rumped Sandpipers typically stay one to five days, but twelve birds have stayed for longer periods, the maximum being thirty and sixty-nine days (mean stay = 3.8 days). Of those that have been aged, twenty-seven have been first-year and seventeen have been adult. Like the Baird's Sandpiper, adult birds of this species occur earlier than first-year birds; only one adult has occurred after 10 September, while only two first-years have occurred in August.

1959–68	1969–78	1979–88	1989–98	1999–08	2009–18
6	14	8	15	14	20

Earliest and latest occurrence in spring and autumn (all records): 15 July–11 November.

Baird's Sandpiper

Calidris bairdii | Gobadán Baird | A

(monotypic)

World	Arctic regions of North America, and Chukotskiy Peninsula
Ireland	Rare autumn visitor
Cork	35 individuals
Conservation	Not globally threatened

Baird's Sandpipers occur in coastal districts, mostly in September (22), but there have also been records in August (7) and October (5), and one in May.

They have occurred at seven sites between Lissagriffin and Ballymacoda. However, Ballycotton has had twenty-three birds, the most for any site. Others have occurred at Sherkin Island, Rosscarbery, Owenahincha and Clonakilty Bay. Baird's Sandpipers typically stay for one to five days, but some stay for periods of up to thirteen days, and one for thirty-two days (mean stay = 5.1 days). Most birds occur singly, but there have been two records of two, both at Ballycotton. Of those that have been aged, fifteen were first-year birds and nine were adult birds. The Baird's Sandpiper occurs earlier than its relative, the White-rumped Sandpiper, and adult birds occur earlier than first-year birds. Only one of eight autumn adults occurred later than 12 September, and only two of fifteen autumn first-years have been seen in August, the earliest date being 25 August.

Baird's Sandpipers have occurred in Cork, and in Ireland, with slightly less than half the frequency of White-rumped Sandpipers. To account for this difference, it is tempting to suggest that vagrant White-rumped Sandpipers might be easier to detect because of their distinctive white rump and call. However, the difference in numbers is probably real and is likely to be related to migratory patterns. Baird's Sandpipers migrate south from the Arctic across the North American prairies, often staging at high-altitude lakes, while the White-rumped Sandpiper migrates south over the west Atlantic. This leaves the Baird's Sandpiper less likely to get carried west across the North Atlantic during the autumn hurricane season.

1959–68	1969–78	1979–88	1989–98	1999–08	2009–18
3	1	9	10	6	6

Earliest and latest occurrence in spring and autumn (all records): 31 May and 1 August–15 October.

Pectoral Sandpiper

Calidris melanotos | Gobadán uchtach | A

(monotypic)

World	Arctic regions of North America and Asia
Ireland	Scarce autumn visitor
Cork	246 individuals
Conservation	Not globally threatened

The Pectoral Sandpiper is the most frequent Nearctic wading bird occurring in Cork, and it has been annual since 1966. More have been recorded at Ballycotton than any other site, with Lissagriffin and Clonakilty Bay also hosting significant numbers, but there is hardly a coastal locality that has not had at least one record. It has also occurred at CCBO and Dursey Island, and inland at the Gearagh, Mallow lagoons, Kilcolman and Charleville lagoons. Six birds have occurred in spring, but most are seen between August and October. Most records have been of one or two birds, but three have been seen together on at least twelve occasions, while four, five and six have also occurred together, the latter at Ballycotton on 11 September 1971. Most birds occur on one or a few days, but small numbers have remained for several weeks, although it is sometimes difficult to determine long-stayers from possible new arrivals on occasions when several have been present at a site. Of nine aged as adult, only one occurred after August, while all

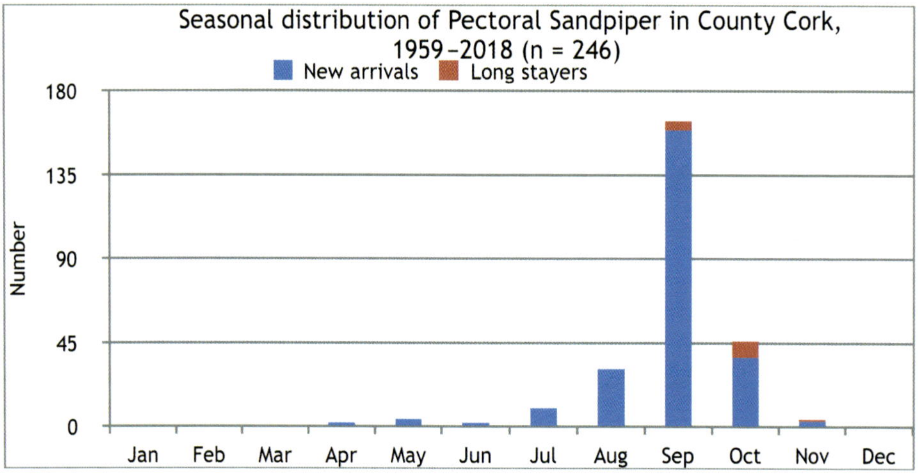

Fig. 7.16 Pectoral Sandpiper.

seventeen first-year birds occurred in August or later. The number occurring per decade has remained remarkably stable over fifty years.

1959–68	1969–78	1979–88	1989–98	1999–08	2009–18
19	47	40	53	47	40

Earliest and latest occurrence in spring and autumn (all records): 13 April–10 November.

Sharp-tailed Sandpiper

Calidris acuminata | Gobadán earr-rinneach | A

(monotypic)

World	Arctic regions of Asia from Lena Delta to Kolyma River
Ireland	9 individuals
Cork	3 individuals
Conservation	Not globally threatened

Adult at Ballycotton on 1 July 1971. Adult at Ballycotton on 27–31 July 2003. First-year at Ballycotton on 4–5 October 2007. The 1971 Ballycotton record was originally believed to be a Pectoral Sandpiper, but recent examination of photographs revealed its identity.

Curlew Sandpiper

Calidris ferruginea | Gobadán crotaigh | A

(monotypic)

World	Arctic regions of Asia from Yamal Peninsula to Chukotskiy Peninsula
Ireland	Uncommon autumn passage migrant
Cork	Scarce passage migrant, mainly in autumn
Conservation	SPEC 1; Red-listed

The Curlew Sandpiper was occasionally seen in large numbers at Cork Harbour, and William Crawford killed sixty (and 10 Dunlin) in one shot with his large strand-gun on 29 October 1847. One wonders how many were present at Cork Harbour, and elsewhere, that day. Crawford shot another on 23 October 1848, and one was shot inland at Kilcolman on 29 October 1884. These occurrences are the only ones known up to 1900.

The Curlew Sandpiper has occurred as a regular autumn passage migrant since 1958. Autumn passage begins in late July, peaks in September and October, and sometimes extends into November. Estuaries and beaches across the county are visited, and occasionally islands such as CCBO and Dursey. Numbers vary greatly between years, with sometimes fewer than twenty, while in others several hundred may occur at widely scattered localities. A minimum of 500 was recorded in 1969, with peaks of 212 at Tivoli on 11 September and 150 at Kinsale on 18 September. Between 300 and 350 occurred in 1988 with a peak of ninety-four at Clonakilty on 18 September. Exceptional weather conditions are likely to be responsible for the occasional westward displacement of large numbers from their normally more easterly migration route.[167;168]

Curlew Sandpipers seldom occur outside the July to November period, although there have been records in every month from December to June.

There have been several autumn inland occurrences at the Gearagh between 2002 and 2018, with a peak of twenty-nine in September. It has also occurred at Kilcolman (September 1972) and Mallow lagoons (September 2003 and February 2004).

Two birds ringed in Norway have been found in Cork, in September 1988 and September 2014.

Stilt Sandpiper

Calidris himantopus | Gobadán scodalach | A

(monotypic)

World	Arctic regions of North America
Ireland	17 individuals
Cork	5 individuals
Conservation	Not globally threatened

Adult at Ballycotton on 14–17 July 1979. Adult at Ballycotton on 6–7 August 1988. Adult at Kinsale Marsh on 14–18 August 1988. First-year at Rosscarbery on 19–20 September 1991, and Inchydoney on 20–1 September. Adult at Lough Beg on 6–11 August 2003.

Purple Sandpiper

Calidris maritima | Gobadán cosbhuí | A

(monotypic)

World	Arctic regions of North America, also Greenland, Iceland, Europe, and Asia
Ireland	Winter visitor
Cork	Scarce winter visitor and spring passage migrant
Conservation	Red-listed

The evidence suggests the Purple Sandpiper has been a scarce and uncertain winter visitor and passage migrant since the 1800s, although the nature of its rocky coast habitat probably means it has been under-recorded, even to the present day. Robert Ball obtained it at Youghal, and seven or eight were seen at Robert's Cove on 9 May 1850. Another flock was at the same place on 17 May 1853, and singles were obtained at Fastnet Rock on 18 May 1888, and on 21 October 1911.

Recent winter atlas surveys confirmed this scarcity with four and nine 10 km squares having records in 1981–4 and 2007–11, respectively. These were widely scattered from the Beara peninsula to Ballymacoda, and no record exceeded ten birds.

A regular but light spring passage occurs from late March or early April to mid-May. This is most obvious at CCBO, Ballycotton and the Beara peninsula. Up to fourteen in a day were recorded at CCBO in the 1960s, and up to eighteen in the 1980s and 1990s. Away from CCBO, sixteen at Ardgroom and fifteen at Owenahincha have been recorded in March. There are records of one or two at CCBO in early June.

Autumn passage, from mid-August to October or early November, is generally lighter. Counts of eighteen at CCBO (November), eighteen at Mizen Head (August) and ten at Dursey Island (September) are noteworthy.

Winter counts of twenty have been made at Owenahincha in December and January, and fourteen at Dursey Island in December, as well as smaller numbers at other locations.

Dunlin

Calidris alpina | Breacóg | A

(subspecies: *schinzii; alpina*)

World	Northern regions of North America, Greenland, Iceland, Europe, and Asia
Ireland	Scarce breeder, common winter visitor, passage migrant
Cork	Common winter visitor, passage migrant
Conservation	SPEC 3; Red-listed

Although considered a common resident in the 1800s, there is no evidence the Dunlin was ever a breeding species in Cork. It was then very common at Cork Harbour and countless flocks used the mudflats between Cork and Cobh where Robert Warren's grandfather once shot 120 with his double-barrel gun.

The Dunlin is a common coastal species, principally at estuaries, from autumn to spring. It is most plentiful at sites from Clonakilty Bay to Youghal Bay, but small populations occur everywhere, including at inland sites on the River Lee reservoirs, Charleville lagoons, Ballyhea gravel pit and Kilcolman.

Autumn passage begins in July and peaks in September. There is a further influx in November. Total counts of the county in January 1992, 1993, 1994 and 1995 gave figures of 16,472, 16,405, 19,820 and 20,513, respectively. Most Dunlin depart in March, although numbers are swollen again by migrants in May. These migrants tend to pass through very rapidly, and the scale of passage varies between years. About 100 Dunlin summered at Ballymacoda in most years in the 1970s and 1980s.

Wintering Dunlin numbers have declined dramatically at all sites since the 1970s and 1980s, especially at Cork Harbour, the most important site in the county. Counts of over 10,000 were regular in the 1970s, but numbers

Dunlin

rarely reached 5,000 during the 2010s. Declines of a similar scale have taken place also at smaller sites, such as Ballymacoda.

Biometric and ring recovery data show that three separate Dunlin subspecies migrate through Britain and Ireland.[169;170;171] The winter population consists largely of birds from northern Europe (*C.a. alpina*). Many of this subspecies moult on the Wash (England) and in the Netherlands, and there are several instances of birds ringed on the Wash in July and August and recovered in Cork later in winter. There have also been each-way recoveries of birds between Sweden, Finland and Cork which are also likely to be this subspecies.

The subspecies which breeds in south-east Greenland, Iceland, and from southern Norway southwards (*C.a. schinzii*) is known to occur on passage from late June to September. Most of this population moves south through France and Iberia to winter in north-west Africa. Dunlin presumed of this subspecies ringed in Cork in autumn have been recovered in Spain, Morocco and Mauritania in autumn and winter, and in spring in known *schinzii* breeding areas such as northern England, Germany and Poland. Some of this subspecies (*schinzii*) also winters here, and one was obtained in Cork on 11 January, while one ringed at Ottenby (Sweden) in July 2016 was captured at Cork Harbour in January 2018.

A third subspecies, the small pale and short-billed Dunlin which breeds in north-east Greenland and winters in West Africa (*C.a. arctica*), has not been identified with certainty in Cork. It would, however, be reasonable to expect this subspecies to accompany *schinzii* on passage, especially in May.

Dunlins (including *alpina*) have often been obtained at lighthouses such as Fastnet Rock and Old Head of Kinsale between August and November.

Broad-billed Sandpiper

Calidris falcinellus | Gobadán gobleathan | A

(subspecies: *falcinellus*)

World	North-east Europe and arctic regions of Asia
Ireland	24 individuals
Cork	5 individuals
Conservation	Not globally threatened

Broad-billed Sandpiper

One at Charleville lagoons on 2 June 1978. One at Ballycotton on 10–14 July 1979. Adult at Ballycotton on 18–20 June 1989. Adult at Ballycotton on 2 May 1996. One at Ballycotton on 2–3 May 2005.

Buff-breasted Sandpiper

Calidris subruficollis | Gobadán broinn-donnbhuí | A
(monotypic)

World	Arctic regions of North America, and north-east Siberia
Ireland	Scarce visitor
Cork	88 individuals
Conservation	Not globally threatened

Most Buff-breasted Sandpipers occur in autumn between August and October, with singles in April, June, July and November. They have occurred at several coastal sites, with two inland records at Kilcolman and Rathduff. Ballycotton (37) is the most important site, followed by Dursey Island (19), Lissagriffin (12), CCBO and Ballymacoda Bay (8 each). One has occurred at Sherkin Island, and one at Kinsale, the only record between Baltimore and Cork Harbour. Most records are of single birds, but there are thirteen occurrences of two or three birds and one each of five and eight. Fifty-eight

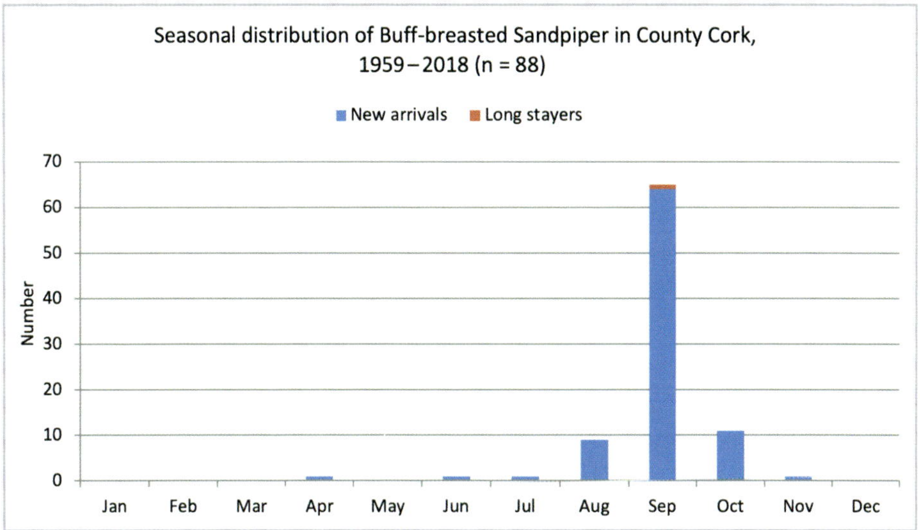

Fig. 7.17 Buff-breasted Sandpiper.

individuals have been seen on one or two days, while thirty others stayed for longer periods, but only four for more than ten days (maximum 18 days). Of thirty-six birds where the age has been published, all except two were first-years.

1959–68	1969–78	1979–88	1989–98	1999–08	2009–18
8	11	16	5	17	31

Earliest and latest occurrence in spring and autumn (all records): 29 April–22 June and (12 July) 3 August–18 October (1 November).

Ruff

Calidris pugnax | Rufachán | A

(monotypic)

World	Northern Europe and northern Asia
Ireland	Scarce winter visitor, uncommon autumn passage migrant
Cork	Scarce winter visitor and passage migrant
Conservation	SPEC 2; Amber-listed

Three Cork occurrences of Ruff (1889, 1896, 1899) were known to Patten.[172] No more were recorded until 1957 and 1958 when one was seen at Cork Harbour in October and seven at Ballycotton in September, respectively.

Since the early 1960s the Ruff has occurred throughout the year, but chiefly in winter and autumn. The pattern of occurrence at Ballycotton illustrates the situation in County Cork. The Ruff increased throughout the 1960s and peaked in the mid-1970s (overall peak count: 70 in September 1976). Numbers declined thereafter, and apart from a brief respite in the 1990s, have remained low. Ruffs have occurred at Ballycotton in all months, and the peak is in September, followed by a winter peak in January and a spring peak in March.[173]

Autumn passage in Cork extends from August to October and at this season Ruffs may be seen at wetlands throughout the county at locations such as Lissagriffin, Clonakilty, Courtmacsherry, Kinsale, Cork Harbour,

Ballycotton, Ballymacoda and Youghal. Inland, the Ruff occurs regularly at sites such as Charleville lagoons, Kilcolman, the Gearagh and Mallow lagoons. Occurrences at other sites follow a similar pattern to that recorded at Ballycotton, although peak counts are usually smaller, but have occasionally reached forty birds. Ruffs have also occurred a few times at CCBO and at Dursey Island.

A male ringed at Ballycotton on 2 October 1976 was captured in the Netherlands on 19 August 1980, and a male ringed in Norway on 21 August 2016 was seen at Timoleague on 13 September 2016.

Jack Snipe

Lymnocryptes minimus | Naoscach bhídeach | A

(monotypic)

World	North-east Europe and northern Asia
Ireland	Winter visitor
Cork	Scarce winter visitor
Conservation	Green-listed

The Jack Snipe was noted as a common winter visitor in the 1800s, but not nearly so much as the Common Snipe, and was not found in wisps, but singly. The winter atlas of 1981–4 showed two concentrations of Jack Snipe. One was in the Roaringwater Bay and Skibbereen area where four contiguous 10 km squares had records. The other involved sixteen contiguous squares in the valleys of the River Lee and River Bride (south), extending south to Kinsale, east to Ballycotton and Ballymacoda and north to Fermoy. The only other records came from the Dunmanway, Kildorrery and Mitchelstown areas. The 2008–11 winter atlas showed records from twenty-two squares in a wide scatter across all regions of the county, including Dursey Island, but with the main concentration of occupied squares centred on and around Cork Harbour.

It usually occurs singly or in small loose flocks of from two to four birds. The peak number appears to be eleven at CCBO on 11 March 1975. Up to six in a day have been seen at Ballycotton (e.g. January and February 1982), and there was a peak of four at Ballyleary Bog in February 1994.

Jack Snipe occasionally occur as early as August (the earliest being at CCBO on 10 August 1980), but peak autumn numbers occur from early October into November. Two have been obtained at Fastnet Rock lighthouse during the autumn migration period.

Departure is typically from late March to early April. The latest records are one at CCBO on 2 May 1994, but one or two were seen there in early June 1994, and one was reported at Castletownbere in July 2003. Occurrences at CCBO in the 1960s indicated a small passage migration in late March and April, and a more prolonged passage between late August and October.

Common Snipe

Gallinago gallinago | Naoscach | A

(subspecies: *gallinago; faeroeensis*)

World	Iceland, Europe, and Asia
Ireland	Common resident breeder, common winter visitor
Cork	Decreasing resident, common winter visitor
Conservation	SPEC 3; Red-listed

The Snipe (*G.g. gallinago*) was a common resident and bred on marshy ground in the 1800s, and it was said to nest in considerable numbers in the vicinity of Skibbereen. The 1968–72 breeding atlas survey revealed it was widespread throughout the county. However, it was curiously absent in a narrow band mainly along the River Blackwater from Millstreet to the Waterford border. Otherwise, there were few gaps of any significance in the breeding range. The 1988–91 survey showed considerable losses across the range of the species, but it remained widespread and there was no significant contraction of range. Density was very low throughout. Further losses occurred by 2008–11 but, surprisingly, there were also gains in a few areas. However, the population was thinly spread at low density across the entire county, but with significant vacant gaps.

Even allowing for differences in coverage between the three atlases, there has clearly been an alarming decline throughout the county, paralleling that for other wetland species such as Moorhen, Common Coot and Northern

Common Snipe

Lapwing, and probably with similar causes. Little is known of breeding density, but up to six pairs bred annually at Ballymacoda, seven pairs at Kilcolman, ten pairs at the Gearagh, four pairs at Sherkin Island, and twelve pairs at CCBO. Snipe ceased to breed at Ballymacoda and at Ballycotton in the 2000s, and the number at CCBO was down to five pairs by 1995.

Fieldwork for the winter atlas surveys of 1981–4 and 2007–11 showed the Snipe was widespread across the whole county. It occurred at the western peninsulas and in many upland areas, although at lower density than lowland habitats. There was a particularly high density of birds along the valley of the River Lee, and this extended eastwards to encompass Cork Harbour and many sites in the east of the county, such as Ballycotton and Ballymacoda.

The Common Snipe, like the Woodcock, presents a challenging shot to the shooting fraternity, and many large bags have been recorded. It was reported that Captain Hungerford shot a total of ninety-eight birds in one day and ninety-nine on another day around Clonakilty, and that twenty-one were killed at one shot by moonlight at a freshwater well near Midleton during hard frost by a person who knew they frequented the area. On another occasion 100 were shot by a single gun in one day. West Cork was said to be famous for Snipe, and instances of large bags include fifty shot by one man in three hours near Skibbereen, and 119 shot by two guns on a November day at Dunmanway, the latter done as a wager. Thirty couple to one gun in a day was not unusual about Bandon, but by 1881 ten to twelve couple in a day was considered good.

Wintering Snipe begin to arrive in August, but numbers are not appreciable until September. One was noted at Fastnet Rock as early as 8 September 1896. Peak numbers are probably not reached until November and December. The Snipe is difficult to count accurately because of its cryptic plumage and fondness for resting up in dense vegetation. Therefore, total counts of all but the smallest wetland sites are likely to be underestimates. Some particularly high counts by recent standards are outlined in Table 7.5. There is an unusual record of eighty at the suburban Cork Lough on 28 March 1978. Hundreds of Snipe flew west at CCBO in January 1963 during very severe weather, but only 2 per cent of all dead birds found there were Snipe.

Site	Date	Number
CCBO	January 1963	1,500+
Ballymacoda	11 February 1976	1,000
Ballymacoda	18 December 1977	410
Ballycotton	Winter 1993/94	Peak of 400
Ballycotton	Winter 1994/95	Peak of 400
Shanagarry	9 February 1982	300
Cork Harbour	30/31 December 1984	219
Ballycotton	31 December 1986	200
Garrettstown	26 January 1989	170
Ballymacoda	November 1989	160

Table 7.5 High counts of Common Snipe in County Cork, 1963–89.

Ring recoveries indicate that many Snipe from England, Scotland and from several north and east European countries occur in Cork in winter (*G.g. gallinago*). Large numbers also come from Iceland (*G.g. faeroeensis*) to winter here with many ring recoveries in support.

Great Snipe

Gallinago media | Naoscach mhór | A

(monotypic)

World	North-east Europe and north-west Asia
Ireland	21+ individuals
Cork	2 individuals
Conservation	Not globally threatened

One obtained at Clonakilty on 17 November 1879. One at Dursey Island on 22 October 1983.

Long-billed Dowitcher

Limnodromus scolopaceus | Guilbnín gobfhada | A

(monotypic)

World	Arctic regions of north-east Asia, and north-west North America
Ireland	Rare visitor
Cork	27 individuals
Conservation	Not globally threatened

The Long-billed Dowitcher had for many years been considered difficult to separate in the field from Short-billed Dowitcher. Consequently, many early records, and the occasional recent one where views are brief or of a bird in flight and where calls are not heard, were of specifically unidentified dowitchers. However, modern optical equipment, the widespread use of photography and an increased knowledge of plumages at different seasons has made the task of identification easier, but not straightforward, for present-day observers. A total of 151 dowitchers had been specifically identified in Ireland to the end of 2018; only three of these (2 per cent) were proven to be Short-billed Dowitchers, while 148 were Long-billed Dowitchers. Forty-one dowitchers occurred in Cork during 1959–2018, the first dating to 1966; twenty-seven of these have been identified as Long-billed Dowitchers with fourteen remaining unidentified.

Long-billed Dowitchers occur in Cork mainly from September to November (23), but there are records of new birds in December, March, May and August (1 each month). They have been recorded at several coastal sites from Lissagriffin to Ballymacoda, with three at the Gearagh, the only inland records. Ballycotton is the most frequented site (10 Long-billed and 8 unidentified), although the occurrence rate there has declined from twelve birds in the twenty-five years from 1966 to 1990 to six birds in the twenty-eight years from 1991 to 2018. Long-billed Dowitchers have occurred also at Cork Harbour (5) at three different sites: Lough Beg (3), Saleen and Harper's Island (1 each). Clonakilty Bay (3) and the Gearagh (3) are the only other sites to have hosted more than one bird. Single birds have occurred at Lissagriffin, Rosscarbery, Timoleague, Garrettstown, Kinsale Marsh and Ballymacoda. Apart from three occasions when two birds were present together (once involving unidentified dowitchers), all other observations have involved single individuals; all three records of two birds were at Ballycotton. Two adult and sixteen first-year birds (Long-billed Dowitcher) have been identified, while one of the unidentified dowitchers at Ballycotton in June 1989 was an adult. Long-billed Dowitchers have generally remained for one to five days (19), with five others remaining between ten and thirty-two days (mean stay = 6 days). However, three individuals (excluded from calculation of mean stay) remained from November 1998 to March 1999, October 1980 to September 1982 and October 1980 to November 1983, respectively – the latter two records involving two birds together for most of the period. Apart from relatively low numbers in the first decade of

1959–68	1969–78	1979–88	1989–98	1999–08	2009–18
1 (2)	0 (7)	5 (5)	6 (0)	10 (0)	5 (0)

Earliest and latest occurrence in spring and autumn (all records): 27 March–18 May and 26 August–18 December.

the present analysis, numbers have fluctuated each decade since then, but overall have remained stable when unidentified birds (14 in parentheses in the decadal table) are included.

Unidentified dowitcher species

Fourteen dowitchers, either Long-billed or Short-billed (*Limnodromus scolopaceus/ griseus*), have been recorded, all probably having been Long-billed Dowitchers. The seasonal pattern of occurrence of these birds matches those that have been identified, with twelve seen from September to November; the geographical spread is also alike, with Ballycotton having had eight (see above) and CCBO and Ballymacoda having had two each, while Lissagriffin and Inchydoney have had one each. Most unidentified dowitchers remained for short periods, but a few remained for longer, while four remained from December 1970 to April 1971, September 1971 to April 1972, October 1979 to April 1980 and November 1979 to April 1980, respectively – the latter two records involving two birds together for most of the period.[174]

Woodcock

Scolopax rusticola | Creabhar | A

(monotypic)

World	Europe and Asia, and Azores, Madeira, and Canary Islands
Ireland	Resident breeder, common winter visitor
Cork	Scarce breeder, common winter visitor
Conservation	Red-listed

The indications are that the Woodcock did not breed in the 1700s. It was known only as a winter visitor and was absent in summer so far south as Cork in the mid-1800s. The first mention of breeding in the county was by Payne-Gallwey who quoted Captain Morgan that it was nesting near Skibbereen.[175] It was a common resident towards the close of the 1800s, having increased of late years throughout Ireland as a breeding species. By the 1930s, the Woodcock was said to have extended its breeding range to every Irish county since 1883 and was nesting commonly in many places.[176]

The Woodcock bred at the Beara peninsula, including woods at Glengarriff, but elsewhere there were only a few records along the River Blackwater and on the border with Limerick during 1968–72, and while commenting on the rarity of breeding in Cork it was suggested there were too few data from the past to decide whether it had declined or was underestimated. There was little change in distribution some twenty years later and the few losses were matched by gains elsewhere. However, breeding was proved in only five 10 km squares in the whole county, three of these being around the Ballyhoura Mountains, and one each at the Beara peninsula and the Clonmult area. The breeding population was equally thinly spread in 2008–11 and extended in an oblique band from the Beara peninsula to the Waterford border. It must be stated that the three breeding atlas surveys probably underestimated the distribution of the Woodcock, as most fieldwork was undertaken by day. Surveying of breeding Woodcock requires fieldwork at dusk and after dark to observe the courtship flight behaviour, and this was apparently not done in the many suitable forestry plantations across the county.[177]

The Woodcock was patchily distributed in 1988–91 during winter and was largely absent near the coast. There were several clusters where good populations were apparently present, such as the Beara peninsula, woodlands in the valley of the River Bride (south) extending east to Cork Harbour, south to Dunmanway and north to Millstreet, and woodlands near the River Blackwater extending west to Newmarket, north to the Ballyhoura Mountains and south-east to the area around Ballincurrig, Ballynoe and Youghal. The winter distribution in 2007–11 showed considerably more occupied squares than in the previous survey, most notably a large block of newly occupied coastal and near coastal squares extending from Bantry to Cork Harbour.

Wintering Woodcock mainly arrive in November.[178] However, they are rarely seen as a migrant at CCBO, for example, where there were fewer than two annually between 1970 and 1988, and the largest numbers occur during hard weather (up to 10 in January 1963, with a peak of 11 on 3 February). Densities in winter may be quite high, with eleven at Minane Bridge on 10 December 1990 and eight at Garrettstown Wood on 17 January 1992, although these numbers are likely to be insignificant in comparison with records from the past.

The Woodcock has long been a favourite sporting bird among the shooting fraternity as a challenging shot and as a delicacy. Because of the large numbers that sometimes occurred in winter, shooters competed to produce the largest bag. Consequently, a literature on the largest number shot in a day developed. The Woodcock was reported as very scarce at Lord Bantry's estate (Bantry and Glengarriff) until 31 December 1849. It then became more plentiful in severe weather in January 1850, so much so that thirty-one were killed in a day by two guns. Cork was said to be very good Woodcock habitat, and a party of four guns killed 141 birds in four days in 1879 in the south of the county. Captain Morgan said he was at the shooting of at least 400 Woodcock in 1880/1, and that the number killed by poachers along the coast of the county in the severe weather of January 1881 was incredible. That month, immense numbers were at Old Head of Kinsale, and local shooters killed about 400 of them, although normally few birds occurred there, and they did not stay long. Payne-Gallwey also noted that 500 birds had been shot in a season at Castlemartyr, and on another occasion 107 were shot at Cool Mountain in three days. The Woodcock was thought to have declined in Cork in winter within living memory, but numbers fluctuated from year to year and even from month to month in the late 1800s.

Large numbers of Woodcock migrate into Ireland each autumn. George Jackson (Lord Bantry's gamekeeper) told Thompson about 1850 they always appeared on the coast near Dursey Island some days before they were found at inland areas. When on autumn migration, the Woodcock has occurred at all lighthouses, such as Bull Rock and Fastnet Rock, and on islands such as Dursey. The Woodcock is also prone to movements during spells of hard frost, and it is said to go to the coasts where it may be met with in sheltered glens.

It has been shown that 13 per cent of winter recoveries of British Woodcock are in Ireland, mostly in the south-west, including many in Cork, having been ringed as nestlings and adults between May and September in Scotland and northern England.[179] No evidence was found that the movements were the result of severe weather, and it was suggested that only a small proportion (about 4 per cent) of British- and Irish-bred birds were truly migratory, but some do go south to France, Spain and Portugal for the winter. Foreign-ringed birds from Fennoscandia and Russia occur commonly in Ireland in winter, but numbers from other European countries are much smaller.

Black-tailed Godwit

Limosa limosa | Guilbneach earrdhubh | A

(subspecies: *islandica; limosa*)

World	Iceland, Europe, and Asia
Ireland	Very rare breeder, winter visitor, passage migrant
Cork	Common winter visitor, passage migrant
Conservation	SPEC 1; Red-listed

The Black-tailed Godwit (*L.l. islandica*) was less numerous than the Bar-tailed Godwit in Cork in the early 1800s, although it was common (relatively) at Cork Harbour in winter, where William Crawford obtained six at one shot in November 1847. It was observed in the same period, though rarely, at Youghal during autumn.

Small flocks continued to occur at Cork Harbour, and at Youghal, in the opening decades of the 1900s, and some were present throughout the winter. There followed a considerable increase during the next few decades, and in the winters of 1938/9 and 1939/40 flocks of up to 120 were observed in Cork Harbour. The increase continued, and flocks of 130 strong were recorded at Cork Harbour where in August 1946 no fewer than 500 were seen in a single flock.

The number of Black-tailed Godwits wintering in Ireland further increased during the 1950s and 1960s following growth in the Icelandic breeding population due to climatic amelioration.[180] The normal winter population at Clonakilty and Rosscarbery in the 1950s was put at 275–300 and 23–7, respectively, while in September 1963 between 290 and 400 birds were recorded at Clonakilty. Summer-plumed birds (25) were also seen at Clonakilty in July 1954.

Cork Harbour was clearly the principal site in Cork with large flocks regularly recorded in the inner harbour areas of Tivoli and Lakelands (850, August 1960; 1,000, August 1963; 1,300, September 1963; 1,000, February 1966). Small numbers were also summering in the same areas of Cork Harbour during these years (up to 240 in July). Several of the principal Irish sites known in the mid-1970s were in Cork (Cork Harbour: almost 4,000 in winter 1974/5; Ballymacoda: over 1,000 in each of the previous three winters with a peak of 1,100; Clonakilty: 500–1,000 birds).[181]

Counts of the county population in January 1992, 1993, 1994 and 1995 gave totals of 2,030, 3,019, 2,672 and 1,985, respectively. Numbers build up gradually after the first arrivals in June and these are swollen by young birds from early August through September. Peak numbers are typically reached between mid-September and November. Cork Harbour holds the largest numbers, with peak winter counts of 2,200–4,200 during the 2010s. Elsewhere, Ballymacoda currently holds a minimum of 1,000 birds, with other important populations at Clonakilty, and smaller numbers at Rosscarbery, Courtmacsherry, Bandon estuary and Ballycotton. Small populations have been recorded at several inland sites: Bateman's Lake, the Gearagh, Mallow lagoons, Kilcolman, and Charleville lagoons.

An exceptionally large flock (peak 5,150) was noted at Youghal Harbour in January and February 2016. This flock was on the River Tourig on the border between Cork and Waterford, and the entire flock fed and roosted on both sides of the river at different times.

The ecology of Black-tailed Godwits was studied at Clonakilty during 1985/6–1988/9.[182] Juveniles arrived on the estuary from early August and peaked at 18 per cent of the population. Birds supplemented their diet by feeding in fields in winter, and they usually roosted in a monospecific flock.

Black-tailed Godwits begin to depart in late January, and in a normal year few remain by March. The spring of 1992 was exceptional, and up to 924 were present at Douglas estuary in late May. Small numbers usually summer at Cork Harbour and at other estuaries, such as Ballymacoda. One was engaged in aggressive defence and mobbed the observer at Harper's Island in June 1996, but breeding was not proved.

Black-tailed Godwits wintering in Cork breed in Iceland, and many ringed birds have been recorded moving between the two countries, as well as visiting other sites on the east coast of Ireland, England, Netherlands, France, Spain and Portugal as part of an international study.[183]

European Black-tailed Godwit *L.l. limosa*: There is one record of this subspecies. Male at Clonakilty on 6–9 April 2017.

Bar-tailed Godwit

Limosa lapponica | Guilbneach stríocearrach | A

(subspecies: *lapponica*)

World	Arctic regions of Europe, also Asia and Alaska
Ireland	Winter visitor, passage migrant
Cork	Winter visitor, passage migrant
Conservation	SPEC 1; Red-listed

The Bar-tailed Godwit was said to be common at Cork Harbour in the 1800s, where it was chiefly known as an autumn visitor, with some staying the winter months. It was much scarcer in bays and estuaries in south and west Cork, and during forty years of punt-gunning, Captain Morgan saw only small parties of five or six birds. It was reported as absent from Glandore and other south-west estuaries in the early 1900s, and four birds in summer plumage were at Rosscarbery in August 1946.

Autumn passage begins in late June, and peaks in August. There is typically a further increase in November and December, with numbers remaining stable thereafter until late February and March. They begin to depart in April, with numbers decreasing gradually through the spring. By June only a few remain on the larger estuaries, although up to 100 have summered at Ballymacoda.

Complete counts in the county in January 1992, 1993, 1994 and 1995 gave totals of 1,417, 1,468, 1,062 and 1,048, respectively. Cork Harbour and Ballymacoda are the main wintering areas. An exceptional count of 3,200 was made at Tivoli in October 1967. However, Cork Harbour totals have declined somewhat from mean peaks of 574 in the late 1970s, although peaks of over 300 often occurred in the 2010s. Numbers also declined at Ballycotton but remained stable at Ballymacoda.

Elsewhere in the county Courtmacsherry and Clonakilty usually hold about 200 birds each (although there appear to have been declines at both sites since the early 1970s), fifty at Rosscarbery, and up to fifty at Bandon estuary. The species has undoubtedly increased in west Cork since the early 1900s.

The Bar-tailed Godwit occurs at sandy coasts and beaches in addition to muddy estuaries, and it does not regularly occur inland, accounting for differences in the range of this species and the Black-tailed Godwit. There are inland records of one at Kilcolman on 1 September 1982, and at the Gearagh on 18 November 2018.

There are two ring recoveries: one ringed at Ballycotton on 6 April 1975 was in the Netherlands on 8 May 1985, and one ringed at Lough Beg on 17 September 1988 was in Guinea-Bissau on 14 March 1993.

Common Whimbrel

Numenius phaeopus | Crotach eanaigh | A

(subspecies: *phaeopus*)

World	Iceland, and northern regions of Europe and Asia
Ireland	Spring and autumn passage migrant, rare in winter
Cork	Passage migrant, rare in winter
Conservation	Green-listed

The Whimbrel was frequently obtained at a newly embanked area near Youghal in the 1800s. It also occurred at Cork Harbour, where it appeared in spring in considerable flocks, although only a few occurred in autumn. The accuracy of this observation (by William Crawford) has been borne out to the present day. Birds were also frequently noted at the Spit Bank in April and May in the late 1800s.

The Whimbrel occurs across the county when on spring migration, and flocks are channelled into estuaries such as Clonakilty, Courtmacsherry, Cork Harbour, Ballycotton and Ballymacoda, where transient flocks are encountered. At this season they have been reported from many lighthouses, including Fastnet Rock and Old Head of Kinsale, where the main movements take place there and at estuaries from the last week of April to the middle of May.

Studies at Cork Harbour have shown a diurnal north-west passage takes place mainly between 20 and 30 April. Passage is heaviest during early morning and late afternoon. Birds also migrate overhead at night with few stopping to feed. A peak of 3,304 was recorded in 1978 with several hundred birds passing daily, reaching 1,875 on one occasion. This passage involves birds returning from African wintering grounds to breeding grounds in Iceland.[184] Numbers vary between years, possibly because the population

Common Whimbrel

migrates on a narrow front with many birds missing Ireland in some years. Some are known to pass north to the west of Ireland during spring.[185]

Away from Cork Harbour, parties may be encountered in spring anywhere along the coast, and some may pause briefly before moving on again. Flocks are usually small, but may sometimes involve large numbers, such as 1,000 near Skibbereen on 26–7 April 1980, 640 at Garrettstown on 27 April 1987, 475 at Dursey Island on 28 April 2000, 474 at Ballymacoda on 23 April 1977, 350 at Ballycotton on 23 April 1998, and 200 inland at Charleville lagoons on 10 May 1991. A total of 339 was seen at CCBO on 30 April 1995. One or two birds remain at some sites during the summer months.

There is an autumn return migration, but numbers are much smaller. However, occasional large flocks do occur, and vast numbers were noted passing south for two hours at Fastnet Rock on 15 September 1888. At the same station many birds were also recorded on 27 July 1889, and on 20 August 1889 about fifty went south-west. Most autumn counts in recent decades involve single figures, but 130 were at Ballycotton on 27 August 1988. It is now known that the main autumn migration route is to the west of Ireland, most birds not making landfall at all.[186] Small numbers continue to occur throughout October and a few birds are still moving through in November. One was recorded at Fastnet Rock on 29 November 1887.

It has been known for a long time that small numbers spend the winter in Cork. Two were reported at Glengarriff on 12 December 1935. A few (1–3) have wintered regularly in various parts of Cork Harbour since at least the late 1940s, with perhaps ten or more in recent times. Winter atlas surveys show an increase, with birds present at coastal sites from the Beara peninsula to Youghal, apparently in response to milder winters.

There have been three ring recoveries: one from Lough Beg on 24 July 1985 was at Guinea-Bissau on 14 March 1986, one from Charleville on 4 May 1990 was at Gortdrum (Tipperary) on 3 May 1992, and one from Charleville on 28 April 1991 was at Loon-Plage (France) on 11 August 1992.

Hudsonian Whimbrel

Numenius hudsonicus | Crotach eanaigh Mheiriceánach | A

(subspecies: undetermined)

World	North America
Ireland	4 individuals
Cork	1 individual
Conservation	Not globally threatened

First-year at Mizen Head on 20–25 September 2011.

Common Curlew

Numenius arquata | Crotach | A

(subspecies: *arquata*)

World	Europe and Asia
Ireland	Resident breeder, common winter visitor, passage migrant
Cork	Former breeder, winter visitor and passage migrant
Conservation	SPEC 1; Red-listed

The Curlew was breeding widely in uplands in the west and north during 1968–72. There were some gaps in the range north of the River Blackwater, and few were breeding south of the Boggeragh Mountains. A major contraction of range was revealed in 1988–91, when breeding was confined to the border with Kerry on the Derrynasaggart and Mullaghareirk ranges, and to the Ballyhoura and Boggeragh ranges. There were no breeding records during 2008–11, but there was one confirmed pair in 2013 and one pair in 2019 in north-west Cork. The decline is believed to be due largely to intensification of hill farming, drainage and increased afforestation.

Passage migrants appear on the coast in July, with peak numbers typically occurring in August. Thereafter, numbers remain relatively stable until late winter when further large numbers may arrive, particularly in hard weather. Total counts for the county in January 1992, 1993, 1994 and 1995 gave figures of 4,627, 7,024, 7,186 and 7,832, respectively. Large flocks occur at the main estuaries during autumn and winter, but numbers have declined at most. Cork Harbour regularly held mean peak counts of over 2,000 birds during the late 1970s, 1980s and early 1990s, but no individual count exceeded 2,000 during the 2010s. Peak counts exceeding 1,000 birds are regular at Ballymacoda and Ballycotton, although numbers are now lower than in the past. Clonakilty and Courtmacsherry regularly hold 500–1,000 birds (with 2,640 at Clonakilty on 7 August 1991), and flocks of 500 or more are not uncommon at smaller sites. Many Curlews spend the day feeding inland on farmland and return to the coast at dusk to roost. This is evident at sites such as Cork Harbour, Ballycotton and Ballymacoda.

The Curlew is one of the most widespread wading bird species at inland sites in winter, although declines are evident, mirroring the breeding season. Flocks of up to 300 are regular at the Gearagh, Kilcolman and Charleville lagoons (500 at Kilcolman on 28 January 1981).

The Curlew regularly visits rocky coasts and islands during migration and has been recorded at Fastnet Rock and Bull Rock. It is a noisy bird at such times and can be heard at night when on the move, particularly in spring, over towns such as Skibbereen and Charleville.

Ring recoveries indicate that many wintering Curlews breed in the uplands of Wales, northern England and Scotland, and birds from many European countries have also been recovered in Cork, especially from Finland.[187] Several birds ringed in Cork and subsequently re-trapped have been notable for their advanced age, but the oldest was a nestling ringed in Germany on 8 June 1981 and found dead at Cobh on 7 January 2010.

Upland Sandpiper

Bartramia longicauda | Gobadán sléibhe | A

(monotypic)

World	North America
Ireland	12 individuals
Cork	2 individuals
Conservation	Not globally threatened

Adult obtained at Newcestown on 4 September 1894. First-year at Dursey Island on 18–24 September 1991.

Common Sandpiper

Actitis hypoleucos | Gobadán coiteann | A

(monotypic)

World	Europe and Asia
Ireland	Summer visiting breeder, scarce in winter
Cork	Scarce breeder and winter visitor
Conservation	SPEC 3; Amber-listed

The Common Sandpiper was breeding in west Cork, and by mountain streams and lakes at Inchigeelagh in the 1800s, but there was no evidence of breeding in east Cork. Breeding took place at the western peninsulas and on the River Lee from Gougane Barra to Dripsey in 1968–72. It was indicated there was a contraction of range in east Cork since the 1950s. However, there seems to be no documentary evidence that the species ever bred there. Losses were recorded throughout the range during 1988–91 when it was confined to the western peninsulas and the River Lee system from Macroom westwards. Further losses occurred on the River Lee during 2008–11, but an apparently new population was recorded in the Rosscarbery to Clonakilty area.

Spring arrival is generally from mid-April to mid-May, and the return movement peaks in July and August. Numbers during autumn migration at many sites, such as Cork Harbour, Ballycotton and Ballymacoda, rarely exceed about fifty birds.

A small wintering population has been known for a long time, one being obtained in 1884 with a further four records from Cork Harbour between 1912 and 1950. There were many winter records through the 1960s and 1970s. The winter atlases have shown an increase in range since the 1980s. Most wintering Common Sandpipers are present at Cork Harbour (about 10 birds) and the River Lee system as far west as Macroom, but there has been an extension in the west between Baltimore and Clonakilty Bay, and to Kilcolman in the north. The increase has been attributed to recent mild winters.

Common Sandpipers do not normally associate with other species, and wintering birds tend to feed alone. However, birds fed over several winters among Common Starlings at a site in Cork Harbour where animal food was dumped.

Spotted Sandpiper

Actitis macularius | Gobadán breac | A

(monotypic)

World	North America
Ireland	52 individuals
Cork	17 individuals
Conservation	Not globally threatened

The Spotted Sandpiper is an autumn visitor, mostly in September and October. It has occurred at several coastal sites from Lissagriffin in the west to Youghal in the east. There has been an even distribution in west Cork and east Cork (the latter including Cork Harbour). CCBO has hosted five birds, but no other site has had more than two. All occurred singly, and there have been one adult and fourteen first-year birds. The single adult was seen in July on Calf Island Middle. Two birds have wintered, at Clonakilty from 7 October 1982 to January 1983, and at Pilmore Strand from 22 November 2015 to 23 February 2016. About half of the other fifteen birds remained for one or two days, while singles have remained for fifteen and sixteen days, respectively (mean stay = 5.0 days). There has been little in the way of a trend in occurrence over the last four decades.

1959–68	1969–78	1979–88	1989–98	1999–08	2009–18
0	1	4	3	4	5

Earliest and latest occurrence in autumn (all records): (11 July) 31 August–27 November.

Green Sandpiper

Tringa ochropus | Gobadán glas | A
(monotypic)

World	Europe and Asia
Ireland	Scarce spring and autumn passage migrant and winter visitor
Cork	Scarce passage migrant and winter visitor
Conservation	Green-listed

The Green Sandpiper occurred by streams in summer, autumn and winter in the 1800s. Robert Ball obtained one near Youghal about October 1822, and it was frequently seen at the River Bandon where it sometimes remained for weeks or months. One was seen inland in west Cork in September 1877. Seven singles occurred between 1927 and 1947. Most of these were at Cork Harbour (4), with others at Glengarriff (2) and Mitchelstown (1), four were in August and September, and three in January, and one remained from August 1945 to April 1946.

The species was reported regularly from about 1950 onwards and it has been annual since 1954. The records indicate an autumn passage migration with some remaining through the winter, and there are records for all months. Green Sandpipers begin to appear from July, with a sparse but steady passage peaking in August and September. At this time individuals may occur in wet habitats anywhere along the coast and especially in the upper reaches of estuaries, in muddy drains, and in freshwater lagoons, small lakes, ponds and streams inland. Birds are increasingly seen in rivers during these months, such as the Rivers Blackwater and Bride (north), and up to twenty-four have occurred at Mallow lagoons, fourteen at Charleville lagoons, and over twenty in the rivers named. It has also occurred at CCBO in autumn, but rarely in spring. Fewer are recorded from late October, and winter records are mainly restricted to inland localities in the valleys of the Rivers Lee and Blackwater.

Evidence of a return passage to Scandinavia in spring is inconclusive, although there is a small spring peak in March. A few birds are present throughout the summer, especially on the River Blackwater.

Solitary Sandpiper

Tringa solitaria | Gobadán aonarach | A

(subspecies: undetermined)

World	North America
Ireland	7 individuals
Cork	3 individuals
Conservation	Not globally threatened

One at Lissagriffin on 5–7 September 1971. One at CCBO on 15–17 September 1974. First-year at CCBO on 27–30 August 2008.

Spotted Redshank

Tringa erythropus | Cosdeargán breac | A

(monotypic)

World	Arctic regions of Europe and Asia
Ireland	Uncommon winter visitor, and autumn passage migrant
Cork	Scarce passage migrant and winter visitor
Conservation	SPEC 3; Amber-listed

Spotted Redshank

William Crawford met with the Spotted Redshank at Cork Harbour in the mid-1800s but did not preserve a specimen. The first authenticated county record was one obtained near Fota Island in December 1898. There were four records of singles, all from Cork Harbour in January, September and December during 1941–6. It has occurred annually since 1964 as a passage migrant and summer visitor.

Autumn passage begins in June when up to eight summer-plumaged birds have been seen at Ballycotton (June 1967) and eighteen at Ballymacoda (June 1970). Peak numbers usually occur in August and September, and it is then widespread at estuarine habitats, especially between Rosscarbery and Ballymacoda, although counts rarely exceed ten (peak of 20 at Ballycotton, August 1977). It has also occurred at CCBO and is rare inland with records only at Charleville lagoons, Bateman's Lake and the Gearagh. Numbers decline at most sites from September onwards, but remain relatively stable at certain favoured locations. There is a slight increase again in February and March, but it is rare everywhere in May.

A decline in numbers in Ireland was noted from the late 1970s. This has led to a contraction of the winter range, which is likely to be caused by climate-driven shifts in populations in north-west Europe. Declines at individual sites such as Cork Harbour, Ballycotton and Ballymacoda have become very evident in the last two decades.

Greater Yellowlegs

Tringa melanoleuca | Ladhrán buí | A

(monotypic)

World	North America
Ireland	14 individuals
Cork	3 individuals
Conservation	Not globally threatened

One obtained at Aghadown on 21 January 1940. One at Ballycotton on 23–4 August 1968. One at Ballycotton on 29 April–6 May 1978.

Greenshank

Tringa nebularia | Laidhrín glas | A

(monotypic)

World	Europe and Asia
Ireland	Has bred, winter visitor, passage migrant
Cork	Winter visitor, passage migrant
Conservation	Green-listed

Small numbers (3–5) occurred at favoured haunts at Cork Harbour and Bantry Bay during the 1800s and early 1900s.

Winter atlas surveys showed a coastal distribution, but small numbers extend up rivers into freshwater habitats, such as the Bandon (to Dunmanway) and Lee (to Gearagh). Greenshanks have occurred at least ten times at Kilcolman, at least five times at Charleville lagoons, and at Mallow lagoons. Up to four birds have also occurred at the Gearagh between July and September.

Greenshanks occur in small numbers on estuaries and sometimes on rocky shores. The earliest to arrive appear in June, with a peak in September and October. Cork Harbour is the most important site, and numbers have increased significantly from a mean peak of sixty-nine during 1984/5–1986/7, to a peak of 150 during 2011/12–2019/20. Other sites (Rosscarbery, Clonakilty, Courtmacsherry, Bandon estuary, Ballycotton and Ballymacoda) hold smaller numbers, typically twenty to forty and rarely over fifty. Most sites hold fewer in winter than during autumn. There is a small spring migration at some sites. Greenshanks depart for their breeding grounds in March and early April, but a pair were displaying at Glandore as late as 24 April.

Ireland is an important wintering area for Scottish-breeding Greenshanks.[188] A nestling ringed in Sutherland in June 1926 was recovered at Inchydoney in October 1926.

Lesser Yellowlegs

Tringa flavipes | Mionladhrán buí | A
(monotypic)

World	North America
Ireland	Rare visitor, mainly in autumn
Cork	56 individuals
Conservation	Not globally threatened

Most Lesser Yellowlegs occur as visitors between August and November, but there have been spring and winter records as well. However, there is evidence of some autumn or early winter arrivals staying for several months and possibly accounting for at least some of the apparently new birds seen in spring. There are three records of single birds overwintering: September 2005 to February 2006 at Clonakilty, October 2006 to May 2007 at Rosscarbery, and October 2007 to April 2008 at Rosscarbery. Another possibly spent the winter at Cork Harbour between December 1969 and April 1970, but there was a gap of three months between sightings. One bird was present at Douglas estuary from May 1992 to April 1993. The

Lesser Yellowlegs has been recorded at coastal sites from Lissagriffin to Youghal – Lissagriffin (4), CCBO (5), Skibbereen (2), Rosscarbery (3), Kilkerran Lake (1), Clonakilty (11), Kinsale Marsh (6), Cork Harbour (6), Ballycotton (13) and Youghal (2) – while three have been recorded inland at the Gearagh.

Most records refer to single birds, but two have occurred on three occasions at Ballycotton, while three have been recorded at CCBO. Because of known records of overwintering, and the relatively long stays made by many other individuals, it has not been possible to calculate a mean length of stay for all birds, but birds that remained for thirty days or less had a mean stay of 5.7 days (n = 47). Where the age was published, seven were adult birds and twenty-three were first-year birds. Higher numbers have occurred in the last three decades (annual mean = 12.7) compared with the previous three decades (annual mean = 6).

1959–68	1969–78	1979–88	1989–98	1999–08	2009–18
2	9	7	16	10	12

Earliest and latest occurrence in spring and autumn (all records): 1 February–17 May and 2 August–12 November.

Marsh Sandpiper

Tringa stagnatilis | Gobadán corraigh | A

(monotypic)

World	Europe and Asia
Ireland	6 individuals
Cork	1 individual
Conservation	Not globally threatened

Adult at Rossmore and Great Island Channel on 20 August 1999.

Wood Sandpiper

Tringa glareola | Gobadán coille | A

(monotypic)

World	Europe and Asia
Ireland	Scarce spring and autumn passage migrant
Cork	Scarce passage migrant
Conservation	SPEC 3; Amber-listed

The first record of Wood Sandpiper was as recently as 1960. Despite assertions that at least one earlier record exists, no documentary evidence can be found. The species has been recorded almost annually since, mainly as a scarce autumn passage migrant. Numbers vary between years, generally with three to ten recorded annually, and a mean of five over the sixty years 1959–2018. It occurred most regularly during the late 1960s to the end of the 1970s, and it has become much scarcer since.

It occurs at a wide range of wetland habitats along the entire coast, as well as at inland wetlands such as Charleville lagoons, Kilcolman, the Gearagh and Mallow lagoons, and it has also occurred at CCBO and Sherkin Island. Most occur in autumn, with 60 per cent in August and 21 per cent in

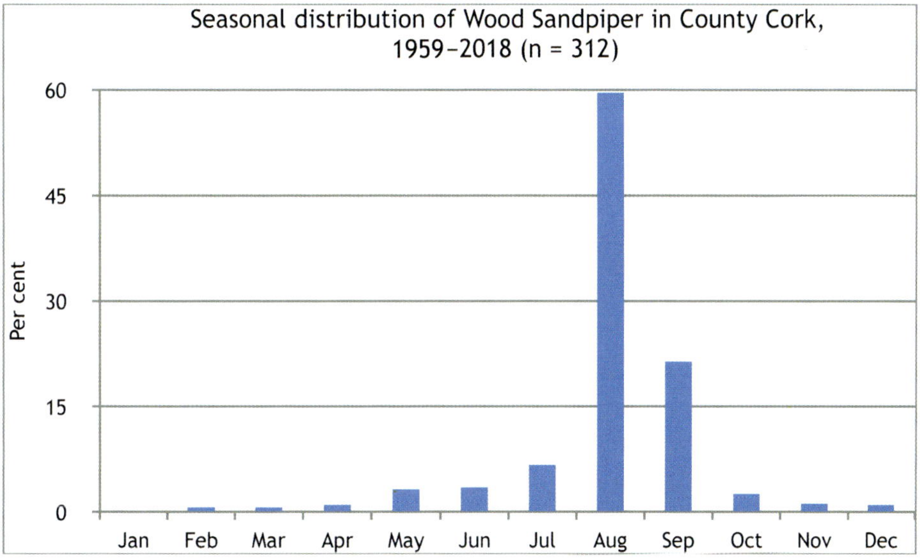

Fig. 7.18 Wood Sandpiper.

September. There are very few spring records, mostly in May and June. One or two have been seen in December, and individuals have been proved (or suspected) to winter on at least three occasions. Most records are of single birds, but in years when the species occurs in above-average numbers, such as 1968 and 1969, flocks of nine and ten have been seen at Ballycotton.

1959–68	1969–78	1979–88	1989–98	1999–08	2009–18
32	84	62	62	42	30

Common Redshank

Tringa totanus | Cosdeargán | A

(subspecies: *totanus; robusta*)

World	Iceland, Europe, Asia, and north Africa
Ireland	Resident breeder, common winter visitor, passage migrant
Cork	Winter visitor, passage migrant
Conservation	SPEC 2; Red-listed

There is no evidence that Redshanks (*T.t. totanus*) bred in west Cork in the 1800s, despite some suggestions that they did. There are two cases involving strong, but circumstantial, evidence of breeding, both in east Cork. The first involved a pair at Shanagarry Marsh in 1967, and the second a pair at Ballymacoda in 1970. No evidence of breeding emerged during atlas surveys, although each produced some birds in suitable habitat at Roaringwater and Bantry Bays.

It has been known since the 1800s that large numbers occur in winter at Cork Harbour. Winter atlas surveys have shown it as distributed along the entire coast, being scarcest at bays in the west and commonest at estuaries from Clonakilty east to Youghal. Redshanks extend into the upper reaches of estuaries, e.g. on the Rivers Lee and Bandon, and into freshwater habitats at sites such as the Gearagh. There are at least eight records for Kilcolman and many from Charleville lagoons. The Charleville records are for most months, with a maximum of seven in January.

Autumn migration begins in mid-June, and numbers build up rapidly in July to peak in August and September. Total counts for the county in January 1992, 1993, 1994 and 1995 gave figures of 2,656, 2,327, 3,417 and 2,551, respectively. Cork Harbour holds the largest population, and peak numbers frequently exceeded 2,000 in the 1980s and early 1990s. Numbers have declined in recent years, and during 2011/12–2019/20 peak counts varied from 1,354 to 1,906. Smaller numbers ranging from about 200–350 occur at Clonakilty, Courtmacsherry, Bandon estuary, Ballycotton and Ballymacoda, although peaks may be higher during migration.

Ringing has shown that some birds from Scotland and northern England move to the south coast of Ireland after breeding. One ringed at Cork Harbour in November 2012 was seen in Spain in December 2013. Large numbers from Iceland (*T.t. robusta*) winter in Cork, although it is unlikely that very many from northern Europe visit Ireland on their autumn migration south.

Turnstone

Arenaria interpres | Piardálaí trá | A

(subspecies: *interpres*)

World	Northern regions of North America, Greenland, Europe, and Asia
Ireland	Common winter visitor, passage migrant
Cork	Winter visitor and passage migrant
Conservation	Amber-listed

The Turnstone was known at Dunmanus Bay, Sovereign Islands, Cork Harbour and Youghal at various seasons during the 1800s. One struck the light at Old Head of Kinsale on 12 August 1910, and it has been seen regularly at the Bull and Calf Rocks in July and August.

Although Turnstones were present along the entire coast during winter atlas surveys, most were recorded at Cork Harbour and eastwards to Youghal. Turnstones are usually found on rocky and seaweed-covered parts of the coast. Although surveys of the county in January 1992, 1993, 1994 and 1995 gave totals of 304, 204, 451 and 402, respectively, these are likely to be underestimates because of the habitats occupied. Numbers at Cork

Harbour vary between winters from a high of 251 in October 1986 to fewer than 100 in several other winters. Autumn migrants begin to arrive at Ballymacoda in July to peak at 201 in October. Numbers at Ballycotton are similar, with a peak of 144 in September. Peak numbers also occur at CCBO in autumn, the highest being thirty-eight during 1959–69. There is an April peak at Ballymacoda and Ballycotton with highs of 201 and 275, respectively. A few birds usually summer at Ballycotton and Ballymacoda. Turnstones (up to 3) have occurred inland at the Gearagh in spring and autumn in recent years.

Turnstones breeding in Greenland and north-east Canada winter in Ireland, including Cork, but it is unclear if birds from the northern European breeding population, which winters in Africa, migrate through Ireland.[189]

Wilson's Phalarope

Phalaropus tricolor | Falaróp Wilson | A

(monotypic)

World	North America
Ireland	89 individuals
Cork	28 individuals
Conservation	Not globally threatened

Wilson's Phalaropes are exclusively autumn vagrants to the coast of Cork, with most occurring during the last ten days of August and in September. They have occurred at most of the important estuarine habitats from Lissagriffin to Ballycotton, with the latter site having had half of all occurrences (14) between 1967 and 1998, but inexplicably, none since then. Cork Harbour (6) and Clonakilty (3) are the only other sites having had more than one individual, while five other sites have had one each, including CCBO. This phalarope occurred much more frequently over two decades in the 1970s and 1980s, but it has reverted to its former extreme rarity during the last thirty years. All occurrences have been of single individuals, and the age of only four has been published, all being first-year birds. Most birds remain for more than one day, and nine have remained for periods of over five days, with a maximum of twenty-one days (mean stay = 5.0 days).

1959–68	1969–78	1979–88	1989–98	1999–08	2009–18
2	7	13	3	1	2

Earliest and latest occurrence in autumn (all records): 1 August–17 October.

Red-necked Phalarope

Phalaropus lobatus | Falaróp gobchaol | A

(monotypic)

World	Northern regions of North America, Greenland, Iceland, Europe, and Asia
Ireland	Rare and sporadic breeder
Cork	8 individuals (6 records)
Conservation	Red-listed

One at CCBO on 16–17 September 1960. Two at Ballycotton on 26 June 1966, joined by a third bird, 27–8 June 1966. First-year at Ballycotton on 7–14 October 1983. First-year at Ballycotton on 29 September–10 October 1992. Male at Rostellan on 6 June 1994. Adult at Galley Head on 11 June 2017. The 1960 CCBO bird occurred during a large movement of Grey Phalaropes at several south-west Cork sites.

Grey Phalarope

Phalaropus fulicarius | Falaróp gobmhór | A

(monotypic)

World	Arctic North America, also Greenland, Iceland, northern Europe and Asia
Ireland	Uncommon autumn passage migrant, scarce in winter
Cork	Uncommon passage migrant
Conservation	Green-listed

Grey Phalaropes appeared in the south of Ireland in winter 1841/2, and one was obtained at Youghal. Another was obtained in Cork in October 1846. Two were obtained in 1886, one at Mitchelstown on 20 October, and another at Castletownshend on 9 November. Only six Cork occurrences were known up to 1900, presumably inclusive of those mentioned. Two further records were added after 1900, one at Power Head on 23 October 1907 and one at Castletownbere in September 1943. One found dead near Whitegate on 6 March 1954 was a rare spring occurrence.

Since 1959, sea watching, mostly at CCBO, has shown the species to be a regular autumn migrant from late August to mid-October, but mainly in late September and early October, although one was at Ballycotton from 30 December 1998 to 3 January 1999. Since most phalaropes pass far out to sea, most remain specifically unidentified. Thus, of about 1,851 phalaropes recorded up to 1998 only about 365 were specifically identified as Grey Phalaropes. However, it is likely that most were this species (see Red-necked Phalarope).

In autumn, Grey Phalaropes appear to migrate south to the west of the Irish coastline, with large numbers occurring inshore only during westerly gales. Annual totals are extremely variable for this reason. In most years fewer than ten birds are recorded, while in sharp contrast, between 600 and 1,000 occurred in 1960, including 320 off CCBO on 22 September. The 1960 invasion (between 14 and 22 September) was without precedent and is believed to have included some Red-necked Phalaropes, although only one was identified. The totals off CCBO may have exceeded 500 on 17 and 20 September, and large flocks were reported off Crookhaven in the same period. Other good years were 1961 (252, including 150 off CCBO on 27 August), 1975 (276, including 148 off CCBO on 1 October) and 1989 (117).

There are records for all parts of the coast from Dursey Island to Ballymacoda Bay. Four have occurred inland, with singles at Charleville lagoons on 15 October 1978 and 5 October 1991, and at the Gearagh on 18 September 2011 and 30 December 2015.

Pomarine Skua

Stercorarius pomarinus | Meirleach pomairíneach | A

(monotypic)

World	Arctic regions of North America, also north-east Europe and Asia
Ireland	Uncommon spring and autumn passage migrant
Cork	Passage migrant
Conservation	Green-listed

Only three occurrences of Pomarine Skua were known in the 1800s, including one off Youghal on 12 October 1834 and one at Cork Harbour in 1899 (location and date not available for third). This remained the situation until the opening of CCBO, where it has been shown to occur regularly in spring and autumn.

Spring passage is concentrated in mid-May. Birds returning to their northern breeding grounds usually pass to the west of Ireland out of sight of land.[190;191;192] Spring totals at CCBO averaged fewer than thirty during 1963–73, although substantially more were recorded in 1978–80, accounted for mainly by an increase in sea watching. Under certain conditions, usually fresh south or south-west winds associated with warm fronts, large numbers are forced close inshore. These typically pass west along the coast as they regain their former track. The most notable movement occurred in spring 1983 when, with persistent south winds, 355 passed west off CCBO.

Autumn passage at CCBO is smaller than in spring and is again westerly. This movement is largely composed of immatures and extends from July to October. Highest autumn totals were forty-two in 1964, forty-five in 1995, seventy in 1991, and 364 in 1992. Apart from these, the highest day counts were twenty-nine at Old Head of Kinsale on 20 August 1989 and twenty-eight at CCBO on 14 August 1979. A few have been recorded in early November, including an unprecedented total of twenty-nine at CCBO on 1 November 1991. There are two winter records, Dursey Island on 8 December 1978 and Ballycotton on 5 January 2000.

Arctic Skua

Arctic Skua

Stercorarius parasiticus | Meirleach Artach | A

(monotypic)

World	Northern regions of North America, Greenland, Iceland, Europe, and Asia
Ireland	Spring and autumn passage migrant
Cork	Passage migrant
Conservation	Green-listed

The Arctic Skua was known in the 1700s for the way in which it chased small gulls and terns to eat their regurgitated food. There were several records during the 1800s. No more were reported until 1960. Sea watching, mainly at CCBO, has since shown that the species is regular on spring and autumn passage. Movements are predominantly westerly at both seasons.

Spring passage typically extends from April to June, peaking in early May. Passage is usually light, and records of more than eight in a day are unusual. The best year was 1983, when a total of sixty-four at CCBO included twenty-two on 5 May, and there was also a count of twenty on 30

April 1995. A total of forty-seven passed Toe Head between 23 April and 8 June 1986, with a maximum of sixteen on 9 May.

Autumn passage is more prolonged, extending from July and peaking in October. Few adults occur, most records referring to immatures. Numbers are occasionally large and in 1967 about 160 passed Old Head of Kinsale on nineteen dates between 9 July and 11 October, with a peak of fifty-two on 5 August, and a total of 361 was recorded in the autumn of 1991. Most mainland records are from Ballycotton, with the species becoming progressively scarcer at more westerly watch points. This includes at CCBO, although most records in any one year are from CCBO due to the relative constancy of sea-watching effort. A count of sixty-two on 2 November 2009 is the highest. By way of explanation, it was suggested that Arctic Skuas, having moved south through the Irish Sea or English Channel, pass west along the Irish coast in decreasing numbers as they veer southwards.[193]

There have been three December records, two at Clonakilty Bay (1978), one at CCBO (1984) and one at Dursey Island (2000). There is an inland record of one at Inishcarra reservoir on 28 August 1992.

Long-tailed Skua

Stercorarius longicaudus | Meirleach earrfhada | A

(subspecies: *longicaudus*)

World	Northern regions of North America, Greenland, Europe, and Asia
Ireland	Scarce passage migrant
Cork	78 individuals
Conservation	Not globally threatened

Most (44 per cent) Long-tailed Skuas have occurred at CCBO, but there are records from almost all main watch points on the coast from Dursey Island to Knockadoon Head, with a reasonably even spread of records from west to east. There is a small spring passage and a larger autumn one, mostly in August and September. Just over 71 per cent of aged birds (n = 56) have been adults. Two birds have been picked up in poor condition, one on the coast between Kilbrittain and Kinsale, and one at Cork Harbour being the victim of oil. Most records refer to one or two birds, but there is one occurrence each of three and four. As would be expected for a passage migrant seabird species, all birds seen from headlands and islands have been seen on a single day.

1959–68	1969–78	1979–88	1989–98	1999–08	2009–18
1	1	5	27	15	29

Earliest and latest occurrence in spring and autumn (all records): 5 May–21 June and 1 July–29 October.

Great Skua

Stercorarius skua | Meirleach mór | A

(monotypic)

World	Iceland, Faeroes, Scotland, east to Kola Peninsula
Ireland	Rare breeder, spring, and autumn passage migrant
Cork	Passage migrant
Conservation	Amber-listed

A Great Skua was obtained near Whiddy Island in winter 1845/6. There were no further records until 1949 when one was seen near Bull Rock in April and one at Fastnet Rock in September.

Although Great Skuas have been recorded annually since 1959, and in every month, they are mainly spring and autumn passage migrants. Spring passage extends from March to June (peaking in April), and autumn passage from late June to November (peaking in September). Most are seen off CCBO, but substantial passage has also been recorded from Dursey Island, Mizen Head, Galley Head and Old Head of Kinsale.

Spring passage is generally light, maximum numbers off CCBO being 137 in 1995, eighty-seven in 1992, eighty-four in 1983, and eighty-three in 1994, with individual counts of up to forty-four on 28 March 1978. Elsewhere, high counts include thirty-nine off Galley Head on 25 April 1977.

Autumn passage is much more substantial although annual totals vary widely. Counts of over 100 in a day have been recorded at CCBO on five occasions in September and October: 1977, 1980, 1982, 1984 and 1996

(124, 119, 118, 105, 103, respectively). Elsewhere, the highest counts are seventy-seven at Old Head of Kinsale on 14 August 2002, and thirty-two at Galley Head the same day. The best autumn on record was 1984, when 1,217 were recorded off CCBO. Numbers on passage on the Cork coast (including CCBO) increased in the 1960s and 1970s but have decreased since then. This may reflect the increase at the time in the Scottish breeding population.[194] Three birds were seen inland at the Gearagh on 7 October 2012.

There has been a recent change in the wintering status of the Great Skua on the Cork coast. Prior to 1982/3 there were four winter records: Ballycotton (December 1968 and February 1973), Bantry Bay (January 1974), and two at Dursey Island (December 1978). A total of fifty have occurred since then, many after southerly gales. These records include twenty in 1982/3 and sixteen in 1989/90. The presence of Icelandic and Norwegian Great Skuas in winter near the west and south coasts of Ireland has recently been established.[195]

A Great Skua ringed as a chick at Foula (Shetland) on 8 July 1977 was found dead at Whitegate on 7 September 1978.

Ivory Gull

Pagophila eburnea | Faoileán eabhartha | A

(monotypic)

World	Arctic Canada, Greenland, and Svalbard to New Siberia Islands
Ireland	19 individuals
Cork	6 individuals
Conservation	Not globally threatened

One obtained at Bantry Bay on 31 January 1852. Adult male found dead at Marina (Cork city) on 16 February 1913. Adult at Ballycotton on 16 October 1969. First-year at Ballycotton on 1–9 January 1980. Adult at Kinsale on 9–10 January 1999. First-year at Baltimore on 3–8 March 2009. The Marina bird was believed to have been killed, and partly eaten, by a Peregrine Falcon. The Baltimore bird had been seen previously in France during 21 January–26 February 2009.

Sabine's Gull.

Sabine's Gull

Xema sabini | Sléibhín Sabine | A

(monotypic)

World	Arctic North America, Greenland, Svalbard to Chukotskiy Peninsula
Ireland	Scarce autumn passage migrant
Cork	Autumn passage migrant
Conservation	Green-listed

Sabine's Gull has been an autumn passage migrant since first recorded in 1962. Numbers vary between years, and occurrences increased through the 1960s and 1970s, but have been stable since around 1990. Birds occur at most headlands and beaches from Dursey Island to Ballymacoda Bay, but most have been at CCBO (39 per cent), with one inland at Charleville lagoons on 14–16 September 1987 and one at the Gearagh on 7 September 2011. The species generally occurs during or after autumnal west or north-west gales, due to displacement from its Greenland to Biscay migration route.[196] Birds at CCBO quickly pass west to regain their previous course while those at Ballycotton and other bays often pause to feed, sometimes remaining for several days. One, presumably the same bird, has occurred at Cork Harbour in eight autumn periods during 2003–13, staying for 12–102 days, the longest being 26 August–5 December 2003. Adults (51 per cent) and juveniles (49 per cent) occur in equal numbers overall, but adults predominate in August and juveniles from September onwards. Most records involve one to three birds, but numbers into high single figures are sometimes recorded.

1959–68	1969–78	1979–88	1989–98	1999–08	2009–18
16	36	84	123	92	129

Earliest and latest occurrence in spring and autumn (all records): 7 April–14 November.

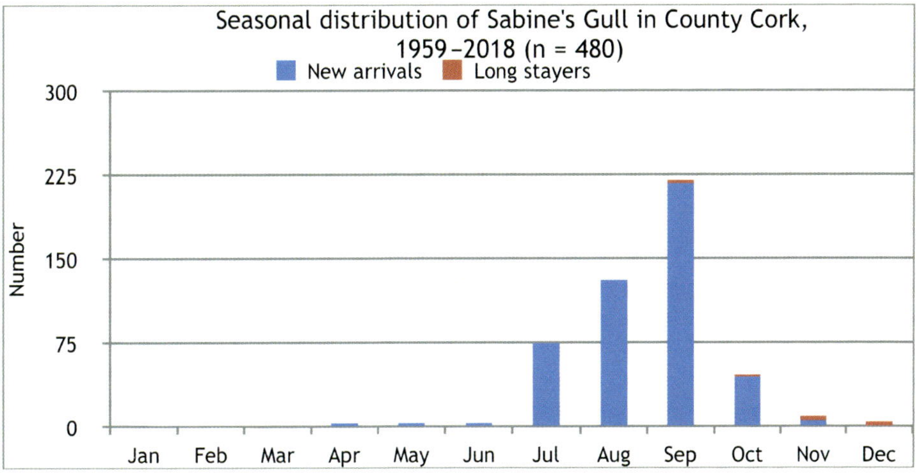

Fig. 7.19 Sabine's Gull.

Kittiwake

Rissa tridactyla | Saidhbhéar | A

(subspecies: *tridactyla*)

World	Northern regions of North America, Greenland, Iceland, Europe, and Asia
Ireland	Common breeder, scarce in winter
Cork	Local breeder, scarce in winter
Conservation	SPEC 3; Red-listed

Kittiwakes were breeding at Bull and Cow Rocks, as well as other unnamed sites in the mid-1800s, and at Old Head of Kinsale in 1907. Kittiwakes increased in Ireland in the 1900s, but the only pre-1969–70 counts are an estimated 300 pairs at Bull Rock in 1884, 200 pairs there in 1955, and 150 pairs at Old Head of Kinsale in 1964, and 550 pairs there in 1967.

Breeding censuses have been carried out regularly during 1969–97; the counting unit used is the Apparently Occupied Nest (AON) (Tables 7.6

Site	1969/70	1971	1985	1989	1993	1997
Bull Rock	642	NC	NC	NC	349	NC
Cow Rock	600	NC	NC	NC	7	NC
Mizen Head	0	NC	76	36	142	NC
Old Head of Kinsale	1,764	NC	2,059	2,056	1,857	1,819
Big Doon	1	15	510	176	76	18
Barry's Head	4	NC	30	0	0	0
Flat Head/Nohoval	14	NC	0	0	NC	0
Robert's Head	0	NC	89	0	0	148

Table 7.6 County Cork totals for breeding Kittiwake, 1969–97 (NC = no count).

Old Head of Kinsale sub-colony	1985	1989	1997
Eastern	48	180	367
Southern	117	143	138
Western (main)	1,894	1,733	1,314
Totals	**2,059**	**2,056**	**1,819**

Table 7.7 Old Head of Kinsale totals for breeding Kittiwake, 1985–97.

and 7.7). Total numbers at Old Head of Kinsale did not change significantly during 1969–97, but there was a 40 per cent decline to 711 AONs in 2015–18.[197]

Kittiwakes are present offshore throughout the year, but observations at CCBO have failed to establish clear movement patterns. Movements are generally heaviest from February to June and again from late October to December when up to 1,000 may pass per hour. The largest movement recorded is of 12,168 going west in one hour in November 1982. Significant offshore passage movements occur at many other headlands, such as Galley Head and Old Head of Kinsale.

The Kittiwake was formerly seldom seen near the coast in winter, and the presence of birds feeding on herring fry at Cobh in January 1948 was considered exceptional. In general, winter occurrences close to shore were rare in Ireland until the early 1960s. In the 1980s there were several large winter influxes to east Cork, each associated with shoals of sprats. In November 1982 an estimated 16,000 Kittiwakes were at Ballycotton Bay and 10,200 at Ballymacoda Bay. In December 1983 an estimated 25,000 were at Ballycotton Bay, and another 3,000–5,000 at Cork Harbour. Numbers were lower in January 1984 with counts of 6,000 at Ballycotton and 3,500 at Cork Harbour. The following winter (1984/5) much smaller numbers were present (600 at Ballycotton, 1,500 at Cork Harbour). Under similar circumstances in 1990, about 50,000 Kittiwakes (all adults) were present at Youghal Bay on 13 December, and there were 25,000 in Cork Harbour on 5 January 1993. Kittiwakes have rarely been seen at the east Cork bays in recent winters, in marked contrast to the 1980s.

Records more than a few kilometres inland are unusual, but one was on the River Lee at Blackrock on 5 April 1976, and one at Cork city on 15 January 1983. One was at Kilcolman on 10 February 1988 and one at the Gearagh on 11 February 2021. Kittiwakes have also been seen at Lough Allua and at Gougane Barra.

There have been ring recoveries of birds from Wexford, Anglesey, Northumberland, France and Iceland.

Bonaparte's Gull

Chroicocephalus philadelphia | Sléibhín Bonaparte | A

(monotypic)

World	North America
Ireland	86 individuals
Cork	17 individuals
Conservation	Not globally threatened

Bonaparte's Gulls have occurred chiefly from December to February, but there are records of new arrivals in March, April, May, July and August. Most have occurred at Cork Harbour (12), with singles at Bantry, Baltimore, Rosscarbery and Ballycotton. One bird has been recorded inland at Mallow lagoons. A bird in Cork city in February 2017 was later identified inland at the Gearagh and Coolcower on the River Lee system near Macroom. Most records refer to single individuals, but two have been seen together on three occasions. An adult bird first recorded at Cobh in winter 2005/6 returned annually until 2010/11, often staying for more than two months, and it was believed to be the same individual that visited Ballybranagan Strand in February 2009. The remaining birds stayed for shorter periods, some for only one to four days, occasionally up to twenty-eight days (mean stay = 7.3 days, excluding the long-staying Cobh bird mentioned above). All have been aged, eleven as adult and six as first-year birds.

1959–68	1969–78	1979–88	1989–98	1999–08	2009–18
0	0	0	3	11	3

Black-headed Gull

Chroicocephalus ridibundus | Sléibhín | A

(monotypic)

World	North-east North America, Greenland, Iceland, Europe, and Asia
Ireland	Common resident breeder, common winter visitor
Cork	Scarce breeder, common winter visitor
Conservation	Amber-listed

Site	Details
Toormore	Usually 2–4 pairs, but up to 30 pairs
Lough Nambrackderg	20–50 pairs in late 1960s; two pairs 1970–92
Lough Gal	25 pairs in 1960s and 1976; none thereafter
The Gearagh	20 pairs in 1965; 12 pairs in 1976; poor success
Kilcolman	30–50 pairs in 1965; 80 pairs in 1995; small colony in 2008; failed attempt in 2009; poor success
Laght, Glanworth	182 pairs in 1991; 52 pairs in 1992; poor success
Shanballymore	11 pairs in 1992; not occupied regularly

Table 7.8 Examples of breeding sites used by Black-headed Gull, 1960s–90s.

The Black-headed Gull was well known in the 1800s, but the only mention of breeding was at an unlikely site, the Cow Rock.[198] Breeding in Cork was not mentioned by Ussher in 1900. However, breeding occurred at several sites in the 1960s (Table 7.8), although it was (and still is) a scarce and local species. Colonies are small and few are regularly occupied for any length of time, although they may be re-occupied many years later. The reason for impermanence is unknown. Most colonies are on freshwater habitats, such as ponds, lakes and reservoirs, but some are in marine habitats. Breeding success inland is often poor, sometimes due to varying water levels, but mostly for unknown reasons.

There was circumstantial evidence in 1965 of a pair breeding at Douglas estuary, and one or two pairs have bred irregularly among Common Terns at Cork Harbour since 1988. In addition to the localities listed, a few pairs breed on islands such as Skiddy and Eyeries, and at others in Glengarriff Harbour. Breeding sites at Ballycotton and Inchydoney have been abandoned since the 1960s.

There have been few consistent counts of Black-headed Gulls. Observations at Ballymacoda indicate that birds are scarcest from March to May, with numbers building slowly to a peak of 8,000 in September. There is typically a decline in October, followed by a second peak (up to 13,200) in November. Thereafter, numbers remain high until February. Large concentrations occur at many other coastal sites, with estimated peaks of

Black-headed Gull

25,000 in Cork Harbour, 20,000 in Clonakilty Bay, and 5,000–10,000 at Courtmacsherry. The maximum count at Ballycotton was 5,400 on 18 February 1994. Black-headed Gulls are widely distributed inland in winter, but usually in small numbers. However, several thousand regularly occur at Kilcolman with a maximum of 5,500 on 1 February 1984. Up to 1,000 have occurred at Charleville lagoons in August, and 1,200 in December 1993. Wintering numbers have declined throughout during the last two decades, and there was a considerable contraction of the inland range during 2007–11, with a large swathe of countryside apparently vacant both north and south of the River Blackwater.

Ring recoveries show movements from Cork to Wexford, Britain, Denmark, Sweden, Finland and France, and movements to Cork from Clare, Mayo, Antrim, Down, Iceland, Britain, Low Countries, Fennoscandia, Germany, Czech Republic, Poland and other Baltic states. Many records result from field observations of ringed birds. One ringed as a nestling in the Netherlands on 27 May 1961 was caught by a dog at Garryvoe on 29 January 1984, when nearly twenty-three years old.[199] Arrivals of birds of all ages begin as early as July, first mostly from the Low Countries, later from further afield.[200]

Little Gull

Hydrocoloeus minutus | Sléibhín beag | A
(monotypic)

World	Great Lakes of North America, Europe, and Asia
Ireland	Winter visitor, passage migrant
Cork	Scarce passage migrant
Conservation	SPEC 3; Amber-listed

The first record of Little Gull in Cork was of two seen at Cobh on 23 February 1888 by William Samuel Mitchell D'Urban, author of *The Birds of Devon* (1892).

The species was not seen again until 13 May 1962 when one was at Clonakilty, and it has occurred annually since 1965. It has been recorded at many points along the coast from Dursey Island to Youghal, with concentrations around Old Head of Kinsale, Cork Harbour and Ballycotton. Birds have been recorded inland at Charleville lagoons (August 1984), Mallow lagoons (July and August 2003) and a few times at Lough Aderry. Some have been recorded in all months, but most occur during August to October. A February peak may reflect poor weather conditions occasionally forcing possible offshore wintering populations to seek refuge in sheltered bays.[201]

There was a considerable increase in numbers during the 1960s, 1970s and 1980s, but a decline set in, and numbers are now back at 1960s levels. Numbers vary greatly between years, peaks occurring in 1978 and 1981. During peak passage it can be difficult to determine exact numbers as there may be an unquantifiable turnover from day to day. Most records are of one to five birds, but flocks of up to twenty-five have been recorded at Ballycotton. For most of the year first-years outnumber adult and second-year birds by about two to one. During peak autumn passage this differential is greater, and first-year birds comprise 85 per cent of records.[202] Ballycotton typically holds the largest concentrations, and while most remain a day or two, some stay for several weeks. Birds seen in winter may also stay for weeks, and occasionally months.

Many records from Ballycotton and Old Head of Kinsale involve birds moving westwards. Some fly into Ballycotton Bay before quickly departing, and a flock of twenty-four flying west on 7 October 1978 was notable. Many occur in light south-east winds, and Hutchinson suggested there may be a more substantial passage at sea normally out of view from land. Most Little Gulls recorded on the Cork coast seem to depart southwards after passing Old Head of Kinsale as usually fewer are seen west of this point. Birds recorded on the south Irish coast in spring probably represent individuals returning to the Baltic for the summer.[203]

1959–68	1969–78	1979–88	1989–98	1999–08	2009–18
41	251	285	147	72	40

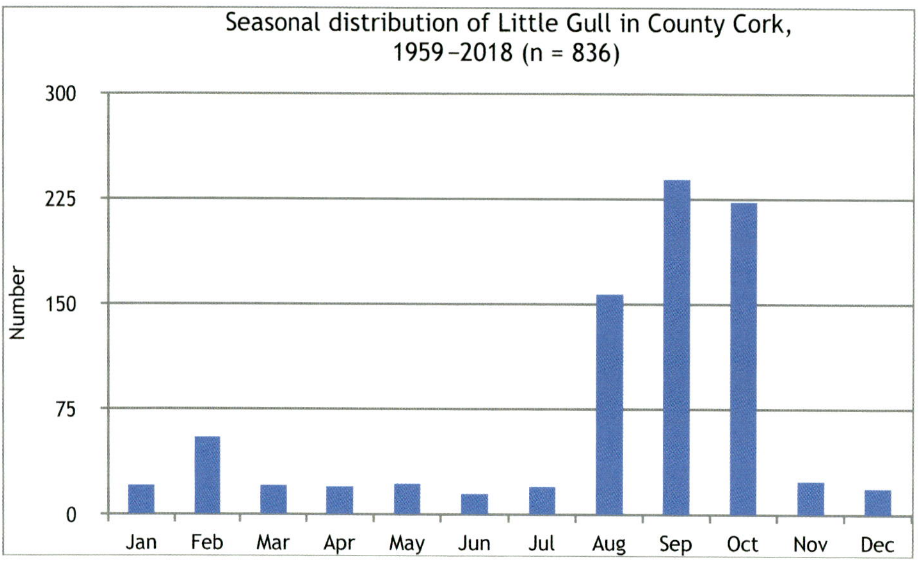

Fig. 7.20 Little Gull.

Ross's Gull

Rhodostethia rosea | Faoileán Ross | A

(monotypic)

World	Arctic Canada and Greenland, and from Taymyr Peninsula to Kolyma River
Ireland	22 individuals
Cork	3 individuals
Conservation	Not globally threatened

First-year at Cobh on 24 February 1985. Adult at CCBO on 24 February 1995. Adult at Kinsale on 9 February–9 March 2014.

Laughing Gull

Leucophaeus atricilla | Sléibhín an gháire | A

(monotypic)

World	North and Central America, and Caribbean islands
Ireland	46 individuals
Cork	17 individuals
Conservation	Not globally threatened

The first Laughing Gull occurred in 1968. There was a cluster of records in 2005 and 2006 associated with Hurricane Wilma, which developed in the Caribbean in early October 2005, and displaced many seabirds. The first record for Cork resulting from 'Wilma' was one at Ballycotton in November 2005, and five more were discovered up to early May 2006. An adult in July 2007 may have arrived earlier. The species has occurred in coastal habitats between Ballydehob and Ballycotton, but most (10) have occurred within Cork Harbour. Most new birds have been found in January (5), but new birds have been found in all months except March and December. Some birds have been recorded on one or two days, but most have remained for extended periods with one present at inner Cork Harbour for eighty days between January and April 1995, and one at Ballycotton Bay from 18 January 2014 to 30 April 2015, although it was apparently absent in April and May 2014. Some birds moved between sites, and one at Cobh and Cork city in June and July 2005 was identified in Galway in late July and August 2005. All birds occurred singly and of sixteen with published ages, thirteen were first-year or second-year birds, and three were adult.

1959–68	1969–78	1979–88	1989–98	1999–08	2009–18
1	0	2	2	11	1

Franklin's Gull

Leucophaeus pipixcan | Sléibhín Franklin | A
(monotypic)

World	North America
Ireland	19 individuals
Cork	3 individuals
Conservation	Not globally threatened

First-year at Kinsale on 11 January 1999. First-year at Blackrock on 22 December 2005, at Carrigaline on 29–30 January 2006, and at Rosscarbery on 12–26 March 2006. Adult at Belgooly on 31 May 2006. The pattern of occurrence during 2005/6 mirrors that of Laughing Gull, both having been displaced off the east coast of the United States of America by Hurricane Wilma.

Mediterranean Gull

Ichthyaetus melanocephalus | Sléibhín Meánmhuirí | A
(monotypic)

World	Western Europe eastwards to Black and Caspian Seas
Ireland	Very rare breeder, recent increase
Cork	Increasing in all months
Conservation	Amber-listed

The first record of Mediterranean Gull in Cork was in October 1964, and from then until 1979 five had occurred. From 1980 onwards it became an annual visitor in ever-increasing numbers, mirroring a national increase.[204] The Irish status change is part of an increase in numbers internationally which led to a westward extension of the breeding range across Europe. Almost 300 individuals had been recorded in Cork to the end of 1999, with reports of twenty or more in several years in the 1990s. Most occurrences were in autumn and early winter, although a few adults summered in Cork Harbour.

The species has continued to increase since 2000 and is present in all months but is scarcest in April to July. In the winters of 2007–11 it was present on the coast from Beara peninsula to Youghal, extending inland to rivers and lakes for 20–30 km. It was most plentiful around Cork Harbour, where a maximum of forty-eight had been counted. Although no coastal site is without a few during autumn, winter and spring, Cork Harbour continues to hold the main population, and peaks of 146 in October 2014 and 152 in October 2019 have been counted, mostly in the Whitegate area, with 160 at Inch Strand in September 2020. There is no evidence of breeding in Cork, although juveniles have occurred as early as July.

There have been sightings of birds ringed in Denmark, Netherlands, Belgium, France, Hungary, Germany (19 birds) and Poland (12 birds). An adult ringed at Cork Harbour in August 2015 was seen in Germany in May 2017, in Belgium in March 2018, in the Netherlands in June 2018, in London in July 2019 and at Cork city in July 2021.

Common Gull

Larus canus | Faoileán bán | A

(subspecies: *canus*)

World	North-west North America, Iceland, Europe, and Asia
Ireland	Resident breeder, common winter visitor
Cork	Has bred, common winter visitor
Conservation	Amber-listed

The Common Gull occurred in winter during the 1800s, but there was no evidence of breeding. The first known breeding was in 1969 with four pairs at Eyeries Island. Four pairs bred at the same site in 1985, and up to five pairs afterwards. Breeding at Roaringwater Bay has occurred since 1982 with one to four pairs at Calf Island Middle and Calf Island West, although young have never been successfully reared.

The Common Gull was rare in Cork before the early 1950s. Numbers considered abnormal visited the south-west during winter 1952/3, an area where few were previously seen, and over the following years it continued to increase. Examples of high counts include about fifty at Rosscarbery in August 1963, 300–400 at Ringabella in November 1964 and eighty-eight at Ballinacurra in September 1965. Terms such as 'extremely numerous' and 'very large numbers' were used to describe occurrences throughout the 1960s.

Common Gulls gather on the coast from July, although peak numbers are not reached until winter. From the 1970s onwards, Ballymacoda and Ballycotton typically held around 5,000 birds. The highest counts at these sites are 15,000 at Ballymacoda in February 1985, and 14,000 at Ballycotton in November 1990. There were 4,000 at Youghal in December 1982, 6,089 at Cork Harbour in December 1995, and 5,043 at Courtmacsherry and Broadstrand Bays in January 1996. Most of these birds depart by late March.

During winter in the early 1980s this species had a more restricted range than the Black-headed Gull, being absent from many inland areas occupied by the latter, and their overall density was much lower, although 365 were at Dunmanway on 14 November 1981. Common Gulls occurred in greatest numbers at the coastal areas described above. There were some losses in winters 2007–11, predominantly inland. There are distinct spring and autumn passage movements at CCBO with peaks in March and October, although numbers are small and rarely exceed twenty in a day.

Common Gulls frequently steal food (kleptoparasitism) from other species by aerial pursuit. This behaviour was studied at Cork city landfill (Kinsale Road).[205] The pursued species was the Black-headed Gull. This is a normal method of feeding for many Common Gulls, and species such as Northern Lapwing are also targeted.

Ring recoveries indicate that wintering Common Gulls originate in Scotland and Scandinavia (particularly Norway), as well as from within Ireland itself. The ringing of nestlings at lakes in north-west Ireland shows that many visit the south coast (including Cork) in autumn and winter.

Ring-billed Gull

Larus delawarensis | Faoileán bandghobach | A

(monotypic)

World	North America
Ireland	Scarce visitor
Cork	Scarce visitor
Conservation	Green-listed

The Ring-billed Gull has been recorded only since 1981 and has occurred annually since then. It is difficult to calculate the number of individuals as many birds have remained for four or five months over the winter period and a few through the summer, some have occurred at the same sites in successive years, and others have moved about between a suite of nearby sites, especially within Cork Harbour. Ring-billed Gulls have occurred along the entire coast from the Beara peninsula to Youghal, but most have been recorded in and around Cork Harbour, with smaller numbers at sites such as Rosscarbery, Clonakilty and Ballycotton. At least two have been seen inland at the Gearagh, while one or two occur in spring at Lough Aderry.

Numbers increased in every decade since the 1980s, but it appears to have become scarcer than previously in the last few years. Most birds are recorded between October and March, but apparent new birds have been seen in most months. Most records are of single birds, but gatherings of four or five are not unusual. All age categories have been recorded, but more than half have been adult.

1959–68	1969–78	1979–88	1989–98	1999–08	2009–18
0	0	48	92	214	269

Lesser Black-backed Gull

Larus fuscus | Droimneach beag | A

(subspecies: *graellsii; intermedius; fuscus*)

World	Iceland, Faeroes, Europe from Iberia to Taymyr Peninsula
Ireland	Summer visiting breeder, winter visitor, recent increase
Cork	Summer visitor, breeder, recent increase in winter
Conservation	Amber-listed

The first definite record (*L.f. graellsii*) of breeding was that of Ussher when he noted smaller and fewer colonies than European Herring Gulls and mentioned Cow Rock and High Island as breeding sites in the late 1800s. By the early 1950s this gull was breeding in small numbers at several islands, but no large colony was known. It was increasing in Ireland in the 1960s, but there were only scattered pairs on the Cork coast and thirty-six pairs at CCBO in 1963. One or two birds were seen at Barry's Head and Robert's Head in July 1963 without evidence of breeding, and a pair nesting at Old Head of Kinsale in 1964 bear out the suggestion of small numbers at that time.

Censuses of the breeding population have been carried out since the late 1960s. Table 7.9 contains figures for the whole county from national censuses. The counting unit used is the Apparently Occupied Nest (AON), and the figures given include subsequent revisions. The Irish population continues to increase, and although numbers at CCBO increased from 103 pairs in 1985–8 to 204 in 1998–2002, numbers decreased to twenty-six pairs in 2015–18. Nevertheless, the total Roaringwater Bay population was 1,288–1,513 pairs in 2016.[206;207]

Year(s)	Number of AONs
1969–70	132
1985–88	339
1998–02	227

Table 7.9 County Cork totals for breeding Lesser Black-backed Gull, 1969–2002.

Breeding in east Cork began, first at Ballycotton Island, during 1988–91. There was further expansion during 2008–11 when Capel Island also had breeding birds. There are no published records of this species breeding inland, or on buildings.

This gull was formerly rare in winter. One was at Cobh in December 1959, with negative reports from several other sites. There were scattered reports of single birds for the months of December and January during the 1960s, and seven were at Rosscarbery on 14 January 1967. Winter numbers increased during the 1970s and early 1980s, with peak counts of 146 at Ballymacoda in December 1981, and 840 at Ballycotton in January 1983.

Status changed dramatically from winter 1982/3 with 1,205 at Ballymacoda, and 1,000 at Ballycotton in December. Over 6,000 were at Cork Harbour in January 1993, 2,700 at Ballycotton in December 1987, and 2,500 at Cobh in December 1990. Away from east Cork there were 2,050 at Courtmacsherry in January 1994. Flocks of up to 200 now regularly occur inland in winter, the peak being 680 at Charleville lagoons in February 1994.[208] Most wintering birds are adult. Flocks have been much smaller in winter and autumn during the last decade, probably due to changes in management at landfill sites.

Autumn flocks can be larger than winter ones in east Cork, and many birds stream back to the coast to roost at dusk having spent the day feeding inland in fields. Flocks often occur inland during autumn, such as counts of 500 at Charleville lagoons in August 1986 and August 1994, and 1,150 in October 1994.

Ring recoveries show movements of Lesser Black-backed Gulls from Cork to the Farne Islands, France, Iberia and Morocco, and to Cork from Skomer Island, Cumbria, Faeroes and Iceland. A nestling ringed at CCBO in June 1991 was seen in Portugal in August 2019, over twenty-eight years later.

European subspecies *L.f. intermedius*: There are eight records (9 individuals) of this subspecies. One at Ballymacoda in October 1981. One at CCBO in October 1986. Two at Cobh in December 1987–January 1988. One at Adrigole in March 1992. One at Ballycotton in September 2000. One at Dursey Island in October 2002. One at Cobh in March 2003. A bird was at Calf Island East in July–August 1989 and May–July 1990; it was paired with a *L.f. graellsii* in 1990 when it was on a nest with one egg, but the outcome was not established.

Baltic subspecies *L.f. fuscus*: There are five records (14 individuals) of this subspecies. Nine at Mizen Head in March 1942. One at CCBO in September 1959. One at CCBO in April 1963. Two at Ballycotton in November 1976. One at Inchydoney in October 1993.

European Herring Gull

Larus argentatus | Faoileán scadán | A

(subspecies: *argenteus; argentatus*)

World	Iceland, Faeroes, Europe east to Kola Peninsula
Ireland	Common resident breeder, winter visitor
Cork	Resident breeder
Conservation	SPEC 2; Amber-listed

The Herring Gull (*L.a. argenteus*) was breeding at Bull and Cow Rocks, Bere Island and CCBO, and between Reannies Bay and Sovereign Islands in the 1800s. There was a considerable colony at Old Head of Kinsale in July 1907.

The Herring Gull underwent a significant population explosion throughout the range from the early 1950s onwards. This was the result of an increasingly affluent society where domestic rubbish was deposited in dumps, often in coastal districts, leading to Herring Gulls becoming a nuisance. Large numbers were culled in some countries for hygiene purposes and to protect other species, such as terns.[209]

Censuses of the breeding population have been carried out since the late 1960s. Table 7.10 contains figures for the whole county from national censuses. The counting unit used is the Apparently Occupied Nest (AON), and the figures given include subsequent revisions. The figures show a considerable reduction in numbers, but recent data indicates the decline has stabilised nationally.[210] However, although severe, the scale of the decline may be somewhat exaggerated due to poor coverage on some parts of the coast, for this and other seabirds. There are no published records of this species breeding inland, or on buildings in coastal towns and fishing ports.

Year(s)	Number of AONs
1969–70	3,511
1985–88	1,990
1998–02	300

Table 7.10 County Cork totals for breeding European Herring Gull, 1969–2002.

The decline of the Herring Gull is consistent with declines elsewhere in Ireland during the same period. Numbers were probably at peak levels in the 1960s and 1970s, but this trend was reversed in the 1980s. The botulism bacterium is believed to be an important factor in the decline. Large numbers of gulls were found dead and dying at colonies in the breeding season, and sometimes in winter. In addition, better maintenance of refuse tips, treatment of sewage and a decrease in fish stocks have led to reduced food availability for gulls.[211]

The scale of the decline can be gauged from the situation at CCBO where there were 662 pairs in 1963, 606 in 1967, 568 in 1983, 176 in 1985–8, 46 in 1998–2002 and 29 in 2015–18. A severe reduction in numbers also took place at Ballycotton Island where there were about 500 pairs in the 1970s and up to 1981, 30 in 1984, 150 in 1985, and 56 in 1995. The picture is not a simple one, and some colonies have suffered less, or even experienced increases. In 1985/6 there were 174 pairs at Sandy Cove Island, and 125 pairs at High and Low Islands combined.

Herring Gulls developed the habit of frequenting inland places at many locations in Ireland at all times of the year with the establishment of bacon factories, and in the Fermoy area numbers of adults were staying throughout the summer by 1943, although there is no exceptional concentration of this species at Fermoy today. Winter atlas surveys showed that most birds were on the coast with fewer inland, but only high ground was completely without them in the early 1980s. However, by 2007–11 there were few birds inland at distances greater than 20 km.

The largest concentrations on the coast in winter occur in towns, estuaries, rubbish tips, and anywhere food is available. Particularly high numbers were recorded in east Cork in the winter of 1982/3, with 12,000 at Ballycotton on 10 January. At Ballymacoda, around 100 is usual, and counts of 960 in December 1982 and 2,000 in February 1985 were exceptional. Since 1985, numbers have fluctuated, but have remained relatively low at this site.

Most nestlings ringed in Cork have been recovered within the county, although there have been movements to Waterford, Wexford, the Irish Sea, Wales and south-west England, and one to Portugal (ringed at Ballycotton in July 1976 and recovered at Oporto in October 1976). Herring Gulls have been recovered in Cork from Galway, Wexford, Dublin, Down, Wales, Isle of Man, northern England, mainland Scotland, Shetland and Iceland. The nominate subspecies (*argentatus*) is considered a winter visitor of uncertain status to Ireland.[212] It has not been proved in Cork from ring recoveries, although several sight records during 2000–5 have been published.[213;214]

Yellow-legged Gull

Larus michahellis | Faoileán cosbhuí | A

(subspecies: *michahellis; atlantis*)

World	North-west Africa and its islands, and Iberia to the Black Sea
Ireland	Scarce visitor
Cork	Minimum of 316 individuals
Conservation	Green-listed

Apart from singles at Rosscarbery in October 1955 and CCBO in September 1966, all records of Yellow-legged Gull (*L.m. michahellis*) have been since 1987. They have occurred in coastal districts from Dursey Island to Youghal, and several times inland on the River Lee reservoirs. However, most occurred at Rosscarbery, Clonakilty, Cork Harbour, Ballycotton and Youghal. Although there are records in every month, most have occurred in August to October and in January. Most have been recorded on a single day, but some have remained for up to two months. Most occurred singly, but up to five have been present at once. In a sample of eighty-two where the age was known, 57 per cent were in adult and 43 per cent in immature plumages.

1959–68	1969–78	1979–88	1989–98	1999–08	2009–18
1	0	2	8	84	220

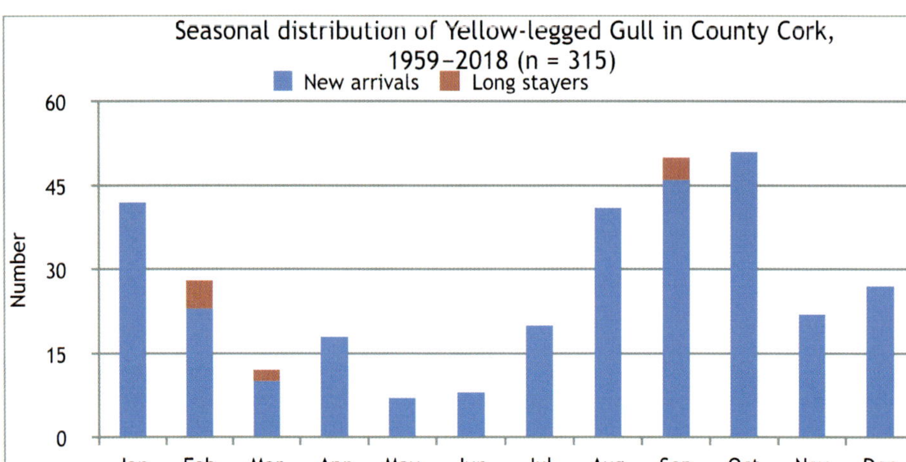

Fig. 7.21 Yellow-legged Gull.

Atlantic subspecies *L.m. atlantis*: There are six records of this subspecies. One at Castletownbere in January 2009. One at Rosscarbery and Owenahincha in August–September 2012. One at Clonakilty in August 2012. One at Clonakilty in September 2012. One at Rosscarbery in October 2013. One at Cork Harbour in September 2015. Records of this subspecies are considered provisional by the Irish Rare Birds Committee, pending further study.

Caspian Gull

Larus cachinnans | Faoileán Chaispeach | A

(monotypic)

World	Eastern Europe and Asia, including Black and Caspian Seas
Ireland	17 individuals
Cork	2 individuals
Conservation	Not globally threatened

First-year at Youghal on 24 February–6 March 2007. Adult at Baltimore on 5 February 2016. The Youghal bird was also seen at Kinsalebeg and Rincrew (Waterford) on 24 and 28 February 2007, respectively.

American Herring Gull

Larus smithsonianus | Faoileán scadán Mheiriceánach | A

(subspecies: *smithsonianus*)

World	North America, and from Taymyr Peninsula to Chukotskiy Peninsula
Ireland	Rare winter visitor
Cork	36 individuals
Conservation	Not globally threatened

The first Cork (and Irish) record of an American Herring Gull was in 1986, and it has been recorded in eighteen of the thirty-three years during 1986–2018. Most records occur from December to March. They have occurred exclusively at coastal sites from the Beara peninsula to Youghal, but most have occurred at Cork Harbour (16), at either Cork city landfill (Kinsale Road), Cork Lough or Cobh. Seven occurred at Ballycotton and five at the Beara peninsula (including 1 at Bantry), three at Baltimore, with two at Red Strand and singles at Clonakilty, Old Head of Kinsale and Youghal. Most records are of singles, but two have occurred on three occasions and three on one occasion. All thirty-six were aged as immature (first-winter, 29; second-winter, 6; third-winter, 1). About half remained one to five days (17), although others stayed for weeks or months, with two at Cobh for more than one year; excluding the latter two, the mean stay has been 12.6 days.

1959–68	1969–78	1979–88	1989–98	1999–08	2009–18
0	0	1	11	16	8

Earliest and latest occurrence in autumn and spring (all records): (31 August) 30 October–29 April.

Iceland Gull

Larus glaucoides | Faoileán Íoslannach | A

(subspecies: *glaucoides; kumlieni*)

World	North-east Canada and Greenland (but not Iceland)
Ireland	Scarce winter visitor
Cork	Scarce winter visitor
Conservation	Green-listed

At least three Iceland Gulls (*L.g. glaucoides*) were recorded at Cork Harbour in January and February 1849, and Ussher and Warren knew of ten occurrences up to 1900, five of which were obtained by Warren himself. One was obtained at Bull Rock in January 1902. Single adults were at Union Quay (Cork city) in December 1946 and February 1948.

It has been recorded increasingly since the 1960s and occurs annually. The ten-year total increased from twenty in 1960–9, forty-seven in 1970–9 to 118 in 1980–9. Numbers vary between years but peaks of over ten have only occurred since the 1980s (overall peak of 33 in 1984). Data are less complete for later decades, but numbers appear to remain around the higher end of the scale. It has been suggested that factors behind the increase may include severe weather on the North American east coast, and a decline in the Icelandic fishing industry.

Iceland Gulls have occurred in all months, but most are winter visitors between November and March. There was a series of records at Tivoli and Cork city in summer 1967. Birds have occurred at all sites on the coast from Dursey Island to Youghal. Favoured localities include Castletownbere, Cork Harbour, Ballycotton and Youghal. Most records are of one to three birds, but up to seven have occurred on more than one occasion. Most are in immature plumage, especially birds in their second year, followed by adult.

Kumlien's Gull *L.g. kumlieni*: There are forty-four records (48 individuals) of this subspecies. The Kumlien's Gull has occurred in all months between August and April, but mostly between December and March at coastal sites between the Beara peninsula and Youghal. Most have been seen at Ballycotton (12), with the Beara peninsula (9) next in importance. One bird returned in four successive winters to Cork Harbour from 1991/2 to 1994/5.

Thayer's Gull

Larus thayeri | Faoileán Thayer | A

(monotypic)

World	Arctic regions of Canada and Greenland
Ireland	9 individuals
Cork	1 individual
Conservation	Not globally threatened

First-year at Cork Lough on 21 February 1990, Cork city landfill (Kinsale Road) on 22 February and 3 March 1990, and Cobh on 26 February–5 March 1990.

Glaucous-winged Gull

Larus glaucescens | Faoileán liath-sciathán | A

(monotypic)

World	Coasts of North Pacific Ocean
Ireland	1 individual
Cork	1 individual
Conservation	Not globally threatened

Sub-adult at Castletownbere on 2 January–2 May 2016.

Glaucous Gull

Larus hyperboreus | Faoileán glas | A

(subspecies: *hyperboreus*)

World	Arctic regions of North America, also Greenland, Iceland, Europe and Asia
Ireland	Uncommon winter visitor
Cork	Scarce winter visitor
Conservation	Green-listed

Glaucous Gulls were recorded on five (or 6) occasions in the 1800s, and there were four records during 1901–50 (Bull Rock, 2 in January 1902, Roche's Point, 1 in March 1913, Rushbrooke, 1 in January 1947), and one at Great Island in August 1958.

It has been recorded increasingly since the 1960s and occurs annually. The ten-year total rose from 36 in 1960–9, to 121 in 1970–9, to 217 in 1980–9. Numbers vary between years with peaks of fourteen (1968), twenty-seven (1977) and forty-three (1983) during the three decades. Data are less complete for later decades, but numbers appear to have returned to about the 1970s level.

There have been records of Glaucous Gulls in all months, but most are winter visitors. Numbers build up from October to peak in January and February. Some remain to spring and a few throughout the summer, but records after April are unusual. One was semi-resident at Cobh from 1983 to 1986. Birds have occurred at all sites on the coast from Dursey Island to Youghal with one inland at Charleville lagoons on 13 March 1994, and one at Lough Aderry on 28 March 1994. Favoured localities are Castletownbere, Clonakilty, Cork Harbour, Ballycotton and Youghal. Most records are of one to three birds, but there were ten at Castletownbere on 11 February 1984; most occur in first-year plumage.

One ringed as a nestling in Iceland on 21 June 1982 was found dead at Ballycotton on 26 March 1983.

Great Black-backed Gull

Larus marinus | Droimneach mór | A

(monotypic)

World	North America, Greenland, Iceland, Europe from France to White Sea
Ireland	Resident breeder, winter visitor
Cork	Common breeder, winter visitor
Conservation	Green-listed

The Great Black-backed Gull was breeding in the 1800s, usually single pairs on top of isolated rocks along the south and west coasts, with about twenty-five pairs at Cow Rock and a colony also at Sovereign Islands. No change in this situation had taken place by the early 1950s, when small colonies and isolated pairs were reported.

Small breeding colonies occur along the Cork coast, almost always in company with European Herring and Lesser Black-backed Gulls. Censuses of the breeding population have been carried out since the late 1960s. Table 7.11 contains figures for the whole county from national censuses. The counting unit used is the Apparently Occupied Nest (AON), and the figures given include subsequent revisions. The figures show a reduction in numbers in Cork, but recent data indicates a slight increase nationally.[215]

Year(s)	Number of AONs
1969–70	381
1985–88	272
1998–02	201

Table 7.11 County Cork totals for breeding Great Black-backed Gull, 1969–2002.

The number of pairs at any colony may vary widely between years. Numbers increased at CCBO from 83 pairs in 1963 to 125 pairs in 1982/3, but then decreased to 45 pairs in 1990. In Roaringwater Bay (excluding CCBO) there were 121 pairs in 1982/3, 92 pairs in 1992, 39 pairs in 1994, and 109 pairs in 2016.[216] Capel Island held a colony of thirty to fifty pairs in the 1970s, but this decreased to about ten pairs in the early 1980s. Other important sites are Sovereign Islands (75 pairs in 1998–2002) and the west side of Mizen Head (27 pairs). There are no published records of breeding inland, or on buildings, although a pair has bred on mooring jetties at Ringaskiddy in recent years. The only data on breeding biology is from CCBO.[217]

Large numbers appear on the coast from late July to February with peak totals often occurring in November and December. The species is scarce inland and rarely goes more than 20 km from the coast. Coastal flocks are usually of fewer than 500, but there have been several records involving larger concentrations at sandy bays. In late October and early November 1971 about 3,300 were recorded between Ballycotton, Harbour View and Clonakilty. The Ballycotton birds departed on the same day as they arrived, and the Clonakilty flock had dispersed within a few days. A more localised concentration occurred in east Cork in the winter of 1982/3 when there were 1,500 at Ballycotton on 23 November and 2,810 at Ballymacoda on 3 December.

There have been several ring recoveries including one from Cork to Clare, and others to Cork from Wexford, Kerry, Galway, Down, Scotland and Norway.

Little Tern

Sternula albifrons | Geabhróg bheag | A

(subspecies: *albifrons*)

World	North America, Caribbean, Venezuela, Europe, Asia, Africa, and Australia
Ireland	Scarce summer visiting breeder
Cork	Has bred, scarce passage migrant
Conservation	SPEC 3; Amber-listed

The earliest reports of the Little Tern are of flocks of six or eight occasionally in Cork Harbour in the 1800s. The first breeding record for Cork (and Munster) was about 1934 when twelve to fifteen pairs were discovered on a small island near Glandore, but the colony ceased to exist before 1945 and has not been reoccupied.

Little Terns bred annually at Ballycotton during 1962–9. Numbers varied between three and ten pairs, apart from 1964 when there were thirty nests. Fifty birds were counted in 1967, but only six pairs bred. During the 1970s the only breeding records were of single pairs in 1976 and 1977, but there were eight birds in June 1977. None have bred since then. The abandonment of this site was probably related to human disturbance and tidal flooding, and the evidence suggests low breeding success.

One pair was present at Ring Strand (Ballymacoda) in July 1964 and June 1965, but breeding was not proved. Five pairs bred nearby at Pilmore Strand in 1967. Breeding success was unknown, and no birds have bred at either site since then. Birds were regularly recorded at Ring and Pilmore Strands as spring migrants in April and May (up to 14) and occasionally in June and July (up to 9) throughout the 1970s and 1980s, but there have been no further breeding attempts.

Sixteen pairs were breeding at Bantry Bay in 1978. Two pairs bred at Roaringwater Bay in 1984, and breeding was also recorded there in 1988–91. Breeding was considered probable there in 2008–11. However, there is no information on breeding continuity between these periods. No Little Terns were recorded breeding in the county during the 1995 survey, and none have bred since that time.

Little Terns are scarce during the spring and autumn migration periods, although individuals may appear at any time between mid-April and October. Most have been recorded in east Cork around the two former breeding sites at Ballycotton and Ballymacoda Bays, but the number and regularity of records in the last two decades has been lower than formerly. Since breeding ceased the spring and autumn peaks at Ballymacoda have been fourteen and nine, respectively, and numbers at Ballycotton have not exceeded these figures.

Further west along the coast they have occurred at Inch Strand, Lough Beg, Fennel's Bay, Old Head of Kinsale, Garrettstown, Courtmacsherry Bay, Clonakilty Bay, Galley Head, Union Hall, Sherkin Island, Mizen Head, Glengarriff and Dursey Island, but they are extremely rare visitors at most sites. They have occurred regularly at CCBO, but not annually, and there are fewer records in recent years than formerly. There is only one spring record from CCBO, three birds in May 1978, and while the autumn peak has been of twenty-two birds on 16 August 1963, most records involve one to three birds. Some birds have been noted as moving westwards off CCBO and at Galley Head.

The earliest spring arrivals involved three at Ballycotton on 11 April 1968 and one on the same date in 1970, and the latest was one at Ballycotton from 12 to 26 November 1977. There are two inland records: two birds at Lough Aderry on 25 April 2009, and one at the Gearagh on 21 June 2015.

Gull-billed Tern

Gelochelidon nilotica | Geabhróg ghobdhubh | A

(subspecies: *nilotica*)

World	North and South America, Europe, Asia, Africa, and Australia
Ireland	22 individuals
Cork	4 individuals
Conservation	Not globally threatened

One at Ballycotton on 29 April 1993. Adult at Courtmacsherry on 26–7 July 1998, and at Kilbrittain on 8 August 1998. Adult at Ballymacoda on 20–1 May 2006. Adult at Clonakilty on 26 April–6 May 2017.

Caspian Tern

Hydroprogne caspia | Geabhróg Chaispeach | A

(monotypic)

World	North America, Europe, Asia, Africa, Australia, and New Zealand
Ireland	11 individuals
Cork	5 individuals (4 records)
Conservation	Not globally threatened

Two adults at Ballycotton on 7–9 August 1988, one to 13 August 1988. Adult at Rosscarbery on 12 July 1991. Adult at Ballycotton on 19–20 June 1998, and at Ballymacoda on 20 June 1998. One at Ballymacoda on 14 August 2000.

Whiskered Tern

Chlidonias hybrida | Geabhróg bhroinndubh | A

(subspecies: *hybrida*)

World	Europe, Asia, Africa, and Australia
Ireland	22 individuals
Cork	7 individuals
Conservation	Not globally threatened

Adult at Ballycotton on 18–19 May 1968. Adult at Ballycotton on 4–15 August 1984. Adult at Kilmacsimon on 28 July–3 August 1985. Adult at Ballycotton on 3–29 May 1988. Adult at Lough Beg on 29–30 April 2007. One at Kilcolman on 5 April–3 May 2008. Adult at Lough Aderry, Ballybutler Lake and Ballyhonock Lake on 4–13 May 2014.

Black Tern

Chlidonias niger | Geabhróg dhubh | A

(subspecies: *niger*)

World	North America, Europe, and Asia
Ireland	Has bred, spring and autumn passage migrant
Cork	Mainly autumn passage migrant, has decreased recently
Conservation	SPEC 3; Amber-listed

Black Terns were said to have bred at a small lake at Roxborough, near Midleton, where Robert Ball observed birds in July for several successive years long before 1834 (about 1819). The breeding record was subsequently considered unsafe, although there is no doubt about the identification. This lake does not exist today.

Apart from these occurrences, the Black Tern had previously been noticed once or twice at Cork Harbour, and one was obtained near Cork city one week before Christmas 1849. Near the end of the 1800s, it was said to be a rare visitor, chiefly in autumn, but the records quoted are for March 1884, about May 1883 and July 1850. Seven occurrences were known for Cork before 1900, one of which was a bird seen in spring at the River Bandon among a flock of Sand Martins. Two were seen at Bandon on 24 April 1954, one at Kinsale on 6 October 1957 and twenty-one at Ballycotton on 8 September 1958.

During 1959–2018 Black Terns occurred as rare spring visitors, and as regular autumn passage migrants. Spring and early autumn birds are mostly adult, while those later in autumn are mostly juvenile. Numbers vary greatly between years and in some none are recorded. Over twenty birds have occurred in five years, 1968 (39), 1974 (30), 1976 (160), 1977 (31) and 1980 (79). The 1976 influx lasted from 23 September to 10 October, and the peak count was fifty-three in the Courtmacsherry Bay area. In 1980, the largest flock involved twenty-three at Inchydoney on 21 September. However, most records involve numbers in single figures. Numbers recorded on passage have declined in line with the species' poor conservation status in Europe.

Black Terns have been recorded on all parts of the coast from Bull Rock to Youghal, but most have been in the large sandy bays of Roaringwater, Clonakilty, Courtmacsherry, Ballycotton and Youghal. There are several inland records, at Castlenalact Lake, the Gearagh, Buttevant, Kilcolman and Charleville lagoons. All except one involved one or two birds, with nine at the Gearagh on 21 September 1980.

1959–68	1969–78	1979–88	1989–98	1999–08	2009–18
86	271	142	51	46	23

Earliest and latest occurrence in spring and autumn (1959–2018): 17 April–1 November.

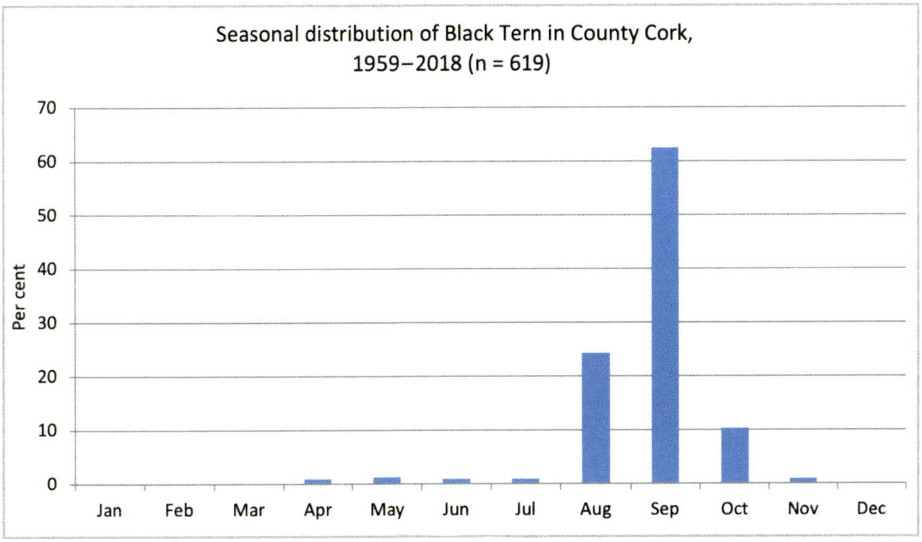

Fig. 7.22 Black Tern.

White-winged Tern

Chlidonias leucopterus | Geabhróg bháneiteach | A
(monotypic)

World	Europe and Asia
Ireland	Rare passage migrant, mainly in autumn
Cork	19 individuals
Conservation	Not globally threatened

There was one record of a White-winged Tern in Cork before 1959, one seen off Glandore in July 1936.

The records since 1959 have been in autumn between July and November (15), with three in May. More than half (10) have been in Ballycotton Bay, nine of these during 1968–83. The other records have been from Cork Harbour (3), CCBO (2), Garrettstown, Castletownbere and Charleville lagoons (1 each), the latter being the only inland record. All records, apart from one, refer to single birds, the exception being two at Ballycotton in September 1975. The age of eleven birds has been published, seven first-year birds and four adults. Most have remained for only one or two days (13), the remainder for up to sixteen days (mean stay = 3.9 days).

1959–68	1969–78	1979–88	1989–98	1999–08	2009–18
1	8	3	1	3	2

Earliest and latest occurrence in spring and autumn (1959–2018): 17 May–23 May and 1 July–16 November.

Elegant Tern

Thalasseus elegans | Geabhróg ghalánta | A

(monotypic)

World	Pacific coast of North America
Ireland	6 individuals
Cork	1 individual
Conservation	Not globally threatened

One at Ballymacoda on 1 August 1982. This bird was present at Carlingford Lough (Down) from 22 June–3 July 1982.

Sandwich Tern

Thalasseus sandvicensis | Geabhróg scothdhubh | A

(monotypic)

World	North and South America, Europe, Asia east to Caspian Sea
Ireland	Summer visiting breeder
Cork	Has bred, passage migrant
Conservation	Amber-listed

The first breeding of Sandwich Terns was in 1980 when eleven pairs were found at Eyeries Island. No further breeding was recorded until 1992 when

Sandwich Tern

ten pairs were at Carrigviglash Island, but they did not return in 1993. [Breeding at Spanish Island (near Baltimore) was incorrectly reported in 1981, the error resulted from confusion with another site named Spanish Island located 2 km inside the Kerry border in Kilmakilloge Harbour.]

The Sandwich Tern is one of the earliest spring migrants to arrive, with the first individuals typically appearing in late March. The earliest record is of one at Great Island on 8 March 2018. Spring passage is generally light and rarely reaches 100 birds (Roche's Point, April 2018), and small numbers occur along the entire coast. There is a period in May and June when few are about, but autumn passage begins in late June when non-breeding birds along with a few adults and early fledged young appear.

Sandwich Terns gather on sandy beaches along the coast from early July to late September. There is a heavy passage in late July and early August at Ballymacoda when flocks of up to 900 have been recorded, but numbers have been smaller in recent years. Considerable numbers were noted by Harvey at Monkstown for a week in October 1852. Cork Harbour (Lough Beg) today holds about 1,000 birds, the largest autumn concentration in Ireland.[218] The high numbers at Lough Beg represent a recent increase as the previous highest count there was 243 birds in August 1987.

Numbers at other coastal sites typically peak between late August and mid-September at 100–200, but there were 400–500 at Ballycotton on 21 August 1995, 471 at Rosscarbery on 10 September 1992 and 348 at Clonakilty on 6 August 1995. The number present at any site often varies significantly from day to day; at Rosscarbery daily counts were 14, 300 and 150 on three successive days in September 1981. Few remain by mid-October, although there are some November records, with the latest at Ballycotton on 9 November 1977. There are several winter records of single birds – at Ballycotton on 6 January 1974, at Bantry Bay on 26 December 2002, at Leap on 28 December 2002, at Roche's Point on 12 January 2003, at Castletownbere on 19 December 2013 – and three at Aghada on 27 January 1999. There is one record of a bird inland, at the Gearagh on 12 May 2017.

There are many ring recoveries of Sandwich Terns, especially at Cork Harbour, of birds from breeding colonies elsewhere in Ireland and Britain, particularly from Lady's Island Lake (Wexford).

Royal Tern

Thalasseus maxima | Geabhróg ríoga | A

(subspecies: undetermined)

World	North and South America, and Atlantic coast of Africa
Ireland	2 individuals
Cork	1 individual
Conservation	Not globally threatened

Adult at Clonakilty on 7 June 2009. This bird was also seen in Caernarfonshire (Wales) from 15 to 20 June 2009.

Lesser Crested Tern

Thalasseus bengalensis | Miongheabhróg chíorach | A

(subspecies: undetermined)

World	Libya, Red Sea, Persian Gulf, and southern Asia to Australia
Ireland	1 individual
Cork	1 individual
Conservation	Not globally threatened

Adult at Ballycotton on 7–8 August 1996. This bird was probably the same as one seen at the Scilly Isles from 2 to 4 August 1996.

Forster's Tern

Sterna forsteri | Geabhróg Forster | A

(monotypic)

World	North America
Ireland	39 individuals
Cork	2 individuals
Conservation	Not globally threatened

First-year at Ballycotton on 11–23 January 2006, at Pilmore Strand on 28–29 January 2006, at Ballycotton on 2 February 2006, at Pilmore and Ring Strands on 12 March 2006, and at Pilmore Strand on 28 January–9 February 2007. First-year at Garrettstown on 14 December 2013. The Ballycotton bird was also present in County Waterford from 17 to 19 March 2006.

Common Tern

Sterna hirundo | Geabhróg | A

(subspecies: *hirundo*)

World	North and South America, Europe, Asia, and north and West Africa
Ireland	Summer visiting breeder
Cork	Summer visitor, breeds east and west Cork
Conservation	Amber-listed

Common Terns bred on islets in Bantry Bay, and at the Sovereign Islands in the 1800s. A flock, including adults and immatures, was noted at a marsh inland of Old Head of Kinsale in July 1907. It was concluded (but not proved) that they may breed nearby. The location was probably Garrylucas Marsh, where breeding has never been documented. It is possible this was a feeding flock from the Sovereign Islands, assuming this site was occupied at the time. The Common Tern was scarcer than the Arctic Tern in the early to mid-1900s, but large numbers bred at Roaringwater and Bantry Bays, and the former colony at Sovereign Islands had disappeared before 1938.

Tern populations and distribution are better known in recent times than formerly. This species often nests in company with Arctic Terns, and it is sometimes impossible to determine the exact number of each species. Terns are also prone to move sites between years, either through disturbance, habitat change or flooding. Table 7.12 contains figures for the whole county from national censuses. The counting unit used is the Apparently Occupied Nest (AON), and the figures given include subsequent revisions. The figures show a reduction in numbers, the cause of which is unknown, but recent data are lacking for the whole county, although Common Terns have increased nationally.[219]

Year(s)	Number of AONs
1969–70	579
1984	149
1995	253

Table 7.12 County Cork totals for breeding Common Tern, 1969–95.

The distribution of breeding colonies ranges from Eyeries Island to islands in Bantry Bay, Dunmanus Bay, Roaringwater Bay and Glandore Harbour. The largest colony in 1969–70 was at Roancarrigmore with 400 pairs, and Eyeries Island with 100 pairs. None was at Roancarrigmore and only a few pairs at Eyeries Island in the mid-1980s, which proves how quickly numbers may change. Common Terns have bred at Roaringwater Bay, but not regularly, and Carrigmore Island has been occupied irregularly, along with Arctic Terns, but with long periods in which none have bred at any site within the bay. Apart from small numbers breeding at Glandore Harbour, none are found east of this site until Cork Harbour is reached.

The earliest record of breeding at Ballycotton dates to 1964 when four pairs bred. Numbers peaked at thirty pairs in 1967, then declined and breeding ceased after 1988 due to drainage and habitat change. Six pairs bred on saltmarsh at Ballymacoda in 1975, but all failed due to tidal flooding, and other breeding attempts also failed. Three to five pairs bred in Cork Harbour in 1969–70, and from 1983 onwards a colony has existed there.[220] Numbers have varied annually and peaked at 157 pairs in 2015. This colony is unique in that nest sites range from natural vegetation to man-made structures, and disturbance, predation and flooding are limiting factors. The colony is the subject of ongoing conservation measures concentrating on the provision of safe floating nest platforms, protection from mammalian and avian predation and reducing human disturbance.

Although Common Terns frequently nest inland in freshwater habitats, no records of inland breeding are known for Cork, and inland records of individuals are rare. At least one occurred at the River Lee reservoirs near Macroom during 1968–72. There were several records at the Gearagh during 2012–18 ranging from one to five birds between May and August, and three records of single birds at Kilcolman, in April 1997, September 1999 and October 2002.

Most Common Terns arrive in May, the earliest being two at Roche's Point on 1 April 1990. An even more extreme date, three at Ballycotton on 15 March 1969 refers to either this species or Arctic Tern. Common and Arctic Terns form feeding and roosting flocks after the breeding season at sites along the coast where breeding does not take place. In the 1800s, Common Terns occasionally appeared in considerable numbers, and flocks were noted in Cork Harbour in July and August. Autumn flocks of Common and Arctic Terns occur mainly on the east Cork coast at the present time. Numbers rarely exceed 100, but in some years 200–500 (mainly adults) are seen from late July to September.[221] Mixed flocks of 5,000 Common and Arctic Terns off Roche's Point in August 1970 and 1,800 at Ballymacoda in August 1977 were exceptional. The latest record concerns one at Roche's Point on 7 November 2004, and there is a record of one at Ballycotton on 26 November 1977 which could be either this species or Arctic Tern.

Nestlings ringed in Cork Harbour have been found in winter quarters on the west African coast in Mauritania, Senegal and Togo, and several have been recorded breeding at the natal site in subsequent years. There has also been interchange between Cork Harbour and Dublin, British and French colonies. One ringed as a nestling in Sweden and captured in Cork Harbour in August was probably on its migration south.

Roseate Tern

Sterna dougallii | Geabhróg rósach | A

(subspecies: *dougallii*)

World	North America and Caribbean, western Europe, Africa, Asia, and Australia
Ireland	Uncommon summer visiting breeder
Cork	Has bred, very scarce passage migrant, mostly in autumn
Conservation	SPEC 3; Amber-listed

The first occurrence of Roseate Tern in Cork was also the first record of breeding, two pairs at Roancarrigmore (Bantry Bay) in 1955.[222] About ten pairs were breeding at this site in 1969. One pair bred at Bantry Bay in 1978. These are the only records of breeding.

The Roseate Tern occurs on the coast as a rare passage migrant, mostly in autumn, with equal numbers in August and September. It has occurred at many sites from Dursey Island to Youghal, with a significant concentration at Ballycotton (57 per cent). High numbers have also occurred at CCBO (23 per cent), but most of these (26) flew west over eight days between 31 August and 11 September 1963. The Roseate Tern typically occurs in numbers of one to three, rarely more, and the largest count was seventeen birds at Ballycotton on 5 September 2016. Three adults and two juveniles were seen at Lough Beg in early July 2003, possibly originating from the

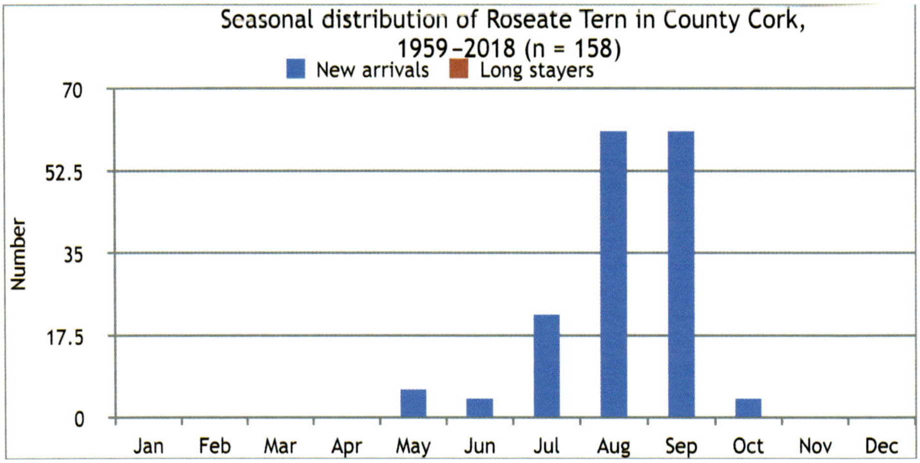

Fig. 7.23 Roseate Tern.

nearest known colony in south Wexford. Strong onward passage is indicated by the fact that only two birds were present for more than one day.

1959–68	1969–78	1979–88	1989–98	1999–08	2009–18
73	9	16	23	15	22

Earliest and latest occurrence in spring and autumn (1959–2018): 4 May–19 October.

Arctic Tern

Sterna paradisaea | Geabhróg Artach | A

(monotypic)

World	Northern regions of North America, Greenland, Iceland, Europe, and Asia
Ireland	Summer visiting breeder
Cork	Summer visitor, breeds in west Cork
Conservation	Amber-listed

There was a large influx of Arctic Terns to the south coast of Ireland beginning on 6 June 1842, and great numbers were reported in Cork. Several pairs were breeding at the Sovereign Islands in 1849. It was also said to breed (when not persecuted) on small islands at Bantry Bay and at other parts of west Cork. However, although once reported as a favoured haunt, it ceased to breed at the Sovereign Islands before 1938. A colony of about 200 pairs at Roancarrigmore about 1950 were mostly Arctic Terns.

From the late 1960s onwards, several national surveys of seabirds, including terns, took place (see Common Tern). Table 7.13 contains figures for the whole county from national censuses. The counting unit used is the Apparently Occupied Nest (AON), and the figures given include all subsequent

Year(s)	Number of AONs
1969–70	115
1984	200
1995	29

Table 7.13 County Cork totals for breeding Arctic Tern, 1969–95.

revisions. Data for the whole county is lacking in recent years, but the figures show a significant reduction in numbers, the cause of which is unknown, although Arctic Terns, mainly a west coast species, have increased in numbers nationally.[223]

Arctic Terns nowadays rarely nest away from west Cork, as was the case in the 1969–70, 1984 and 1995 surveys when they were recorded from Roaringwater Bay westwards to the Beara peninsula. There was one nest at Ballycotton in 1964, seven nests in 1977 and one nest in 1980, and one pair bred at Ballymacoda in 1975 but the nest was destroyed by tidal flooding. Breeding also took place at Ballycotton at least once during 1968–72, and three pairs bred at Glandore Harbour during 1985–7. Breeding has been recorded at about seventeen sites, although only about half of these are used in any one year, and there has been a slow decline. There were 200 pairs at Roancarrigmore in July 1955, more than the total recorded for the whole county in 1969–70. There were 155 pairs at five islands in Roaringwater Bay in 1982 with 127 of these on Calf Island Middle, although all apparently failed to produce young.

Most colonies recorded in the 1984 survey were small, but there were fifty-six pairs at Whiddy Island and forty pairs at Adrigole. Arctic Tern colonies are notoriously unstable, however, and in 1985 there were 120 pairs at Whiddy Island but none at Adrigole. About fifteen pairs bred at Roaringwater Bay in 1992, and twenty pairs at Carrigmore Island in 1993 (where they also bred during 1963–7). Only twenty-nine pairs were counted in the survey of 1995, although at least thirty-four pairs bred at Roaringwater Bay in 2005.

The first Arctic Terns generally arrive in mid-April, the earliest definite record being one at CCBO on 9 April 1991. At CCBO, where this is the commonest tern species, passage generally peaks in mid-May. Autumn passage peaks in early September, the number of birds recorded off CCBO suggesting that the species may often be overlooked along the mainland. At this locality, autumn concentrations of up to 500 Common or Arctic Terns have been found to consist of, on average, 77 per cent Arctic. A peak of 530 Arctic Terns was recorded there on 21 August 1990. Arctic Terns may be seen at many parts of the coast in autumn, but numbers are usually in single figures. The latest record has been one at Ballycotton on 5 November 2005. Single Arctic Terns have been recorded inland at the Gearagh on five occasions in 2016, 2017 (2) and 2018 (2) in the months of April, May, July and August.

Common Guillemot

Uria aalge | Foracha | A

(subspecies: *albionis; aalge*)

World	North Atlantic, North Pacific and Arctic Oceans, and Baltic Sea
Ireland	Common resident breeder
Cork	Common resident breeder
Conservation	SPEC 3; Amber-listed

The Guillemot (*U.a. albionis*) bred at Bull Rock, Cow Rock, Old Head of Kinsale and Reanies Bay in the 1800s. There was no attempt at quantifying range or numbers until the late 1960s. It is known that Guillemots breed (or have bred) at ten different colonies from Bull Rock to Robert's Cove. The Cow Rock and Old Head of Kinsale are the most important sites, but the lack of complete counts makes it difficult to discern a population trend at some sites. Counts at CCBO have been erratic, but the trend is one of decline.

Several national surveys have been carried out, but not all colonies have been visited each time. The results are shown in Table 7.14, the counting unit used is the individual bird, and the figures given include subsequent revisions. The figures show a consistent reduction. However, numbers nationally have increased recently, as has been the case at Old Head of Kinsale where there were 4,157 individuals during the 2015–18 period.[224] About 2 per cent of Guillemots at Bull and Cow Rocks are of the 'bridled' form.[225]

Year(s)	Number of birds
1969–70	6,075
1985–88	4,745
1998–02	3,763

Table 7.14 County Cork totals for breeding Common Guillemot, 1969–2002.

A light spring passage is usually recorded from mid-April to mid-May (maximum count of 1,446 west per hour at CCBO on 8 April 1991). Autumn passage is heaviest in late October and November, the peak rate at CCBO being 4,238 west per hour on 23 November 1991.

Guillemots disperse widely but thinly along the coast in winter. However, December passage at CCBO may reach up to 2,000 per hour. During the winter of 1983/4 there was an exceptional influx, and an estimated 8,000–10,000 auks (Guillemot and Razorbill) were present off Ballycotton.

Guillemots occasionally return to their breeding cliffs for brief periods in mid-winter; hundreds may be present at Old Head of Kinsale on such occasions. Winter atlas surveys show concentrations of birds around the tip of the Beara peninsula, at Roaringwater Bay and around Courtmacsherry Bay and Old Head of Kinsale, and between Cork Harbour and Youghal.

Birds from almost all parts of Scotland (*aalge*) – especially the island of Canna on the west coast but also from the east coast Isle of May – England and Wales (*albionis*), and from Great Saltee Island (Wexford), occur along the south coast of Ireland in winter.

Auks (Guillemot and Razorbill) have been beset by many problems, such as oil contamination and trapping in fishing nets. In 1977 and 1978 there were unconfirmed reports of large auk catches in drift nets set for salmon off the Cork coast.[226] Shooting of auks at the Old Head of Kinsale colony also took place in the 1970s.

The iron ore carrier *Kowloon Bridge* went aground on the Stags Rock off Toe Head in November 1986. Fuel oil leaked from the wreck and spread in an easterly direction. A total of 536 oiled birds (all species) were killed by oil between Cork Harbour and Youghal Harbour, and a further 570 birds were alive but with oil on their plumage. Over 1,500 Guillemots may have been killed and this species accounted for 85 per cent of all birds killed in east Cork.[227]

Razorbill

Alca torda | Crosán | A

(subspecies: *islandica; torda*)

World	North Atlantic Ocean, and Baltic and Barents Seas
Ireland	Common resident breeder
Cork	Common resident breeder
Conservation	SPEC 1; Red-listed

The Razorbill (*A.t. islandica*) was breeding at Bull Rock, Cow Rock, Stags of Castlehaven (i.e. Stags Rock), High Island, Old Head of Kinsale and Reanies Bay in the 1800s. There was no attempt at quantifying range or numbers until the late 1960s. Razorbills are known to breed (or have bred) at nine different colonies from Bull Rock to Robert's Cove. The Bull Rock and Cow Rock are the most important sites, but the lack of complete counts makes it difficult to discern a population trend at some sites. Counts at CCBO show a continuing decline.

Several national censuses have been carried out, but not all colonies have been surveyed each time. The results are shown in Table 7.15, the counting unit used is the individual bird, and the figures given include subsequent revisions. The figures show a significant reduction. However, numbers nationally have been increasing during the 2015–18 period.[228]

Year(s)	Number of birds
1969–70	2,938
1985–88	717
1998–02	149

Table 7.15 County Cork totals for breeding Razorbill, 1969–2002.

Peak passage of Razorbills occurs in April and is dominantly westerly. The species is, given its local breeding status, more numerous than Common Guillemot at this time. Numbers at CCBO usually peak at under 1,000 per hour, but exceptionally reach 10,000 per hour in calm conditions following rough weather. These may be Razorbills that have returned to Irish waters after wintering further south and are beginning to visit their colonies while also undertaking long feeding trips. A lull in passage at CCBO from July to early August coincides with the moult period of adults and is followed by a light autumn passage (e.g. 1,500 on 30 October 1994).

Winter atlas surveys showed a wider distribution for the Razorbill than the Common Guillemot, although fewer Razorbills were present at any given place. There were small concentrations around the tip of the Beara peninsula, Roaringwater Bay, Old Head of Kinsale, and Ballycotton and Youghal Bays. Occasionally, Razorbills outnumber Common Guillemots, as in east Cork and Bantry Bay.[229;230] There were 300 in Cuskinny Bay, Cork Harbour, on 26 February 1994.

Winter ring recoveries in Cork involve birds from Wexford and Kerry, from Wales and the Isle of Man, and from Scotland as far north as Shetland. A nestling ringed at Bull Rock in July was recovered in Spain the following December.

Northern Razorbill *A.t. torda*: Nine of this subspecies were recorded based on measurements among a sample of 228 Razorbills that had drowned in fishing nets on the east Cork coast in January and February 1983.[231]

Black Guillemot

Cepphus grylle | Foracha dhubh | A

(subspecies: *arcticus*)

World	North Atlantic and Arctic Oceans, and Baltic Sea
Ireland	Resident breeder
Cork	Resident breeder
Conservation	SPEC 2; Amber-listed

The coast between Robert's Cove and Sovereign Islands held several breeding pairs, including three or four at Reanies Bay, and the coast near Castletownshend was also a breeding haunt during the 1800s. It was reported from Fastnet Rock in April 1889, but not as a breeder.

The Black Guillemot is the most widespread, but not the commonest, auk breeding on the Cork coast due to its habit of nesting beneath boulders and in holes in cliffs that are lower than those favoured by other auk species. The breeding range during 1968–72 extended from Dursey Island to Capel Island, but with gaps where nesting habitat was absent. They were commonest at the western peninsulas and bays, and scarcest in the east. There was little overall change in distribution during 1988–91, but significant losses were recorded around Dursey Island and the western peninsulas, although some losses may be due to low (or no) survey effort. Elsewhere, losses were equalled by gains, but they were apparently absent from Galley Head and Seven Heads. However, east of Cork Harbour breeding no longer took place at Power Head, and the only occupied site was Knockadoon Head. The distribution in 2008–11 was almost the same as in previous surveys, although there were gains in the western bays and peninsulas, and breeding was recorded at Capel Island. Birds bred at three east Cork sites in the 1970s and 1980s (Power Head, Knockadoon Head and Capel Island), and a pair bred at Ballycotton Island in one year, possibly 1977. Breeding has since ceased at all east Cork sites.

It is impossible to compare directly the totals obtained during the various seabird censuses carried out since 1969 due to the early surveys being carried out in June, when many birds are on nests. Pre-breeding surveys produce better results in April and May. It is now clear that surveys in the 1970s and 1980s underestimated the population. During the 2000 survey a total of 679 individuals was recorded. Largest concentrations occur at Roaringwater Bay, especially from February to April when individuals are displaying and visiting the breeding cliffs. A maximum count of 146 was made between Baltimore and Schull on 14 April 1989. This may represent a breeding population of about seventy pairs. At CCBO there were forty-four pairs in 1963, sixteen pairs in 1967, thirty-nine pairs in 1969, thirty-one pairs in 1976, ninety-nine individuals in 1990, and 111 individuals in 2000. There have been few counts of breeding numbers away from CCBO, so it is difficult to comment on population level changes.

Cork Harbour has been a wintering site for many years, two being recorded there in February 1849. The recent peak in Cork Harbour has been forty at Cobh on 10 January 1993. Recent winter atlas surveys have shown the Black Guillemot population to be concentrated at the sheltered waters of Bantry, Dunmanus and Roaringwater Bays, and at Glandore Harbour. There was also a small concentration of birds at Cork Harbour, with scattered individuals extending west to Kinsale and east to Youghal Bay. Highest densities were in the bays around the western peninsulas. At least some breeding sites are deserted soon after the young fledge. An exodus takes place from CCBO as early as the beginning of August. Presumably some of these birds spend the winter in the inner part of Roaringwater Bay (peak count of 82 on 15 September 1988). They are scarce at CCBO between August and February except when calm conditions and bright weather bring flocks to the island to visit their nesting cliffs. Sea passage off CCBO is almost non-existent.

The Black Guillemot is a sedentary species, making only localised movements from breeding sites to sheltered waters of estuaries and bays for the winter. There has been a single ring recovery; one ringed as a nestling at Rockabill Island (Dublin) on 13 July 1991 was found at Ballydehob on 21 October 1991.

Little Auk

Alle alle | Falcóg bheag | A

(subspecies: *alle*)

World	Arctic waters of North Atlantic, North Pacific and Arctic Oceans
Ireland	Scarce winter visitor
Cork	Scarce winter visitor, mainly in November
Conservation	Green-listed

Two Little Auks were recorded at Cork Harbour, where Harvey said it was frequently seen in the early 1800s. It was later described as an occasional but uncertain winter visitor, but only one record was cited, a bird at Castletownbere in January 1869. Only three occurrences for Cork, presumably inclusive of those cited, were known up to 1900.

Large numbers of storm-blown Little Auks sometimes occur, such an event being known as a 'wreck'. A major 'wreck' took place on the Cork coast (and elsewhere) following south-west gales in early February 1950.[232] At least fifty birds were documented as blown ashore around Goleen, although one report stated there were hundreds in one cove east of Goleen. Little Auks were also recorded at Baltimore, near Skibbereen, Castlehaven and Clonakilty, inland as far as Drimoleague and east to Midleton. An exact number was not computed, but it is possible that as many as 500 birds may have died across the county based on the number recorded in the Goleen area.

Between 1959 and 2018 at least 509 birds were recorded. Most (67 per cent) occurred in November, with smaller numbers in other months, except June and July, the only months with no records. Little Auk records are dependent on the extent of sea watching in different months over the years. It is known that sea-watching effort has varied, and the seasonal proportions may not truly reflect abundance at any given time. Migrants have been recorded passing headlands and islands on the Cork coast from Dursey Island to Knockadoon Head, but CCBO alone has accounted for 84 per cent of the total. Many records involve from one to ten birds observed predominantly flying west during short sea-watching sessions. However, exceptional numbers were recorded at CCBO on 7–8 November 1991 when totals of 191 and 58 passed west, respectively. The difficulty of interpreting these data is demonstrated by the fact that following the peak year of 1991

(263 birds), the next ten years (1992–2001) produced only eighteen birds. This is most likely to be the result of a lack of sea-watching effort, especially in late autumn and early winter, rather than an absence of birds.

In addition to the birds seen passing headlands, a small number of singles have been recorded at many sites from Dursey Island to Pilmore Strand in inshore waters, often in harbours, estuaries or on fresh or brackish waters adjoining the coast. These birds often appear in distress, probably resulting from exhaustion and starvation following difficult feeding conditions at sea during severe weather. At least twenty-eight such birds can be identified from the data since 1959. A further thirteen have been recorded as dead or dying on beaches, as well as two more at inland sites (Ballydesmond and Doneraile). Such birds have occurred from October to April, with most of them between November and March. However, despite many gales occurring since 1959, no serious wrecks have been recorded in Cork. Apart from the large numbers recorded during 1991, the temporal distribution of occurrences has not changed significantly.

1959–68	1969–78	1979–88	1989–98	1999–08	2009–18
12	39	87	328	30	13

Earliest and latest occurrence in autumn and spring (1959–2018): 3 August–11 May.

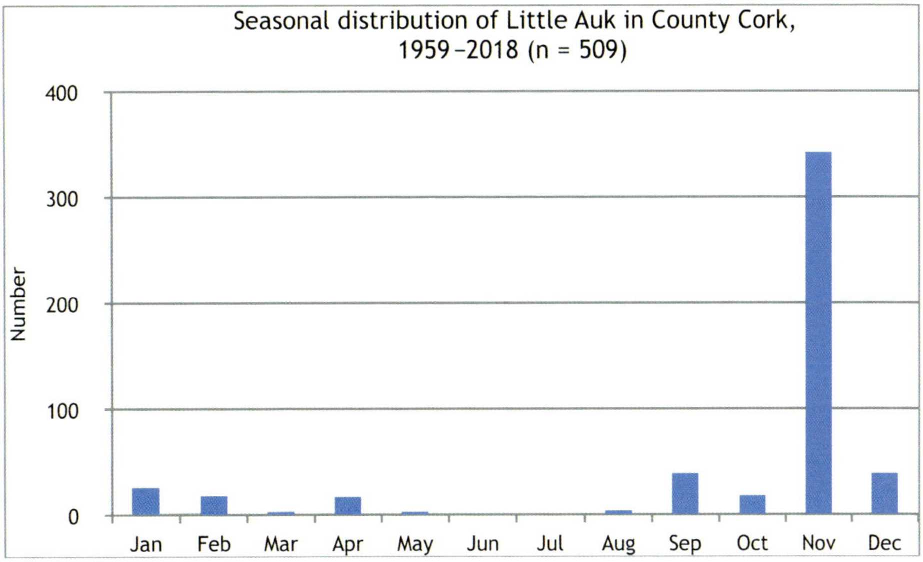

Fig. 7.24 Little Auk.

Puffin

Fratercula arctica | Puifín | A

(monotypic)

World	North Atlantic and Arctic Oceans
Ireland	Common summer visiting breeder, rare in winter
Cork	Summer visitor and breeder, very rare in winter
Conservation	SPEC 1; Red-listed

There are two references to breeding of Puffins which cannot be verified. The first is supposed breeding at Ballycotton Island in the 1700s.[233] The second is that large numbers had nests at Old Head of Kinsale in July 1907.[234] Old Head of Kinsale is a mainland site easy to observe by naturalists, but there are no further indications of breeding.

Various authors reported breeding by Puffins at Bull Rock, Cow Rock, Stags of Castlehaven (i.e. Stags Rock), High Island and at other unnamed parts of the coast during the 1800s. The only certain sites occupied today are Bull Rock and Cow Rock. An estimated 200 pairs were at each of these sites in 1969, and around 500 pairs in the 1980s. A colony of about thirty pairs was at CCBO in the 1960s but it has been extinct since the 1990s. The colony at the Stags Rock has long since vanished. High Island has for many years been known as a minor breeding site and it is probable that small numbers still breed there. Four were seen on the sea and on High Island in May and one in June 1985 by Pat Smiddy and John Earley. One was later found dead in a fishing net in the area. A pair bred at Mizen Head in 1985, and in June 1994 six adults were reported on the sea near Ballycotton Island.

Puffins are common on offshore passage in spring (late March to early May) and autumn (October to early December). The highest numbers off CCBO occurred in early April 1962, when over 6,000 passed west in twenty-one hours (peak rate 293 per hour). A total of 530 passed CCBO on 7 April 1983, with up to 400 per hour going west there in late April 1971. The highest passage at a mainland site was 240 west per hour off Galley Head on 12 April 1977. Autumn counts at CCBO include 1,800 per hour on 26 November 1970, and sixty-nine going west in one and a half hours off Galley Head on 2 October 1991.

Puffin

Winter records are more unusual as Puffins (unlike other auks) disperse into the Atlantic rather than remaining inshore at this season. Single Puffins were recorded on most winter sea watches at CCBO from 1981/2 to 1983/4, and a count of 191 west per hour in early January 1964 was exceptional.

There have been a few ring recoveries involving individuals from Puffin Island (Kerry), St Kilda (west Scotland), Isle of May (east Scotland) and Orkney Islands; the bird from Isle of May was found at Ballycotton in December 1978.

Pallas's Sandgrouse

Syrrhaptes paradoxus | Gaineamhchearc Pallas | B

(monotypic)

World	Asia from Aral Sea eastwards
Ireland	Rare irruptive visitor
Cork	About 6 individuals (1 record)
Conservation	Not globally threatened

About six at Mallow in June 1888. The occurrence reported here coincided with an invasion to Europe and Britain, with some occurring in all Irish provinces. It has occurred in Ireland in two invasions, in 1863 and 1888.

Rock Dove and Feral Pigeon

Columba livia | Colm aille | A, C1

(subspecies: *livia*)

World	Almost worldwide, introduced and domesticated widely
Ireland	Resident breeder
Cork	Resident breeder, feral population
Conservation	Green-listed

There was said to be a great variety of these birds in the 1700s. This was a clear reference to what we now call the Feral Pigeon, while of the Rock Dove it was said they bred very commonly on the coast. Rock Doves have evidently been resident and common coastal birds for a very long time. A decrease in numbers was reported in certain coastal regions of Cork, notably around Glengarriff, but in the south-west it was generally considered plentiful in the early decades of the twentieth century.[235] The genetic purity of Rock Doves has been diluted through interbreeding with escaped birds and most of those seen on the coast exhibit colours other than the true type, even on the coast of south-west Cork.

Rock Doves nest colonially in crevices and caves in sea cliffs. Almost every stretch of coast has a few resident breeding pairs, although the species is in decline at some western locations alongside a decrease in the growing of arable and cereal crops in the immediate vicinity of cliffs. In general, the Rock Dove population becomes purer with increasing distance from Cork city, where the main concentration of Feral Pigeons occurs. At CCBO, the population (34 pairs in 1963, 20 pairs in 1967, 14 pairs in 1995) was reported to be relatively pure in the late 1960s, and while the proportion of Feral Pigeons has increased since then, there are apparently some pure birds still present. Eight pairs were breeding at Sherkin Island in 1985, but by 1994 there were none there. Although there are small flocks of apparently true Rock Doves on the coast of the Beara peninsula, it ceased to breed at Dursey Island in the early 1990s.

Largest numbers of Rock Doves occur at CCBO in September and October, and again in December and January. These peaks may result from autumn and winter flocking of local birds, but it has been suggested that birds from adjoining sections of coast arrive on the island in autumn, with winter immigrants swelling numbers in December and January. In autumn and winter, flocks of up to fifty are regularly seen on the coast, and 300–400 have been counted at Old Head of Kinsale. Peak counts at CCBO typically involve seventy-five to eighty-five birds, but up to 252 in a day were recorded in the autumns of 1985 and 1986.

During the 1968–72 breeding atlas it was reported there was likely to be less contact between Rock Doves and Feral Pigeons on the coast from Waterford westwards. However, even at that time many Rock Doves at Knockadoon Head, in the east, showed the mixed colours of feral and escaped birds. Nevertheless, the situation may have been different elsewhere, especially at the western peninsulas. In any case, Rock Doves were shown as breeding along the entire coastline from Dursey Island to Capel Island. The 1968–72 survey also showed the Feral Pigeon as breeding only at Cork city, but this form may have been under-recorded. There was then the nucleus of Feral Pigeon populations in the east at grain stores at Mogeely and Ballinacurra, and at Youghal where there was an active pigeon racing club.

The 1988–91 breeding atlas showed the coastal situation regarding Rock Doves had hardly changed, but there was a major range expansion of Feral Pigeons at inland sites. Feral Pigeons had pushed north to Fermoy and west to Dunmanway, but density was low except around Cork city. They were nesting at a wide range of sites, including in bridges (Midleton) and caves (Carrigacrump). The range of the Feral Pigeon expanded further by 2008–11. The whole county was occupied except for the extreme north and some parts of the west, the unoccupied areas being largely upland and probably unsuitable for Feral Pigeons.

The coastal winter distribution of Rock Doves does not differ markedly from that of the breeding season, and birds were present from Dursey Island to Knockadoon Head. Inland, Feral Pigeons had spread to many new areas by the mid-1980s. Populations can be traced to towns such as Charleville, Mitchelstown, Kanturk, Mallow, Fermoy, Ballymakeery, Bandon and Midleton, as well as Cork city. These towns have grain storage sites providing a ready source of food. There was a noticeable concentration of birds along parts of the Rivers Blackwater and Bandon, encompassing some of the mentioned towns. At least some of these records (in these intensive farming areas) are attributable to cattle and pig farms with animal feeding facilities. By 2007–11 the winter range of the Feral Pigeon had expanded inland in a similar way to that observed during the breeding season. Many small populations had developed around individual farms, often where new buildings had been erected for domestic animals which provided the birds with a combination of feeding facilities, shelter, and nest sites. Almost the whole county now hosts a very high density of Feral Pigeons during the breeding and winter seasons.

Stock Dove

Columba oenas | Colm gorm | A

(subspecies: *oenas*)

World	Europe, Asia, and north-west Africa
Ireland	Resident breeder
Cork	Resident breeder
Conservation	Red-listed

It is not known precisely when the Stock Dove colonised Cork as a breeding species. In 1926 it was said to have nested in the Mallow area within the previous three years. It was believed to be breeding as far west as Timoleague in 1925, in 1943 it was thought to be widespread, but in 1956 it was rarely seen around Union Hall.

The 1968–72 breeding atlas showed several gaps in distribution across the county. It was largely absent from the south-west and the western peninsulas, and from a band along the east to west section of the Rivers Blackwater and Bride (north). Although many records were not of proved breeding, their distribution showed the potential for increase. A pair bred at CCBO for the first time in 1963, and ten pairs did so in 1965, and eight pairs bred in the Ballintubbrid area in 1976.

The 1988–91 breeding atlas showed the Stock Dove had suffered considerable losses north of the River Blackwater and in the west. The main area of distribution was in the east where there was a relatively stable situation. However, by 2008–11 the situation was reversed, and breeding birds had pushed into areas of north Cork where they were apparently absent in 1988–91. They were largely absent from higher ground and were completely absent from the western peninsulas. The Stock Dove no longer breeds at CCBO, where it has been described as almost a rarity in recent years. This is consistent with its current absence from the western peninsulas.

The winter atlas survey of 1981–4 showed that a considerable contraction of range had taken place since the early 1970s. By the early 1980s hardly any were recorded west of a line from Killavullen to Kinsale or north of the River Blackwater. The winter distribution was almost confined to the east of the county, with the densest population around Cork Harbour. The winter distribution in 2007–11 nearly mirrored that recorded in the breeding season and showed an expansion into parts of coastal west Cork and the north-west of the county. There is no evidence that Irish birds emigrate in winter.

Flocks of from ten to twenty birds are regularly recorded outside the breeding season, especially around Cork Harbour and in east Cork. The largest concentrations on record are 137 near Mallow racecourse in March 1992, sixty-one at Cobh in January 1984, fifty-one at Ballintubbrid in September 1976 and fifty at Ballycotton in October 1991.

Woodpigeon

Columba palumbus | Colm coille | A

(subspecies: *palumbus*)

World	Europe, Asia, north Africa, and Azores
Ireland	Very common resident breeder
Cork	Common resident breeder
Conservation	Green-listed

The Woodpigeon was breeding throughout the county during all three atlas surveys. It was common inland and in coastal districts, but density was lowest at the western peninsulas and in upland areas. Winter atlas surveys showed there was no difference compared to the breeding season, and density was highest in the most intensive farming areas.

Woodpigeons breed on some islands, and on Capel Island, where there are no trees or bushes, a nest was found on the ground in 1977. CCBO was colonised as recently as 1963 when three pairs bred, and where eight to ten pairs continue to do so. At Sherkin Island, twenty to forty pairs also breed.

Woodpigeons may cause damage to agricultural crops, especially to ripening cereals, legumes and clovers. A total of 89,094 was shot in County Cork in 1982/3 and 1983/4 as a crop protection measure. However, this level of culling appears to have had little effect on the overall population. They will eat a range of wild plant species also, and Robert Warren noted them taking the roots of silverweed on ploughed ground at Castle Warren during the winter.

Large flocks of Woodpigeons sometimes occur in autumn and winter, some of which may be migrants. At least 15,000 were present at Castlemartyr in January 1975. Other high counts include at least 5,000 at Oysterhaven in November 1964, with smaller flocks of 2,000 at Templenacarriga in November 1991 and 1,000 at Garrettstown in November 1993.

There is little information available about passage movements, although flocks of 500–800 birds may occasionally be seen flying south-west over Cork Harbour in late autumn, several thousand were at Galley Head on 15 October 1994, and Woodpigeons were reported among a rush of birds at Fastnet Rock lighthouse in October 1914. However, ringing has shown the sedentary nature of British and Irish Woodpigeons, with only 1 per cent of ringed birds having been recovered overseas. The evidence for many continental birds in Ireland is equally slight, and although one nestling from the Netherlands ringed in May 1969 was recovered in Cork city in February 1970, there is somewhat more evidence that some British birds visit Ireland.

Collared Dove

Streptopelia decaocto | Fearán baicdhubh | A

(subspecies: *decaocto*)

World	Europe, Asia, and north-east Africa, introduced elsewhere
Ireland	Common resident breeder
Cork	Resident breeder
Conservation	Green-listed

The first record of Collared Doves in Cork was in July 1962 when two pairs were seen at Ballinacurra, and they bred there the same year. In 1964, there were twenty birds at Ballinacurra and two pairs bred, eight birds at Ballinhassig and one pair bred, and one pair bred at Douglas Road (Cork city) (total of 4 breeding pairs). During 1964 and 1965 birds were seen at several Cork city suburbs, such as the Mardyke, Monaghan Road, Ballintemple and Blackrock, while outside the city birds were recorded at Cobh. CCBO had its first record in June 1964, and Old Head of Kinsale in August 1965, while breeding took place at Timoleague also in 1965. Eight birds at a grain store near Midleton railway station in 1965 had increased to sixty or seventy birds by November 1966, and one was seen at Mallow in May 1967.[236]

The main concentrations in 1970 were summarised as breeding around Cork city, Ballinhassig, Cobh, Midleton, Youghal, Kinsale and Bandon. They were local away from the south-east but were breeding at Mallow and near Millstreet and probably also around Kanturk, Skibbereen, Grenagh and Castletownbere, and a few were present in summer elsewhere.[237]

Collared Doves were widely distributed during the 1968–72 breeding atlas, but especially in a coastal band from Glandore to Youghal. Some had reached the western peninsulas and several inland towns such as Mallow and Charleville. The range had not changed significantly by 1988–91, but increases were evident in the centre and north-east. The increase and spread continued through 2008–11, by which time there were very few vacant areas, and those few that existed were mainly on higher ground.

The winter distribution of the Collared Dove in 1981–4 showed there were hardly any west of a line from Millstreet to Glandore, apart from a few in the Bantry and Skibbereen areas, but they were widely spread across the rest of the county. The winter distribution in 2007–11 was very similar to that of the breeding seasons of 2008–11. Away from the main urban and cereal-growing areas of the south and east this species is rather scarce. Highest counts involve 204 at a grain store at Carrigtwohill in September 1992. Grain storage areas have repeatedly been sites for Collared Dove flocks, especially during the autumn and winter.

Collared Doves disperse in all compass directions, but mostly to the west and north-west. There are several records of birds ringed in England and recovered in Cork proving this tendency.

Turtle Dove

Streptopelia turtur | Fearán | A

(subspecies: *turtur*)

World	Europe, Asia, north Africa, also Madeira and Canary Islands
Ireland	Has bred, scarce spring and autumn passage migrant
Cork	Has bred, scarce spring and autumn passage migrant
Conservation	SPEC 1; Red-listed

Turtle Doves were observed and obtained at several locations in the 1800s, such as Bantry, Castlefreke, Carrigaline, Castle Warren and Youghal. Several were obtained at lighthouses between 1889 and 1913, with records for April, May, August, September and October at Bull Rock, Berehaven, Fastnet Rock and Old Head of Kinsale. It was noted as increasing as a visitor in 1909, and Cork was regarded as probably the most visited Irish county, but no breeding record was known.

One Turtle Dove was at a demesne near Doneraile in June 1954, and it was said to be an annual visitor to the Glandore district. The first case of breeding was in 1968 when young birds were seen being fed by adults near Whitegate, and they may have bred at the same site in 1966 and 1967. Turtle Doves also probably bred near Allihies and near Courtmacsherry Bay during the 1968–72 breeding atlas, and a few birds, probably migrants, were seen at a few other coastal localities. A pair was at Ballintubbrid for over a month in summer 1979, and two downy young were seen soliciting food from an adult at Crookhaven on 30 September 1990. There are no further records of breeding, although birds occur in coastal districts nearly every year in spring.

Spring migration extends from mid-April to June, when small numbers occur regularly at coastal localities. It is generally a very light passage, although peaks of fifty-nine and fifty-three were recorded at CCBO in May 1965 and May 1994, respectively. Counts exceeding twenty birds have been obtained at CCBO on a few other occasions, always in spring. Elsewhere in Cork the highest spring count is of twelve at Old Head of Kinsale in May 1994. Most other counts refer to only one or two birds. Single birds often remain for only one day, although small flocks may remain for a week or more. At CCBO, numbers often dwindle over a week following the arrival of a flock.

Turtle Doves are less numerous on autumn passage, which extends from August to November. Maximum counts at this season are always much lower and rarely exceed five. Most Turtle Doves, in both seasons, are recorded at coastal sites from Dursey Island to Youghal, although small numbers may filter inland, some to north of the River Blackwater where they may go unrecorded.

There are four winter records of single birds, at Midleton from 16 December 2004 to 2 January 2005, at Ballymacoda from late November to at least 3 December 2008, at Skibbereen from 25 January to 30 March 2010, and at Clonakilty from 24 December 2016 to 2 January 2017.

1959–68	1969–78	1979–88	1989–98	1999–08	2009–18
251	167	249	282	115	148

Earliest and latest occurrence in spring and autumn (1959–2018): 1 April–25 November.

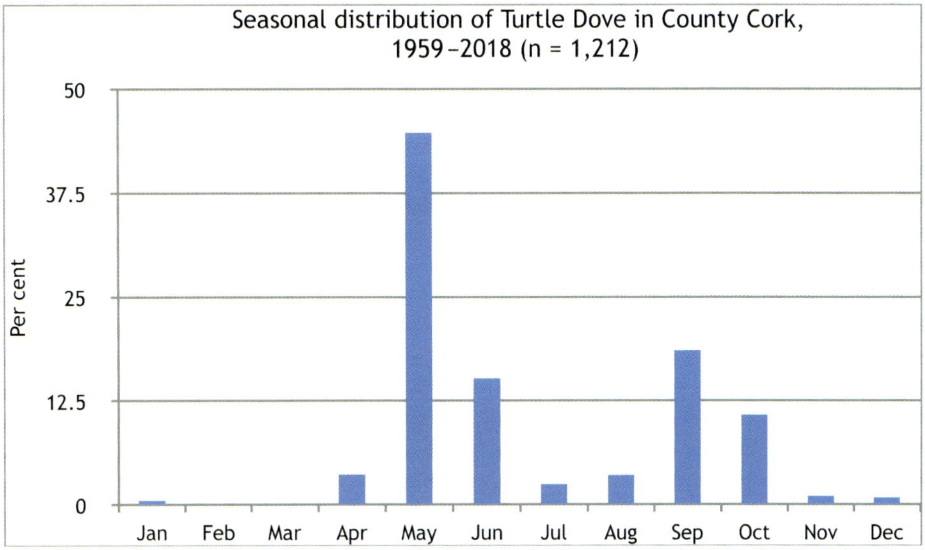

Fig. 7.25 Turtle Dove.

Mourning Dove

Zenaida macroura | Colm caoineadh | A

(subspecies: undetermined)

World	North and Central America, and Caribbean islands
Ireland	3 individuals
Cork	1 individual
Conservation	Not globally threatened

One at Garinish on 25 October 2009.

Great Spotted Cuckoo

Clamator glandarius | Mórchuach bhreac | A

(monotypic)

World	Southern Europe, south-west Asia, and Africa
Ireland	8 individuals
Cork	1 individual
Conservation	Not globally threatened

First-year at Ringaskiddy on 15 February 2009.

Common Cuckoo

Cuculus canorus | Cuach | A

(subspecies: *canorus*)

World	Europe, Asia, and north Africa
Ireland	Summer visiting breeder, recent decrease
Cork	Scarce summer visitor and breeder, recent decrease
Conservation	Green-listed

The Cuckoo was common in the late 1800s, arriving in April, usually leaving in July, but with young remaining to August. Ussher mentioned numbers of birds roosting in favoured coverts in bare tracts of ground, but he did not elaborate. The breeding habits of the Cuckoo are well known but are often the subject of ill-informed opinion.[238]

In Cork, there are records of two species having been parasitised by the Cuckoo. The Meadow Pipit has been parasitised several times at CCBO, while a young Cuckoo was observed in a nest at Power Head in 1982. The only other record is that of Robert Warren who saw a young Cuckoo in the nest of a Dunnock at Monkstown, probably in the 1840s.[239]

During the breeding atlas survey of 1968–72 it was widespread with very few vacant areas. Twenty years later, a major decline in range was recorded. Cuckoos had almost completely disappeared from many areas and only in the west was there a continuity of population. Further losses occurred by 2008–11 by which time the main range was west of Galley Head, there being significant losses in all other regions.

Reasons for these decreases have been speculated upon, including a decline in its main host, the Meadow Pipit, and a decline in its main food source, large caterpillars. However, the range of the Cuckoo has declined significantly more than that of the Meadow Pipit in Ireland. It has also been speculated that the Common Reed Warbler, a recent Irish colonist, might act as a host of the Cuckoo since it does so in Britain. However, it is known that Cuckoos which specialise on a particular host are effectively different subspecies, and may not be readily able to switch to a new host. The Cuckoo is one of several trans-Saharan migrants in general decline, therefore the causal factors may, at least partly, lie outside the breeding range.[240]

At CCBO a few Cuckoos bred in most years in the 1960s and early 1970s, but they have been noticeably scarcer since 1977 and breeding has not been proven in recent years. The decline at CCBO since the 1970s has been described as dramatic. However, on Sherkin Island the breeding population appears to have been more constant at four or five pairs in most years during 1977–94.

Although individuals have been recorded at coastal districts in early April on several occasions (the earliest being one at Castlemartyr on 3 April 1982), most arrive from the last week of April and they quickly filter inland and to upland areas. Most records concern single birds, although there are counts of up to twenty-one at CCBO from 27 to 30 May 1967 (including a flock of 14 seen together in flight), with fifteen there on 14 May 1965. Song has usually ceased by late June, although one was heard sing on 2 July. Most subsequent sightings, peaking in early August, are of coastal migrants, often of juveniles. September records are rare, with the latest at Ballyshane Strand on 6 October 1996. The Cuckoo has been obtained at lighthouses on spring and autumn migration, and there are records from Fastnet Rock (May 1901), Old Head of Kinsale (August 1902, July 1905) and Roche's Point (July 1911, August 1913).

There are two colour morphs of the female Cuckoo, the grey morph and the so-called hepatic or brown morph. One hepatic female was seen at Ballintubbrid on 17 August 1979.

Yellow-billed Cuckoo

Coccyzus americanus | Cuach ghob-bhuí | A

(monotypic)

World	North America and Caribbean islands
Ireland	10 individuals
Cork	3 individuals
Conservation	Not globally threatened

One obtained at Youghal in autumn 1825. One found dead at CCBO on 13 October 1969. One at CCBO on 30 October 1986.

Barn Owl

Tyto alba | Scréachóg reilige | A

(subspecies: *alba*)

World	North and South America, Europe, Asia, Africa, and Australia
Ireland	Uncommon resident breeder, has decreased
Cork	Scarce resident breeder, has decreased, but recovering locally of late
Conservation	SPEC 3; Red-listed

The Barn Owl was relatively common in the 1700s and 1800s, and frequented ruined buildings, chimneys, lofts and hollow trees. It bred about

Barn Owl

Youghal, including the chancel of Saint Mary's Church, and at Castle Warren near Ringaskiddy, the ancestral home of Robert Warren.

The Barn Owl appeared to be scarce during the 1968–72 breeding atlas, with most of the county having no records, although it is likely it was under-recorded, especially since the adjoining county of Kerry had a healthy population. An optimistic population estimate for Cork may have been about 100 pairs at that time. By 1988–91 further declines had taken place, although some new 10 km squares were occupied. Winter atlas fieldwork also under-represented the distribution, although that of 2007–11 showed a distribution not unlike the breeding seasons of 2008–11.

Subsequent intensive fieldwork has clarified the situation and provided new information. Contrary to the distribution suggested in 1988–91, Barn Owls were present throughout the county during 1991–5 wherever the land was lower than 150 m above sea level. The total number of pairs was estimated at between sixty and ninety, most likely towards the lower end of the scale. The estimate was revised in 2006 (40–60 pairs).[241] Distribution included the west, with sightings from all the peninsulas. They were present in the Clonakilty, Rosscarbery, Skibbereen and Drimoleague areas, and along some of the smaller and larger river valleys.

However, despite the relatively wide distribution, Barn Owls occur at low density, with rarely more than one pair per 10 km square in suitable habitat. Ruined buildings of one kind or another are the most frequently used nest sites, with a minority in trees.

Dedicated fieldwork and conservation effort has continued throughout the county, particularly in the Duhallow area, and this work has highlighted what is probably close to the true status of the Barn Owl at present (2020). While there are significant areas without records, the population is spread across the county from Glengarriff to Killeagh and from the coast to the Limerick border, with the main populations in the north, north-west, south-west and central areas of the county.

Conservation measures have included the provision of nest boxes, which are readily used by resident owls. During 2010–20, eighty to ninety-seven sites have been occupied at least once and in 2020 alone fifty-six sites were occupied, nineteen involving nest boxes. Productivity was 3.1 chicks fledged per breeding attempt (n = 36).[242]

Declines in the Irish population have been reported since the 1950s. Declines have occurred in Cork over the same timeframe, but especially since 1970. There are several contributary factors, such as agricultural intensification leading to reduced food supply and hunting areas, and loss of nest sites through demolition or renovation of old buildings. However, there is evidence of a recovery in some regions in the last decade, although road deaths are an important cause of mortality.[243]

The diet of the Barn Owl has been examined in detail, and the results of twenty-nine Irish studies reviewed.[244] Small rodents and shrews are the most important prey species. The field mouse is the most important individual prey species, followed by bank vole, pygmy shrew, house mouse, and brown rat. Birds, bats and common frog are taken less frequently. The greater white-toothed shrew, a recently introduced species, has a restricted distribution in Cork, but is taken in high numbers by Barn Owls where it is present, and this may be partly responsible for the recently reported increase in the population.[245]

The Barn Owl is a sedentary species, but ringing has shown that young birds disperse during their first four or five months of life, usually within 50 km of the natal area. After this there is little additional movement, although one ringed at Blarney on 27 July 1976 was found at Callan (Kilkenny) (100 km) on 16 February 1978. Other recent movements have been within the county and over much shorter distances.

The Barn Owl is a vagrant to CCBO, with five records of single birds during 1959–96 in the months of April, August, September and October (twice).

Scops Owl

Otus scops | Ulchabhán scopach | A

(subspecies: *scops*)

World	Europe, Asia, and north Africa
Ireland	16 individuals
Cork	4 individuals
Conservation	Not globally threatened

Male obtained at Fastnet Rock on 6 May 1907. One found injured at Inchydoney on 27 April 1993, died on 10 May. Adult male at CCBO on 17–19 May 1999. One found injured (subsequently died) at Crookhaven on 4 November 2005.

Snowy Owl

Bubo scandiacus | Ulchabhán sneachtúil | A

(monotypic)

World	Arctic regions of North America, also Greenland, Europe, and Asia
Ireland	Has bred, 87 individuals
Cork	2 individuals
Conservation	SPEC 3; Red-listed

One at Inchigeelagh in September 1827. Female at Bere Island on 5 September 2019.

Long-eared Owl

Asio otus | Ceann cait | A

(subspecies: *otus*)

World	North America, Europe, Asia, and north Africa
Ireland	Resident breeder
Cork	Resident breeder
Conservation	Green-listed

The Long-eared Owl bred in old nests of Woodpigeon, Magpie, Rook and Hooded Crow in the 1800s, and it was the commonest owl in wooded districts, especially in demesnes near Cork Harbour, notably Coolmore.

Like the Barn Owl, this species was under-recorded during the 1968–72 breeding atlas. Comparison with Kerry, where survey effort was good, serves to illustrate the extent of the likely under-recording. It was still apparently scarce in 1988–91, but there were more records than previously, and it was more frequent in the south and east. Nevertheless, the 1988–91 distribution was considered an underestimate and it was said to be widespread, with pairs in most plantations and old orchards. Five birds were calling simultaneously at Doolieve on 5 November 1987, suggesting that at least locally, breeding densities were high. A considerable expansion of range had taken place by 2008–11 when it was found almost throughout the county with only uplands in the north-west and parts of the western peninsulas vacant. The apparent expansion may have been, at least partly, the result of an increase in conifer plantations, but most of the increase can be explained by intensive night-time coverage by some surveyors. A county population estimate in 2006 suggested at least 300 breeding pairs.[246]

The winter atlas survey of 1981–4 provided few records of this nocturnal species. There is no doubt that it was under-recorded as few observers made survey visits at night. While there were some new winter records in 2007–11, it was very patchily distributed. However, it is difficult to accept that this apparent winter scarcity is real, given its widespread nature in the breeding season (2008–11), and it is likely that while considerable local effort was mustered in the breeding seasons, winter effort was lower.

The diet of the Long-eared Owl has recently been reviewed based on published studies.[247] Small mammals, especially field mouse, are important prey, but even in areas where the bank vole is common, they are not taken in high numbers. All other mammals, and birds, are taken less frequently.

Although there is believed to be an immigration of birds from the continent in autumn, its extent is unknown. No birds appear to have been taken at Cork lighthouses. At CCBO, Long-eared Owls are mainly scarce autumn visitors, and apart from isolated occurrences in March, May and July, the records are distributed from August to December. Exceptional numbers occurred in 1975 and 1991, with up to five and seven in October, respectively. It is not known if these were of Irish or continental origin. It was suggested that two birds which appeared at the end of October on a gorse-covered hillside devoid of trees in a bog near Skibbereen, where the species was little known, were migrants. Birds are also occasionally recorded at Dursey Island, Galley Head and Old Head of Kinsale in autumn.

Short-eared Owl

Asio flammeus | Ulchabhán réisc | A

(subspecies: *flammeus*)

World	North and South America, Iceland, Europe, and Asia
Ireland	Very rare breeder, uncommon winter visitor
Cork	Scarce winter visitor
Conservation	SPEC 3; Amber-listed

In the 1800s, the Short-eared Owl was not rare in low marshy ground in winter. It was even said to be abundant at Ballyvergan Marsh, and it occurred more commonly in some winters than others (e.g. October and November 1903). Birds were obtained at Ballycotton in November 1907 and at Youghal in March 1914. In the first half of the 1900s it was described as a scarce winter visitor about Fermoy, it was plentiful in the Schull district, and was occasionally found in marshes near Whitegate. The status of the Short-eared Owl at the present time does not differ significantly from these accounts.

It has not been proved to breed in Cork, although it has bred in the adjoining counties of Kerry and Limerick. A single bird was seen in breeding habitat in Cork near the border with Limerick in June 1977. Two pairs were also reported in breeding habitat in Cork in 1977, but it is not clear if this was at a different site to the one reported in June. One bird was recorded in the breeding season during 1988–91, and again in 2008–11, in the north-west. There have been several subsequent reports of Short-eared Owls in breeding habitat in the breeding season in the north-west and north (e.g. 1990s, 2002–6, 2009, 2016), but without positive evidence of breeding. Some have been noted to linger in coastal (and other) wintering areas at least to early May, and there have been occasional June and July records. Breeding may have been overlooked and is a possibility in the future.

The species is most regular as an autumn migrant and winter visitor. Although the numbers involved are small, it is the most frequently reported owl in Cork due largely to the open nature of the habitats occupied, and to its partly diurnal habits. Some birds arrive in August, but most not until September, October and November and they usually remain until April and occasionally May. It generally occupies low-lying coastal marshes and dunes, and several such sites are used each winter, favoured ones being around Clonakilty, Cork Harbour, and from Ballycotton to Youghal. It has occurred inland, e.g. around wetlands such as the Gearagh and Kilcolman, although it may be under-recorded away from the coast. On arrival, it is frequently seen at islands and exposed headlands where it generally does not spend the winter, such as Dursey Island, CCBO, Old Head of Kinsale, Barry's Head and Knockadoon Head. Most records concern only one or two individuals, although the Clonakilty and Ballymacoda areas have each held up to seven birds.

Studies of diet have been carried out at two sites, Clonakilty and Ballymacoda. All published data have been reviewed recently.[248] At these sites the diet is largely an avian one, with over 50 per cent by number and biomass consisting of birds. Small mammals are also taken, with the most important being field mouse and bank vole. Among the birds, Dunlin, Common Snipe and Common Redshank are the most important prey items. One bird at the Bull Rock on 25 May 1977 subsisted largely on European Storm Petrels.

Birds from Scottish and Fennoscandian breeding grounds have been recovered in Cork. One ringed as a nestling in May 1956 in south-west Scotland was recovered at Meelin in January 1963; this bird, at six years and almost eight months of age, became the oldest individual recorded from British and Irish ringing. One ringed as a nestling in Finland in June 1988 was recovered at Bull Rock in April 1989.

Nightjar

Caprimulgus europaeus | Tuirne lín | A

(subspecies: *europaeus*)

World	Europe, Asia, and north Africa
Ireland	Very rare summer visiting breeder, recent major decrease
Cork	Former summer visitor and breeder, now believed extinct
Conservation	SPEC 3; Red-listed

The Nightjar was a summer visitor in the mid-1800s, and it was reportedly occasionally shot around Bandon. Towards the end of the 1800s it was said to seldom appear before the middle of May, and it was common about heaths and uplands, especially near plantations. In the first half of the 1900s it was described as more numerous and widely distributed in Munster than other provinces, although Cork was not specifically mentioned.

In 1957–8 it was uncommon or locally common with no decrease in much of the mid, east, and north of the county, but it did not breed annually in the west.[249] Just over ten years later in 1968–72 it was found in only four 10 km squares. All related to cases of probable breeding, and the sites ranged from the west to the south and north. There had been a serious decline which had occurred unnoticed and almost without comment. There were no records during 1988–91, and only one record during 2008–11 in the north-west on the border with Kerry, but without proof of breeding. Traditional sites in west Cork known to Bernard O'Regan which held breeding pairs in the 1940s and 1950s were abandoned by the early 1980s. Many of these sites were checked without success in 1982.

Records of birds in breeding habitat during the breeding season since 1965 are shown in Table 7.16. While there is no certainty that birds bred at all the listed sites, it is likely that breeding took place at most, at least through the 1970s. Maurice Smiddy heard a male singing at Ballymacoda at night in early May 1972; this male was killed on the road on 13 May, and no further birds were seen there. The site was a bracken-covered hillside since converted to pasture. It is believed that at least one pair bred at Ballyhoura Mountains in 2019, although an extensive survey in 2020 was negative.[250] This is the first confirmed breeding record for Cork for at least thirty years since it is considered unlikely that breeding took place at Sherkin Island in 2001 (Table 7.16).

Year	Details
August 1965	Singing pair, Midleton
1965	Singing, Dungourney
1969–70	Singing, Ballyhoura Mountains
1969–70	Singing, Nagles Mountains
1969–70	Singing, Ballygarvan
June 1970	Singing birds, four sites between Castletownbere and Glengarriff
May 1972	One singing, Ballymacoda
July 1977	One singing, Dunmanus Bay
June/July 1991	Pair, north Cork
May/June 2001	One singing, Sherkin Island (present for three weeks)
June 2012	Male, near Timoleague
July 2019	Two or three singing, Ballyhoura Mountains (Cork side)

Table 7.16 Records of Nightjar in breeding habitat, 1965–2019.

Month	Year(s)
May	1964, 1967, 1994, 2005, 2006
June	1988, 1998 (2), 2009, 2012
July	2014
August	2009
September	1977, 2016
October	1968, 1969, 2002, 2005, 2012, 2016

Table 7.17 Records of Nightjar at migration watch points and outside the breeding season, 1964–2016.

Single birds have also been recorded on twenty occasions at migration watch points on the coast, and occasionally inland outside the breeding season and at sites where the presence of breeding habitat could not be confirmed. The records are summarised in Table 7.17.

The Nightjar is a long-distance migrant, wintering in Africa south of the Sahara Desert. Birds have been obtained at Daunt's Rock lightship on 24 September 1905, and on the mainland at Ballycotton on 6 September 1907.

Common Nighthawk

Chordeiles minor | Éan oíche | A

(subspecies: undetermined)

World	North and Central America
Ireland	2 individuals
Cork	1 individual
Conservation	Not globally threatened

First-year at Ballydonegan on 24 October 1999.

Chimney Swift

Chaetura pelagica | Gabhlán simléar | A

(monotypic)

World	North America
Ireland	18 individuals
Cork	12 individuals (7 records)
Conservation	Not globally threatened

One at CCBO on 23 October 1999. One at Ballydonegan on 23 October 1999. One at Gyleen on 25 October 1999. One at CCBO on 29–30 October 2005, two at Sherkin Island on 29 October, and one at Baltimore on 30 October; this series of sightings involved at least three individuals. One at Clonakilty on 30 October 2005. Three at Courtmacsherry and Broadstrand on 31 October–3 November 2005, with four on 1 November, and three at Galley Head on 2 November; this series of sightings involved at least four individuals. One at Mizen Head on 26 October 2015.

Needletail Swift

Hirundapus caudacutus | Gabhlán earrspíonach | A
(subspecies: undetermined)

World	Asia
Ireland	1 individual
Cork	1 individual
Conservation	Not globally threatened

One at CCBO on 20 June 1964.

Common Swift

Apus apus | Gabhlán gaoithe | A
(subspecies: *apus*)

World	Europe, Asia, and north Africa
Ireland	Common summer visiting breeder
Cork	Common summer visitor and breeder, mostly in towns
Conservation	SPEC 3; Red-listed

The Swift bred widely in the 1800s wherever lofty buildings or ruins afforded holes for nesting. It was widespread across the county during 1968–72, with a few gaps at the western peninsulas and at uplands and areas without suitable breeding sites. A general thinning of positive 10 km squares was evident in 1988–91, with inland areas affected most. There were some further losses in 2008–11, almost all of which were in the west. However, it was widely distributed, with the highest density centred around Cork city and in north Cork. Swifts breed exclusively in man-made structures in built-up areas, and especially in older buildings in villages and towns.

Earliest migrants typically appear in late April or the first few days of May, but there are several records for early April, and one was at Carrigaline on 13 March 1983. Maximum numbers occur during mid-summer, when feeding flocks of up to 1,000 have occurred at Youghal. There were 1,500 at Charleville on 22 July 1986. CCBO also experiences its highest numbers in late July (250 on 30 July 1975). Most have departed by late August although small numbers occur each year in September, but in October records are rare. There are a handful of records for November, with the latest modern one at CCBO on 15 November 1975, although one was obtained at Clonakilty some days before 27 November 1907.

The Swift is a long-distance migrant, wintering in Africa well south of the Sahara Desert. Swifts have been reported and obtained at Fastnet Rock during autumn migration between June and August, the latest date being 19 August. A juvenile obtained there on 31 July and a juvenile obtained at Old Head of Kinsale lighthouse on 12 August indicates the rapid departure of young birds after fledging. A mid-season movement of parties going westwards or arriving from the south in June and July on the Cork coast has been noted. However, these movements may be related to weather conditions rather than to newly arriving birds.

Little Swift

Apus affinis | Gabhlán beag | A

(subspecies: undetermined)

World	Africa and southern Asia
Ireland	2 individuals
Cork	1 individual
Conservation	Not globally threatened

One at CCBO on 12 June 1967.

Alpine Swift

Tachymarptis melba | Gabhlán Alpach | A

(subspecies: *melba*)

World	Southern Europe, southern Asia, Africa, and Madagascar
Ireland	86 individuals
Cork	22 individuals
Conservation	Not globally threatened

There were two records of Alpine Swift before 1959, one obtained 15 km off CCBO about mid-summer 1829, and one obtained near Doneraile in June 1844 or 1845.

Twenty individuals have occurred since 1959, all singly except one record involving five birds in Cork city in March 2002. Most (13) occurred in spring between March and June, with seven in autumn between July and October. All records in 1959–2018 have been at or near the coast, the Doneraile record (above) being the only truly inland one. They occurred at ten sites ranging from Dursey Island to Ballycotton, but only CCBO (5), Old Head of Kinsale (2), Cork city (5) and Ballycotton (2) have had more than one individual. All records relate to birds seen on one or two days, apart from one of the Ballycotton birds which remained for twelve days (mean stay = 1.8 days).

1959–68	1969–78	1979–88	1989–98	1999–08	2009–18
2	2	4	0	8	4

Earliest and latest occurrence in spring and autumn (1959–2018): 10 March–2 June and 27 July–6 October.

Common Kingfisher

Alcedo atthis | Cruidín | A

(subspecies: *ispida*)

World	Europe and north Africa, Asia south and east to Indonesia
Ireland	Resident breeder
Cork	Scarce resident breeder
Conservation	SPEC 3; Amber-listed

The Kingfisher was found at Cork Lough in the 1700s, and at a small lake at Dower. In the late 1800s it was said that it would be more common if not shot so ruthlessly on the shore in winter. By the mid-1900s it was rare around Skibbereen but was breeding near Clonakilty and possibly near Leap. Cork had a good breeding population in 1967 and had apparently not been badly affected by the cold winters of 1961/2 and 1962/3.

Although widely distributed across the county during 1968–72, there were many gaps in the breeding range. There was a concentration along the River Blackwater, and they were absent, as expected, from upland areas. There were some losses in 1988–91, especially at the River Blackwater and Beara peninsula. There were gains throughout the county in 2008–11, including at the River Blackwater, although some of these gains may reflect better coverage compared with 1988–91.

Breeding density on the River Blackwater and several tributaries was surveyed in 2008 and 2010, and the River Ilen in 2010.[251] Density on the River Blackwater and tributaries was 0.07 territories per km in 2008, and 0.05 territories per km in 2010. However, birds were not evenly distributed, and there were large stretches of river without Kingfishers. Within the Blackwater system, the River Bride (north) consistently had higher numbers. The River Ilen held higher numbers than the River Blackwater system at 0.5 nests per km. There was no evidence the cold winter of 2009/10 affected numbers in the breeding season of 2010. A survey in 1992 of a 14 km stretch of the River Bride (south) between Farnanes and Classis found a minimum of five pairs (0.36 territories per km), but only two pairs (0.14 territories per km) during a repeat survey in 1994.

Winter atlas surveys in 1981–4 and 2007–11 showed that Kingfishers were present at rivers and lakes from the River Lee southwards to the coast. They were also present at many wetlands, lagoons, estuaries and inlets along most of the coast from Beara peninsula to Youghal. The Kingfisher is largely sedentary, but there is a movement towards the coast in autumn, and many birds recorded at that season and in winter are probably dispersing juveniles. While most birds occur as single individuals, twos and threes occur at many sites. Singles occur at several sites around Cork Harbour, and if all were occupied at once they would account for about fifteen birds. A total of sixteen Kingfishers were recorded at CCBO during 1959–96, fifteen occurred between 7 August and 5 October, and one on 18–31 March. Kingfishers have also occurred at Sherkin Island in autumn. One was killed striking the light at Bull Rock in September 1904.

Bee-eater

Merops apiaster | Beachadóir Eorpach | A

(monotypic)

World	Southern Europe, south-west Asia, and north and South Africa
Ireland	66 individuals
Cork	22 individuals
Conservation	Not globally threatened

There were three records of Bee-eater before 1959, including a flock of seven together (5 obtained) in the Whitegate and Trabolgan area in late April–early May 1888. The second and third records involved one inland at Aghern in August 1937, and two near Rosscarbery in April 1955.

Twelve records involving single individuals occurred during 1959–2018, ten in spring and two in autumn. Eleven of these twelve occurred at the western peninsulas and islands – Dursey Island (4), Beara peninsula (2), Mizen peninsula (1) and CCBO (4) – with the remaining record from Crosshaven in April and May 1964. Most birds (10) have remained for one or two days, and two for seven and ten days, respectively (mean stay = 2.5 days). One has been aged as a first-year bird and one as an adult.

1959–68	1969–78	1979–88	1989–98	1999–08	2009–18
2	0	2	1	2	5

Earliest and latest occurrence in spring and autumn (1959–2018): 19 April–4 June and 30 September–17 October.

Roller

Coracias garrulus | Rollóir | A

(subspecies: *garrulus*)

World	Europe, Asia, and north Africa
Ireland	20 individuals
Cork	4 individuals
Conservation	Not globally threatened

One obtained at Dunmanway in September 1851. First-year male obtained at Skibbereen on 29 October 1883. One at Aghada on 26 September 2011. First-year at Manch, Dunmanway on 16 November 2014.

Hoopoe

Upupa epops | Húpú | A

(subspecies: *epops*)

World	Europe, Asia, and Africa, also Canary Islands and Madagascar
Ireland	Scarce spring and autumn migrant
Cork	Rare spring and autumn migrant
Conservation	Not globally threatened

The Hoopoe was a rare bird in the mid-1700s. It occurred regularly in the 1800s, mostly in spring, but with records from January to October from coastal districts such as CCBO, Old Head of Kinsale and Youghal, but also inland at Bandon and Enniskean. Ussher said it occurred so frequently some might remain to breed if not so often shot, and he knew of thirty-six occurrences in the county. Although it was noted that Hoopoes occurred very frequently in Cork, only one new record was added during 1901–50, one killed, probably by a Sparrowhawk, on the beach near Shanagarry in April 1948.

There was an influx of Hoopoes in spring 1954 involving twenty-one birds. Two were seen in March, with seventeen in April and two in May. All were near the coast, with eleven at Whitegate. One was at Castletownshend in October 1954, followed by a smaller influx in April 1955 involving five birds, four of which were seen in the area extending from Rosscarbery to Toe Head, with one at Castletownbere. One was seen at the Mardyke in November 1957, one near Skibbereen in March 1958 and one at a Cork city suburb in October 1958.

A minimum of 263 Hoopoes occurred in the sixty years 1959–2018. While numbers have varied from zero in some years to thirty in others, the decadal mean has been forty-four birds, and four per annum, showing a remarkable consistency across thc period. There has been an approximately even distribution of records along the coast from Dursey Island to Youghal

Hoopoe

at many of the main migration watch points, such as Galley Head and Old Head of Kinsale, with many also at intervening points. About 22 per cent have occurred at CCBO and 26 per cent at east Cork sites. Hoopoes arrive on a broad front, as in years when there are many birds some occur in widely separated locations from west to east, the best years having been 1965 (30), 1980 (14) and 2005 (10). In the 1965 influx all birds occurred over the few days between 29 March and 5 April, while in 1980, nine were recorded between 11 and 17 April. On arrival, some birds tend to filter inland for distances of from 5 to 15 km, with the farthest from the coast being singles at Castlelyons (March 1975) and Macroom (2005).

There have been three definite spring arrivals in east Cork in late February. One has occurred in December, two in January and three in February in the milder climate of west Cork, indicating that winter survival is possible. Most have occurred on only one day, but individuals have stayed at one location for several weeks. Most records are of single individuals, although two and three have occurred together on a few occasions. The largest number together in the sixty-year study period refers to eight birds at Churchtown South in late March or early April 1968.

1959–68	1969–78	1979–88	1989–98	1999–08	2009–18
63	26	38	38	41	57

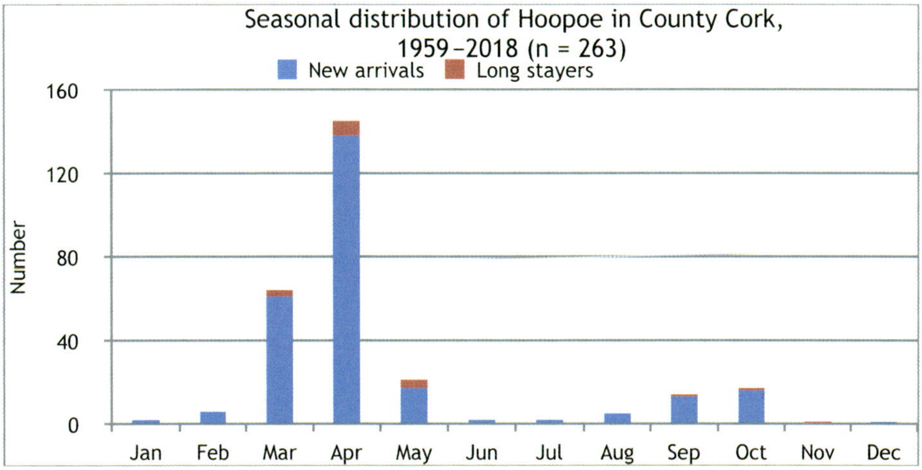

Fig. 7.26 Hoopoe.

Wryneck

Jynx torquilla | Cam-mhuin | A

(subspecies: *torquilla*)

World	Europe, Asia, and north Africa
Ireland	Rare passage migrant, mainly in autumn
Cork	232 individuals
Conservation	SPEC 3; Amber-listed

Two Wrynecks occurred in Cork before 1959, and both were males. The first was found dead at Fastnet Rock on 17 September 1898 and the second was obtained at Ballyhooly on 14 November 1925. The Ballyhooly Wryneck is the only inland Cork occurrence, and is also the latest date, and one wonders if it was attempting to overwinter.

Wrynecks have occurred as autumn passage migrants since 1959 (216), with a few in spring (14). They occurred at many sites on the coast from Dursey Island to Ballyvergan Marsh, but most occurred from Clonakilty west to Dursey Island (216). CCBO has had the most (94), followed by Dursey Island and the Beara peninsula combined (56), the Mizen peninsula (37) and the area around and including Galley Head (19). Old Head of Kinsale has had very few (2), and numbers thin out on reaching east Cork, where eleven birds have been recorded at four sites (Power Head, Ballycotton, Knockadoon Head, Ballyvergan Marsh). Most records are of one bird, but two, three and four occurred together on ten, three and three occasions, respectively. This is a difficult species to age in the field, and only one has been assigned a first-year bird. Wrynecks occurring in Ireland in autumn are likely to be displaced Scandinavian breeders as they migrate south-west, while those in spring are likely to be overshooting from southern Europe.

1959–68	1969–78	1979–88	1989–98	1999–08	2009–18
7	12	26	29	64	92

Earliest and latest occurrence in spring and autumn (1959–2018): 4 April–27 May and 13 August–10 November.

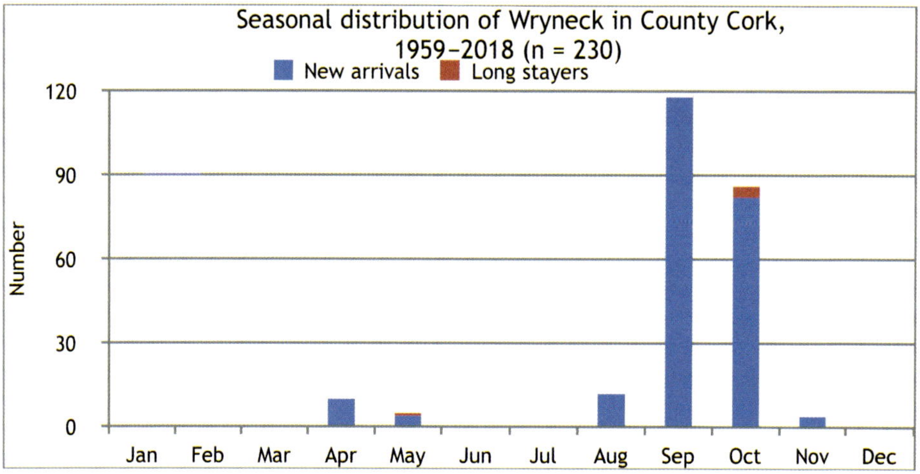

Fig. 7.27 Wryneck.

Yellow-bellied Sapsucker

Sphyrapicus varius | Súdhiúlaí tarrbhuí | A

(monotypic)

World	North America
Ireland	1 individual
Cork	1 individual
Conservation	Not globally threatened

First-year female at CCBO on 16–19 October 1988.

Great Spotted Woodpecker

Dendrocopos major | Mórchnagaire breac | A

(subspecies: *anglicus*)

World	Europe, Asia, north Africa, and Canary Islands
Ireland	Rare breeder, recent increase, rare winter visitor
Cork	Rare visitor to 2008, currently showing signs of colonisation
Conservation	Green-listed

The Great Spotted Woodpecker occurred singly in Cork on three occasions before 1959, obtained at an undisclosed location about 64 km from Cork city in 1862, obtained near Doneraile in October or November 1889 and seen near Aghinagh in early February 1950.

There were no further records until one was seen at CCBO on 19–20 April 2008. There have been several other records since (up to January 2021), all in potential breeding habitats (Leap, Bandon, Doneraile, Blarney, Carrigtwohill, Fota and Midleton). Drumming and test hole drilling has been noted at Doneraile, Blarney, Fota and at two woodland sites near Midleton. Given the rapid development of breeding populations elsewhere in Ireland, it is likely that Cork will soon have breeding Great Spotted Woodpeckers also.

Irish breeding woodpeckers are likely to have colonised from the British population based on mitochondrial DNA analysis.[252] Colonisation of Ireland took place following major range expansion in the British population, a trait identified in several other colonists.

Philadelphia Vireo

Vireo philadelphicus | Glaséan Philadelphia | A
(monotypic)

World	North America
Ireland	2 individuals
Cork	1 individual
Conservation	Not globally threatened

One at Galley Head on 6–17 October 1985.

Red-eyed Vireo

Vireo olivaceus | Glaséan súildearg | A
(subspecies: undetermined)

World	North and South America
Ireland	70 individuals
Cork	37 individuals
Conservation	Not globally threatened

Red-eyed Vireos occurred in autumn at coastal headlands and islands from Dursey Island to Knockadoon Head, peaking in the first ten days of October. Four occurred at east Cork sites, the rest being at Toe Head (1), Baltimore (1), but with most at CCBO (15), Mizen peninsula (8), Beara peninsula (5) and Dursey Island (3). All records refer to singles, but two overlapped at CCBO in 1988. Just under half were present for only one day, with the remainder staying for several days, and singles for nine and ten days, respectively (mean stay = 3.2 days).

1959–68	1969–78	1979–88	1989–98	1999–08	2009–18
1	1	6	12	5	12

Earliest and latest occurrence in spring and autumn (all records): 5 September–30 October.

Golden Oriole

Oriolus oriolus | Óiréal órga | A

(subspecies: *oriolus*)

World	Europe east to central Asia, and north Africa
Ireland	Rare migrant
Cork	108 individuals
Conservation	Not globally threatened

There were thirteen records (18 individuals) of Golden Orioles up to 1958, but only two after 1900 (1909 and 1952). Apart from one record of six near Bantry in April 1870, all were of single individuals, and all dated records were from April to June. Six records were from inland sites at Dunmanway, Bandon, Cloyne, Castlemartyr and near Midleton (2). The remainder were at coastal sites between Bantry and Youghal.

The Golden Oriole has occurred as an over-shooting spring migrant (86) since 1959 with most in May (69) and four in autumn. It has occurred in thirty-eight of the sixty years between 1959 and 2018. Most records were of one or two birds, but there is one record each of three and four (Old Head of Kinsale and CCBO, respectively), and up to fourteen were at CCBO in 1994. In only four years have five or more birds been recorded (21 in 1994, and five each in 1997, 2005 and 2006). Birds have occurred at many sites from Dursey Island to Knockadoon Head, and all except one relate

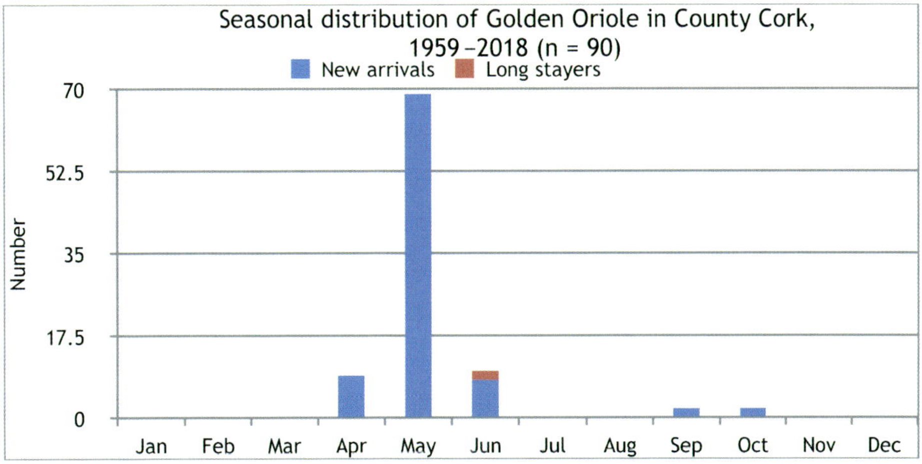

Fig. 7.28 Golden Oriole.

to coastal locations, this being one found dead at Castlemartyr in April. CCBO has provided records of most individuals (62), with a further eleven at Dursey Island and Beara peninsula. Only one other site has hosted more than two or three birds, with a total of five at Old Head of Kinsale. Most birds have been recorded on only one day (43), but others have remained for up to nine days, and one for eighteen days (mean stay = 2.5 days). Of those that have been aged and sexed, ten were adult, thirteen were first-year, twenty-seven were male and twelve were female.

1959–68	1969–78	1979–88	1989–98	1999–08	2009–18
8	5	13	39	17	8

Earliest and latest occurrence in spring and autumn (1959–2018): 17 April–23 June and 8 September–24 October.

Isabelline Shrike

Lanius isabellinus | Scréachán gainimh | A

(subspecies: *isabellinus*)

World	South-west and central Asia
Ireland	6 individuals
Cork	3 individuals
Conservation	Not globally threatened

One at Old Head of Kinsale on 17–20 October 2006. Adult female at Mizen Head on 19–21 October 2007. One at Toe Head on 31 October–1 November 2016. The Mizen Head bird has been assigned to subspecies *isabellinus*. The others remain undetermined.

Red-backed Shrike

Lanius collurio | Scréachán droimrua | A

(subspecies: *collurio*)

World	Europe east to central Asia
Ireland	Rare migrant
Cork	122 individuals
Conservation	Not globally threatened

There were three records of Red-backed Shrikes before 1959, all first-year birds obtained at Fastnet Rock in autumn 1910, 1930 and 1931.

Records since 1959 indicate an autumn passage migrant (105), with a smaller passage in spring (14). All birds occurred at coastal sites from Dursey Island to Youghal. More have been recorded at CCBO (60) than at other sites. The Mizen peninsula (21) and Dursey Island and Beara peninsula combined (19) have been the next most frequented areas. Smaller numbers have been recorded in the Galley Head area (6), Old Head of Kinsale (4), Ballycotton (4) and Knockadoon Head (3), with singles at Baltimore and Youghal. Most records are of single individuals, but four and two occurred together at CCBO (1976 and 2010), and two at Mizen Head (2006). Many birds have occurred on only one day (57), but many remained from two to five days (45), while seventeen birds were present for longer periods with a maximum of fourteen days (mean stay = 3.0 days). Of those that have been aged and sexed, seventy-four were first-year and eight were adult birds, while nine males and eight females were identified.

The peak autumn migration is late September and October and the birds involved may be reverse migrants from a southerly European breeding

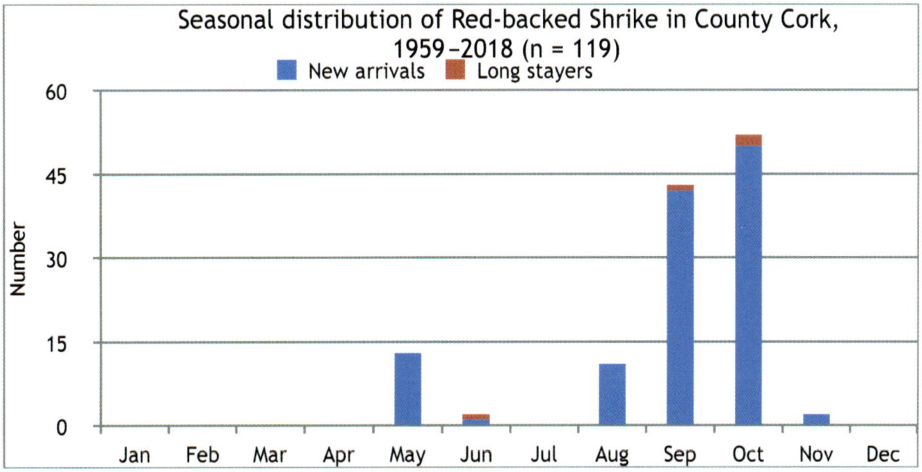

Fig. 7.29 Red-backed Shrike.

range, while records in August may involve birds with a more central European origin displaced by easterly winds.[253]

1959–68	1969–78	1979–88	1989–98	1999–08	2009–18
13	15	17	24	22	28

Earliest and latest occurrence in spring and autumn (1959–2018): 1 May–10 June and 6 August–24 November.

Unidentified shrike species

One shrike, either Red-backed or Woodchat (*Lanius collurio/senator*), was recorded at CCBO on 27 September 1961.

Lesser Grey Shrike

Lanius minor | Mionscréachán liath | A

(monotypic)

World	Southern and eastern Europe east to central Asia
Ireland	6 individuals
Cork	2 individuals
Conservation	Not globally threatened

First-year at Ballycotton on 6 September 1985. First-year at CCBO on 18 September 2005.

Great Grey Shrike

Lanius excubitor | Mórscréachán liath | A

(subspecies: *excubitor*)

World	North America, Europe, and Asia
Ireland	48 individuals
Cork	4 individuals
Conservation	Not assessed

One obtained near Cork city in 1834. One obtained at Carrigaline in late October 1844. One obtained at Carrigaline in early August 1845. One at Courtmacsherry on 28 October 1991.

Unidentified shrike species

One shrike, either Great Grey or Lesser Grey (*Lanius excubitor/minor*), was recorded at Cloyne on 12 December 1966.

Woodchat Shrike

Lanius senator | Scréachán coille | A

(subspecies: *senator*; *badius*)

World	Southern Europe, south-west Asia, and north Africa
Ireland	Rare spring and autumn migrant
Cork	47 individuals
Conservation	Not globally threatened

There was one record of a Woodchat Shrike (*L.s. senator*) before 1959, an adult female found dead at Whitegate in May 1951.

Occurrences during 1959–2018 indicate a spring (31) and autumn migrant (15). All records are of single individuals, and all except one (7 km inland at Castlemartyr) occurred on the coast. Most have been recorded at CCBO (17), the Mizen peninsula (9), Old Head of Kinsale (3) and Knockadoon Head (3), but there are records of one or two birds at many sites from Dursey Island to Pilmore Strand (14). Of those that have been aged, fourteen were first-year and fifteen were adult, and there were fifteen males and four females. Sixteen birds remained for only one day, but seventeen remained for longer than five days, one for twenty-three days (mean stay = 5.0 days).

Balearic Woodchat Shrike *L.s. badius*: There is one record of this subspecies. Adult at Mizen Head on 3–5 June 2002.

1959–68	1969–78	1979–88	1989–98	1999–08	2009–18
3	4	4	7	11	17

Earliest and latest occurrence in spring and autumn (1959–2018): 27 March–23 June and 12 July–20 October.

Chough

Chough

Pyrrhocorax pyrrhocorax | Cág cosdearg | A

(subspecies: *pyrrhocorax*)

World	West and south Europe, Asia, north Africa, Ethiopia, and Canary Islands
Ireland	Uncommon resident breeder
Cork	Local, mostly coastal resident and breeder
Conservation	SPEC 3; Amber-listed

The Chough was common in the 1700s and frequented rocks, old castles and ruins on the coast. In east Cork, Robert Ball and William Thompson saw Choughs at Capel Island in summer 1834, and they knew it frequented other parts of the Cork coast. However, the Chough became scarce and then disappeared from the east shortly before 1900, but it remained common in the west, especially on the promontories and islands. The cause of the decline in east Cork is unknown. It occurred throughout the year at Dursey Island, Crookhaven (very numerous), Old Head of Kinsale and Ballycotton

Island in the late 1800s, and up to ten in winter, and over forty on one occasion, were reported from Bull Rock.

The Chough is characteristic of the Cork coast, and between Ardgroom and Capel Island there are few cliffs without a pair. A survey put the Cork population at between seventy-three and ninety-eight pairs in 1960–5.[254;255] This survey was incomplete, but at that time the population was believed to be larger than previously. However, the severe winter of 1962/3 was thought to have reduced numbers between Kinsale and Crosshaven, although populations at CCBO were not affected. Numbers were said to be holding up, if not increasing, in the south of the county.

Three comprehensive surveys have subsequently been carried out, and the results are summarised in Table 7.18. Inland breeders were all in west Cork and ranged from 3 to 24 km from the sea. Many non-natural nest sites were used including signal towers, quarry buildings, houses, a lighthouse, an old fort, the underside of a bridge, mine shafts and associated buildings, but most pairs used caves and crevices in sheer rock cliffs. There is evidence from recent work that farm buildings, including modern ones, are increasingly used as nest sites. These studies emphasised the importance of short-grazed pasture in which Choughs can dig for food, and the presence of insect-rich undisturbed soils, factors which might render the species vulnerable to future agricultural changes. In all surveys, density was greatest at the western peninsulas and islands. Exceptionally large post-breeding flocks of seventy-seven at Old Head of Kinsale in July 2003 and sixty-seven at Dursey Island in July 2000 have occurred.

Outside the breeding season Choughs visit saltmarshes, beaches and sand dunes in search of food. Pairs or small groups may fly up to 10 km inland to feed on farmland, and this behaviour has been widely observed around Kinsale, Riverstick, Minane Bridge, Cork Harbour, Ladysbridge

Year	Breeding (pairs)	Non-breeding (birds)	Islands (pairs)	Inland (pairs)
1982	153	171	34	5
1992	282	292	39	18
2002–03	257	251	30	12

Table 7.18 Breeding censuses of Chough in County Cork, 1982–2003.

and Ballymacoda. One was at the Gearagh on 28 July 2018. Largest flocks are seen in west Cork. Flocks of seventy-eight and sixty-five have occurred at Dursey Island in September, sixty-seven at Garinish (Beara peninsula) in September, sixty at Barley Cove (Mizen peninsula) in September, fifty at CCBO in November, fifty at Old Head of Kinsale in November, and forty-two at Seven Heads in September. A cliff-roosting flock of forty-five birds was noted at Allihies in August 2020.

Choughs feed mainly in coastal heathland and on cliff sides, as well as in adjoining semi-improved grassland. However, they show a level of flexibility and some feed in newly mown silage fields among other crow species. In winter, they sometimes feed in improved pastures, as well as in arable fields where root crops have been harvested. They sometimes feed opportunistically where prey is abundant or easy to obtain, and observations have been made of a bird feeding on grasshoppers, of a family feeding on Lepidoptera larvae, and of a flock feeding in burned heathland.

Magpie

Pica pica | Snag breac | A

(subspecies: *pica*)

World	Europe, Asia, and north Africa
Ireland	Very common resident breeder
Cork	Common resident and breeder
Conservation	Green-listed

The Magpie was unknown in Ireland about 1680 but was common by 1750. In the late 1800s Magpie hawking, using Peregrine Falcons, was practised around Fermoy. It was exceedingly common at Castlemary up to 1945 but had become scarce by 1957, and in 1964 it was rare. The breeding population at CCBO varied from eight to fifteen pairs in the 1960s, with sixteen pairs in 1986. The number of breeding pairs at Sherkin Island varied from five to sixteen pairs during 1977–94, while one or two pairs breed regularly on the almost treeless Dursey Island since the 1990s.

The Magpie was present throughout the county during the three breeding atlas surveys. Density was highest in the agricultural regions of the south and east. Large numbers are shot each year as a game protection measure, and 12,905 were killed during the seasons of 1982/3–1983/4, although populations remain at a high level. Winter season distribution mirrors the breeding season and includes bare coastal districts.

Magpies are sedentary, making short and local dispersive movements. These movements generally occur in autumn and early winter, and flocks are often seen. Flock movements are obvious at coastal headlands and islands and often exceed 100 birds, which usually remain at any location from a few hours to a few days, e.g. 500 at Mizen Head in October 1995, 320 at CCBO in November 1990 and 110 at Old Head of Kinsale in November 1991. Smaller flocks occur at Dursey Island and other coastal sites. A flock of forty-one flew in over the sea at Old Head of Kinsale in October 1983 (weather conditions at the time were cyclonic), and one was at Bull Rock in October 1975.

Jay

Garrulus glandarius | Scréachóg | A

(subspecies: *hibernicus*)

World	Europe, Asia, and north Africa
Ireland	Resident breeder
Cork	Widespread resident, nowhere numerous, increasing of late years
Conservation	Green-listed

Robert Ball said the Jay bred near Youghal in 1837, but his view was that it was rare. Reports in 1839 that it had recently become common around Bandon due to protection by Lord Bandon were not borne out by his successor, who stated in 1893 that he had never heard of Jays being seen there. It was said to have bred at Castlehyde but was absent by the early 1890s. In 1900, Ussher described it as formerly present in the River Blackwater valley (Cork), but it had been exterminated due to killing for its wing feathers, which were used in making salmon fishing flies. At that time (1900) it was believed to be absent as a resident from Cork, although it irregularly wandered from adjoining counties where it was breeding. Three records of birds obtained about this time have been detailed: 1888 from Doneraile, 1898 from Rostellan, and 1900 from Upton.[256]

The reason why the Jay disappeared from the River Blackwater valley has been debated and described as apocryphal.[257] These authors argue that shooting of Jays for making fishing flies was not the reason for its disappearance; they accepted that shooting for fishing flies was taking place, but that the more likely cause was habitat destruction, i.e. woodland felling.

Cork was recolonised in the early 1910s, but the subsequent expansion has been poorly documented. Jays became established about 1912 at Convamore and Ballyhooly woods and numbered six to eight birds in 1915. A pair appeared at Longueville about 1921, and by 1926 it was said to be very numerous there; in September that year a flock of twenty was seen, and later a larger flock, but these soon disappeared, leaving the local population unchanged. It was very scarce in east Cork, especially south of Midleton, and one at Castlemary in April 1964 was noteworthy. In November 1964 about six individuals were present near Whitegate, an area where they were normally seldom seen. Jays were seen regularly in 1964, but not in any numbers, in the wooded valleys of the Rivers Leamlara and Glashaboy, at Brooklodge, Fota and Ballyhooly, all in east Cork, and it was extending its range in new plantations.

Between 1968 and 2011 Jays expanded from populations at Glengarriff and the valleys of the Rivers Blackwater, Lee and Bandon to populate almost the entire county, except for some coastal regions and upland areas, where woods are isolated or absent. There is almost no difference between winter and breeding season distribution, and Jays now occur regularly in suburban gardens at Blackrock and Sunday's Well in Cork city.

The Jay is sedentary in Ireland and Britain, but populations in northern Europe are irruptive. It has not been recorded in sixty years at CCBO, but it has occurred twice at Sherkin Island, on 26 October 1997 and 22 October 2005. One occurred at Dursey Island on 23 October 1997. It has also occurred near the western end of the Beara peninsula, and seven were seen on 15 October 2011. It has occurred several times at Mizen Head in October. One flew out to sea at Roche's Point in February 2001, and seven flew high over Cobh going west on 7 November 2005. Whether any of these were birds of European origin is unknown.

Jackdaw

Coloeus monedula | Cág | A

(subspecies: *spermologus*; *monedula/soemmerringii*)

World	Europe east to central Asia, and north-west Africa
Ireland	Common resident breeder
Cork	Common resident breeder
Conservation	Green-listed

The Jackdaw (*C.m. spermologus*) was resident and common throughout the county, breeding in sea cliffs (Reanies Bay) and chimneys in the 1800s. They become extremely common at Cobh in the late 1800s, following the roofing of the cathedral.

The range of the Jackdaw extended across the county, on coasts and inland, during all three breeding atlas surveys, with density appearing greatest in the north-west, and in coastal parts of the south and east. Winter surveys showed the distribution was identical to the breeding season. The sparsest populations were generally in areas away from human activity and settlements. Many pairs breed along the coast, where their preference for crumbling cliffs reduces competition with Choughs. Inland, most nests are in chimneys and in ruined buildings, with some also using quarries and trees. The Jackdaw is loosely colonial with many small groups of five to ten pairs, with larger numbers occurring at cliffs and extensive ruins, although many breed as single pairs.

Jackdaws did not breed at CCBO during the 1960s but have colonised since the late 1970s and are increasing (10 pairs bred in 1986). Jackdaws also nest at Sherkin Island (though numbers declined from about sixty to sixteen pairs between 1977 and 1994), but do not do so at Dursey Island.

Flocks are formed after the breeding season; at Cobh up to 3,000 collect to roost in late summer and autumn, following well-defined narrow flight paths from Midleton and Carrigaline. These, and accompanying Rooks, move to a larger roost at Fota in winter. Flocks of 1,000 birds have occurred at Rochestown in December, and of 400 at Doolieve Wood in January, and 200 were roosting on the island in Cork Lough in December 2006.

Evidence of migration among Irish birds is slight, but some birds arrive from Britain and the continent for the winter (see below). Large numbers occur at coastal headlands and islands in autumn. It is believed these flocks (including Rooks) move out to the islands and beyond in high-pressure systems, returning to the mainland at dusk. Largest numbers are present at CCBO in October when flocks arrive daily from the mainland, and 500–2,500 birds are often involved in such movements. Smaller flocks visit Dursey Island during similar weather conditions, peaking at 380 in October 2010. Elsewhere on the coast, a flock of 1,500 (with Rooks) was recorded at Rathbarry (near Galley Head) in October 2005. The Jackdaw was once reported at Fastnet Rock in October.

Northern/Eastern Jackdaw *C.m. monedula/ soemmerringii*: Jackdaws from northern and eastern European subspecies have been identified in Ireland, mostly in winter, and five records have been documented for Cork: CCBO October 2000; Rochestown November 2001; Ballycotton December 2004; Sherkin Island April 2005; Sherkin Island October to November 2005.

Rook

Corvus frugilegus | Rúcach | A

(subspecies: *frugilegus*)

World	Europe and Asia, introduced elsewhere
Ireland	Very common resident breeder
Cork	Very common resident breeder
Conservation	Green-listed

The Rook was regarded as a pest in the 1700s and destroyed much grain, but despite thousands being killed it remained very numerous. Rooks take a wide range of food items, from the shore and from land. They commonly take marine invertebrates at estuaries, beaches and rocky shores. About Christmas 1846 Robert Warren saw Rooks feeding on dead or dying sprats taken from the surface of Cork Harbour. Many dead Rooks were once washed ashore at Cork Harbour, during a dense fog in the 1800s.

The range of the Rook extended across the county during all three breeding atlas surveys. Breeding density was mostly high, but was lower at the western peninsulas, presumably where nesting and feeding opportunities are least favourable. There was no difference between breeding and winter season distribution with the densest populations in the agricultural east and north.

There is often a resumption of breeding behaviour in the autumn (this also applies to some other species), and some even begin to build nests, but this seldom leads to breeding. However, fresh eggs are occasionally reported in autumn, and a nestling was seen on 17 November, following a spell of very mild weather. A statement that Rooks nest on Sherkin Island is incorrect.[258]

Although there is no evidence that the Rook is migratory in Ireland, birds have been recorded in autumn at Fastnet Rock. Flocks of over 100 occasionally occur at CCBO, mainly in autumn, many arriving from the mainland in the morning during high-pressure systems and departing again at dusk. Rooks tend to be scarcer at Dursey Island in autumn, and they are often absent, about 100 birds being recorded in October 2010. There was a mixed flock of 3,000 Rooks and Jackdaws at Mizen Head in October 1995. A record of thousands of Rooks flying north-west at Spit Bank in November 1890 suggests local movement rather than true migration, perhaps involving food-seeking or a roost flight-line.

Rooks (often combined with Jackdaws) form flocks after the breeding season. Such flocks may roost at the breeding rookery or at other sites. They then often follow defined flight-lines to favoured feeding areas, before returning to the roost at dusk. Feeding areas include pastures, recently cut silage fields, cereal stubble, and cultivated land. Roost flocks may sometimes be very large, possibly involving tens of thousands of birds, but usually the numbers are smaller. There were combined roosts of 14,000 at Fota in December 1989, and of 8,000–9,000 at Doneraile in December 1994. There were roosts of 2,000 Rooks at Rushbrooke in summer 1989, and of 1,000 Rooks at Cuskinny in December 1989.

Carrion Crow

Corvus corone | Carróg dubh | A

(subspecies: *corone*)

Ireland	South-west Europe, and Asia
Ireland	Has bred, scarce spring and autumn visitor
Cork	Rare spring and autumn passage migrant
Conservation	Not globally threatened

The Carrion Crow was rare in Cork in the 1700s. Although Harvey said it was resident (clearly, he was mistaken), Robert Ball and Joseph Stopford never encountered it at Youghal or in west Cork, respectively, in the 1800s. William Corbet claimed to have shot old and young birds at Trabolgan, and to have seen it at Ballycotton. Ussher and Warren knew of a Cork specimen in the British Museum, and one (stated to be paired to a Hooded Crow) was reported shot near Rostellan in March 1884, with others shot there in previous years. However, the foregoing cannot be taken as referring to confirmed breeding. The only occurrence in the 1900s before the opening of CCBO appears to be two birds seen at Lombardstown on 6 August 1956.

Winter atlas surveys show a scatter of occurrences from CCBO to Midleton, and north to Newtownshandrum. The impression gained is of a straggler rather than a species with a constant presence in the county.

Records since 1959 indicate an autumn passage migrant, with a much smaller spring passage. A few birds occurred in winter months.[259] Most Carrion Crows occur at CCBO (68 per cent), and the remainder at coastal sites from Dursey Island to Youghal. There are a few inland records (3) apart from those alluded to in the last paragraph. Most records are of birds seen on one day, although some have stayed for longer, some possibly for several months over the course of a winter. One bird remained near Cloyne from October 1967 to the breeding season of 1968 when it was apparently paired to a Hooded Crow, but breeding was not proved. Another stayed throughout 1976 at Roche's Point. Most records are of one to five or six birds, but twenty were at CCBO in October 1975 and ten in October 1977. It is now much rarer at CCBO than formerly.

1959–68	1969–78	1979–88	1989–98	1999–08	2009–18
53	99	51	12	31	30

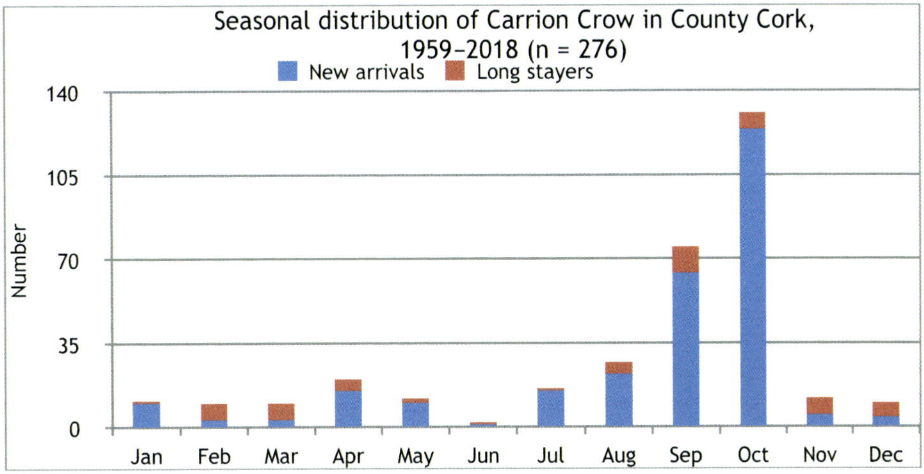

Fig. 7.30 Carrion Crow.

Hooded Crow

Corvus cornix | Caróg liath | A

(subspecies: *cornix*)

World	Western Europe to central Asia, and north-east Africa
Ireland	Very common resident breeder
Cork	Common resident breeder
Conservation	Green-listed

The Hooded Crow was a common bird in the 1700s. It nested in sea cliffs as well as in trees in the 1800s. Warren recorded six nests in a demesne of 24.3 ha near Cork (probably Coolmore) and seven nests in one season at Ballybricken. As many as fifteen or sixteen nests were recorded on the Myross peninsula in west Cork within an area of 5.6 km by 3.2 km. It remains a common species, and eleven to twenty-five pairs breed annually at CCBO, mostly in sea cliffs. About eight to ten pairs breed at Sherkin Island. Despite its importance as a predator, very little has been published on the breeding biology of the Hooded Crow. A mean of two chicks fledged from a total of fourteen nests at Lough Hyne over three years 1988–90, where a maximum of five pairs held territories that included part of the

Hooded Crow

lough.[260] Although it is generally a shy bird, and will often shun built-up areas, a pair nested on a tree in South Mall, Cork city in 2016.

The range of the Hooded Crow extended across the county in all three breeding atlas surveys. Density was high throughout, especially in the agricultural east and north. Winter atlas surveys showed little difference in distribution from the breeding season.

Hooded Crows frequently occur in large flocks where food is plentiful, often in coastal districts on mudflats and sandflats, and around landfill sites (e.g. flocks ranging from 62 to 180 have been recorded at CCBO, Sherkin Island, Rosscarbery, Currabinny Wood, Ballycotton, Youghal, Mallow and Charleville). The largest flock on record is of 240 at Cork Cricket Club (Mardyke) on 17 December 2018. Hooded Crows are frequently killed as a game protection measure, and 22,300 were reported shot in the winters of 1982/3 and 1983/4, but it has shown no signs of decline.

Hooded Crows are important foragers and predators on intertidal molluscs and other invertebrates, and their impact, diet, food-caching behaviour, and distribution have been studied at Lough Hyne and other west Cork estuarine habitats.[261;262;263] They are opportunist in nature and will exploit prey that is occasionally available. Although Hooded Crows do not migrate, they frequently visit Fastnet Rock in quest of food.

Raven

Corvus corax | Fiach dubh | A

(subspecies: *corax*)

World	North America, Greenland, Iceland, Europe, Asia, and north Africa
Ireland	Resident breeder, recent increase
Cork	Thinly spread resident breeder, recent increase
Conservation	Green-listed

The Raven was well known in the 1700s. It nested in tall trees and old towers, and there was a ravenry at Ightermurragh Castle where they nested annually. A pair nested at Reanies Bay in 1848, where Robert Warren saw them predating the nests of Great Cormorants. A pair also nested on a Scots pine at Ballybricken, beside Monkstown Creek, before 1851. In the fifty years 1850–1900 the Raven was much reduced in numbers throughout Ireland, especially at inland sites, due to systematic shooting, trapping and poisoning by gamekeepers and shepherds. By the early 1890s it was very scarce and local in Cork and bred only at a few remote sea cliffs, those in west Cork being the most frequented, although Derek Ratcliffe suggested some must have persisted at wild and little-trodden districts at inland sites.

There was a slow recovery in the early to mid-1900s. Between 1960 and 1968 breeding took place in five coastal 10 km squares ranging from CCBO to Knockadoon Head. Breeding also took place in two inland 10 km squares, but neither of these apparently referred to tree nesting. However, the extent of the range in the 1960s was probably underestimated.

During 1968–72 the breeding distribution was mostly confined to the coast. It was rare inland, being commonest in uplands on the border with Kerry, but with outliers at Crossbarry and Ballyhoura Mountains. The surveys of 1988–91 and 2008–11 showed a major extension of range to almost all parts of the county, inland as well as coastal. The indications are that the 1970s to the 2000s was a good period for Ravens, with pairs nesting at most coastal cliffs, and inland at quarries, ruined buildings and trees. Although usually a shy bird, it occasionally occurs over Cork city and a pair has nested near University College. Winter distribution since the 1970s has mirrored the breeding season, with indications of higher density at the western peninsulas and in the north-west.

One or two pairs have bred (or have attempted to breed) annually at CCBO since 1959, although four pairs bred in 1983 and 1984, and six pairs in 1986. Only two pairs bred in 1990 and this was apparently due to an outbreak of myxomatosis in the rabbit population, therefore food is likely a limiting factor. There is usually one breeding pair and one non-breeding pair at Sherkin Island, while one or two pairs breed in most years at Dursey Island. A study area of approximately 840 km^2 in east Cork and west Waterford had forty-five nesting sites in the late 1980s and early 1990s distributed on coastal cliffs (17), trees (22), rock quarries (3) and ruined buildings (3).

Post-breeding flocks are common in autumn (August to October), especially at coastal sites, and where food is plentiful. A flock of seventy-six at Dursey Island in October 1986 is the largest on record, and there are several high counts from CCBO ranging from forty to sixty-one during August to October. A flock of thirty-two was seen at Mizen Head in September 1988. Flocks also occur inland, and there were fifty-eight near Midleton in June 1995 and forty at the Gearagh in January 1995. Flocks may also form at winter roosts (50 at Skeagh forest in December 1992) and at quarry and dump sites (40–5 at Benduff slate quarry and dump in 1986, 32 at Kildorrery dump in October 1995, 23 at a quarry near Midleton in April 1995 and 15 at Glengarriff dump in 1992).

The diet of Ravens in Cork has been studied from pellet analysis at two locations, two nest sites at Roaringwater Bay and one nest site at Ballycotton.[264;265] At Roaringwater Bay, vertebrate remains occurred in 85 per cent of pellets, believed to be from sheep and other large species which had been fed upon as carrion. Rabbit was the only wild mammal identified. Marine species, insects, eggshell and bird remains were also identified. At Ballycotton, diet included a range of wild mammals, such as fox, rabbit, brown rat and field mouse, as well as common frog, birds, insects, eggshell and a variety of marine species, but there was no evidence of feeding on large domestic animals.

There are no data relating to the movements of Ravens in Cork, but elsewhere there is a dispersal of young birds away from the natal area, although in general it is a sedentary species. Elsewhere, movements of over 50 km may be made daily between feeding and roosting areas outside the breeding season.

Goldcrest

Goldcrest

Regulus regulus | Cíorbhuí | A

(subspecies: *regulus*)

World	Europe, Asia, Azores, and Canary Islands
Ireland	Common resident breeder, passage migrant and winter visitor
Cork	Resident breeder, passage migrant and winter visitor
Conservation	SPEC 2; Amber-listed

Goldcrests were present in every spruce plantation in the mid-1800s, and many arrived in winter from northern Europe. The breeding range extended throughout the county during all three atlas surveys. Density was high except for coastal areas of west Cork. There are no differences between breeding and winter season distribution. One to three pairs bred regularly at Sherkin Island during 1977–94, while it bred for the first time at CCBO in 1986.

Spring passage is generally very light, and at CCBO this is typically reflected by counts of up to ten in late March and April. The exception was an arrival of at least 600 on 3 April 1988. Autumn passage is usually much heavier, both at CCBO and on the mainland. It extends from late August through to November, peaking in mid-October. Numbers are highest at CCBO, with an average of over 1,000 bird-days annually in September and October, and influxes on 26 September 1998 (400) and 4 October 1987 (300). Elsewhere, counts exceeding 100 birds are rare.

Because of its diminutive size, the Goldcrest suffers high mortality during cold winter weather such as 1961/2 and 1962/3. In the autumns of 1962 and 1963 there were zero and ten bird-days, respectively, at CCBO. These figures can be compared with the autumns of 1959–61 (average of 180 bird-days per annum) and 1964 (341 bird-days), which indicates how quickly the population recovered.

There are autumn records at Bull Rock, Fastnet Rock, Crookhaven, Old Head of Kinsale and Ballycotton lighthouses. Goldcrests come to Ireland and Britain in winter from northern Europe and from France. Ringing has shown movements between Cork and Wexford, west Wales and southern England.

Firecrest

Regulus ignicapilla | Lasairchíor | A

(subspecies: *ignicapilla*)

World	Europe, south-west Asia, and north Africa
Ireland	Scarce autumn visitor, very rare in spring and winter
Cork	Scarce visitor in autumn, rare in spring and winter
Conservation	Green-listed

There was one record for Cork before 1959, one found freshly dead near Glengarriff on 7 December 1943.

The next records were in October 1959 when seven were seen at CCBO. Since then it has occurred in very variable numbers, with none in 1963, 1965, 1970, 1974 or 1992. On the other hand, thirty or more occurred in 1978, 1980, 1989, 1990, 2000, 2003 and 2005. This increase is in line with an expansion in range and numbers in northern Europe. It occurs predominantly in autumn (95.5 per cent). Birds typically arrive during south-east winds resulting in westerly displacement of continental migrants. A few have been seen in winter and spring, and at least some are believed to have overwintered, e.g. at CCBO in 1975/6 and 1982/3. Although most birds have been recorded on only one day, some remain for longer periods.

During influxes, numbers tend to decline from a peak over the following week or two as birds move on. Turnover at some sites makes it difficult to accurately assess the total in years when numbers are high. Most occurrences are of single birds, but low double figures have been seen at CCBO, peaking at thirty-three on 5 October 2005.

Most birds have been recorded at CCBO, but other west Cork sites have also hosted high numbers, such as Dursey Island, Beara peninsula, Mizen peninsula, Galley Head and Old Head of Kinsale. Smaller numbers occurred at east Cork sites, such as Power Head, Ballycotton and Knockadoon Head. Apart from a few in Cork city, there are only two inland records, singles at the Gearagh in April and at Kealkill in January.

1959–68	1969–78	1979–88	1989–98	1999–08	2009–18
32	84	145	150	215	132

Earliest and latest occurrence in autumn and spring (1959–2018): 2 September–11 April.

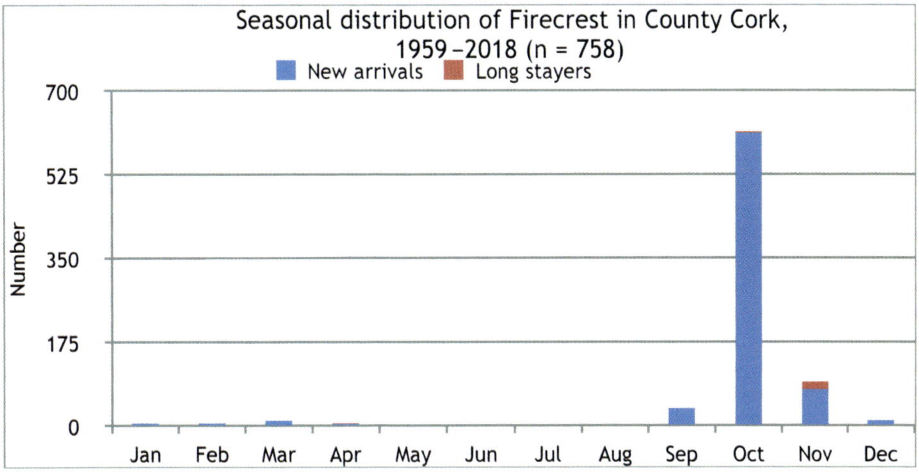

Fig. 7.31 Firecrest.

Ruby-crowned Kinglet

Regulus calendula | Rí beag deargchorónach | A

(subspecies: undetermined)

World	North America
Ireland	1 individual
Cork	1 individual
Conservation	Not globally threatened

First-year female at CCBO on 27 October 2013.

Blue Tit

Cyanistes caeruleus | Meantán gorm | A

(subspecies: *obscurus*)

World	Europe, western Asia, and north Africa
Ireland	Very common resident breeder
Cork	Common resident and breeder
Conservation	Green-listed

Blue Tits were distributed throughout the county during all three breeding atlas surveys, and there was no difference between the breeding and winter season. It bred irregularly at CCBO in the 1960s, ten pairs bred in 1986 and a small population is now established. About ten pairs (occasionally more) breed at Sherkin Island, but it does not breed at Dursey Island.

Breeding biology was studied in farmland in east Cork.[266] Breeding success (proportion of eggs that produced fledged young) at 80 per cent was higher than in woodland at Coleraine, and considerably higher than in urban Belfast (Northern Ireland). The mean egg-laying date was 5 May (n = 96), and the mean clutch size was 7.75 eggs (n = 100).

Irish Blue Tits are sedentary. An arrival of birds occurs at CCBO each year between August and October, and numbers present have increased in recent years, but this may partly reflect the increased breeding population, and probably involves birds dispersing from the mainland and Sherkin

Island. Significant autumn numbers have not been noted at other sites, although there were seventeen at Galley Head in October 1991. One was at Bull Rock on 18 October 1945 after a south-east gale, and one ringed at Bardsey Island (Wales) on 4 October 2003 was recovered at CCBO on 23 October 2003.

Great Tit

Parus major | Meantán mór | A

(subspecies: *newtoni*)

World	Europe and Asia south to Timor, and north Africa
Ireland	Very common resident breeder
Cork	Common resident and breeder
Conservation	Green-listed

Great Tits were present throughout the county during all three breeding atlas surveys, and there was no difference between the breeding and winter season. A few breeding attempts took place at CCBO in the 1960s, but successful breeding did not occur until 1971, and there were four pairs by 1986. There has been a sustained increase in occurrences since about 1980, and the breeding population is now established. There is a breeding population of up to ten pairs at Sherkin Island.

Breeding biology was studied in farmland in east Cork.[267] Breeding success (proportion of eggs that produced fledged young) at 81 per cent was higher than in woodland at Coleraine, and considerably higher than in urban Belfast (Northern Ireland). The mean egg-laying date was 7 May (n = 81), and the mean clutch size was 6.25 eggs (n = 80).

Irish Great Tits are sedentary. There is typically an autumn influx to CCBO, presumably of birds dispersing from the mainland, and a first-year bird ringed there in October 2018 was captured near Clonmel (Tipperary) in February 2019. Significant autumn influxes have not been noted at other sites. One was obtained at Fastnet Rock on 16 October 1899.

Coal Tit

Periparus ater | Meantán dubh | A

(subspecies: *hibernicus*; *ater*)

World	Europe, Asia, and north Africa
Ireland	Very common resident breeder
Cork	Common resident breeder
Conservation	Green-listed

The Coal Tit (*P.a. hibernicus*) was distributed throughout the county during all three breeding atlas surveys, but was scarce at the western peninsulas, and there was no difference between the winter and breeding season.

Irish Coal Tits are birds of coniferous woodland, where Blue and Great Tits may be scarce, and where winter counts may reach forty-five, as at Doolieve Wood in January 1990. They are also found in deciduous woodland, parks and gardens. In winter, they often accompany other tit species, Goldcrests and Treecreepers.

Coal Tits are sedentary, but occasionally occur at offshore islands. Although there were no records at CCBO (see below) before 1970, there were regular occurrences of from one to twenty individuals up to 1986. The increase at CCBO has continued, and a maximum of 132 birds occurred on 12 October 1997. Coal Tits have occasionally bred at Sherkin Island, and small autumn influxes have occurred there and at Dursey Island.

Coal Tits are occasionally seen at mainland headlands, mostly in autumn. Unusually heavy passage was recorded at Galley Head in 1991, with twenty-nine there and twenty-three at nearby Sands Cove on 11 October, and forty-five at the latter on 12 October.

European Coal Tit *P.a. ater*: There are two records of this subspecies. One at CCBO on 6–11 October 1959. One at CCBO on 13 October 1961.

Bearded Tit

Panurus biarmicus | Meantán croiméalach | A

(subspecies: undetermined)

World	Southern Europe and southern Asia
Ireland	Rare breeder and rare migrant to south and east coast counties
Cork	Minimum of 15 individuals (3 records)
Conservation	Amber-listed

Eight (including 3 males) at CCBO on 13 October 1972. Six at Ballycotton on 28 October–early November 1979, one male and two females on 28 September 1980 and one or two to December 1980, one male from 11 January 1981 onwards, and five in May 1981; various sightings of singles (including 1 male) to 28 June 1981, one male on 27 January 1982. Male at Ballymacoda on 15 April 1981.

It is likely the birds present at Ballycotton between October 1979 and January 1982 refer to the same group (peak of 6), and although there were significant gaps between some sightings, all observations are treated as referring to one record. Juveniles were never seen at Ballycotton, and there was no confirmed evidence of breeding there.[268;269]

Greater Short-toed Lark

Calandrella brachydactyla | Fuiseog ladharghearr | A

(subspecies: undetermined)

World	Europe, Asia, and north Africa
Ireland	83 individuals
Cork	38 individuals
Conservation	Not globally threatened

Short-toed Larks have occurred in coastal districts in spring (17) and autumn (21). The first ten occurrences (1962–89) were all in autumn, but there has since been a reversal of this trend with seventeen in spring and eleven in autumn. CCBO (18) and Dursey Island (11) have hosted most, with others at Ballycotton (4), Knockadoon Head (1) and Pilmore Strand (1) in the east, while the Mizen peninsula (2) and Galley Head (1) are the other sites in the west. All records are of single individuals, although two briefly overlapped at CCBO in October 2011. They tend to remain for several days and although fourteen have been recorded on one day, most have remained for longer periods, some for up to sixteen days with one for twenty-five days (mean stay = 5.1 days). The age of only two birds has been published; both were adults.

1959–68	1969–78	1979–88	1989–98	1999–08	2009–18
2	0	7	10	9	10

Earliest and latest occurrence in spring and autumn (all records): 25 April–5 June and 28 August–24 October.

Woodlark

Lullula arborea | Fuiseog choille | A

(subspecies: *arborea*)

World	Europe, western Asia, and north Africa
Ireland	Former breeder, now very rare visitor
Cork	Former breeder, now rare vagrant
Conservation	Not globally threatened

The Woodlark was a resident in the 1700s and 1800s. Robert Ball said it was frequent and sought by bird-catchers, which possibly led to its extinction. No information on range was available before 1875.

William Corbet said it formerly bred at Doneraile and near Castlehyde, and a nest was found at Castlehyde about 1887. Corbet obtained two at Rathcormack in January 1887 and five at Trabolgan, probably also in the 1880s. Woodlarks bred on the River Blackwater valley from Castlehyde (Cork) to Cappoquin (Waterford), but it was almost extinct by about 1870.

From the foregoing, it appears that Woodlarks bred regularly in certain parts of the valleys of the Rivers Blackwater and Awbeg, and possibly also in the valley of the River Bride (north) (Rathcormack area) and perhaps in the Trabolgan area, but breeding had ceased, or almost ceased, by the 1880s.

There were no further records of Woodlarks until 1954, when a pair bred near Rosscarbery. Four young, not fully able to fly, and two adults were seen on 17 June. Two other adults were present, song was heard, but no nest was located, and no further young were seen. No birds could be found there in subsequent years.

Five individuals have been recorded since 1959. One at CCBO on 2–7 September 1965. One at CCBO on 1–6 September 1966. One at Cobh on 12 December 1991. One at Dursey Island on 20–1 October 2007. One at CCBO on 13 October 2016.

Skylark

Alauda arvensis | Fuiseog | A

(subspecies: *arvensis; intermedia*)

World	Europe, Asia, and north Africa, introduced elsewhere
Ireland	Common resident breeder and winter visitor
Cork	Declining resident breeder and winter visitor
Conservation	SPEC 3; Amber-listed

The breeding range of the Skylark (*A.a. arvensis*) extended across the county during 1968–72. Gaps had developed by 1988–91, and by 2008–11 these had enlarged and extended from Bandon south to the coast, and to north and east Cork, all areas of intensive agriculture.

The breeding population at CCBO was probably seventy pairs in 1965. There were ten pairs in 1969, twenty-nine pairs in 1986, twelve pairs in 1990, but only three pairs in 2016–17. There were forty pairs in 1985 at Sherkin Island, declining to six or eight pairs by 1994. It was a common breeding species at Dursey Island in 1993. The trend is downward, and this has been reflected over much of the county as farming has intensified.

In the 1960s and 1970s Skylarks were breeding widely on the east Cork coast, in dunes as at Ballycotton, Ballymacoda, Pilmore and Redbarn, and heathland as at Power Head and Knockadoon Head, with ten pairs or more at each site. These sites host, at most, one or two pairs (and none at some) since the early 2000s. Grazing marsh on the estuary of the River Womanagh has been abandoned as grazing intensified and conifer plantations developed. Skylarks were never a breeding species of autumn- or spring-sown cereal fields in east Cork, as is the case in Britain.[270] While individuals are often heard singing over such fields in spring, they do not remain to breed. The Skylark is nowadays relatively common only in low-intensity farmland on the peninsulas and islands, and in unforested uplands.

Winter atlas surveys revealed a patchy distribution during 1981–4 and 2007–11, with highest densities along the coast from Galley Head to Youghal, reflecting use of weed-infested littoral and saltmarsh habitats, dunes and adjoining stubble fields. Stubble fields are an important winter habitat, but there has been a shift towards winter-sown cereals in east Cork, removing an important feeding habitat after ploughing in October. Many fields where formerly 80–100 Skylarks wintered now hold none. Flocks of up to 500 birds were recorded at Great Island in the mid-1980s. Skylarks appear to shun areas of intensive pasture in winter.

Skylarks make a series of complex movements, the extent of which are not fully understood. Large movements occur on the Cork coast in autumn and winter, and in cold weather flocks move towards the milder west coast. During the severe winter of 2010/11 over 1,000 Skylarks were present at Ballymacoda during January. Skylarks have been obtained at Fastnet Rock from September to November (and once in March), and large numbers have sometimes occurred, and others have been obtained at Bull Rock in November and December. Skylarks have also been obtained at Old Head of Kinsale lighthouse in spring and autumn.

Migrant flocks pass over and fly out to sea in a south-west direction in October at CCBO, and over 8,000 were recorded during 22–4 October 1990, but large numbers do not occur every year, possibly because birds leaving the south coast move south before reaching so far west at CCBO. Autumn movements take place on a broad front and have been noticed at Dursey Island, Old Head of Kinsale, Roche's Point and Ballycotton where several hundred have been counted in a day. The highest count at Ballycotton involved 525 moving west in October 1968. Movements inland are rare (perhaps under-recorded), but 110 birds were flying north over Kilcolman in October 1992.

Eastern Skylark *A.a. intermedia*: There is one record of this subspecies. One obtained at Old Head of Kinsale on 7 October 1910.

Shore Lark

Eremophila alpestris | Fuiseog adharcach | A
(subspecies: *flava*)

World	North America, north-west South America, Europe, Asia, and Morocco
Ireland	21 individuals
Cork	2 individuals (1 record)
Conservation	Not globally threatened

Adult male at CCBO on 13–22 April 2007, second adult male on 20–22 April.

Sand Martin

Riparia riparia | Gabhlán gainimh | A
(subspecies: *riparia*)

World	North America, Europe, Asia, and north-east Africa
Ireland	Abundant summer visiting breeder
Cork	Common summer visiting breeder
Conservation	SPEC 3; Amber-listed

The Sand Martin was widely distributed across the county during 1968–72. The range contracted during 1988–91 but expanded again by 2008–11. These changes were related to declines in Irish and British breeding populations during 1968/9, 1983/4 and 1990/1 associated with drought in the wintering grounds in the Sahel region of Africa.[271]

Sand Martins use a variety of natural and artificial nest sites, although natural sites are used less frequently. There are several colonies on the River Blackwater east of Fermoy and some west of Mallow, as well as many small colonies on other rivers. There is a history of breeding on sea cliffs of glacial till in east Cork, and elsewhere. There are (or were) colonies at Claycastle, Ring Strand, Ardnahinch Strand, Ballybranagan Strand and White Bay. Sea cliff colonies have also been seen at Red Strand, Long Strand and Garrettstown, and at other west Cork sites since 2000.

Sand Martin

Birds occasionally nest in holes in walls, usually near water, and there are many records of nesting in stone bridges, and a few of nesting in buildings. Since the Sand Martin is an opportunistic nester, such sites are not unusual.

Most colonies, including the largest ones, are in sand and gravel pits. It is probably true to say that wherever there is (or was) a sand pit, Sand Martins nest in it. Extensive sand pit workings on the north and east shores of Cork Harbour, and in the wider east Cork area, have all been sites of large colonies. One colony was in an urban environment on the north side of Cork city when a housing estate was built in a former sand pit, but the colony was deserted before 2021. Colonies have also been recorded in stored piles of sand and ground limestone, and once in a road cutting at Ballinacurra.

During the 'Celtic Tiger' building boom period (1995–2008) numerous small sand pits were operating in east Cork, along with some larger established ones. All had breeding Sand Martins. All small and some larger sand pits suddenly closed from 2008 onwards following the collapse of the building industry. Sand pits remain suitable for nesting only if new cliffs are being exposed by extraction, and smaller ones, especially, quickly become unsuitable due to wind, rain and rabbit erosion.

Returning migrants arrive from mid-March, occasionally earlier (Mizen Head on 2 March 1989), but the main arrival takes place in April. Numbers build up rapidly and large flocks are often seen feeding over lakes, reservoirs and rivers. Sand Martins occur at CCBO in spring, with a double peak in autumn involving early first brood juveniles and later adults and second brood juveniles. The decline following the 1968/9 crash was reflected at CCBO by reduced passage numbers in succeeding years.

Numbers on the mainland are generally higher, especially in east Cork, but a count of over 6,000 at Ballycotton on 20 August 1995 was particularly high. Flocks of about 500 birds, mostly juveniles, were observed at Lough Aderry in August 2019 and 2020.

Most Sand Martins have departed by late September, but there are many October and a few November sightings. A few occur in winter (1 at Ballycotton on 8 December 1973, 2 at Castlemartyr on 1 December 1974, 1 at Ballycotton on 16 December 1974, 1 at Cobh on 14 December 1991 and 1 at Ovens on 12 January 1992).

Ringing results at breeding colonies, and autumn roosts at Ballycotton and Ballyvergan, have shown considerable interchange within Cork and adjacent counties; one adult from Ballymacoda in July 1987 was in central Scotland in July 1989. Sand Martins begin their autumn migration by moving eastwards to cross the Irish Sea. They join British birds before moving south through France and Iberia, and then to the wintering area in the Sahel region of Africa.[272;273] Several recoveries of Cork-ringed Sand Martins have been obtained in Senegal, and one ringed in Senegal on 21 March 1990 was captured at Midleton on 17 June 1990. One ringed at Mourneabbey in July 1973 was recovered in Malta in April 1974; the same bird was captured in France in September 1976.

Sand Martins roost in reedbeds in wetlands after the breeding season. Flock size at a roost at Ballyvergan was highest in July and declined in August and September, and it was estimated that at least 9,000 birds used the site during that season.[274] Reedbeds are also used as roosts in spring and summer but the numbers are smaller than in autumn, although a flock of 300 was observed on 6 June. The first juveniles attended the Ballyvergan roost on 27 June. Small numbers (maximum 50) roost in maize crops along with Barn Swallows on agricultural land in east Cork during July to September.

Barn Swallow

Hirundo rustica | Fáinleog | A

(subspecies: *rustica*)

World	North America, Europe, Asia, and north Africa
Ireland	Very common summer visiting breeder
Cork	Common summer visiting breeder
Conservation	SPEC 3; Amber-listed

The Swallow was present and breeding throughout the county during all three atlas surveys. Density was high across most of the county, especially in south and east coastal areas, in association with intensive agriculture.

Most Swallows today build their nests in association with humans inside buildings. Sites range from garden sheds to farm and industrial buildings. Other site types are also used, but less frequently. Some nest in open porches and a few under the eaves of modern houses. A small segment of the population nests beneath river bridges. Swallows nested in a marine cave at Schull Harbour for over fifty years. A pair nested in a marine cave at Roche's Point, and a pair nested in an inland cave near Cloyne during the 1980s. Nests have also been found beneath shipping quays at Youghal, in mooring dolphins at Ringaskiddy docks, and at the entrance to a burial vault at Knockraha.

About thirty pairs were breeding at CCBO in 1965, declining to ten pairs in 1990, increasing again to eighteen pairs in 1995. Numbers breeding at Sherkin Island in the late 1970s and 1980s ranged from thirty-five to forty-five pairs, but this declined to ten to thirty pairs in the first half of the 1990s.

Breeding biology and nest site characteristics have been studied in east Cork. Swallows are multi-brooded and nest in small, loose colonies (maximum 8 nests), but in 68 per cent of cases pairs nested alone. Laying dates varied from 29 April to 18 August for all nesting attempts. The mean clutch size was 4.56 and the mean number of chicks fledged was 4.04. Following fledging, juvenile Swallows return to the nest at night, and often roost within it, for periods ranging from six to twenty-two nights (mean stay = 13 nights).[275]

Swallows roost communally in reedbeds during spring, summer and autumn. This behaviour was studied at Ballyvergan in the early 2000s. Some used the reedbed immediately following spring arrival, and use continued throughout spring and summer, although numbers were small and peaked at 175 on 7 May. The first juveniles attended the roost on 16 June. Roosting patterns and behaviour have been studied intensively during autumn. Roosting time varied between eleven and thirty-five minutes after sunset (mean 22.2 minutes), and variation may be explained by prevailing weather conditions. There was a significant turnover of birds at the Ballyvergan roost from day to day, and only one bird was re-trapped out of 2,484 captures. It was estimated that at least 36,000 Swallows used the site during the autumn period. Swallows also roost in maize crops on agricultural land, and in one study numbers per night varied from fifteen to 1,200 birds.[276;277]

There are isolated records of Swallows in January and February, but the earliest arrivals are in March, with the main arrivals in mid-May, although passage is rather diffuse and large concentrations are seldom reported.

Autumn passage begins in late July and is usually heaviest in east Cork. Reedbed roosts at this time may hold several thousand; there were 15,000 at Ballycotton on 18 August 1986, and 10,000 at Ballyvergan on 9 September 1990. The autumn peak at CCBO is usually between 150 and 400, although 5,000 were present on 27 August 1991. Most have departed by late September or early October, but a few remain to November, and December records are not uncommon. During winter atlas surveys a scatter of records occurred along the coast from the Beara peninsula to Youghal, and while most were late migrants, at least some were probably genuine overwintering birds.

Pre-migratory and migratory movements of Swallows in Britain and Ireland have been studied by analysis of ring recoveries.[278] First-year birds make wandering movements in August which become increasingly oriented to the south-east from September. On the south Irish coast a daytime eastwards movement can be observed during fine and settled weather from late August. Birds ringed in Cork are frequently captured on the south coast of England and in France and Iberia, before they cross the Mediterranean and make their way south, some going to the tip of South Africa.

House Martin

Delichon urbicum | Gabhlán binne | A

(subspecies: *urbicum*)

World	Europe, Asia, and north Africa
Ireland	Common summer visiting breeder
Cork	Summer visitor and breeder
Conservation	SPEC 2; Amber-listed

A colony was recorded on sea cliffs at Reanies Bay in the mid-1800s, but birds also nested on houses. Nesting on sea cliffs was known at only two localities in Cork in the 1940s, but this is probably an underestimate.

The breeding range extended across the county during 1968–72, but there were certain gaps. These included an absence around Cork city. Although there were losses in 1988–91, it was present throughout the county in 2008–11, representing the most widespread distribution ever recorded. However, a wide distribution can mask significant population changes.

Several sea cliff colonies were known in the mid-1960s: Power Head (peak of 50 pairs), Ballycotton (no figure given) and Frower Point (6 pairs). In the same period, four pairs were nesting on a fish-freezing plant at Ballycotton. In 1969, sea cliff nesting was widespread, and there were colonies at Old Head of Kinsale, Newfoundland Bay, Rocky Bay, Robert's Cove, Power Head and Ballycotton (total of 53 pairs). About twelve pairs nested on sea cliffs at CCBO in the 1960s, but in 1995 and 1996 only one pair was nesting (on a building), sea cliff nesting having ceased. Unusual nest sites are sometimes used. An ovoid nest with the opening at the top was seen beneath a stone railway viaduct near Cork city before 1951, and a pair nested in a concrete cavity building block in the ceiling of an open storage area at Castletownbere in 1974.

There were many small colonies on sea cliffs by the 1980s, especially in east Cork, where there were about seventy pairs at seven colonies. Nine small sea cliff colonies were recorded in 1985 during a partial census from Cork Harbour to Rosscarbery. Sea cliff breeding in scattered colonies is relatively common from Knockadoon Head to Rosscarbery, but further west colonies are scarce, although there are small ones as far west as the

Beara peninsula. Nesting at inland rock quarries appears to be unusual but has occurred at Leamlara.

One of the largest inland colonies in the county, on a disused water tower near Killeagh, declined from around 120 nests in the early 1980s to forty-five in 1994 and 1995, and has since disappeared, the reason being unclear. On Great Island, there were fifteen to twenty pairs in Cobh town centre and up to 100 pairs at Carrignafoy up to 1980, but there have been few since then. There has been a general increase in small colonies on houses in the countryside since the 1970s, but not necessarily an increase in the overall number of pairs (note the demise of some large colonies, and perhaps their re-distribution at smaller ones). Nesting beneath bridges by House Martins is much rarer than among Barn Swallows.

The House Martin is the latest hirundine to arrive, the earliest being one at Little Island on 20 February 1988. Records in March are unusual, and the species is scarce until the first half of May. Spring passage is light with a maximum daily count of 100 at CCBO in May. Autumn passage is usually recorded in August and September, but it does not form large roosts and counts seldom exceed fifty birds. Maximum counts at CCBO are of about 100 birds in August (1967, 1969 and 1976), and the highest count at Ballycotton is of 120 in August 1970. It is regularly recorded into late October and November, with three occurrences in December, and one at Midleton from November 1975 to 14 January 1976.

Red-rumped Swallow

Cecropis daurica | Fáinleog ruaphrompach | A

(subspecies: *rufula*)

World	Southern Europe, Asia, and Africa
Ireland	55 individuals
Cork	27 individuals
Conservation	Not globally threatened

Red-rumped Swallows occur as overshooting spring migrants (22), mostly in April and May, with the remainder in September and October (5). Most occurred between Firkeel and Baltimore (16), and between Owenahincha and Cork Harbour (6), with a further five at east Cork sites. The Mizen peninsula (7) and CCBO (6) have hosted the greatest number, with one 9 km inland at Lough Aderry. All records relate to single birds, and four have been aged as adult. Most (21) have been seen on only one day, and the remainder (6) on two or three days (mean stay = 1.3 days). There has been an increase in frequency in the most recent three decades (there were 4 in 2011 alone), which may be related to the northward expansion of Spanish and French breeding populations.

1959–68	1969–78	1979–88	1989–98	1999–08	2009–18
0	0	2	5	8	12

Earliest and latest occurrence in spring and autumn (all records): 19 February–12 June and 4 September–25 October.

Cetti's Warbler

Cettia cetti | Ceolaire Cetti | A

(subspecies: undetermined)

World	Southern Europe, western Asia, and north Africa
Ireland	4 individuals
Cork	1 individual
Conservation	Not globally threatened

Presumed male (in song) at Ballymacoda on 21 May 2001.

Long-tailed Tit

Aegithalos caudatus | Meantán earrfhada | A

(subspecies: *rosaceus*)

World	Europe and Asia
Ireland	Common resident and breeder
Cork	Common resident and breeder
Conservation	Green-listed

Long-tailed Tit

The Long-tailed Tit was rare and was unknown to Robert Ball around Youghal in the mid-1800s. It was common at the end of the century, but whether it increased or was overlooked earlier is unknown. Long-tailed Tits occurred widely during all three breeding atlas surveys, being absent only from high ground and from exposed coastal districts. The winter range is continuous across the county and does not differ significantly from the breeding season. However, this is a thinly spread species at all seasons.

Long-tailed Tits are typically found in small flocks outside the breeding season. Flocks rarely exceed twenty individuals but counts ranging from twenty to twenty-eight (n = 7) have been published for September to January. However, such counts are probably biased, reporting of large flocks being more likely than of small flocks. Long-tailed Tits are badly affected during cold winter weather, although they have the capacity to recover quickly.

Flock size was systematically studied over four winters during 2007–11. The first two winters were typically mild followed by two which were atypically cold.[279] The mean flock size across all four winters and across the four months from November to February was 6.1 individuals (range = 1–15; n = 114). Mean flock size progressively reduced across the four months from 7.6 in November to 3.6 in February and was not affected by the cold winters. However, a lower proportion of sites held Long-tailed Tit flocks in the latter months of the second cold winter compared with the same months of the mild winters. From this it is hypothesised that whole flocks may have been lost during the cold winters, rather than individuals within flocks being affected.

The first record at CCBO was not until 1968, although the observatory existed since 1959. Small flocks have since been recorded in autumn between August and November. They were reported to be increasing in Cork in 1965 and the increase in records at CCBO perhaps reflected a recovery in the wider population following the severe winter of 1962/3.

Greenish Warbler

Phylloscopus trochiloides | Ceolaire scothghlas | A
(subspecies: *viridanus*)

World	North-east Europe, and central and southern Asia
Ireland	43 individuals
Cork	29 individuals
Conservation	Not globally threatened

The Greenish Warbler is an autumn migrant, with only one spring record at CCBO on 30 May 2012. It occurs at coastal sites, chiefly in west Cork, especially at CCBO (16) and Mizen peninsula (4). There have also been records from Dursey Island (2), Beara peninsula (1), Baltimore (1), Galley Head (2), Old Head of Kinsale (1), Barry's Head (1) and Ballycotton (1), the latter the only east Cork record. All occurred singly, and where they have been aged, one was adult and seven were first-year birds. Almost half (14) stayed for only one day, while the remainder stayed from two to nine days, with one staying for twenty days at CCBO (mean stay = 3.5 days).

1959–68	1969–78	1979–88	1989–98	1999–08	2009–18
2	0	3	2	14	8

Earliest and latest occurrence in spring and autumn (all records): 30 May and 29 August–19 October.

Two-barred Warbler

Phylloscopus plumbeitarsus | Ceolaire báneiteach | A
(monotypic)

World	East central Asia
Ireland	1 individual
Cork	1 individual
Conservation	Not assessed

One at Dursey Island on 26 October 2019.

Arctic Warbler

Phylloscopus borealis | Ceolaire Artach | A
(monotypic)

World	Northern Europe and Asia
Ireland	14 individuals
Cork	8 individuals
Conservation	Not globally threatened

One at CCBO on 8–10 September 1968. One at CCBO on 20–2 October 1977. One at CCBO on 5 October 1981. One at Toe Head on 20–5 October 1985. First-year at CCBO on 9–18 October 2009. One at Garinish on 6–7 September 2010. One at Dursey Island on 13 September 2013. One at Mizen Head on 21 September 2014.

Pallas's Leaf Warbler

Phylloscopus proregulus | Ceolaire Pallas | A

(monotypic)

World	Central and eastern Asia
Ireland	42 individuals
Cork	26 individuals
Conservation	Not globally threatened

This is a late autumn migrant to coastal districts. Eighteen birds have occurred during 22 October–10 November. Most have occurred at CCBO (11), and others at Dursey Island (2), Galley Head (5), Power Head (2) and Knockadoon Head (4), while singles have occurred at Sherkin Island and Ballycotton. The high proportion of birds recorded at east Cork sites (7) is in contrast with some other Asiatic and Siberian vagrants. Nearly half (12) were seen on only one day, but some remained for several days and one for seven days (mean stay = 2.2 days). Three records refer to two birds together, the rest to singles.

1959–68	1969–78	1979–88	1989–98	1999–08	2009–18
1	0	3	7	8	7

Earliest and latest occurrence in spring and autumn (all records): 12 October–13 November.

Yellow-browed Warbler

Phylloscopus inornatus | Ceolaire buímhalach | A

(monotypic)

World	North-east Europe and Asia
Ireland	Scarce autumn visitor, subject to occasional influxes
Cork	Scarce autumn visitor, with recent major influxes
Conservation	Not globally threatened

Yellow-browed Warbler

The first Yellow-browed Warbler was recorded in 1959, at CCBO. Occurrences have been annual since except in 1962–5 and 1972. All but three individuals occurred in autumn, 91.5 per cent in October. One occurred in March 1993 at Schull, one in April 2004 at University College Cork and one in February 2017 at Harty's Quay. These birds may have arrived in previous autumns. There was no obvious change in status up to 1984 (average of 2.8 birds per annum). There was a major influx in 1985 (at least 70), and a smaller one in 1986 (at least 56). The numbers increased between 1987 and 2004 (average of 12.8 birds per annum), but there were differences between years, ranging from a high of thirty-four (2003) to a low of four (1997). In 2005, 2006, 2007 and 2008, minimum totals of seventy, fifty-four, fifty-six and fifty-six birds occurred, respectively. During the final decade in the study period an average of fifty-seven per annum was documented, almost as many as had occurred in the previous five decades.

Most birds occurred at CCBO, Dursey Island, Beara peninsula and Mizen peninsula, with declining numbers at watch points east to Galley Head and Old Head of Kinsale, and smaller numbers extending to the east Cork sites of Ballycotton and Knockadoon Head. While most records refer to one to three birds, five to ten (sometimes more) have frequently occurred at west Cork sites.

Major influxes have always been part of wider influxes to other parts of Europe and to Britain.[280] The reasons behind these influxes have been debated, but the origin and destination of the birds involved is unclear.[281;282] Reverse migration (i.e. flying in the opposite direction to the usual route) has been hypothesised, but as these influxes involve large numbers there is a high chance that some survive and return to their breeding grounds, and contribute genetically to the pioneering of new wintering areas.[283]

1959–68	1969–78	1979–88	1989–98	1999–08	2009–18
26	26	180	117	318	573

Earliest and latest occurrence in spring and autumn (all records): 11 February–2 April and 15 September–16 November.

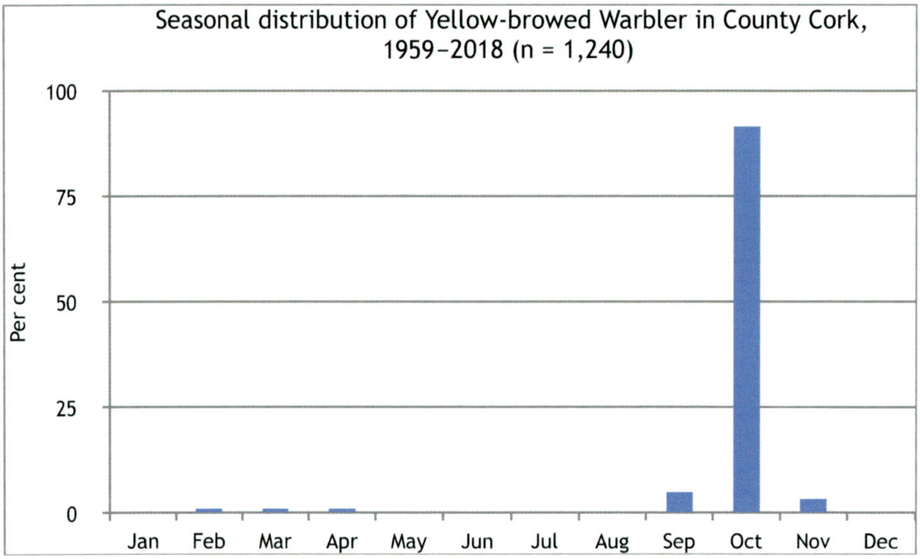

Fig. 7.32 Yellow-browed Warbler.

Hume's Leaf Warbler

Phylloscopus humei | Ceolaire Hume | A

(subspecies: *humei*)

World	South-central Asia
Ireland	3 individuals
Cork	2 individuals
Conservation	Not globally threatened

One at Knockadoon Head on 18 December 2003–6 January 2004. One at CCBO on 23 October 2016.

Radde's Warbler

Phylloscopus schwarzi | Ceolaire Radde | A

(monotypic)

World	Central and eastern Asia
Ireland	22 individuals
Cork	12 individuals
Conservation	Not globally threatened

Radde's Warbler is a rare late autumn migrant, and all have occurred between 14 October and 1 November. While most birds occur singly, there were two together at Barry's Head in October 2016, one of which had been ringed at Brownstown Head (Waterford) six days previously, prompting the question of whether some small passerine migrants might visit more than one headland and be recorded, unknowingly, as different individuals. Occurrences have been at coastal districts from Crookhaven to Barry's Head, with CCBO (5) hosting the most records. Two have also occurred at Old Head of Kinsale, with singles at three other sites (Crookhaven, Ballymacrown and Galley Head). Half of the individuals have stayed for only one day, but others have stayed longer, six days being the maximum (mean stay = 2.5 days).

1959–68	1969–78	1979–88	1989–98	1999–08	2009–18
0	0	2	1	5	4

Earliest and latest occurrence in spring and autumn (all records): 14 October–1 November.

Dusky Warbler

Phylloscopus fuscatus | Ceolaire breacdhorcha | A

(subspecies: undetermined)

World	Central and eastern Asia
Ireland	12 individuals
Cork	9 individuals
Conservation	Not globally threatened

One at Old Head of Kinsale on 31 October–1 November 1987. One at CCBO on 1–15 November 1987. One at CCBO on 19–30 October 1995. One at CCBO on 26 September 1998. One at CCBO on 14 October 2007. One at Ballycotton on 22–3 October 2007. One at Power Head on 22–5 October 2009. One at Knockadoon Head on 28–31 October 2015. One at Dursey Island on 29 October 2016.

Western Bonelli's Warbler

Phylloscopus bonelli | Ceolaire Bonelli | A

(monotypic)

World	South-west Europe and north Africa
Ireland	20 individuals
Cork	13 individuals
Conservation	Not globally threatened

This species is an autumn migrant between August and October. Most have occurred in mid-September, with fewer thereafter, but there have been three records in late October. It has occurred at only four coastal sites, always singly: Beara peninsula (2), Mizen Head (2), CCBO (6) and Galley Head (3). In contrast to many vagrant passerines, this warbler tends to remain for extended periods. Only four have been seen on only one day, and singles have stayed for eight, fourteen and twenty-nine days (mean stay = 5.7 days). Only in 2015 (2) and 2016 (3) has more than one individual been recorded in a year.

1959–68	1969–78	1979–88	1989–98	1999–08	2009–18
1	1	2	0	2	7

Earliest and latest occurrence in spring and autumn (all records): 26 August–28 October.

Unidentified Bonelli's warbler species

Five Bonelli's warblers, either Western or Eastern (*Phylloscopus bonelli/orientalis*), have been recorded. One at CCBO on 28 August 1968. One at Old Head of Kinsale on 22 August 1984. One at Lissard on 12 September 1990. One at Crookhaven on 1 September 1991. One at CCBO on 10 September 2014.

Wood Warbler

Phylloscopus sibilatrix | Ceolaire coille | A

(monotypic)

World	Europe east to central Asia
Ireland	Rare summer visiting breeder, rare migrant in spring and autumn
Cork	One breeding record, 183 individuals as migrants
Conservation	Red-listed

The Wood Warbler bred in Cork on one occasion when a pair and nest containing eggs was discovered at Glengarriff in June 1938.

This species has occurred in all but ten of the sixty years during 1959–2018. Spring migrants account for 23 per cent of passage, while autumn migrants account for 77 per cent. More migrants occurred at CCBO (146) than at any other site. Smaller numbers occurred at Mizen peninsula (15), and Dursey Island and Beara peninsula combined (8). Galley Head (5) and Old Head of Kinsale (4) are the only other sites which have hosted more than one bird. Four sites from Sherkin Island to Rosscarbery have each had one bird, while the only occurrence in east Cork is of one at Knockadoon Head. Most records are of single birds, but there have been multiple occurrences on several occasions at CCBO. Most birds have been present for only one day, but many have remained between two and ten days, and one for thirteen days (mean stay = 2.3 days). Numbers peaked in the third decade of this analysis but have subsequently remained stable at a lower level.

It was suggested that the small number of spring migrants in Ireland conform to the theory of overshooting from a southerly breeding population, but the possibility that some were returning Irish breeders could not be excluded. Autumn occurrences may be drift migrants of Scandinavian origin, reverse migrants from more southerly populations, or departing Irish breeders. The Wood Warbler is one of a few migrants breeding in Britain and Ireland to take a south-east route through Italy and across the Mediterranean Sea and Sahara Desert to their wintering grounds in Africa, but the return migration in spring is to the west of the autumn route.

1959–68	1969–78	1979–88	1989–98	1999–08	2009–18
14	33	53	34	24	25

Earliest and latest occurrence in spring and autumn (1959–2018): 9 April–3 June and 26 July–26 October.

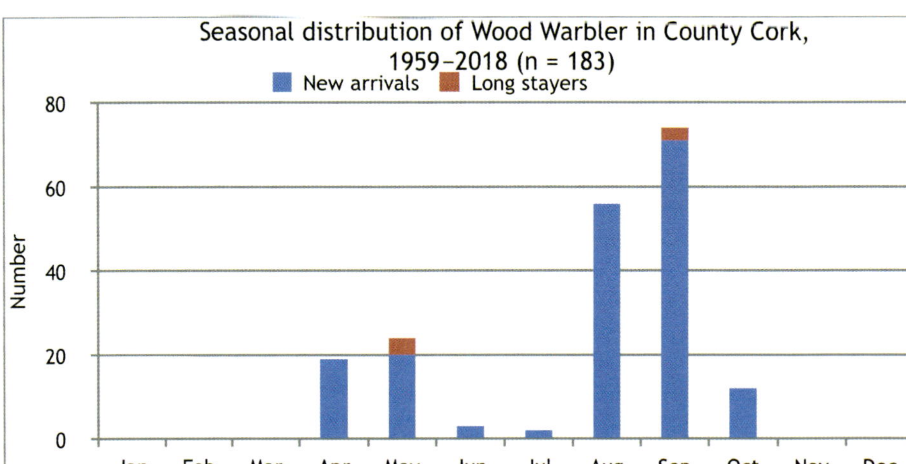

Fig. 7.33 Wood Warbler.

Common Chiffchaff

Phylloscopus collybita | Tiuf-teaf | A

(subspecies: *collybita*; *abietinus*; *fulvescens*; *tristis*)

World	Europe and Asia
Ireland	Very common summer visiting breeder, scarce winter visitor
Cork	Common summer visiting breeder, scarce winter visitor
Conservation	Green-listed

The Chiffchaff (*P.c. collybita*) has been a summer visitor and breeder for over two centuries, but it was almost certainly more restricted in distribution in the past than at present.[284] It bred throughout the county during all three atlas surveys at moderate to high density. Breeding took place at CCBO for the first time in 1986, and occurrences there are increasing. Breeding took place for the first time at Sherkin Island in 1994.

Chiffchaffs are one of the earliest summer migrants to arrive. The first usually appear around mid-March, and a moderate spring passage peaks in mid-April. At CCBO there are generally fewer than forty birds in a day, and an arrival of 170 on 3 April 1988 was exceptional. This figure may have been exceeded on 27 April 1962 when 600 *Phylloscopus* warblers were present. Although numbers are usually much lower, forty to fifty were at Dursey Island in late March 2012. Passage at Old Head of Kinsale does not usually exceed twenty birds, although 200 *Phylloscopus* warblers were present on 21 April 1968 and 100 on 15 May 1970.

Autumn passage extends from early August to early November, usually peaking in October. The typical autumn maximum at CCBO is around fifty birds in a day, although there were 250 on 22 October 1990, and almost 100 have occurred several times. Elsewhere, autumn counts of over twenty are rare. Intensive coverage at the Beara peninsula since 1986 rarely produces this figure, while at Old Head of Kinsale the highest count was thirty-two on 1 October 1989. Up to forty birds were present at Crookhaven in October 1994. Chiffchaffs have been obtained at Fastnet Rock and Old Head of Kinsale lighthouses during spring and autumn migration.

Chiffchaffs have occurred in winter in small numbers since the 1800s, and five records between November and February were noted in the mild Cork climate. One was caught on 25 January 1913, and two were seen at

Glengarriff each winter from 1925 to 1928, and they sang on warm days. In December 1939 and January 1940 the Reverend Patrick Kennedy observed several birds, and he counted eight together on one occasion; birds were seen in the same area in January 1941.

Chiffchaffs were recorded in winter in slowly increasing frequency until the mid-1970s, since when it has become more numerous at this season. In most winters since 1976 up to twenty have been recorded annually, and a total of about 150 was reported in 1980/1. A minimum of 250–300 Chiffchaffs (perhaps as many as 600–700) may have wintered in the county each year since the early 1980s. During winters 1981–4 Chiffchaffs were mostly distributed along the coast, the main wintering area being a block of contiguous 10 km squares around Cork city and harbour, extending eastwards to Youghal. The population thinned out in a westerly direction, but some reached the western peninsulas. There were few inland records, and none north of the River Blackwater. There was a significant increase in the winter range by 2007–11, and while most records were within 20–30 km of the coast, there was an extension into the River Blackwater valley.

The Chiffchaff is largely a trans-Saharan migrant, but there are no winter recoveries of birds ringed as nestlings to indicate the precise origin of birds wintering in Ireland. However, various data suggests a wide area of origin. A small proportion are likely to be Irish breeding stock (*collybita*) based on ringed birds known to be present in the breeding season, and in winter, but a larger proportion are likely to be *collybita* of continental origin. However, many birds present here in winter are known to be subspecies *tristis* (see below), and while *abietinus* might be considered a candidate for frequent occurrence based on its closer geographical range, the evidence suggests that it might be rarer.

Northern and Eastern Chiffchaff *P.c. abietinus/ fulvescens/ tristis*: Subspecific identity of the Chiffchaff is difficult. There are many records of birds showing characters of one or other of these subspecies, but publication of the records has often been couched in the language of probability, and some have been referred to as northern or eastern Chiffchaffs without mention of subspecies. Records of three subspecies (in addition to the nominate *collybita*) have been published.

One record has been published as *fulvescens* (which breeds in central Asia), one at CCBO on 12–16 October 1970. The validity of *fulvescens* has been questioned, and it may be an intergrade due to hybridisation in the overlap zone between *tristis* and *abietinus*.

Four records have been published as *abietinus* (which breeds in Fennoscandia), single birds at CCBO on 2 September, 23 September and 10 October 1962, and at Ballycotton from early November to 6 December 1975. One of the CCBO birds remained for six weeks.

Eleven records involving twenty-five individuals have been published as *tristis*, or probable *tristis* (which breeds in Siberia east of *abietinus*) during 1968–82. All have occurred at CCBO and all have been in October (19) and November (6). Most records refer to single birds but two have occurred together twice, three once, and eleven once.

No records of *tristis* were published for Cork in the years 1983–2006. Identification was clearly difficult in the past, but there is now a better understanding of field characters, plumage and especially vocalisation than previously. A clearer understanding of its status is emerging, assisted by the trend towards digital recording.

Between 2007 and 2018 there were fifty records of *tristis* involving sixty-three individuals. One record involved a presumed returning individual in a subsequent winter. Most have occurred as single birds, but two together have been seen five times, and four and six on one occasion each. Most have occurred in October (37) with an even spread of records thereafter to March. Most birds have been seen on one or a few days, but a few have remained for up to two months. The geographical spread of records extends along the coast from Dursey Island to Knockadoon Head, with one inland at the Gearagh. Most occurred at CCBO (16), and at various locations in and around Cork city and harbour (13).

Willow Warbler

Phylloscopus trochilus | Ceolaire sailí | A

(subspecies: *trochilus*; *acredula*)

World	Europe and Asia
Ireland	Very common summer visiting breeder
Cork	Common summer visitor and breeder
Conservation	SPEC 3; Amber-listed

The Willow Warbler (*P.t. trochilus*) was a common summer visitor in the 1800s. It bred throughout the county during all three atlas surveys. Density was higher in the west than elsewhere. It did not breed at CCBO in the 1960s, but a small breeding population is now present in most years. The breeding population at Sherkin Island was one pair in the late 1970s and early 1980s, but a significant increase to twenty-five to thirty pairs took place by 1994.

There are many late March arrival dates, the earliest being individuals at Kilcolman on 19 March 1986 and at Carrigrohane on 20 March 1990. The main arrivals are in April and early May. Spring numbers at coastal watch points seldom exceed 100 in a day, but arrivals of 600 *Phylloscopus* warblers at CCBO on 27 April 1962, and 200 at Old Head of Kinsale on 21 April 1968 and on 9 April 1995 were predominantly of this species. Elsewhere, numbers are smaller, e.g. thirty to forty birds at Dursey Island in late March 2012.

Autumn passage is typically lighter than spring, but is generally more protracted, extending from late July to late October, peaking in early September. There are occasional early November records, the latest involving singles at Ballymacoda on 6 November 1977 and at CCBO on 6 November 1982. At this season high counts at CCBO include 100 on 30 August 1976, and ninety on 15–16 September 1986. Numbers at other sites rarely exceed twenty in a day. Willow Warblers have been obtained at Bull Rock, Fastnet Rock and Old Head of Kinsale lighthouses during spring and autumn migration. Willow Warblers are trans-Saharan migrants, and one bird ringed at Doneraile on 30 June 1991 was recovered in the Ivory Coast on 14 February 1992. Two birds ringed in southern Norway have been captured at CCBO in September 2017 and July 2019, respectively.

Northern Willow Warbler *P.t. acredula*: There are five records of this subspecies. It was recorded at CCBO on four occasions with one to five birds, once each in April, May, September and October. One was recorded at Old Head of Kinsale on 25 September 1976. This subspecies is probably overlooked since it is difficult to separate in the field from the nominate *trochilus*.

Blackcap

Sylvia atricapilla | Caipín dubh | A

(subspecies: *atricapilla*)

World	Europe east to central Asia, north Africa, and islands off north-west Africa
Ireland	Summer visiting breeder, recent increase, passage migrant, winter visitor
Cork	Increasing summer visiting breeder, passage migrant, and winter visitor
Conservation	Green-listed

The Blackcap was a summer visitor in the 1800s. There was no indication of breeding in Cork in the early 1950s at a time when it was believed unlikely to be breeding outside Leinster. Most records of singing males were believed to be unmated or were birds which had wintered but had not yet left, as suggested for one at Glengarriff in March 1942.

There were many records of Blackcaps during the 1960s, including at CCBO, but most related to wintering or to spring and autumn migrants. One was singing near Mitchelstown in June 1956, but breeding was not proved. The first proof of breeding was in 1969 when birds were singing at two sites near Kilworth on 23 June, and the next day one was carrying food near Skibbereen, and at least one pair, perhaps three, were proved breeding. These records preceded what was to be a dramatic expansion of the breeding range across the county in the following decades.

Fieldwork on the 1968–72 breeding atlas, including the records quoted above, showed the Blackcap was scarce with a few isolated populations. A decline was reported in the late 1970s, but it was increasing in the early 1980s, especially in east Cork. The range continued to expand over the following four decades, and apart from being scarce or absent from parts of the western peninsulas, it was present across the county by 2008–11. It now breeds in many parts of Cork city, such as Sunday's Well and off the Lower Glanmire Road, and a few pairs have recently colonised CCBO.

The Blackcap has been a migrant at CCBO since 1959, mainly in autumn.[285] Spring migration is light, with numbers rarely exceeding ten in a day. It is usually scarce away from CCBO in spring, with numbers rarely reaching ten birds, e.g. Dursey Island on 10 April 1995.

Autumn migration at CCBO is strong and protracted, and numbers are higher than at any mainland site. A significant increase in autumn passage has taken place there since the 1960s. The peak recorded in a day between 1959 and 1969 was eleven birds, and this rose to twenty-six in October 1973, and twenty-eight in October 1979. Counts have increased further since, notably fifty in October 1990, and fifty-three in October 1993. Small numbers occur everywhere along the coast, with numbers frequently reaching low double figures, e.g. thirty at Old Head of Kinsale on 30–1 October 1993.

Analysis of Blackcap numbers at bird observatories in Ireland and Britain suggests that increases are due to more regular easterly airflows bringing more birds westwards from Europe, a total population increase, a change in habits, or to a combination of all three.[286]

Harvey said Blackcaps probably wintered occasionally in the 1800s, and he knew of several occurrences and obtained two taken in November 1839. Robert Ball saw Blackcaps in winter 1833, probably at Youghal, and one was found dead at Youghal in January 1838. Barrington had no record from a Cork lighthouse, but a male was obtained on 8 December 1896 on the mainland opposite Fastnet Rock (presumably Crookhaven). Specimens were obtained at Ballycotton and Old Head of Kinsale lighthouses in November 1910 and October 1911, respectively, and a male was obtained at Old Head of Kinsale on 20 November 1965.

When commenting on winter occurrences, Barrington suggested it was more likely these birds were winter immigrants, rather than birds from the then very small breeding population staying behind for the winter. This was the first indication of the possible origin of birds wintering in Ireland.[287]

Irish (and British) Blackcaps leave their breeding grounds in September and go south to wintering areas mainly in southern Iberia and north-west Africa. Birds ringed at CCBO in September and October have been recovered in France, Spain and Portugal in November, December and February. Birds passing through coastal sites, such as CCBO, in October and November are presumed to be of continental origin, and it is from these that the wintering population is derived, largely coming from west-central Europe, and a link has been drawn between increases in the European breeding population and increased wintering numbers.[288]

There have been several surveys of wintering Blackcaps in Ireland and Britain. The first showed no birds in Cork, but it is known that some had been overlooked. Some continued to be recorded in small numbers during the 1960s and 1970s, and by 1969/70 and 1970/1 wintering was widespread. Another survey during 1978/9 also recorded none in Cork. However, exceptional circumstances (a postal strike) prevented data from reaching the organiser showing that some were indeed present in Cork that winter.

The winter distribution of Blackcaps was mapped for the first time during 1981–4, when it was scarce. The main wintering area was in a block of contiguous 10 km squares around Cork city extending eastwards to Youghal, with a wide scatter of records elsewhere. The winter range had increased considerably by 2007–11 with squares occupied across the county, apart from on high ground. Many birds are now reported at bird feeders and berried shrubs in suburban gardens around Cork city and in east Cork. Up to ten were present at Great Island in January 1991 and at least twenty-eight were there in winter 1993/4. There has clearly been a significant recent increase in the number of Blackcaps breeding and wintering in Ireland and Britain.[289]

Garden Warbler

Sylvia borin | Ceolaire garraí | A

(subspecies: *borin*)

World	Europe east to north-central Asia
Ireland	Scarce summer visiting breeder, spring and autumn passage migrant
Cork	Has bred, scarce passage migrant, mainly in autumn
Conservation	Green-listed

The Garden Warbler occurred at Sunday's Well for several years leading up to 1852, and Ussher was satisfied it bred. A male was noted at Cuskinny about 1876 and another at Passage West in 1888, but no females were seen. A pair was observed at Monkstown during 1893, and the male was heard singing. Song ceased about 24 June, and it was believed breeding took place.[290;291] The only record of a bird in breeding habitat in the first half of the 1900s was a singing male at Glengarriff during 12–16 June 1941.

Two birds were singing near Lough Allua on 11 September 1957, and one was seen near Charleville on 25 September 1957. These records are too late as proof of breeding and may be birds on migration, but they would have been either in or near suitable nesting habitat. There was one case of breeding by a pair near Inishannon in June 1969, and one case of probable breeding at Glengarriff between 1968 and 1972. The 1988–91 atlas showed a cluster of three 10 km squares with Garden Warbler records centred on Glengarriff, but only one involved proved breeding, and there was a single sighting of one bird at Tracton on 25 June 1990. There were no records for Cork during the 2008–11 survey.

One or two birds were at Kilcolman in July 1994, and two were observed at Ryecourt Wood in May 2005, but breeding was not proved. There were also several sightings at inland locations with suitable nesting habitat of single birds on single days, e.g. Kilcolman July 2006, Fermoy July 2014, Ovens June 2014, April 2015 and 2016, and Macroom September 2010 and April 2011. It is possible some of these refer to migrating birds, but they could also indicate nearby breeding.

Migrating Garden Warblers were obtained at Fastnet Rock on 18 November 1910 and 20 September 1912, and they also occurred there during a rush of birds on 28–30 September 1913. Birds in potential breeding habitat, as discussed above, have been excluded from the decadal table below. A light passage occurs at CCBO in spring, and a larger passage in autumn. Spring migration peaks in May, and autumn migration occurs between August and early November. Although most migrants are recorded at CCBO, there is a light scatter of birds at almost every coastal location between Dursey Island and Youghal. The numbers at CCBO vary between years, but often reach double figures, with a peak of twenty-eight birds on 5 October 1981. Counts exceeding three birds at other sites are rare. Most birds are seen on one day and stays of longer than three or four days are unusual. Apart from a decline in the fifth decade, numbers have been remarkably stable over the sixty years of this study.

1959–68	1969–78	1979–88	1989–98	1999–08	2009–18
98	147	154	183	86	122

Earliest and latest occurrence in spring and autumn (1959–2018): 6 March–29 November.

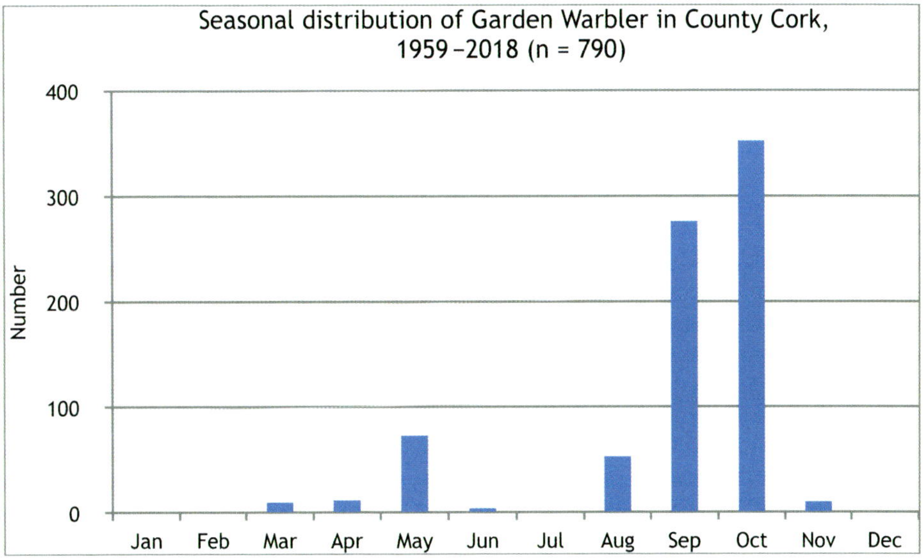

Fig. 7.34 Garden Warbler.

Barred Warbler

Curruca nisoria | Ceolaire barrach | A

(subspecies: undetermined)

World	Europe east to central Asia
Ireland	Rare autumn passage migrant; never recorded in spring
Cork	123 individuals
Conservation	Not globally threatened

All occurrences of Barred Warblers have been at coastal sites between Dursey Island and Ballymacoda, and all have been in autumn. Most birds occurred at CCBO (51), with Mizen peninsula (22), Beara peninsula (19) and Dursey Island (13) also providing numbers in double figures. Four other west Cork sites have had smaller numbers: Old Head of Kinsale (5), Galley Head (4), with singles at Sherkin Island and Baltimore. Seven birds have occurred at four east Cork sites: Power Head (1), Ballycotton (2), Knockadoon Head (3) and Ballymacoda (1). All but two records refer to one bird, the exceptions being two at CCBO in 1971 and 1995. The ages of fifty-nine have been published, all first-year birds. Over two-thirds (84) were recorded on only one day, and twelve birds remained longer than five days, nineteen days being the maximum (mean stay = 2.2 days). The Barred Warbler has occurred in Cork in forty-two of the sixty years during 1959–2018 and has been annual since 2004. The decadal trend over the period 1969–78 to 1999–2008 was steady at a mean of almost seventeen birds per decade, but in the most recent decade this number increased to fifty birds.

1959–68	1969–78	1979–88	1989–98	1999–08	2009–18
6	14	12	19	22	50

Earliest and latest occurrence in spring and autumn (all records): 29 August–13 November.

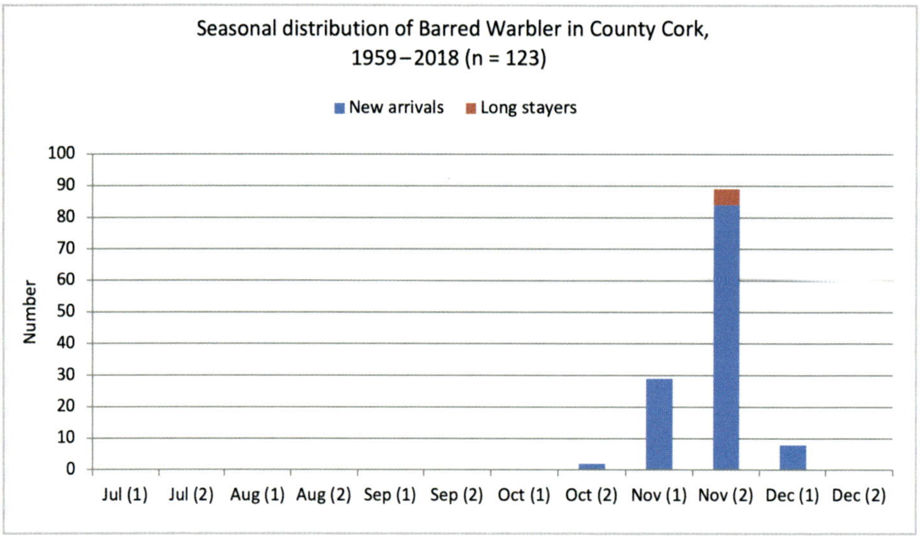

Fig. 7.35 Barred Warbler.

Lesser Whitethroat

Curruca curruca | Gilphíb bheag | A

(subspecies: *curruca*)

World	Europe and Asia
Ireland	Rare summer visiting breeder, scarce spring and autumn passage migrant
Cork	Scarce passage migrant, mainly in autumn
Conservation	Not globally threatened

The Lesser Whitethroat has been a regular autumn (92 per cent of records) passage migrant since the first record in 1959, with smaller numbers in spring (8 per cent of records). There is one winter record of a bird at Carrigaline on 17–21 January 2015. Only two years (1967 and 1974) have been without a record during the 1959–2018 period. Most records are of single birds, but up to five have occurred at CCBO on the same day. Most birds are present for only one day, but some stayed for several days, but stays of more than ten days are unusual. CCBO recorded more birds than any other site. Dursey Island, Beara peninsula, Mizen peninsula, Sherkin Island, Galley Head and Old Head of Kinsale have provided most of the rest. All records are from coastal headlands and islands, but very few have been recorded east of Old Head of Kinsale at sites such as Inch Strand, Ballycotton and Knockadoon Head. There has been an increase in sightings since 1989, when over ten birds occurred for the first time in one year.

It was suggested by Clive Hutchinson that the small number of spring migrants in Ireland resulted from overshooting, and that autumn occurrences resulted from reversed migration, the autumn migration of this species being south-east across Europe to wintering grounds in north-east Africa. The range of the Lesser Whitethroat has increased by 32 per cent in Britain over the forty years from 1968 to 2008. With breeding now taking place on the coast of Wales, this species is a likely candidate for colonisation in Ireland, although attempts to date have not been successful.[292] However, there are recent indications that this might be changing, and birds have been heard in song at Ballinadee and Ballynoe in May 2019, and north of Cork city in May and June 2020.

One ringed in Yorkshire on 8 October 2016 was recovered at CCBO on 12 October 2016.

1959–68	1969–78	1979–88	1989–98	1999–08	2009–18
20	25	36	102	125	141

Earliest and latest occurrence in spring and autumn (all records): 20 April–23 June and 21 August–11 November (17 January).

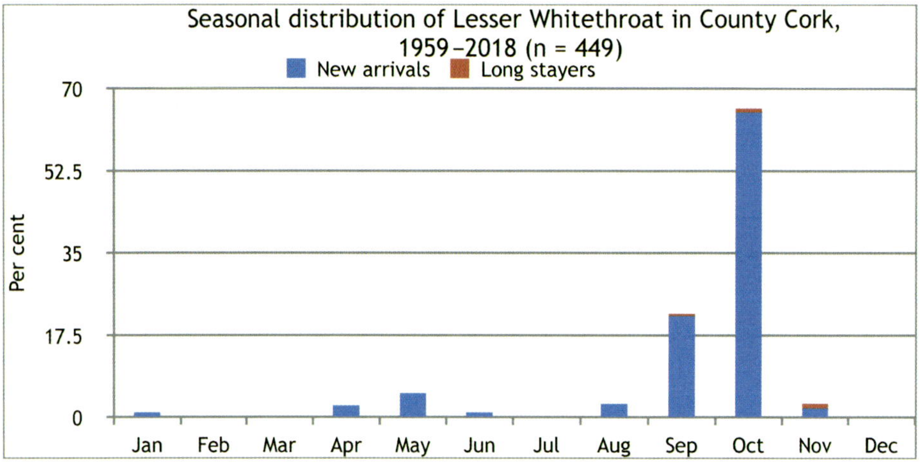

Fig. 7.36 Lesser Whitethroat.

Common Whitethroat

Curruca communis | Gilphíb | A

(subspecies: *communis*)

World	Europe east to central Asia, and north Africa
Ireland	Common summer visiting breeder
Cork	Summer visitor and breeder
Conservation	Green-listed

This species was a common summer visitor throughout the 1800s. Although present across the county during 1968–72, there were several gaps in the range. The 1988–91 survey showed that nearly all inland breeders had been lost, and apart from a few scattered records, most were in a band along the coast and around Cork Harbour. There was some expansion by 2008–11, but the main areas without birds were confined to the centre and north-west of the county.

Whitethroat populations in Britain crashed dramatically in 1969, the result of climatic deterioration in winter quarters in the Sahel region of Africa.[293] Numbers also declined in Ireland that year, and breeding populations at CCBO dropped from forty pairs in 1965 to five pairs in 1969. Thereafter, numbers increased to twenty-five to thirty pairs in 1992. Numbers on autumn passage at CCBO also recovered after 1974. Breeding populations at Sherkin Island have been erratic, declining from sixteen pairs in 1977 to six pairs in 1989, and increasing again to twenty-nine to thirty-nine pairs in 1994.

The first individuals usually arrive on the coast in the second half of April (the earliest was one at CCBO on 14 April 1984), but spring passage does not peak until mid-May. Up to fifty have been recorded at CCBO in spring (25 May 1962), although this was prior to the 1969 population crash. Numbers since the early 1970s have typically been lower and usually below twenty birds. The highest recent count on the island was about forty birds from 16 to 20 May 1998.

Autumn passage occurs along the entire coast and peaks in late August and early September. Before 1969, up to thirty had been recorded at CCBO at this time, with a maximum of thirty-three on 1 September 1963. Peak autumn numbers remained below twenty for most of the 1970s and 1980s but were higher in the 1990s with up to forty in autumn 1991.[294] October records are scarce, and the latest individual was one at Ballycotton on 11 November 1981. The Whitethroat has often been obtained at lighthouses such as Fastnet Rock and Old Head of Kinsale during spring and autumn migration periods.

There are two mid-winter records of Whitethroats, two near Bantry and two at nearby Whiddy Island, each from November 1974 into 1975.

Dartford Warbler

Curruca undata | Ceolaire fraoigh | A

(subspecies: *dartfordiensis*)

World	South-west Europe and north-west Africa
Ireland	10 individuals
Cork	7 individuals
Conservation	Not globally threatened

One at CCBO on 27–8 October 1968. One at CCBO on 15 August 1972. One at CCBO on 19 October 1972. Adult at CCBO on 25–6 October 1975. One at CCBO on 16 October 1987. Adult male at Dursey Island on 16–22 May 1999. One at Brow Head on 31 March 2003.

Western Subalpine Warbler

Curruca iberiae | Ceolaire Fo-Alpach | A

(subspecies: *iberiae* or *inornata*)

World	Southern Europe and north Africa
Ireland	58 individuals
Cork	24 individuals
Conservation	Not globally threatened

The Subalpine Warbler occurs in spring (13) and autumn (11). All occurred in coastal districts, mostly in west Cork, with a total of eighteen at CCBO, Mizen peninsula and Dursey Island, combined. Three occurred at Old Head of Kinsale and one each at Toe Head, Inch Strand and Knockadoon Head, the latter two the only occurrences for east Cork. All have occurred singly, except one record of two together (male and female) at Mizen Head in 2008. Of those where the sex was known, thirteen were male and four were female, while three adult and seven first-year birds were identified. One of the males (Mizen Head, 30 April 2018) was in song. Half stayed for only one day, while the remainder stayed from two to nine days, with one remaining at CCBO for forty-seven days in August and September 1990 while it underwent primary moult (mean stay = 1.9 days, excluding the long-staying CCBO bird).[295]

1959–68	1969–78	1979–88	1989–98	1999–08	2009–18
1	0	2	3	9	9

Earliest and latest occurrence in spring and autumn (all records): 10 April–16 May and (2 August) 25 September–26 October.

Sardinian Warbler

Curruca melanocephala | Ceolaire Sairdíneach | A

(subspecies: undetermined)

World	Southern Europe, south-west Asia, north Africa, and Canary Islands
Ireland	3 individuals
Cork	3 individuals
Conservation	Not globally threatened

Male at CCBO on 10–12 April 1993. Male at Knockadoon Head on 14–21 April 1993. Male at Dursey Island on 20 April–14 May 2014.

Pallas's Grasshopper Warbler

Helopsaltes certhiola | Ceolaire casarnaí Pallas | A

(subspecies: undetermined)

World	Central and eastern Asia
Ireland	2 individuals
Cork	1 individual
Conservation	Not globally threatened

First-year at CCBO on 8 October 1990.

Common Grasshopper Warbler

Locustella naevia | Ceolaire casarnaí | A

(subspecies: *naevia*)

World	Europe, and western and central Asia
Ireland	Summer visiting breeder
Cork	Scarce breeder, but numbers probably underestimated
Conservation	Green-listed

The song of this warbler was heard by Robert Ball around Youghal. It occasionally bred near Enniskean, song was heard near Bandon, and heaths, young plantations and other areas of long herbage were frequented in the 1800s.

The Grasshopper Warbler, which favours damp and dry grassy places with rank vegetation, was patchily distributed as a breeding species during 1968–72. The main area of continuity was at the Beara peninsula, and the uplands south and east of Bantry extending northwards along the Cork/Kerry border. Surprise was expressed at its absence from areas with suitable habitat. There were losses throughout the range in 1988–91. The range in 2008–11 was more extensive than during either of the two previous surveys, although there were significant areas without birds in the centre, north and east of the county. This species is undoubtedly thinly spread and when mated sings only at dusk and dawn. Consequently, it is easy to overlook in extensive surveys, but unmated birds may sing throughout the summer.

A few occur at coastal sites on spring migration, and are usually encountered in ones and twos, rarely up to five or six in number, but twenty-one and fifteen at Old Head of Kinsale on 21 April 1968 and 15 May 1970, respectively, were exceptional. It is rare at CCBO, the normal spring migration route not extending so far west, although it has bred in recent years. Its normal arrival time is from the second half of April, but one was at Garryvoe on 9 April 1980 and one at Dursey Island on 8 April 2000. Autumn migrants are even rarer at coastal watch points, and the few records are scattered from late July through to late September. There are only a few October records, the latest being one at CCBO on 24–5 October 1990.

The Grasshopper Warbler is a trans-Saharan migrant. One was obtained at Old Head of Kinsale on 22 September 1909 and two were obtained at Fastnet Rock on 30 September 1913 following a rush of birds. One ringed at Ballyvergan Marsh on 8 September 2002 was recovered in Sussex (England) on 19 September 2002.

Savi's Warbler

Locustella luscinioides | Ceolaire Savi | A

(subspecies: undetermined)

World	Europe, western and central Asia, and north Africa
Ireland	11 individuals
Cork	3 individuals (2 records)
Conservation	Not globally threatened

Male (in song) at Ballyvergan Marsh on 17–23 June 1985, joined by second bird (presumed female) on 19 June 1985. Male (in song) at Ballycotton on 3–15 May 1988. The record for 10 km square W98 is an error; it refers to W96, the Ballycotton occurrence given above.[296]

Eastern Olivaceous Warbler

Iduna pallida | Ceolaire bánlíoch | A

(subspecies: undetermined)

World	South-east Europe, south-west and central Asia, and north Africa
Ireland	3 individuals
Cork	3 individuals
Conservation	Not globally threatened

One at Dursey Island on 16 September 1977. First-year at CCBO on 18 September–9 October 1999. One at CCBO on 24 September–1 October 2006. The Dursey Island bird was exhausted and easily captured and was later released on the mainland.

Booted Warbler

Iduna caligata | Ceolaire tosaithe | A

(monotypic)

World	North-east Europe east to central Asia
Ireland	7 individuals
Cork	2 individuals
Conservation	Not globally threatened

First-year at Ballycotton on 2 September 2004. One at Firkeel on 7 October 2016.

Sykes's Warbler

Iduna rama | Ceolaire Sykes | A

(monotypic)

World	Central and south-west Asia
Ireland	2 individuals
Cork	2 individuals
Conservation	Not globally threatened

One at CCBO on 17 October 1990. One at Garinish on 2 October 2013.

Icterine Warbler

Hippolais icterina | Ceolaire ictireach | A

(monotypic)

World	Northern and central Europe east to north-central Asia
Ireland	Rare passage migrant, mainly in autumn
Cork	169 individuals
Conservation	Not globally threatened

The Icterine Warbler is an autumn migrant between August and October to coastal headlands and islands from Dursey Island to Knockadoon Head. There are three spring occurrences, all at CCBO. It occurred in all but nine of the sixty years during 1959–2018, although eight of the blank years have been since 2000. Most have occurred at CCBO (117), followed by Dursey Island and Beara peninsula combined (22). The Mizen peninsula, Galley Head and Old Head of Kinsale have each had eight birds, while the east Cork sites of Ballycotton (4), Gyleen and Knockadoon Head (1 each) account for the rest. Most records refer to single birds, but there are fifteen instances of two and three together, and one when five were present at the same time at CCBO (August 1984). Typically, Icterine Warblers remain for only one or two days (126), but a small number have remained for longer periods, one for seventeen days and another for more than twenty-two days (mean stay = 2.5 days). The ages of twelve individuals have been published, eleven of which were first-year birds and one an adult.

The number of individuals occurring over the sixty years of this study has declined. The mean number per decade in the first three decades was thirty-nine, while this decreased to eighteen in the most recent three decades. The decline is probably due to population changes in Europe, where there has been a decrease in breeding numbers in the south-west of its range in France, but an increase in Sweden and elsewhere.

1959–68	1969–78	1979–88	1989–98	1999–08	2009–18
38	32	46	26	16	11

Earliest and latest occurrence in spring and autumn (all records): 11 May–8 June and 6 August–7 November.

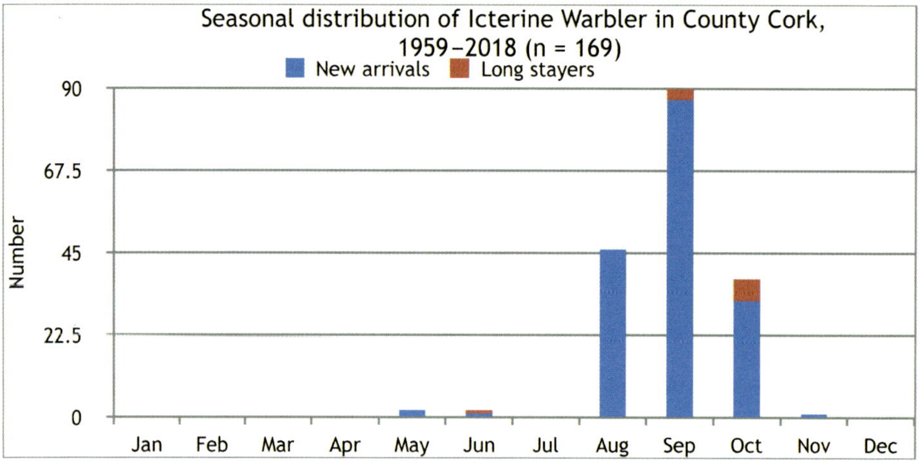

Fig. 7.37 Icterine Warbler.

Melodious Warbler

Hippolais polyglotta | Ceolaire binn | A

(monotypic)

World	South-west Europe and north Africa
Ireland	Rare passage migrant, mainly in autumn
Cork	135 individuals
Conservation	Not globally threatened

The first record of a Melodious Warbler was of one obtained at Old Head of Kinsale on 23 September 1905. The next was not until 1959, since when it has become a regular autumn migrant with one spring record (Knockadoon Head, May 2000). It has been absent in eleven of the sixty years during 1959–2018 and has occurred at many coastal headlands and islands from Dursey Island to Knockadoon Head. Most have occurred at CCBO (87), while Mizen peninsula (13) and Dursey Island and Beara peninsula combined (12) provide most of the rest. Galley Head (10), Old Head of Kinsale (6), Sherkin Island (2) and Barry's Head (1) are the only other west Cork sites where it has occurred. Three have occurred in east Cork, two at Knockadoon Head and one at Inch Strand. Most records are of single individuals, but two occurred together on seven occasions. Only three have been aged, all as first-year birds. Most have been recorded on only one day, but many have remained for longer periods, with nine remaining for more than ten days, the longest stay being nineteen days (mean stay = 2.9 days).

Unlike its relative the Icterine Warbler, this species, with a south-west breeding range in Europe which has expanded, has not declined over the sixty years of this study.

1959–68	1969–78	1979–88	1989–98	1999–08	2009–18
26	14	27	28	19	20

Earliest and latest occurrence in spring and autumn (1959–2018): 7 May and 4 August–31 October.

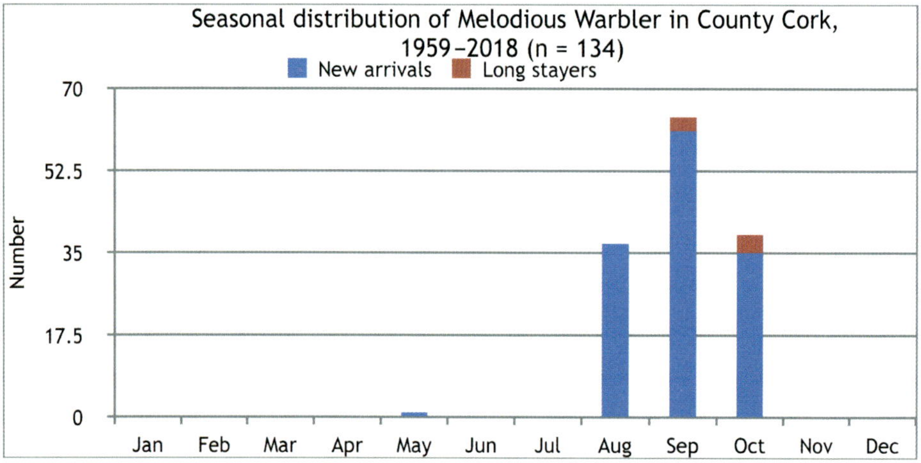

Fig. 7.38 Melodious Warbler.

Unidentified *Hippolais* warbler species

Twenty-six *Hippolais* warblers, either Icterine or Melodious (*Hippolais icterina/ polyglotta*), have been recorded during 1959–2018. Most of these have occurred at CCBO, with one each at the Beara peninsula and Knockadoon Head; all have occurred in the autumn, and all fall within the date ranges documented for both species.

Aquatic Warbler

Acrocephalus paludicola | Ceolaire uisce | A

(monotypic)

World	Central Europe east to central Asia
Ireland	13 individuals
Cork	8 individuals
Conservation	Vulnerable

First-year male obtained at Bull Rock on 20 September 1903. One at CCBO on 31 August 1961. One at CCBO on 17–24 September 1963. One at CCBO on 14–15 October 1973. One at CCBO on 29–30 September 1976. One at CCBO on 11–12 October 1976. One at CCBO on 3 September 1981. One at CCBO on 26 September 1989.

Sedge Warbler

Acrocephalus schoenobaenus | Ceolaire cíbe | A

(monotypic)

World	Europe, and Asia east to River Yenisey
Ireland	Common summer visiting breeder
Cork	Common summer visitor and breeder
Conservation	Green-listed

Early authors described the Sedge Warbler as common, usually arriving in May. It bred throughout the county during 1968–72. Although there were losses in 1988–91, by 2008–11 most had been made good. The range extended across the county, with the main absences in the north, and highest densities in coastal areas.

Sedge Warblers breed in marshy habitats. Reedbeds are important, but it is not confined to reedbeds, and other wet vegetation types are utilised. Some nest in dry habitats, especially young conifer plantations, but hedgerows on farmland and bramble thickets, often with a trickle of water nearby, are also used.

Breeding at CCBO has been erratic, at least to the 1990s. Seven pairs were breeding in 1965, and six pairs in 1969, but it did not breed every year. It now breeds at several locations, and occurrences have been increasing since the 1990s. Numbers at Sherkin Island increased from seven pairs in 1977 to thirty pairs in 1992, but then declined to seventeen to twenty pairs in 1994, with some moving into drier habitats.

The earliest Sedge Warbler was one at Cobh on 17 March 1983, but few are present by late April and the main arrival is in May. Migration at CCBO is normally heavier in spring and peaks in mid-May when up to ten birds have occurred, with a maximum of fifty on 16 May 1997. Autumn migration peaks between late July and mid-August. Numbers are lower by early September, and few remain by the end of the month. Autumn 1964 was exceptional at CCBO, with atypically high counts of twenty-six and thirty-five on 2 and 3 September, respectively. The latest record is of one at CCBO on 21 October 1959. It has been noted that fewer migrant Sedge Warblers passed through CCBO in the period 1972–86 compared with the 1960s.

Sedge Warbler

Ringing has shown that reedbeds hold greater numbers in late summer and early autumn than the local breeding population and their offspring combined, indicating a throughput of birds from other parts of Ireland. Studies at Ballycotton and Ballyvergan Marsh in the 1980s, 1990s and 2000s involved trapping and ringing thousands of birds. These have produced a series of recoveries from Donegal in the north to Senegal in the south, including one ringed at Ballycotton on 22 August 1987 and recovered in Devon the next day. The recoveries illustrate the autumn migration route, with Cork-ringed Sedge Warblers being found in Wexford, Wales, south coast of England, Belgium, France, Iberia, Morocco and Senegal. One ringed at Ballycotton on 3 August 1984 was controlled in Avon (England) on 9 August and found dead in Morocco on its return migration on 20 April 1985. Many Sedge Warblers have been obtained in spring and autumn at Fastnet Rock and Old Head of Kinsale lighthouses.

Paddyfield Warbler

Acrocephalus agricola | Ceolaire gort ríse | A

(monotypic)

World	Eastern Europe east to central Asia
Ireland	5 individuals
Cork	2 individuals
Conservation	Not globally threatened

First-year found dead at Galley Head on 13 October 1991. One at Garinish on 5 October 2000.

Blyth's Reed Warbler

Acrocephalus dumetorum | Ceolaire Blyth | A

(monotypic)

World	North-east Europe east to central Asia
Ireland	16 individuals
Cork	8 individuals
Conservation	Not globally threatened

One at CCBO on 20 October 2006. One at Mizen Head on 10–15 October 2007. One at Dursey Island on 26–7 September 2010. One at CCBO on 3–5 October 2011. One at Ballynacarriga (Beara) on 8 October 2012. One at Garinish on 5–6 October 2015. One at CCBO on 14–16 October 2015. First-year at Knockadoon Head on 7–8 November 2018.

Marsh Warbler

Acrocephalus palustris | Ceolaire corraigh | A

(monotypic)

World	Europe and western Asia
Ireland	8 individuals
Cork	5 individuals
Conservation	Not globally threatened

Adult at Ballyvergan Marsh on 5 August 1991. One at Galley Head on 30 September–3 October 2004. One at CCBO on 25–6 September 2009. Male (in song) at Old Head of Kinsale on 9 June 2014. First-year at Crookhaven on 7–10 October 2017. The CCBO bird had been ringed in Norway as a Common Reed Warbler.

Common Reed Warbler

Acrocephalus scirpaceus | Ceolaire giolcaí | A

(subspecies: *scirpaceus*)

World	Europe east to central Asia, and north Africa
Ireland	Scarce breeder, recent increase, passage migrant
Cork	Scarce breeder, passage migrant
Conservation	Green-listed

Reed Warblers were first identified in Cork in September 1960, although there were records of unidentified *Acrocephalus* warblers in autumn 1959. The species has occurred annually as a passage migrant since, and recently as a breeder. Unidentified *Acrocephalus* warblers are treated here as Reed Warblers.[297]

The earliest indication of breeding was of two seen in suitable habitat at Ballycotton in August 1977, and another in July 1979. During August and September 1980, seven juveniles were trapped, and several more seen. Breeding was first proved in 1981, when at least five males were singing through the summer, food-carrying was seen, and seventeen were trapped.

Breeding has been almost annual since 1981, with five to ten pairs in most years, although there was a hiatus following habitat change from 1990/1 which reduced the area of reedbed.

Breeding was proved at Ballyvergan Marsh in 1983, when up to eleven were present in June and July. Breeding has since been annual, with an estimated twenty pairs in 1994 and twenty-five to fifty pairs by 1996, although in recent years censusing has been sporadic.

Apart from the above, the only records indicating possible breeding up to 1998 were three singing males at Flaxfort Marsh (Courtmacsherry Bay) in May 1991, a juvenile at Inchydoney in July 1991, a male singing at Pilmore Strand in June 1995 and another singing at Flaxfort Marsh in June 1998.

Fieldwork for the 2008–11 breeding atlas showed the range had extended west to Castlefreke. However, breeding took place at only two 10 km squares (Rostellan, W86 and Ballyvergan, X07). The three squares occupied west of Cork Harbour all referred to probable breeding, and while breeding was not proved at Ballycotton in 2008–11, a small population may have been overlooked.

Migrant Reed Warblers have occurred annually since 1960 at CCBO, as well as at other coastal sites, although the presence of breeding birds since the 1980s has clouded the accuracy of migrant records. In collating the data for the decadal table birds in breeding habitats were excluded. Spring migration is meagre and accounts for about 4 per cent of records. There were no spring records up to 1969 at CCBO, and it has rarely occurred there since at this season. An earlier national analysis showed 5 per cent of migrants occurred in spring.[298]

Reed Warblers have occurred at almost every site on the coast from Dursey Island to Youghal during autumn migration. CCBO has had more records than any other site, but otherwise the spread of occurrences is reasonably even. Migrant numbers have varied annually although there has been a sustained decadal increase over the sixty years of this study. Most records involve one to three or four birds. Easterly and westerly movements of juveniles occur on the south English coast in autumn, and those recorded in Cork may be a combination of wandering local and English birds.[299] Birds occurring from late September onwards are more likely to be reverse migrants from east European populations as they occur at a time when most west European birds have already departed.[300] Ring recoveries support this theory since birds from the Netherlands, Belgium and Poland have been recovered at CCBO in September (1985, 1996, 1998).

The Reed Warbler is a trans-Saharan migrant, and while no Irish ringed birds have been found in the wintering area, data from ringing in Cork suggests that they move east to the south coast of England, then to France and Morocco and down the west coast of Africa. A juvenile ringed at Ballycotton in August 1995 was recovered alive at Ballyvergan Marsh seven years later in July 2002.

1959–68	1969–78	1979–88	1989–98	1999–08	2009–18
48	62	99	191	204	250

Earliest and latest occurrence in spring and autumn (all records): 24 April–4 November.

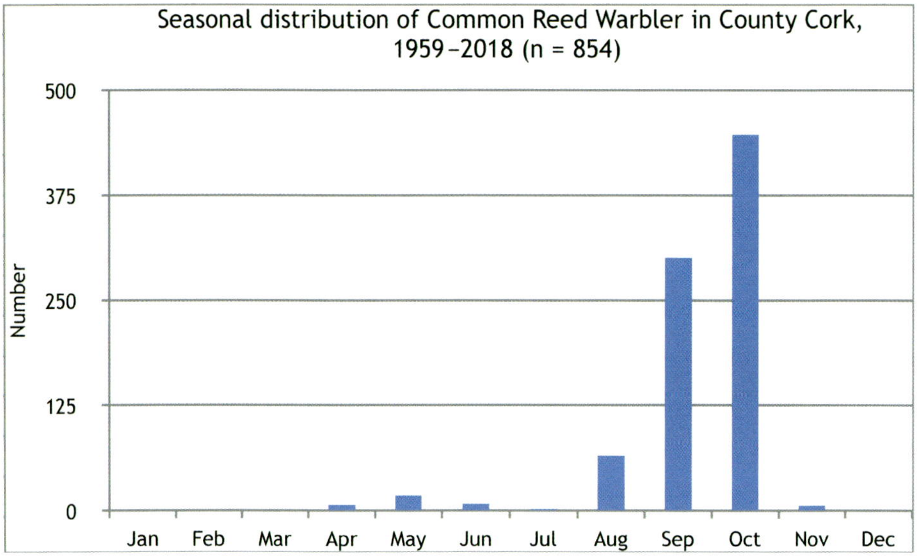

Fig. 7.39 Common Reed Warbler.

Great Reed Warbler

Acrocephalus arundinaceus | Mórcheolaire giolcaí | A

(subspecies: *arundinaceus*)

World	Europe east to central Asia, and north Africa
Ireland	4 individuals
Cork	3 individuals
Conservation	Not globally threatened

One found dead at Castletownshend on 16 May 1920. Male (in song) at CCBO on 10–26 June 1964. One at CCBO on 18 May 1979

Fan-tailed Warbler

Cisticola juncidis | Ceolaire earrfheanach | A

(subspecies: undetermined)

World	Southern Europe, southern Asia south to north Australia, and Africa
Ireland	2 individuals
Cork	2 individuals
Conservation	Not globally threatened

One at CCBO on 23 April 1962. One at CCBO on 18 April 1985.

Bohemian Waxwing

Bombycilla garrulus | Síodeiteach boihéamach | A

(subspecies: *garrulus*)

World	Northern regions of Europe, Asia, and North America
Ireland	Winter visitor, occasionally irruptive
Cork	Scarce irruptive winter visitor
Conservation	Green-listed

The Waxwing is usually rarer in southern Irish counties, and the first record for Cork was one obtained by Robert Ball at Castlemartyr in 1820. Three or four were seen (1 obtained) near Carrigaline in January 1850, a winter when an influx took place in Ireland. Another was obtained in March 1937, and there were three records during winter 1946/7 (including 1 obtained in May 1947), another winter when an influx took place in Ireland.

Waxwings have occurred in Cork in twelve winters since 1959, but the number involved in individual winters has varied from one bird to hundreds. In Table 7.19, the numbers given attempt to take account of the likelihood of duplication, and for analysis Ballincollig is included in Cork city, while Carrigaline, Glanmire, Little Island, Cobh and Carrigtwohill are included in Cork county. It is probable that higher numbers than those given occurred in some years, and the likelihood of the same mobile flock visiting several tree-lined suburban roads must be considered.

During the first influx in 1965/6 most birds occurred at sites within Cork city, with three in the county at Trabolgan. More recently, while significant numbers occurred within the city, large flocks also occurred in county towns. In winters when large numbers occurred, the first birds appeared in late October or early November. Typically, numbers remained

Winter/year	Number (city/county)	Extreme dates
1959	?	January
1965/66	41 (38/3)	8 November–29 March
1968	4 (4/0)	1–3 January
1969	1 (0/1)	7–8 March
1970/71	6 (0/6)	11 October–6 February
1996	14 (7/7)	12 February–9 March
2004/05	629 (395/234)	23 October–23 April
2005/06	76 (25/51)	31 October–22 March
2008/09	33 (0/33)	14 November–17 March
2010/11	9 (5/4)	30 November–7 January
2012/13	194 (67/127)	2 November–26 February
2017	37 (0/37)	11–12 February

Table 7.19 Numbers of Bohemian Waxwing in County Cork, 1959–2017.

high through to February or March, some remaining to April. Apart from birds in Cork city, almost all parts of the county were visited, including Goleen, Baltimore, Skibbereen, Clonakilty, Kinsale, Cobh and Ladysbridge, and inland locations, such as Kealkill, Macroom, Mallow and Castletownroche. Waxwings were generally seen in groups ranging from a few to twenty or thirty, but during 2004/5 there were loose gatherings of at least 100 birds.

Waxwings have also occurred at coastal migration watch points. They have occurred several times in October and November, and once in January at Dursey Island. It has been seen at Mizen Head in October and December, and at Ballycotton lighthouse in October and March.

Unidentified waxwing species

One waxwing, either Bohemian or Cedar (*Bombycilla garrulus/ cedrorum*), was recorded at CCBO on 14 May 2002.

Treecreeper

Certhia familiaris | Snag | A

(subspecies: *britannica*)

World	Europe and Asia
Ireland	Common resident breeder
Cork	Scarce but widespread resident
Conservation	Green-listed

The Treecreeper was common in the 1800s and it was present wherever there were trees. Robert Ball saw it at Youghal, and Robert Warren said it bred in holes on a wall at Castle Warren. It was widely distributed as a breeder during 1968–72. The 1988–91 and 2008–11 atlas surveys indicated little change, with absences only from coastal sites and high ground. Records suggest a thinly spread species which is absent only where there are no trees, but one which is easily overlooked. Winter atlas surveys showed a similarly thin distribution.

Most sightings of Treecreepers are of single birds or pairs and family parties during the breeding season, and six birds at Rostellan on 3 March 1989 was a high number. They stay on or close to trees most of the time, and most foraging is done on tree trunks, although individuals forage for invertebrates on cement walls, dry stone walls, moss-covered bridges, wooden fencing and wooden telephone poles where these structures occur close to trees, and also on the ground, and grit is probably taken in some cases.

Treecreepers have occurred at least twenty-one times at CCBO between 20 June and 25 October, although few have been recorded after August, perhaps reflecting post-breeding dispersal. One spring record has occurred at CCBO. Dispersal may also explain the presence of one at Crookhaven on 9 October 1992.

Wren

Troglodytes troglodytes | Dreoilín | A

(subspecies: *indigenus*)

World	North America, Iceland, Faeroes, Europe, Asia, and north Africa
Ireland	Very common resident breeder
Cork	Common resident breeder
Conservation	Green-listed

Wrens were distributed widely during all three breeding atlas surveys. Birds were present in all habitat types, including at poorly vegetated uplands and islands. Distribution in winter was identical to the breeding season, and density was high throughout.

The Wren is one of the commonest breeding birds at CCBO. Numbers declined following the cold winters of 1961/2 and 1962/3. It was estimated that 750–1,000 pairs were present in 1959–61. This figure dropped to 150–200 pairs in 1962, and seventy-five pairs in 1963. Numbers increased to 391 pairs in 1986, well below original estimates, which were probably too high. Wrens breed commonly at other islands, including Sherkin and Dursey, as well as smaller ones where a few pairs breed.

Wren

Irish Wrens are believed to be sedentary. No regular movements have been recorded concerning Cork birds. However, some have occurred at places that give rise to the possibility that migrants occur. Two were obtained at Fastnet Rock on 30 October 1897, and the rock was visited again in October 1914. One was found dead at Bull Rock on 29 October 1914. One was circling the M.V. *Inishfallen* for twenty minutes from 5.45 a.m. when about 5 km from Cork Harbour on 13 April 1962. There is little evidence of movement at CCBO, although this would be difficult to detect given the large resident population.

Wrens often make their nests inside buildings, and inactive Barn Swallow nests are sometimes occupied. During a study of Barn Swallow nests in 2005 and 2006, 8.2 per cent (n = 159) and 7.5 per cent (n = 146), respectively, were occupied by Wrens. Wrens also make their nests inside inactive Dipper nests beneath river bridges. Wrens roost communally during winter, and sites used include natural ones among vegetation, Barn Swallow and Dipper nests, and nest boxes erected for tits.

Wrens (occasionally Dunnock) are associated with the tradition of *Hunting the Wren* on St Stephen's Day (26 December). In the 1700s, Smith said to hunt and kill the Wren on this day was an ancient custom. Killing the

Wren was forbidden by the lord mayor of Cork, Richard Dowden, from the end of 1845. Although the Wren Boys tradition goes on to the present, the Wren has seldom been killed in recent decades. The Wren is now protected by the Wildlife Act, 1976.

Grey Catbird

Dumetella carolinensis | Catéan liath | A

(monotypic)

World	North America and Bermuda
Ireland	1 individual
Cork	1 individual
Conservation	Not globally threatened

One at CCBO on 4 November 1986.

Common Starling

Sturnus vulgaris | Druid | A

(subspecies: *vulgaris*)

World	Iceland, Europe, Asia, Azores, and Canary Islands, introduced elsewhere
Ireland	Very common resident breeder, very common winter visitor
Cork	Common resident breeder, and winter visitor
Conservation	SPEC 3; Amber-listed

The Starling was present in winter and summer in the 1700s, and probably had a breeding range confined to the coast. It was breeding in sea cliffs at Reanies Bay and in eaves of houses in the mid-1800s. It was increasing as a breeder throughout the country at the end of the 1800s but was mainly a winter visitor to Cork and was unknown in the west during summer.

The Starling continued to extend its breeding range, although the west was not reached until the 1930s. It was breeding at Bantry in 1935 and was well established ten years later, during which decade a few bred at Skibbereen and Castletownbere. CCBO was colonised in about 1955, and the population was 80–100 pairs during 1963–86. The breeding population at Sherkin Island declined from 85–100 pairs in 1977 to twenty-eight to thirty-three pairs in 1994, possibly due to loss of open grassland. Starlings were breeding throughout the county during all breeding atlas surveys.

A large immigration from the continent occurs in autumn. Later in winter, especially during cold weather, further birds arrive from Scandinavia, Germany, Poland and Britain. Starlings have occurred in October and November at Bull Rock, Fastnet Rock and Galley Head.

Winter surveys show the Starling as widely distributed, with indications of higher densities in the north. Starlings forage in a variety of habitats during winter and at other times. Pasture fields, including where silage has been harvested, are important feeding areas as flocks roam from field to field, often in company with crows. Seaweed-strewn beaches are also utilised as feeding sites. Large flocks can be seen at these times, such as 10,000 near Cork airport on 23 November 1969, and 3,500 at Ballycotton on 21 October 1969.

Starlings roost communally after the breeding season, but it is during the winter that spectacular flocks occur, although surprisingly few records have been documented. A roost at Newcastle (near Grenagh) in a forest plantation held at least 100,000 in January 1992, the largest gathering of birds of any species on record for Cork. About 50,000 roosted in a reedbed at Ballyvergan in November and December 1990, and 30,000 in a forest plantation near Killeagh in January 1990. A pre-roost flock of 5,000–10,000 occurred at Timoleague in January 2018.

Starling flocks were believed to foretell disaster, especially in urban settings. A battle between opposing flocks at Cork city in October 1621 has been dubbed the 'Battle of the Birds' and was said to foretell the Burning of Cork which happened on 30 May 1622.[301;302] Another battle apparently took place in Fermoy in November 1930, this time between Starlings and Rooks.[303]

Rosy Starling

Pastor roseus | Druid rósach | A

(monotypic)

World	South-east Europe, and south-west and central Asia
Ireland	Rare passage migrant, mainly in autumn
Cork	46 individuals
Conservation	Not globally threatened

The Rosy Starling is a spring and autumn passage migrant to coastal districts from Dursey Island to Knockadoon Head, but one has occurred in winter at Cobh (January and February 2004). Most occur at the western peninsulas and islands, CCBO (15), Dursey Island and Beara peninsula (7), Mizen peninsula (7), Galley Head (4) and Old Head of Kinsale (6). Seven birds occurred from Cork Harbour eastwards: Carrigaline (1), Cobh (1), Power Head (1), Ballycotton (3) and Knockadoon Head (1). Of those that have been aged, ten were adult and thirty-four were first-year birds. Apart from two and three together at CCBO in 2002 and 2011, all other records are of single birds. While nearly half (21) have been recorded on only one day, there is a tendency for some to make protracted stays and sixteen individuals have remained for periods ranging from ten to fifty-five days (mean stay = 8.4 days). There has been an increase in occurrences over the last three decades of this study, greater than might be predicted of a species whose breeding range is east of the Balkans.

1959–68	1969–78	1979–88	1989–98	1999–08	2009–18
1	1	1	9	12	22

Earliest and latest occurrence in spring and autumn (all records): 15 May–28 June and 20 July–6 November (20 January).

Dipper

Cinclus cinclus | Gabha dubh | A

(subspecies: *hibernicus*)

World	Europe, Asia, and north Africa
Ireland	Resident breeder
Cork	Resident breeder
Conservation	Green-listed

The Dipper frequented stony rivers in the 1700s. In the 1800s it was described as a bird of rocky mountain streams, where it was common, but it also inhabited lowland streams. Today it nests from mountain streams almost down to the coast, and on slow-flowing rivers provided there is some broken water.

Breeding atlas surveys have shown it is widely distributed, in winter as well as the breeding season, although its status in the south-west has been commented upon in the past. In 1948 it was described as occasional in streams in the south-west and south and was apparently absent from much suitable habitat there in 1956. If these observations are correct then it is difficult to explain the apparent change in status, but birds there possibly suffered more than elsewhere during cold winters, of which there were several in the 1940s, 1950s and 1960s. However, a more likely explanation is underestimation of the range, and there is evidence that the first two breeding atlases poorly represented the situation. Independent surveys in 1994 and 1995 showed that a single surveyor confirmed breeding in nineteen previously blank 10 km squares, with breeding evidence found in a further ten squares in which Dippers had been noted as present without evidence of breeding in 1988–91. Dipper studies in lowland streams and rivers north and west of Cork Harbour and in east Cork showed no change in status over the period from the mid-1980s to the late 2010s.[304;305] Some breed in urban streams at Blackpool and Glen Park in Cork city.

The breeding biology of Dippers has been studied by examining 501 nest histories.[306] Most pairs are single brooded, only 8.2 per cent attempting a second clutch. The mean clutch size was 4.16 eggs, and 3.48 young fledged per successful pair. Dippers occur at a rate of about one pair per km in good-quality rivers, but in poor habitats density is lower. Lack of nest sites may be a limiting factor in some rivers and numbers may be increased by providing nesting platforms or boxes.

Dipper movements have been studied by ringing.[307] Juveniles begin dispersal from natal territories in June, when fledged less than two months. One male, fledged for forty-five days, had travelled 35.2 km, and one female, fledged for seventy days, had travelled 37.5 km. Both birds had crossed watersheds. Females moved greater distances than males, and birds of both sexes showed strong fidelity to breeding sites. Dippers have been recorded at CCBO on three occasions (6 October 1994, 16 June 2006, 1 June 2009), but never at Sherkin or Dursey Islands, although they have been seen near the tip of the Beara peninsula. The occurrence of a juvenile at CCBO in June 2009 confirms the dispersal pattern based on ringing results. Ringing has revealed the maximum lifespan of some birds, the oldest in Cork being eight years, nine months and four days.

Dipper diet was studied at the Rivers Araglin, Douglas and Glashaboy.[308;309] A wide variety of aquatic insect taxa was taken, but studies during spate conditions revealed an increase in insect prey of terrestrial origin indicating that birds were forced to seek food away from the river corridor during such times. Dippers have been seen at the edge of the estuary at Douglas in winter, and on the estuary of the River Glashaboy during the low tide period, indicating a degree of flexibility in foraging strategy.

White's Thrush

Zoothera dauma | Smólach White | B

(subspecies: *aurea*)

World	Eastern Europe and Asia south to Sumbawa
Ireland	5 individuals
Cork	1 individual
Conservation	Not globally threatened

One obtained at Bandon in early December 1842.

Siberian Thrush

Geokichla sibirica | Smólach Sibéarach | A

(subspecies: undetermined)

World	Eastern Asia
Ireland	2 individuals
Cork	1 individual
Conservation	Not globally threatened

Probable first-year female at CCBO on 18 October 1985.

Hermit Thrush

Catharus guttatus | Smólach ruaphrompach | A

(subspecies: undetermined)

World	North America
Ireland	2 individuals
Cork	2 individuals
Conservation	Not globally threatened

First-year at Galley Head on 25–6 October 1998. First-year at CCBO on 19–20 October 2006.

Swainson's Thrush

Catharus ustulatus | Smólach Swainson | A

(subspecies: undetermined)

World	North America
Ireland	8 individuals
Cork	6 individuals
Conservation	Not globally threatened

One at CCBO on 14–16 October 1968. One at CCBO on 8 October 1990. One at Garinish on 11–12 October 1999. First-year at Galley Head on 11 October 2008. One at CCBO on 14 October 2017. First-year at CCBO on 17 October 2018.

Grey-cheeked Thrush

Catharus minimus | Smólach glasleicneach | A

(subspecies: undetermined)

World	Northern regions of North America and north-east Siberia
Ireland	13 individuals
Cork	11 individuals
Conservation	Not globally threatened

All eleven individuals were recorded between 3 and 29 October, peaking in the middle third of the month. All occurred singly at Dursey Island (2), CCBO (5), Rosscarbery (1), Galley Head (2) and Old Head of Kinsale (1). Some stayed for several days, the maximum being nine (mean stay = 3.2 days). The first record was in 1982, and the species has occurred with increasing frequency in the last two decades. In 2017, three different birds occurred in a small geographical area between Rosscarbery and Red Strand, the most recorded in any year. Of seven that have been aged, all have been first-year birds.

1959–68	1969–78	1979–88	1989–98	1999–08	2009–18
0	0	2	1	3	5

Earliest and latest occurrence in spring and autumn (all records): 3 October–29 October.

Veery

Catharus fuscescens | Smólach Wilson | A
(subspecies: undetermined)

World	North America
Ireland	1 individual
Cork	1 individual
Conservation	Not globally threatened

First-year at CCBO on 17–18 October 2018.

Ring Ouzel

Turdus torquatus | Lon creige | A
(subspecies: *torquatus*)

World	Europe, south-west Asia, and north Africa
Ireland	Rare summer visiting breeder, scarce passage migrant
Cork	Has bred, scarce passage migrant
Conservation	Red-listed

The Ring Ouzel was scarce in the 1700s and was found in the mountains of 'Ivelary' (the parish near Inchigeelagh, now called Iveleary). Unpublished notes by Ussher (about 1900) detail breeding sites south of Mallow and at 'Coomashesh' (5 km from Millstreet), where two pairs bred each year. The reference to south of Mallow is probably to the eastern end of the Boggeragh Mountains, while the site near Millstreet may refer to Comeenatrush. A major decrease was reported in Ireland during the first half of the 1900s, and it was almost unknown in Cork by the early 1950s.

The Ring Ouzel was probably always a scarce breeding bird in Cork due to habitat limitations. None was found during 1968–72 or 1988–91. However, some potential breeding sites were not surveyed. Two pairs were seen near Millstreet (Caherbarnagh range) in May 1977 and two or three were singing west of Glengarriff (Caha Mountains) in June and July 1977. Birds held territory in the north and west of the county in 1986 and 1987.[310]

A bird was singing at Beara peninsula in June 1991, but a search of the Caherbarnagh range in June 1995 produced a negative result. It is likely that 1977 was the last year of breeding in Cork (Caherbarnagh), but with the possibility of breeding continuing to the early 1990s at the Beara peninsula. None was found in breeding habitats during 2008–11, and records from four coastal 10 km squares referred to passage migrants.

Three Ring Ouzels were recorded at Fastnet Rock (22 September 1887, 3 April 1891 and 30 October 1897), and others were obtained there on 20 October 1911, and three during a rush of birds on 28–30 September 1913. There are also records from Old Head of Kinsale in spring and autumn (1906–10), with an early occurrence at Ballycotton on 29 March 1949.

It is typically rare in spring but is more regular in autumn, peaking in October. At CCBO, one or two occur regularly, with occasionally up to twelve in a day. The highest autumn total recorded there was an exceptional fifty-two birds on 16 October 1973, with thirty-two there on 17 October 2006. Most records are of one to three birds and counts in double figures are rare away from CCBO, but fourteen were at Dursey Island on 15 October 2014. CCBO, Dursey Island and Old Head of Kinsale are the most regularly visited sites, but records extend along the coast from Dursey Island to Ballymacoda, although east Cork occurrences are rare. Despite an assertion that it was a regular migrant at Roche's Point, records during 1959–2018 do not support this.[311] Apart from records at breeding areas, there appears to be only one other inland occurrence, one at Charleville on 13 November 2008. The origin of Irish south coast migrants is unknown but could involve Scottish and Scandinavian birds.

Apart from breeding season records, which are excluded from the decadal table, the Ring Ouzel has occurred annually in Cork over the sixty years of this study. It is ironic that the number of migrants appears to be at its highest during the most recent two decades, at a time when the Irish breeding population is at its lowest level on record.

1959–68	1969–78	1979–88	1989–98	1999–08	2009–18
89	107	43	67	158	198

Earliest and latest occurrence in spring and autumn (1959–2018): 19 March–13 November.

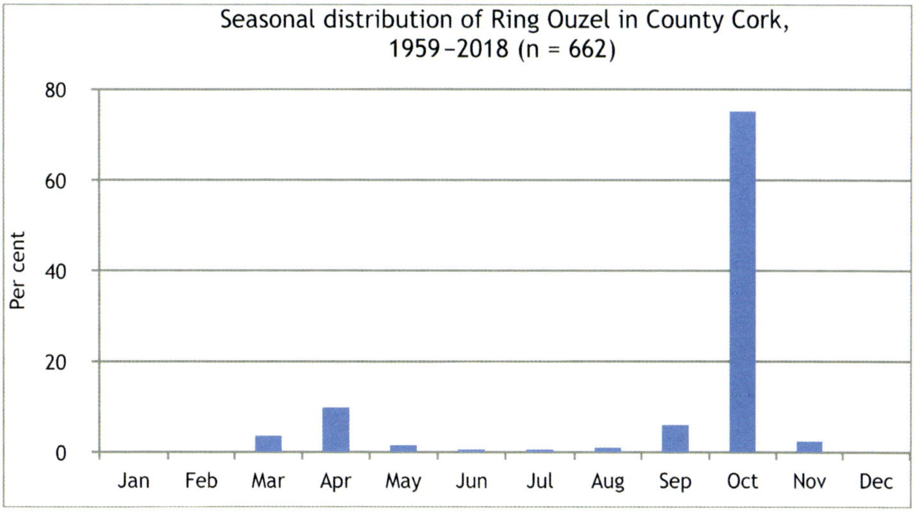

Fig. 7.40 Ring Ouzel.

Blackbird

Turdus merula | Lon dubh | A

(subspecies: *merula*)

World	Europe, Asia, north Africa and its islands, introduced elsewhere
Ireland	Common resident breeder, winter visitor, passage migrant
Cork	Common resident, winter visitor, passage migrant
Conservation	Green-listed

The Blackbird was present throughout the county during all three breeding atlas surveys, but density was low relative to the east of Ireland. It is among the commonest breeding passerine at CCBO, and despite two cold winters (1961/2 and 1962/3), numbers remained stable when censuses in 1965 and 1986 produced figures of 137 and 135 breeding pairs, respectively. A mean of sixty-eight pairs per annum bred at Sherkin Island during 1977–94.

Winter atlas surveys showed that distribution mirrored the breeding season. There is an immigration in October and November, and large numbers have been recorded at lighthouses, including Bull Rock and Fastnet Rock. Blackbirds, unlike other thrushes, appear to be less affected by cold weather and are usually not involved in movements at the onset of a cold spell. At CCBO, flighty flocks of migrants are seen in most (but not all) years in October, and numbers are highest in November (up to 147 in a day). Up to 150 have been recorded at Dursey Island in October, but numbers are usually smaller.

Irish- and British-bred Blackbirds are mainly sedentary, but some from northern England and Scotland winter in Ireland. Migrants from north European countries also winter in Ireland (and Cork), especially from Fennoscandia, Denmark, Germany and the Netherlands.

Fieldfare

Turdus pilaris | Sacán | A

(monotypic)

World	Iceland, Europe, and Asia
Ireland	Common winter visitor
Cork	Common winter visitor
Conservation	Green-listed

Fieldfares were less frequently seen in the south of Ireland than Redwings in the 1800s, but the reverse was the case at exposed sites like Dursey Island. Winter atlas surveys showed that it occurred throughout the county. Fieldfares are less affected than some other thrushes by cold weather, but the severe winter of 1813 reduced numbers about Youghal.

Fieldfares are recorded at CCBO from late September to mid-April but are regular only in October and November when up to 280 have been seen in a day. Largest numbers occur on the island during cold weather movements, and thousands were reported there in January 1962 during such conditions. Westerly movements also took place at CCBO in December 1962 and January 1963, another very cold winter. Modest numbers occur at Dursey Island in most years with sometimes up to 400 in October. During autumn, Fieldfares have been recorded at Bull Rock, Fastnet Rock and Old Head of Kinsale lighthouses.

Numbers vary greatly between and within winters and can change quickly at any site as mobile flocks roam the countryside in search of

Fieldfare

feeding opportunities at berry-bearing hedgerows and orchards. Timing of departure from Scandinavian breeding grounds depends on the availability of rowan berries, and the numbers wintering in Ireland are dependent on weather conditions in Britain and Europe. Because of these factors, the timing of arrivals and the numbers are extremely variable. Few large counts have been recorded, but in January 1982 there were up to 10,000 at Ballybraher and 5,000 at Midleton. A total of 3,000 flew over Cork Harbour on 11 January 1982, although most flocks involve fifty to several hundred birds.

Fieldfares often roost overnight on low bushes and trees in marshy areas (e.g. at Ballyvergan Marsh and Ballymacoda), and Robert Ball saw about 500 roosting together in a spruce tree near Youghal in the early 1800s.

Fieldfares arrive later than Redwings and are usually the scarcer of the two. They are rarely seen before the beginning of October, arriving chiefly in November, although three were seen near Bandon on 17 August 1948. There was one at CCBO on 1 September 1972, and in 1994 a passage movement began there as early as 11 September, with forty-six birds seen on 17 September. Fieldfares often remain late in spring, and four were seen at Banduff Bog on 8 April 1965, when courtship display was noted. Most have left by early to mid-April, but small numbers may remain to early May. Fieldfares wintering in Ireland come from Norway and Sweden, and possibly from further east in Siberia. A nestling ringed at Dovre (Norway) on 20 June 1946 was recovered at Whiddy Island in March 1947.

Song Thrush

Turdus philomelos | Smólach ceoil | A

(subspecies: *clarkei*; *philomelos*)

World	Europe and Asia, introduced elsewhere
Ireland	Very common resident breeder, common winter visitor, passage migrant
Cork	Very common resident, winter visitor and passage migrant
Conservation	Green-listed

The Song Thrush (*T.p. clarkei*) was common and widely distributed during all three breeding atlas surveys. Winter distribution mirrored that of the breeding season.

The Song Thrush was a winter visitor to CCBO until the late 1800s or early 1900s, since when it has become a breeding species. It suffers badly during cold winter weather. The breeding population at CCBO in 1959 and 1960 was about fifty pairs but declined to eight pairs in 1963 following two cold winters. Numbers increased to sixteen pairs in 1964, eighteen pairs in 1965 and forty-one pairs in 1986. At Sherkin Island, the number of breeding pairs increased from fifteen in 1977 to thirty-three in 1994.

Breeding biology was studied in 2001–3 based on analysis of 100 nesting attempts.[312] The mean clutch size was 4.12 eggs, and the mean number of young at fledging was 3.69. Clutch size was highest in April and lowest in June. Of ninety-three nests where the outcome was known, 45 per cent succeeded in fledging at least one young.

The influence of nesting habitat on breeding ecology was examined at three study sites.[313] The presence of dense vegetation was found to be a significant factor when selecting a nest site and this should be an important consideration in conservation measures aimed at maintaining or improving populations.

There is a large winter immigration of Song Thrushes (*T.p. philomelos*), described over 100 years ago at lighthouses, especially Fastnet Rock. This movement takes place mostly in October and November, and other stations are also visited, such as Bull Rock, Dursey Island and Galley Head. Immediately after the breeding season, Song Thrushes become scarce or absent in July and August, and this has been ascribed to local movements.

Autumn migration from mid-October takes place at CCBO, and up to 180 birds have occurred in a day, but there is no evidence of spring migration. Over 100 birds in a day have occurred at Dursey Island in autumn. Further large movements caused by cold weather may occur at any time during any winter, and at CCBO up to 140 occur during such influxes in January and February. Cold weather movements usually involve birds from Scotland and northern England, and during such times flocks are observed heading south-west in search of milder conditions.

Snails form an important part of the diet of Song Thrushes in winter, and it is common to see piles of broken snail shells in sand dune systems at that season.

Redwing

Turdus iliacus | Deargán sneachta | A

(subspecies: *iliacus*; *coburni*)

World	Iceland, Faeroes, Europe, and Asia
Ireland	Common winter visitor, passage migrant
Cork	Common winter visitor
Conservation	SPEC 1; Red-listed

The cold winter of 1813 reduced Redwing numbers about Youghal. Large numbers moved westwards across the country during cold weather in the 1800s, and the same happens today. A westerly movement during cold weather took place at CCBO in December 1962 and January 1963. Large numbers also occurred there during cold weather in 1969 (e.g. 320 on

20 February). Numbers can vary greatly between winters depending on severity, the cold winter of 1981/2 producing many more birds than the mild winter of 1982/3.

Winter atlas surveys showed the Redwing was distributed throughout the county. There are no breeding records, but some can be seen late in spring and one or two such records in the north of the county in 2008–11 refer to late migrants.

Redwings occur at CCBO from mid-September to mid-May. Most occur in October and November, when up to 300 in a day have been recorded. Larger numbers sometimes occur, and over 2,300 were present on 13 October 1972 and 1,600 on 13 October 1973. One was seen at CCBO on 19 July 1977, but the earliest autumn record is of one on 4 September 1982, although few occur before early October. Redwings have been obtained at Fastnet Rock in October and November, and at Bull Rock in December.

Elsewhere, high autumn and early winter counts include 1,500 at Dursey Island in October 2012, 600 at Kilcolman in November 1989, and several flocks of 500+ in the Dunmanus Bay area in October 1971. Mid-winter numbers may increase rapidly as birds arrive to escape cold weather in Europe. Under such conditions in 1982, over 3,000 flew south-west over Cork Harbour on 11 January, and there was an estimated 10,000 at Ballybraher at the same time. In contrast to the large numbers recorded in autumn, spring migration is negligible. Redwings leave Cork in March, with the last stragglers usually departing by mid-April, but a few sometimes remain to mid-May.

Most racially identified Redwings are the continental subspecies (*T.i. iliacus*), and one ringed as a nestling in Finland was at CCBO in January 1963. The Icelandic subspecies (*T.i. coburni*) has been identified only a few times at CCBO, and once at Dursey Island on 4 November 2017. One ringed as a nestling in Iceland in May 1958 was at Douglas in February 1962. Of forty killed at Old Head of Kinsale lighthouse in October 1976, one was identified as *coburni* and thirty-seven as *iliacus*. Nevertheless, it is likely that continental birds are scarcer in Ireland than Icelandic ones. There are more ring recoveries in Ireland (several in Cork) of Icelandic than of continental birds.[314]

Mistle Thrush

Turdus viscivorus | Liatráisc | A

(subspecies: *viscivorus*)

World	Europe, Asia, and north Africa
Ireland	Common resident breeder
Cork	Common resident, occasional migratory movements
Conservation	Green-listed

The Mistle Thrush colonised Ireland in the 1800s, and by 1900 it was breeding in all counties. The first record for Cork was one obtained by Robert Ball, probably at Youghal, in 1818. It was breeding in Cork for some years before 1845, probably beginning between 1820 and 1840. It was numerous in the late 1800s, and flocks formed from June and July onwards. A preference for coastal clifftops in Cork was noted for July.

The Mistle Thrush was widely distributed during 1968–72, and it sometimes bred in exposed coastal districts, such as Seven Heads and Old Head of Kinsale. The range remained the same during 1988–91 and 2008–11. It breeds occasionally at CCBO, usually nesting in dry stone walls. Two pairs bred in 1960, three pairs in 1961 and one pair in 1969.

Post-breeding flocks occur in August and September, the largest being 100 at Model Farm Road (Cork city) from 15 September to 7 October 1993, fifty at Youghal on 16 August 1976 and fifty at Ballymacoda on 19 August 1976.

Winter atlas surveys show that it has a wide distribution. Flocks are exceptional in winter when individuals become territorial at good feeding areas, but there is a record of 100 near Inchigeelagh on 3 November 1985.

Most British-born birds are sedentary, but some move south to the continent. Observations indicate that some continental birds arrive on the British east coast, but there is no evidence that any reach Ireland. Large immigrations have occasionally been noted on the south coast in October and November, but only one bird has been obtained at Fastnet Rock in autumn (6 November 1891). Numbers rarely exceed ten birds at CCBO, but on one occasion in 1959 an extraordinary movement was observed when 84, 313 and 292 birds passed over the island on 28–30 October, respectively. Small numbers also occur at Dursey Island, but always in

American Robin

Turdus migratorius | Smólach imirce | A

(subspecies: undetermined)

World	North America
Ireland	10 individuals
Cork	1 individual
Conservation	Not globally threatened

One at Glengarriff on 16 January 1977.

Spotted Flycatcher

Muscicapa striata | Cuilire liath | A

(subspecies: *striata*)

World	Europe, Asia, and north Africa
Ireland	Common, decreasing summer visiting breeder
Cork	Decreasing summer visitor and breeder
Conservation	SPEC 2; Amber-listed

The Spotted Flycatcher was widely distributed but was not numerous in the 1800s. It bred throughout the county during 1968–72 and 2008–11. The field experience of surveyors in the latter period was that only one or two pairs were present in the minimum of eight tetrads surveyed per 10 km square.

Although there are records of two at Cobh on 5 April 1989 and one at Crossbarry on 6 April 1969, Spotted Flycatchers begin to arrive in early May. At CCBO, spring passage is light with rarely more than five in a day, although there were up to forty-eight from 10 to 18 May 1967 and thirty on 16 May 1995. An equally light spring passage occurs at mainland sites. At Old Head of Kinsale, where one or two are present during May, there were counts of fourteen on 10 May 1981 and twelve on 19 May 1990.

Autumn passage is heavier and occurs from August to October, peaking in September. Maximum counts at CCBO are fifty on 26 September 1976 and thirty-five on 3 September 1964, but peaks of fewer than twenty are usual. Up to five are recorded at mainland coastal sites from Dursey Island to Ballymacoda. The latest individuals were one at CCBO on 3 November 1968 and one at Old Head of Kinsale on 6 November 1982.

There are two ring recoveries from Bardsey Island (Wales); one on 8 August 1965 was at Schull on 26 May 1967, and one on 20 September 1976 was at Buttevant on 28 July 1977. Birds have also been obtained at Fastnet Rock during spring and autumn migration.

Rufous Bush Robin

Cercotrichas galactotes | Torspideog ruadhonn | A

(subspecies: *galactotes*)

World	Southern Europe, south-west Asia, and Africa
Ireland	3 individuals
Cork	2 individuals
Conservation	Not globally threatened

One obtained at Old Head of Kinsale on 23 September 1876. One at CCBO on 20 April 1968. The 1876 bird was misidentified as a Common Nightingale, although its identity and subspecies (*galactotes*) were later established. The CCBO bird was not identified to subspecies.

European Robin

Erithacus rubecula | Spideog | A

(subspecies: *melophilus*; *rubecula*)

World	Europe, Asia, north Africa, Azores, Madeira, and Canary Islands
Ireland	Very common resident breeder
Cork	Very common resident breeder
Conservation	Green-listed

The Robin (*E.r. melophilus*) was common and widely distributed at high density in all habitats during all three breeding atlas surveys. Its distribution in winter was equally widespread.

It occurs commonly at the larger islands. About 160 pairs were nesting at CCBO in 1965, and there was an increase to 413 pairs in 1986. Numbers breeding at Sherkin Island varied from 90 to 130 pairs between 1977 and 1994. Robins often nest in unusual sites, particularly around (and in) houses and gardens, and are said to have a particular fondness for nesting on walls clad in ivy in County Cork.

The breeding territory density of Robins has been studied at several sites.[315] Densities varied between habitat types, with highest densities in semi-improved grassland and lowest at suburban gardens. Density on farmland ranged from 22 to 263 territories per km^2. A density of fifty-five territories per km^2, also on farmland, was recorded at Kilcolman.[316]

Autumn passage movements on the south Irish coast are difficult to detect due to the large resident population. Autumn counts of up to 100 are occasionally reported at coastal sites. Between fifty and one hundred arrived at Sherkin Island on 22 October 1990, and many hundreds arrived there in mid-October 1994. A large influx to the south coast was mentioned for October 1987 but was not reported in detail.[317] Robins have occurred at Fastnet Rock in March, and in September to November, seven birds on one occasion in November, and it has also occurred at Bull Rock in July. Some British-bred Robins are known to go south to the continent in winter, but it is unknown if Irish birds do the same. One juvenile ringed in Scotland in August was recovered at Clonakilty the following March.

European subspecies *E.r. rubecula*: There is one record of this subspecies. About fifty birds were at CCBO on 9 October 1959, but it is likely to be under-recorded.

Thrush Nightingale

Luscinia luscinia | Filiméala smólaigh | A

(monotypic)

World	Northern and eastern Europe, and western Asia
Ireland	4 individuals

Cork	4 individuals
Conservation	Not globally threatened

One at CCBO on 29 October–1 November 1989. First-year at CCBO on 26 October 1990. First-year at CCBO on 15 October 1999. One at Dursey Island on 22 October 2013.

Common Nightingale

Luscinia megarhynchos | Filiméala | A

(subspecies: *megarhynchos*)

World	Europe, western Asia, and north Africa
Ireland	36 individuals
Cork	13 individuals
Conservation	Not globally threatened

The Common Nightingale has occurred in spring (4) and autumn (9). Most records are of single birds, although two were at CCBO in October 1976. All but one has been recorded at the western peninsulas and islands: CCBO (8), Beara peninsula (3), Mizen Head (1) and Ballycotton (1). Most have stayed for only one day, but two remained for five days (mean stay = 1.9 days). Although it has been suggested that Nightingales might have formerly bred at the Gearagh, there is no convincing evidence that they have done so.[318]

1959–68	1969–78	1979–88	1989–98	1999–08	2009–18
1	3	1	2	2	4

Earliest and latest occurrence in spring and autumn (all records): 16 April–9 May and 20 August–11 October.

Unidentified nightingale species

One nightingale, either Common or Thrush (*Luscinia megarhynchos/luscinia*), was recorded at CCBO on 11 October 1959.

Bluethroat

Luscinia svecica | Gormphíb | A

(subspecies: *svecica*)

World	Europe, Asia, Alaska, and north-west Canada
Ireland	45 individuals
Cork	22 individuals
Conservation	Not globally threatened

The Bluethroat is an autumn passage migrant (20), but two have occurred in spring. It has only occurred singly, with most staying for only one day (mean stay = 2.5 days), but one stayed for sixteen days at Ballycotton in November 2009. Most occurred at CCBO (14), and others at Dursey Island (2), Mizen Head (2), Galley Head (1) and Ballycotton (2). The only inland record has been at Kilcolman (see Red-spotted Bluethroat below). Three males and two females have been recorded, and where ages have been published the breakdown is two adult and three first-year birds.

Red-spotted Bluethroat *L.s. svecica*: There is one record of this subspecies. Male at Kilcolman on 6–14 April 1995. This is the only individual where the subspecies has been identified.

1959–68	1969–78	1979–88	1989–98	1999–08	2009–18
5	7	1	3	3	3

Earliest and latest occurrence in spring and autumn (all records): 6 April–26 May and 28 August–8 November.

Red-flanked Bluetail

Tarsiger cyanurus | An t-earrghorm rua-chliathánach | A

(monotypic)

World	North-east Europe and Asia
Ireland	6 individuals
Cork	5 individuals
Conservation	Not globally threatened

First-year at Dursey Island on 10 November 2009. One at CCBO on 12 October 2010. One at Galley Head on 26 March 2012. First-year at Mizen Head on 20 October 2015. One at Lissagriffin on 22–3 October 2016.

Black Redstart

Phoenicurus ochruros | Earrdheargán dubh | A

(subspecies: *gibraltariensis*)

World	Europe, Asia, and north-west Africa
Ireland	Scarce winter visitor, uncommon passage migrant
Cork	Scarce spring and autumn migrant
Conservation	Green-listed

Robert Ball knew the Black Redstart around Youghal in 1818. Over the next few years several were obtained or seen, including an adult male at a cliff in January 1843. One was obtained at Castlefreke in November 1845. The Bull Rock, Fastnet Rock, Old Head of Kinsale, Power Head and Ballycotton produced many specimens obtained between 1887 and 1912, mostly in October, but also in November to January, and in March and April, with one on 15 June 1903. A flock was reported at Fastnet Rock in October 1924. Three immature birds were seen near Cork city in December 1943, two adult males were found dead at Ballycotton lighthouse on 5 March 1948, and an adult male was at Castlemary in November and December 1957.

There have been records in all years since 1959, most frequently in autumn, but also in winter and spring. Early autumn records occur from mid-October. Most arrive on south-east winds, and they may appear anywhere on the coast. Most move on in a week or two, but some remain through the winter. One ringed at Ballymacoda on 17 October 1977 was recovered in Portugal on 13 January 1978, indicating the onward passage south of some. Highest numbers are recorded at this time, e.g. sixty at CCBO in October 1990, fifty at Mizen Head in October 2004 and fifty at Knockadoon Head in November 1994.

Spring numbers are smaller, and rarely reach double figures, fourteen at Roche's Point in late March or early April 1969 being the highest. There has been no real temporal trend over the sixty years of this study; numbers vary greatly between years.

Most records are from coastal areas. Winter atlas surveys in 1981–4 and 2007–11 showed a scatter of records along or near the coast from Dursey Island to Youghal, with concentrations in and around Cork city and harbour, and at sites east to Youghal. The total number in the winter population in any year probably does not exceed thirty birds.

Inland records came from the Kildorrery and Mitchelstown area. However, it is likely that small numbers filter inland to towns, villages, quarries and industrial sites; two were reported at Doneraile on 25 November 2014. Beaches, cliffs and man-made structures, such as car parks and buildings, are favoured wintering sites at or near the coast. Most winter records refer to one to three birds, and it is exceptional to find more than five at any site.

There are no confirmed breeding records, although a first-summer male was singing at Oliver Plunkett Street, Cork city, on 1 June and 17 July 2010. Some wintering birds stay late into spring, and these or late spring migrants may account for some records in the 2008–11 breeding atlas. There are a few summer records, e.g. at Bull Rock on 15 June 1903, a female at Cobh from late July to 4 August 1987, a female at Dursey Island on 15 June 1996, and a male there on 17 June 1996.

Common Redstart

Phoenicurus phoenicurus | Earrdheargán | A

(subspecies: *phoenicurus*)

World	Europe, Asia, and north-west Africa
Ireland	Rare summer visiting breeder, spring and autumn migrant
Cork	Scarce spring and autumn migrant
Conservation	Red-listed

The first records of Common Redstart were single birds obtained at Fastnet Rock in October 1887, September 1896 and September 1909. Old Head of Kinsale lighthouse produced three birds in September 1909 and singles in October 1910 and September 1913. The spring of 1949 was remarkable for the numbers seen and the early date at Fastnet Rock (up to 5 on on 24–5 March). One was seen near Rosscarbery on 24 April 1957.

Spring occurrences are rare, and the only June records were in 1971 with singles at Goleen and CCBO, each present for one day. Autumn passage occurs from August to October. CCBO accounts for most migration at both seasons, but away from this site Dursey Island and Beara peninsula, Old Head of Kinsale and Ballycotton have had repeated occurrences, but records extend along the coast from Dursey Island to Ballymacoda Bay. There have been inland records at Mount Gabriel in August 1977, Kilcolman in October 1993 and near Blarney in May 2013. Numbers rarely exceed three or four in a day at CCBO, peak counts being ten in October 1959 and October 1989. Six at Old Head of Kinsale in September 1969 is the mainland peak.

There is one winter occurrence, a male at Oysterhaven on 19 December 1965. This record has been questioned, but there are credible winter reports in Europe as far north as Germany.[319]

1959–68	1969–78	1979–88	1989–98	1999–08	2009–18
105	99	108	101	56	129

Earliest and latest occurrence in spring and autumn (1959–2018): 26 March–22 June and 9 August–17 November.

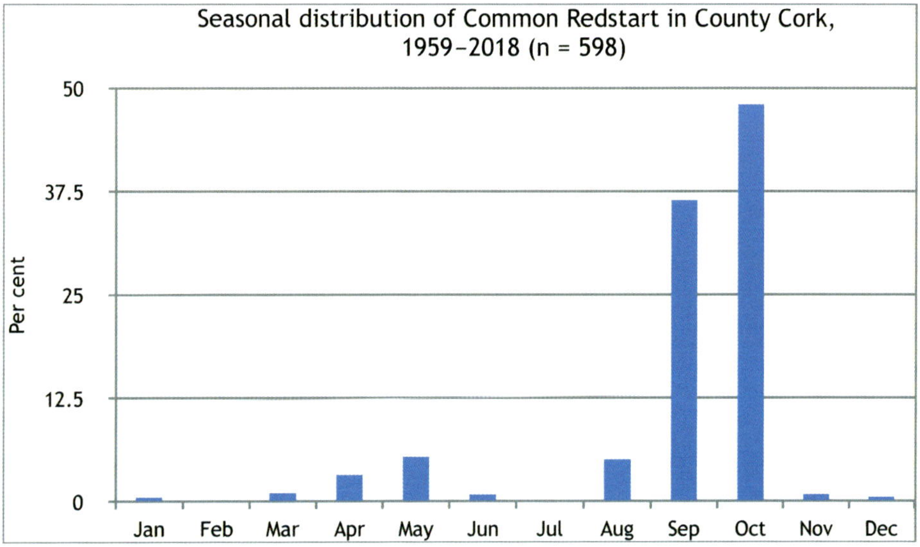

Fig. 7.41 Common Redstart.

Whinchat

Saxicola rubetra | Caislín aitinn | A

(monotypic)

World	Europe and western Asia
Ireland	Scarce summer visiting breeder
Cork	Has bred, spring and autumn passage migrant
Conservation	SPEC 2; Red-listed

The Whinchat was a rare summer visitor in the 1800s, but breeding was not mentioned. One pair nested at the south side of Bantry Bay in 1911, and a pair near Macroom in July 1954 was believed to be breeding, but proof was lacking. None was recorded during the breeding atlas of 1968–72. In late June 1970 single birds were seen in potential breeding habitat near Dursey Sound and near Eyeries. Records at two coastal 10 km squares in west Cork during 2008–11 referred to migrants.

There is a light spring passage and a stronger autumn one, with two obtained at Fastnet Rock on 20 September 1912. The greatest proportion of passage at both seasons takes place at CCBO, and it varies from year to year from a few birds to double figures, e.g. peak of twenty-nine on 26 September 1976. Away from CCBO, counts have seldom exceeded ten birds, e.g. fifteen at Ballycotton on 17 September 1969. There is no evidence for a change in status across the sixty years of this study. Whinchats have occurred at almost all coastal (and near coastal) sites from Dursey Island to Ballymacoda Bay, the only inland occurrences during 1959–2018 being two at Kilcolman in October 1992 and one at the same place in August and September 1997.

There are winter records of singles at Slatty Bridge on 20 February 1980 and at Cork city in January and February 2001.

1959–68	1969–78	1979–88	1989–98	1999–08	2009–18
111	258	204	211	155	227

Earliest and latest occurrence in spring and autumn (1959–2018): 28 March–27 November.

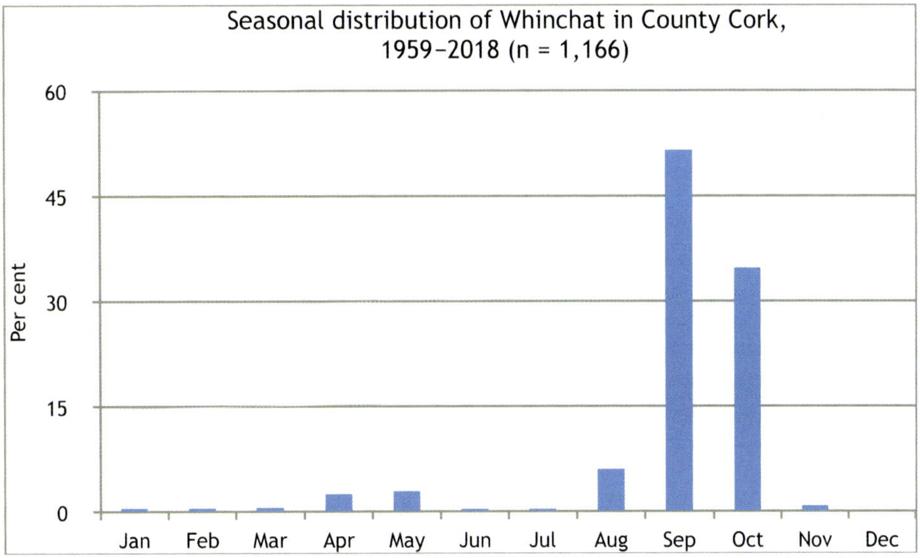

Fig. 7.42 Whinchat.

European Stonechat

Saxicola rubicola | Caislín cloch | A

(subspecies: *hibernans*)

World	Europe and north-west Africa
Ireland	Common resident breeder
Cork	Common resident breeder
Conservation	Green-listed

The Stonechat was a common resident in coastal districts of Munster in the late 1800s, the province where it was most abundant. On the coast, it nests in heathland above cliffs, in dunes and in patches of rough ground at the margins of fields. Young forest plantations with gorse, ling heather, bell heather, bramble and rough grass are inland habitats.

During 1968–72, the Stonechat was breeding throughout the county, the only gaps being inland. There were considerable losses at inland areas during 1988–91, but these were made good by 2008–11. Winter distribution in 1981–4 showed a strong coastal bias, but the range expanded at inland areas and only a few 10 km squares in the west and north were devoid of birds in 2007–11.

The breeding population at CCBO has been surveyed at intervals during 1959–98. Estimates in the early years put the population at 50–150 breeding pairs. Following the severe winters of 1961/2 and 1962/3 breeding numbers declined to three pairs. The population then increased until 1986, when it declined again after the severe winter of 1985/6. During the period 1965–98, a mean of twenty-three pairs per annum bred, with the peak being thirty-three pairs. Given the stability of the population since 1965, it was suggested that previous estimates of over 100 pairs were exaggerated.[320] Breeding population counts at Sherkin Island gave a mean of twenty-seven pairs per annum during 1977–94.

Breeding biology has been studied based on data from about 100 nests.[321] It is multi-brooded, frequently having three nesting attempts in a season, and a mean overall clutch size of 5.17 eggs which fledge a mean of 4.6 young.

The diet of nestling Stonechats has been studied at east Cork sites.[322] Young birds were fed an exclusively invertebrate diet, with Coleoptera, Hymenoptera, Scarabeidae, Arachnida and terrestrial larvae of other species accounting for 94 per cent of the diet.

A movement of Stonechats takes place at CCBO in autumn, and an estimated 1,000 birds were there in August 1959. Many may have been dispersing birds from the mainland during times of high population levels, but there is evidence that some Irish birds leave the country in winter. A first-year bird ringed at CCBO on 19 September 1961 was recovered near Cadiz (Spain) on 1 November 1961, and one was obtained at Fastnet Rock on 20 October 1887.

Siberian Stonechat

Saxicola maurus | Caislín cloch Sibéarach | A

(subspecies: *maurus* or *stejnegeri*)

World	Northern Europe and Asia
Ireland	10 individuals
Cork	8 individuals
Conservation	Not globally threatened

Female or first-year at Old Head of Kinsale on 27 October–1 November 1988. One at Galley Head on 20 October 1990. Female or first-year at Crookhaven on 5 October 1992. First-year at Galley Head on 23 October 1993. Male at CCBO on 24 September 1998. First-year male at Galley Head on 8 October 2010. First-year at Firkeel on 1 October 2012. One at Barry's Head on 11–17 October 2016. These birds have not been identified to subspecies, but five have been published as either *maurus* or *stejnegeri*, while a sixth has been published as belonging to one of the eastern races.

Isabelline Wheatear

Oenanthe isabellina | Clochrán gainimh | A

(monotypic)

World	South-east Europe and Asia
Ireland	1 individual
Cork	1 individual
Conservation	Not globally threatened

One at Mizen Head on 10–17 October 1992.

Northern Wheatear

Oenanthe oenanthe | Clochrán | A

(subspecies: *oenanthe*; *leucorhoa*)

World	Canada, Greenland, Iceland, Europe, Asia, Alaska, and north Africa
Ireland	Summer visiting breeder, scarce passage migrant in spring and autumn
Cork	Summer visitor, breeding mainly in west, passage migrant
Conservation	SPEC 3; Amber-listed

During 1968–72 the Wheatear (*O.o. oenanthe*) was breeding at the western peninsulas and islands, and intermittently on the coast eastwards to Old Head of Kinsale. It also bred in the uplands on the border with Kerry, with

Northern Wheatear

an outlier breeding station at Ballyhoura Mountains. The range has since contracted, and none now breed east of Galley Head, or inland east of Macroom.

At least thirty-four pairs were breeding at CCBO in 1963, fifty pairs in 1965, but only nineteen pairs in 1986. The population has continued at a low level, ranging from thirteen to twenty-three pairs up to 1998. The decline has been linked to an increase in shrubby vegetation in previously open areas. The breeding population at Sherkin Island remained stable at about ten to fifteen pairs during 1977–94.

Early returning Wheatears appear in mid-March (earliest were at Coomhola Mountain on 25 February 2016, at Knockadoon Head on 5 March 2010 and at Garryvoe on 6 March 1977), with spring passage peaking in mid-April. Highest daily totals at CCBO approach 100 birds, and patterns and numbers are similar at Dursey Island. Spring migration occurs along the entire mainland coast, but numbers are lower.

Autumn passage extends from August to October, peaking in September. Up to thirty are commonly recorded at this time, but sometimes more, e.g. 100 at Pilmore Strand on 20 September 1980 and 100 at Ballycotton on 18

September 2008. Highest autumn counts at CCBO and Dursey Island usually fall short of 100 birds. Small numbers are observed at inland locations such as Charleville lagoons, Kilcolman and Moorepark. Individuals occur at coastal sites to mid-November, and the latest record is one at Youghal on 21 December 1989. The Wheatear occurs on spring and autumn migration at Fastnet Rock, where sixty-two were killed during a rush of birds on 28–30 September 1913. A series of birds have been obtained at Old Head of Kinsale lighthouse at both seasons.

Greenland Wheatear *O.o. leucorhoa*: The Greenland subspecies is under-recorded, and it is probably a regular passage migrant in spring and autumn. In the past, it has been obtained at Fastnet Rock and Old Head of Kinsale in April, May, September and October, and once at Bull Rock on 7 December. Recent examples have occurred at CCBO, most often in September, at Sherkin Island in May, and at Ballycotton in September. It has occurred in spring at Dursey Island, where most are of this subspecies from late September.

Pied Wheatear

Oenanthe pleschanka | Clochrán alabhreac | A

(monotypic)

World	South-east Europe and Asia
Ireland	4 individuals
Cork	2 individuals
Conservation	Not globally threatened

First-year male at Knockadoon Head on 8–16 November 1980. First-year female at Ballymacoda on 11–15 December 2017.

Black-eared Wheatear

Oenanthe hispanica | Clochrán cluasdubh | A

(subspecies: *hispanica*)

World	Southern Europe, south-west Asia, and north Africa
Ireland	5 individuals
Cork	1 individual
Conservation	Not globally threatened

First-year male at CCBO on 26–7 May 1992.

Desert Wheatear

Oenanthe deserti | Clochrán fásaigh | A

(subspecies: undetermined)

World	South-west and central Asia, and north Africa
Ireland	8 individuals
Cork	2 individuals
Conservation	Not globally threatened

First-year male at Red Strand on 27 October–2 November 1990. First-year male at Garryvoe on 26–7 November 1995.

Red-breasted Flycatcher

Ficedula parva | Cuilire broinnrua | A

(monotypic)

World	Europe and Asia
Ireland	Rare passage migrant, mainly in autumn
Cork	213 individuals
Conservation	Not globally threatened

One Red-breasted Flycatcher was obtained at Bull Rock on 18 November 1903. Records since 1959 show that it is an autumn migrant to coastal headlands and islands between Dursey Island and Knockadoon Head. Four birds have occurred in spring, two in August, one in December, but the majority in October. Most have been at CCBO (131), with others at Dursey Island and Beara peninsula combined (24), and at Mizen peninsula (24). Smaller numbers have occurred at Galley Head (12) and Old Head of Kinsale (8), and seven birds have been shared by four other west Cork sites. Six birds occurred at two east Cork sites. It has been absent only in 1982, 1992, 1996, 1998 and 2004 during 1959–2018. It has increased in frequency across the sixty years of this study. Most records refer to single individuals, but two together have been recorded on twelve occasions. Three have been recorded at CCBO on three occasions, and four together once at Mizen Head. The majority are present for only one or two days, but some remain for up to seven days (mean stay = 1.7 days). Of those individuals where their age was published, first-years predominated (70) over adults (10).

1959–68	1969–78	1979–88	1989–98	1999–08	2009–18
27	29	37	18	48	53

Earliest and latest occurrence in spring and autumn (1959–2018): 19 April–26 May and 29 August–22 November (2 December).

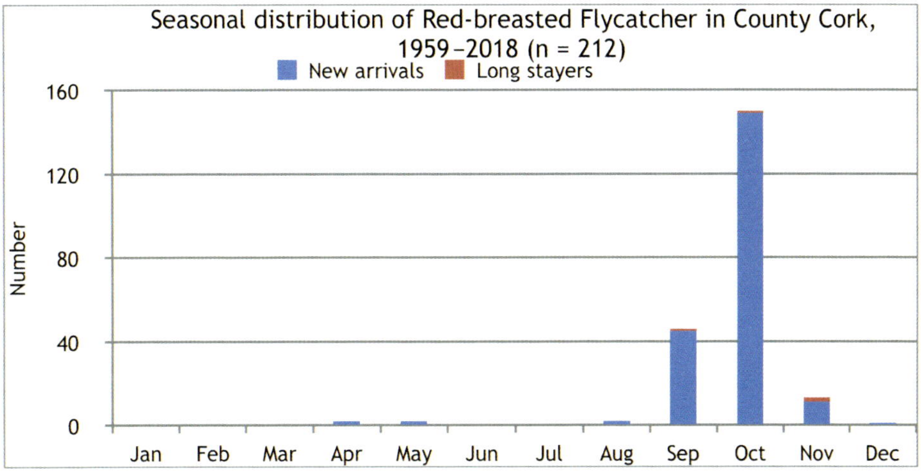

Fig. 7.43 Red-breasted Flycatcher.

Taiga Flycatcher

Ficedula albicilla | Cuilire Taiga | A

(monotypic)

World	Ural Mountains to Kamchatka Peninsula
Ireland	1 individual
Cork	1 individual
Conservation	Not globally threatened

First-year at Galley Head on 21–5 October 2018.

Pied Flycatcher

Ficedula hypoleuca | Cuilire alabhreac | A

(subspecies: *hypoleuca*)

World	Europe, Asia, and north Africa
Ireland	Has bred, scarce spring and autumn passage migrant
Cork	Scarce passage migrant in spring and autumn
Conservation	Amber-listed

Four Pied Flycatchers were obtained at Fastnet Rock in September and October between 1886 and 1899. Fourteen occurrences at Cork lighthouses were later reported, one being obtained at Old Head of Kinsale (September 1909), and one at Ballycotton on 19 April 1914. Unprecedented numbers occurred on the Irish coast in September 1949 during south-east winds, including three killed and others seen at Fastnet Rock.

It has occurred annually since 1959 at almost every coastal (or near coastal) site from Dursey Island to Ballymacoda. A passage takes place at CCBO, and at other well-watched sites, most migrants moving on very quickly and rarely staying for more than one day. It occurs predominantly in autumn and is rare in spring (3.5 per cent of individuals; n = 1,581). Spring records usually involve one or two birds in April or May. Daily counts at CCBO in autumn often exceed twenty birds, with a peak of thirty-six on 3 September 1966. Elsewhere, highest counts have been of seventeen at Crookhaven in September 1990 and ten each at Old Head of Kinsale and Galley Head in

September 1976. One ringed as a nestling in the Netherlands in May 2000 was recovered at CCBO in May 2001.

Numbers occurring in each decade in this analysis appear to have remained relatively stable, although an examination of data reveals large differences between years, some having much higher totals than others.

1959–68	1969–78	1979–88	1989–98	1999–08	2009–18
200	188	277	318	337	261

Earliest and latest occurrence in spring and autumn (1959–2018): 18 March–21 June and 11 July–10 November.

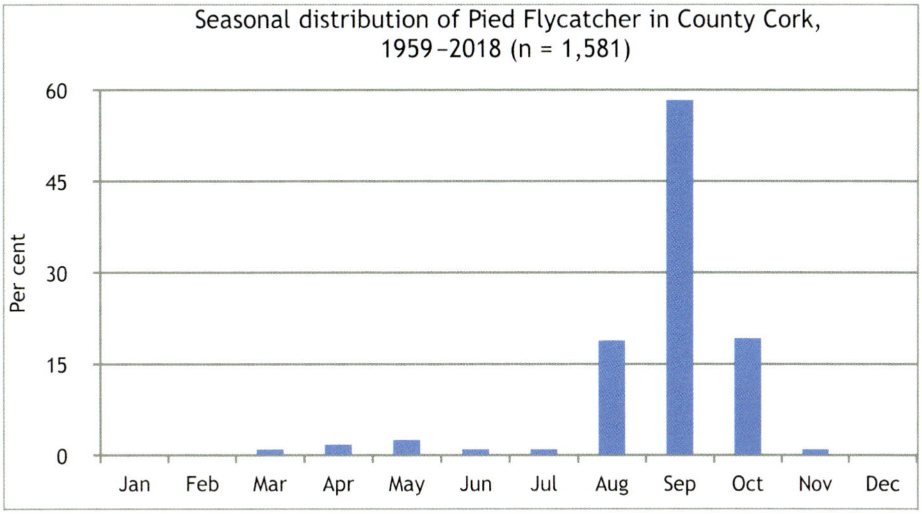

Fig. 7.44 Pied Flycatcher.

Dunnock

Prunella modularis | Donnóg | A

(subspecies: *hebridium*)

World	Europe and Asia, introduced elsewhere
Ireland	Very common resident breeder
Cork	Common resident and breeder
Conservation	Green-listed

The Dunnock was distributed widely during all three breeding atlas surveys, and there was an area of high density to the west, north and east of Cork Harbour. Winter surveys mirrored that of the breeding season. Dunnocks occur in a wide range of habitats, from hedgerows to woodland, and gardens of built-up areas.

The Dunnock breeds on some islands. At CCBO, where it is one of the commonest breeding species, it was scarcely affected by the severe winters of 1961/2 and 1962/3, and the number of breeding pairs increased from 243 in 1965 to 310 in 1986. Between 100 and 200 pairs have been recorded breeding at Sherkin Island.

Although Dunnocks have been obtained from offshore Irish light-stations, no records were obtained from stations off the Cork coast. Evidence of passage migration or immigration is almost non-existent at CCBO. Neither is there evidence from ring recoveries of immigration into Ireland, and post-breeding dispersal of adults and post-natal dispersal of juveniles can be measured in hundreds of metres rather than kilometres.

House Sparrow

Passer domesticus | Gealbhan binne | A

(subspecies: *domesticus*)

World	Iceland, Europe, Asia, and north Africa, introduced elsewhere
Ireland	Very common resident breeder
Cork	Common resident breeder
Conservation	SPEC 3; Amber-listed

House Sparrows were present throughout the county during all three breeding atlas surveys. Winter distribution did not differ from the breeding season. Density was highest in the agricultural areas of the south and east.

House Sparrows nest in loose colonies ranging from a few to several dozens of pairs. With the decline of thatch as a roofing material, House Sparrows switched to nesting beneath tiles and slates of dwellings as well as in holes and crevices in farm buildings. Some breed in nest boxes set up for tits, and some have usurped tits and House Martins from their nests. A nest was built in the base of an occupied Magpie nest at Ballymacoda and in the

House Sparrow

base of an occupied Rook nest at Ballintubbrid, and a nest was also built in an old Barn Swallow nest inside a building. Sparrow flocks consisting of adults and their offspring are largest in autumn when they move to fields when cereal grain is available, but they return to the vicinity of their nesting sites to roost at night.

A colony at Ballymacoda has increased to about 400 birds, probably due to food provisioning. About 100–150 pairs bred at CCBO during the 1960s, and up to 200 frequented stubble fields after the breeding season. About 150 pairs were continuing to breed there in 1986. At Sherkin Island, the breeding population declined from about forty to sixty pairs in 1977 to about ten pairs in 1994 following a decline in arable farming. House Sparrows have not bred at Dursey Island since 1994, but they are common near the tip of the Beara peninsula.

House Sparrows also occur in gardens in built-up areas, where they benefit from food provisioning. Grain stores are also occupied as are locations within towns and cities. Up to 150 occur in autumn near Kent railway station in Cork city, and about 200 and 250 at two locations near

Cobh. There are many small colonies in Cork city and its suburbs, but they are usually absent from the city centre.

There is no evidence of migration, and the few observations reported from Fastnet Rock may have been misidentified Tree Sparrows. Occasional records at CCBO of birds flying out to sea are likely to be local rather than migratory movements. Ringing data indicates that House Sparrows make short movements, with birds at Cobh staying within 0.5 km of the ringing site over four consecutive winters.

Tree Sparrow

Passer montanus | Gealbhan crainn | A

(subspecies: *montanus*)

World	Europe, and Asia south to Lesser Sunda Islands, introduced elsewhere
Ireland	Resident breeder
Cork	Former breeder, scarce visitor
Conservation	SPEC 3; Amber-listed

The first record of a Tree Sparrow was one found dead at Fastnet Rock on 2 October 1911. Two were present near Skibbereen from December 1945 to February 1946, and two were at the same place for some days in July 1954. There were no further records until 1960 when it was recorded at CCBO.

A search for breeding sites was conducted in 1964 following a sighting of two birds at ruins near Midleton in April that year. These did not stay, and no breeding site was located. A colony at Roche's Point was discovered in 1965 and breeding continued to 1977. Birds bred in ruins (1 pair in a tree) at two separate sub-colonies. In 1965 there were at least seven occupied nests. The number breeding increased in 1966, and a third sub-colony was found. Numbers decreased in 1967 when four or five pairs bred. In June 1968 there were more Tree Sparrows in the Whitegate and Roche's Point area than could be accounted for by the number of breeding pairs. Three, five and three pairs bred in 1968, 1969 and 1970, respectively. Two pairs bred in 1977 but breeding then ceased. A few birds continued in the area and the last record was two birds on 12 March 1980.

A colony was discovered at a farmyard at Old Head of Kinsale in 1982 and birds bred there until 1992, but not thereafter, although one bird was present in March 1993. Up to ten pairs bred, and the number of individuals showed annual changes typical of this species.[323] The peak was twenty-four in August 1983, then declining before another peak at thirty in December 1989, finally going into terminal decline. Nest boxes were installed in 1994, but the colony was not reoccupied.

The history of the Tree Sparrow in breeding and winter atlases is worth exploring. It is shown as breeding during 1968–72 at two 10 km squares in east Cork, one of which is the described Roche's Point colony. The other square (W96), which extends eastwards from Power Head to Ballycotton, also showed proved breeding. This record is of interest as there is no documentary evidence of a colony in the literature. The winter atlas of 1981–4 showed presence in two 10 km squares in north Cork where there was no previous history of presence (R40 and W58). Whether presence in these squares represented new populations is unknown, but none was recorded during the breeding or winter seasons of 2007–11. The only winter season records for Cork during 2007–11 came from three contiguous squares in east Cork, no doubt stray birds from a breeding population in west Waterford.

Apart from the two breeding colonies, which are excluded from the decadal table, Tree Sparrows have occurred in spring, with smaller numbers in autumn and a few in winter. Most have occurred at CCBO, followed by Dursey Island. Records have been evenly spread along the coast, sometimes several kilometres inland, eastwards to Ballymacoda with only two proper inland occurrences, at Dripsey and Newmarket. Most records are of birds seen on only one day, but occasionally some linger for several days or even weeks. Most occurrences involve fewer than ten birds, but eleven and thirteen have occurred at CCBO in June. Apart from the demise of the breeding colonies, numbers have been stable for the last thirty years.

The Tree Sparrow is largely sedentary. Two birds were seen flying in from the sea at Old Head of Kinsale in May 1994, and movement through Dursey Island in spring is rapid, involving birds leaving within hours of arrival. It was suggested by Clive Hutchinson that occurrences in spring at islands and headlands are of adults seeking suitable breeding sites. If this is the case, then they have been relatively unsuccessful at finding such sites in Cork.

1959–68	1969–78	1979–88	1989–98	1999–08	2009–18
9	38	11	35	47	45

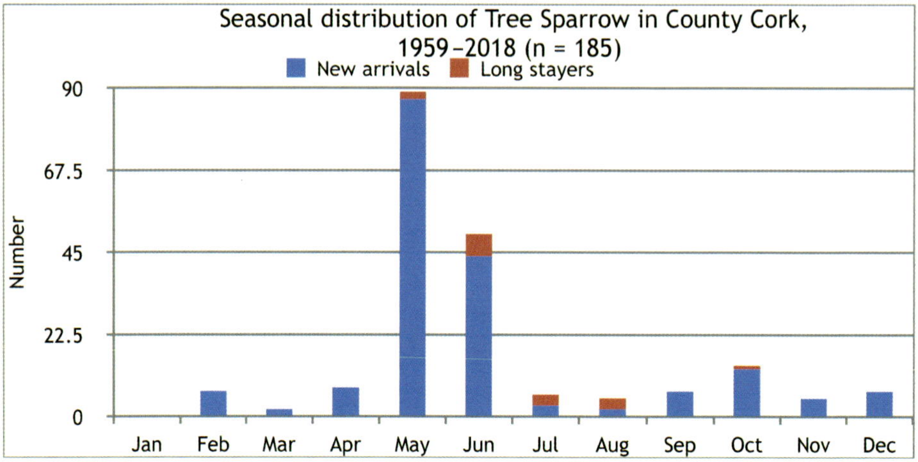

Fig. 7.45 Tree Sparrow.

Western Yellow Wagtail

Motacilla flava | Glasóg bhuí | A

(subspecies: *flavissima*; *flava*; *thunbergi*)

World	Europe, Asia, north Africa, Alaska, and north-west Canada
Ireland	Rare summer visiting breeder, scarce passage migrant
Cork	Has bred, scarce passage migrant
Conservation	SPEC 3; Amber-listed

The first records of Yellow Wagtail (*M.f. flavissima*) were at Old Head of Kinsale in August 1909 and September 1913. The next were three seen near Skibbereen in September 1956. Another (possibly 2) was at the Marina (Cork city) in April 1957. After the establishment of CCBO in 1959, the status of the Yellow Wagtail in Cork was revealed.

One pair bred at Ballinwilling, on the north shore of Ballycotton Bay, in 1965 and 1966. A male was seen in June 1968, but no further breeding attempt took place there, or at any other Cork site.

The Yellow Wagtail is a spring passage migrant to the coast from Dursey Island to Youghal Bay and occurs annually in varying numbers. It is very rare inland, with records at the Gearagh in October 1964 and Charleville lagoons in May 1994. It was rare in spring but less so in autumn at CCBO in the 1960s, with up to eight birds occurring in a day. Yellow Wagtails were more numerous in the 1960s and 1970s than in the last four decades. They were regularly recorded during that period at Ballycotton, and occasionally at sites in west Cork such as Crookhaven, Galley Head and Clonakilty. It was said of autumn 1968 that birds were present at many scattered localities in small numbers, but no details were given.

There was an exceptional influx in autumn 1976 when at least fifty birds passed through CCBO in late September. There were peak counts of seven at Galley Head, six at Old Head of Kinsale and ten at Ballycotton, with reports of one or two at many other sites. Numbers peaked at eight birds at Ballycotton on 29 September 1977. There are two winter records, both in 1968, at Ballycotton on 8 December and Crookhaven on 29 December.

Because of difficulties establishing reasonably accurate numbers for the 1960s and 1970s, the period analysed here is the forty years from 1979 to 2018. Although numbers across the four decades have not changed (mean = 105), numbers within individual years have varied from two to four up to at least thirty birds. They occur annually, but not at any one site, and during the forty-year period the peak count was fifteen birds at CCBO in September 1996. No other record has exceeded ten birds. There has been a tendency towards a steady increase in spring occurrences across all four decades from 22 per cent in 1979–88 to 33 per cent in 2009–18. The geographical spread has included the entire coastline, with concentrations at well-watched sites such as Dursey Island, CCBO and Ballycotton.

There have been suspected cases of other subspecies of the Yellow Wagtail, but no subspecies other than *flavissima*, *flava* and *thunbergi* have been proved to occur. Suspected cases, and birds not assigned to subspecies, have been included in the analysis of *flavissima*.

1979–88	1989–98	1999–08	2009–18
109	114	96	100

Earliest and latest occurrence in spring and autumn (1979–2018): 6 April–10 November.

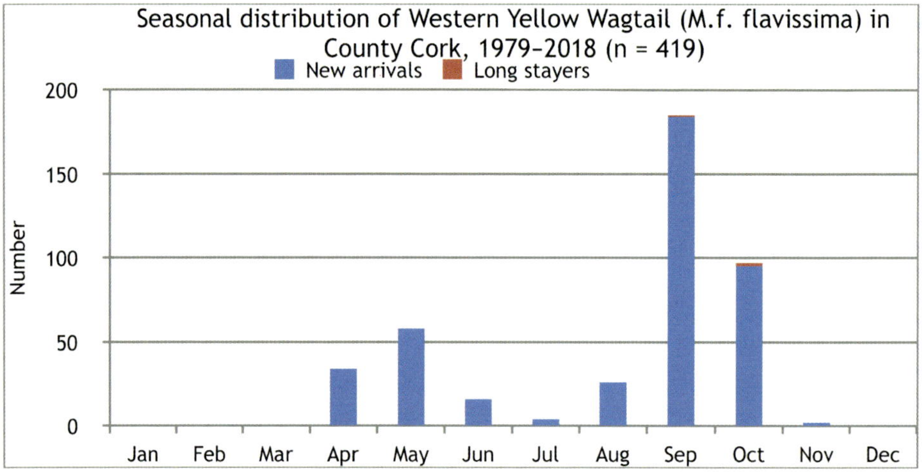

Fig. 7.46 Western Yellow Wagtail.

Blue-headed Wagtail *M.f. flava*: There are thirty-five records (43 individuals) of this subspecies. It is a spring (49 per cent) and autumn (51 per cent) passage migrant to coastal sites between Dursey Island and Ballycotton. Most occurred at CCBO (23) and Ballycotton (9). Sixteen were identified as male and two as female. Most were adults, and one a first-year bird. Most records are of singles, but five on one occasion, and two on four occasions have occurred at CCBO. Birds typically remain for one day (31), but some remained from five to seven days (11), with one for thirty-seven days at CCBO (mean stay = 2.9 days).

Grey-headed Wagtail *M.f. thunbergi*: There is one record of this subspecies. Male at Cloyne on 9 May 1984.

Eastern Yellow Wagtail

Motacilla tschutschensis | Glasóg oírthearach | A

(subspecies: undetermined)

World	North-east Siberia and north-west North America
Ireland	3 individuals
Cork	1 individual
Conservation	Not assessed

First-year at Dursey Island on 3 November 2017.

Citrine Wagtail

Motacilla citreola | Glasóg chiotrónach | A

(subspecies: undetermined)

World	Eastern Europe and Asia
Ireland	34 individuals
Cork	9 individuals
Conservation	Not globally threatened

First-year at Ballycotton on 15–16 October 1968. First-year at Ballycotton on 7–24 September 1980. First-year at Ballycotton on 21 September 1993. First-year at Ballycotton on 14 September 2003. First-year at Ballycotton on 2 September 2004. First-year at Lissagriffin on 3 September 2005. First-year at Lissagriffin on 22 September 2007. First-year at Ballycotton on 3 October 2015. One at Lissagriffin on 23 August 2016.

Grey Wagtail

Motacilla cinerea | Glasóg liath | A

(subspecies: *cinerea*)

World	Europe, Asia, north-west Africa, and Azores, Madeira, and Canary Islands
Ireland	Resident breeder
Cork	Common resident breeder
Conservation	Red-listed

The Grey Wagtail was present throughout the county during all three breeding atlas surveys, although it was often scarce or absent in coastal 10 km squares. Breeding density was high at upland and lowland areas.

Surveys of winter distribution showed presence throughout the county, including on the coast and at the western peninsulas. Ones and twos occur at all types of wetland habitats, including puddles, around farmyards where cattle are fed or where dung heaps are present. A winter increase was noted in west Cork in the late 1800s. This species suffers declines in severe winter weather, and this can be seen in lower numbers at breeding streams in the following summer.[324]

Grey Wagtail

Breeding biology of the Grey Wagtail has been studied in east and north-east Cork.[325] Egg-laying peaks in the second week of April, with some evidence of a later second peak. Most clutches contained five eggs (mean of 4.79), and the mean brood size at fledging was four young per successful brood.

An autumn passage takes place at CCBO and peaks in September at usually fewer than thirty birds. Counts of seventy on 31 August 1976 and fifty on 11 September 1983 were exceptional. The mean numbers migrating through the island in autumn 1963 declined to 84 per cent of that during 1959–67, attributable to the effects of the severe winters of 1961/2 and 1962/3, although Irish birds were considered to have fared better than populations elsewhere.[326]

Autumn passage has been noted at other coastal locations, but generally numbers of ten or fewer are involved. A light passage takes place at Ballyvergan Marsh and Ballymacoda, where low-flying birds pass by in ones and twos, mostly in an easterly direction. A single bird was obtained at Fastnet Rock during a rush of birds in September 1913, and four were obtained at Old Head of Kinsale in September and October. The origin of birds seen on these movements is unknown, but they are probably predominantly Irish, and may involve some from Britain.

Pied Wagtail

Motacilla alba | Glasóg shráide | A

(subspecies: *yarrellii*; *alba*)

World	Europe, Asia, north-west Africa, east Greenland, Iceland, and Alaska
Ireland	Common resident breeder
Cork	Common resident breeder
Conservation	Green-listed

The Pied Wagtail (*M.a. yarrellii*) was a common resident in the 1800s. The range extended throughout the county during all three breeding atlases, with highest densities in east Cork. At CCBO, seventeen pairs were breeding in 1965, and fourteen pairs in 1986, and at Sherkin Island numbers declined from nineteen pairs in the 1980s to four or five pairs in 1994. Several pairs also breed at Dursey Island.

Winter range does not differ from the breeding season, with high density throughout. Flocks commonly occur in autumn, often at beaches, and communal roosting takes place then, and in winter. Robert Ball observed flocks about Youghal during October, and an influx was observed into west Cork in winter in the 1800s.

The best-known communal roost is in Cork city centre, with separate roosts on trees at South Mall, Grand Parade and City Hall, and estimates of up to 550 birds have been made. Pied Wagtails forage by day in open spaces in the suburbs and in fields outside the city and fly into the city centre before dark to avail of the higher ambient temperature. Up to 350 birds gather on flat roofs at Cork University Hospital in late autumn and winter before heading into the city centre before dark.

Roosts are also present at Cobh (100–200 at the Promenade), Midleton (about 350) and Youghal (about 50), and probably at many other towns throughout the county. A roost at an industrial building at Little Island contained 350 birds in January 1990. A roost of up to sixty-five occurs in bushes at CCBO during autumn. Roosts also occur at reedbeds, especially in autumn, such as Shanagarry and Ballyvergan. Roosts in reedbeds at Ballinamona Lake in the 1980s held about eighty birds, another at Inch Strand in October 1994 held forty-two birds, both containing some White Wagtails (*M.a. alba*).

Small numbers of passage migrants occur in spring from mid-March to April. Autumn migration is more substantial, and loose flocks of up to 250 are a feature of coastal districts. Autumn migration is noticeable from Dursey Island to Youghal.

Pied Wagtails visit Fastnet Rock regularly and have been obtained in February, March, September and October, but five killed during a rush of birds on 28–30 September 1913 were not identified to subspecies. Too few Irish birds have been ringed to know if any move to the continent in winter. However, birds from northern Britain do move south, some to the continent, and some probably visit Ireland.

Pied Wagtails often take small flying insects, but one was seen to chase, catch and eat a spring hawker dragonfly at Lough Aderry.

White Wagtail *M.a. alba*: White Wagtails, which breed in Iceland and on the continent, are scarce spring migrants but are more common in autumn. One seen at the River Lee on 20 April 1899 was the first mention of this subspecies in Cork. It is rare in spring at CCBO, but more regular in autumn, although more than ten in a day seldom occur. An influx of 100 took place on 1 September 1963. This subspecies occurs on the coast from Dursey Island to Youghal, most commonly at beaches in the east of the county. Numbers usually range from single figures to about fifty. There is a long series of records for Ballycotton dating from the 1960s through to the present of twenty to fifty birds in September.

Richard's Pipit

Anthus richardi | Riabhóg Richard | A

(monotypic)

World	Central and eastern Asia
Ireland	Rare passage migrant, mainly in autumn
Cork	80 individuals
Conservation	Not globally threatened

Richard's Pipits occur in coastal districts in autumn, mostly in October (62). There is one spring record (Old Head of Kinsale, 21 April 1992), and one inland record (Kilcolman, 1–3 December 1994). There is one other December record, at Galley Head in 2014. They have occurred at several sites from Dursey Island to Pilmore Strand. Most records are from CCBO (33), with nineteen at Dursey Island and Beara peninsula and seven at Galley Head. Several other west Cork sites have had smaller numbers, with twelve at five east Cork sites, Ballycotton having the most at six birds. All except three records are of single birds, three and two at CCBO and two at Ballycotton. Most (64) have been recorded on only one day, with fifteen for two to six days and one for twelve days (mean stay = 1.5 days). Richard's Pipits occurred in thirty-eight of the sixty years since 1959.

1959–68	1969–78	1979–88	1989–98	1999–08	2009–18
7	4	13	17	19	20

Earliest and latest occurrence in spring and autumn (all records): 21 April and 13 September–27 December.

Tawny Pipit

Anthus campestris | Riabhóg dhonn | A
(subspecies: undetermined)

World	Europe, Asia, and north Africa
Ireland	42 individuals
Cork	18 individuals
Conservation	Not globally threatened

Tawny Pipits are spring (4) and autumn (14) migrants to coastal sites. They occurred singly except on one occasion when two were present at CCBO in September 1976. Most have been at CCBO (10), and others at Dursey Island (3) and Ballycotton (3), with singles at Pilmore Strand and Old Head of Kinsale. Most birds have stayed for only one day, but individuals have stayed for four, five and six days (mean stay = 1.7 days). Nine occurred at CCBO between 1959 and 1976, but only one has occurred there since, in 2004.

1959–68	1969–78	1979–88	1989–98	1999–08	2009–18
6	6	0	1	3	2

Earliest and latest occurrence in spring and autumn (all records): 23 April–18 May and 3 September–17 October.

Olive-backed Pipit

Anthus hodgsoni | Riabhóg dhroimghlas | A

(subspecies: undetermined)

World	Asia
Ireland	11 individuals
Cork	8 individuals
Conservation	Not globally threatened

One at CCBO on 13–14 October 1990. One at Dursey Island on 14 October 1990. One at CCBO on 24 October 1990. One at Cobh on 23–4 January 1991. One at Dursey Island on 5 October 1992. One at CCBO on 17 October 1993. One at Dursey Island on 10 October 2016. One at Firkeel on 19 October 2016.

Tree Pipit

Anthus trivialis | Riabhóg choille | A

(subspecies: *trivialis*)

World	Europe and Asia
Ireland	Scarce spring and autumn passage migrant
Cork	Scarce spring and autumn passage migrant
Conservation	SPEC 3; Amber-listed

There were no records of Tree Pipits until the establishment of CCBO in 1959. It has since been recorded in all but four of the sixty years to 2018 (1963, 1970, 1971 and 1974). Most have been recorded at CCBO (68 per cent). Dursey Island and Mizen peninsula come next with 11 per cent and 7 per cent, respectively. The remaining records are from almost every prominence and beach between Allihies and Ballymacoda Bay, but only 8 per cent have been east of Galley Head, with two birds marginally inland (Cork city and Midleton).

Tree Pipits are passage migrant in autumn (75 per cent) and spring (25 per cent). Most are recorded on only one day (many birds fly over at CCBO and Dursey Island), with some remaining for two to four days. One bird remained on farmland at Midleton from 7 November 2017 to 25 February 2018.[327] Most records are of one to three birds, but four (twice) and five (once) have occurred together. Numbers have not changed over the six decades of this study, but a higher proportion have occurred in spring since 1994 than previously.

1959–68	1969–78	1979–88	1989–98	1999–08	2009–18
44	33	54	47	49	64

Earliest and latest occurrence in spring and autumn (all records): 21 March–2 June and 8 August–28 October (7 November).

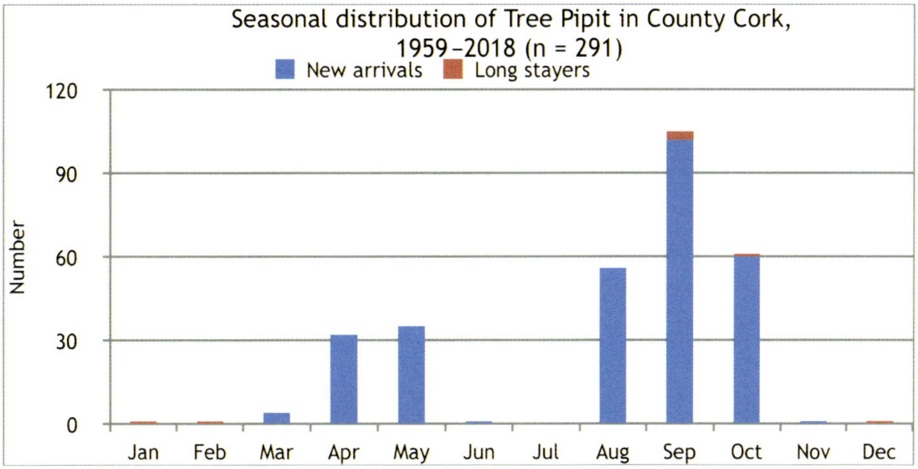

Fig. 7.47 Tree Pipit.

Pechora Pipit

Anthus gustavi | Riabhóg Pechora | A

(subspecies: undetermined)

World	North-east Europe and northern Asia
Ireland	2 individuals
Cork	1 individual
Conservation	Not globally threatened

One at Garinish on 27–8 September 1990.

Meadow Pipit

Anthus pratensis | Riabhóg mhóna | A

(subspecies: *whistleri*; *pratensis*)

World	Greenland, Iceland, Europe, and north-west Asia
Ireland	Very common resident breeder, winter visitor, passage migrant
Cork	Common resident breeder, winter visitor and passage migrant
Conservation	SPEC 1; Red-listed

The Meadow Pipit (*A.p. whistleri*) was breeding throughout the county during 1968–72. It was also widespread in 2008–11 but there were gaps in the range, and density was low apart from the Beara peninsula.

The breeding population at CCBO was censused in 1965 (355 pairs) and 1986 (370 pairs). A mean of 153 pairs (range = 110–200) bred at Sherkin Island during 1977–94. It breeds commonly at Dursey Island, but numbers have not been quantified. Two pairs at Bull Rock in 1889 were presumably breeding, although none was seen during visits in 1955, 1969 and 1970. Meadow Pipits breed on coastal heaths, dunes and unimproved upland grassland. They may sometimes sing over intensive lowland grassland in spring, but they desert before it is harvested for silage.

Meadow Pipits are equally widespread during winter, and there is a relatively high density throughout. They are frequently found on intensive pastures which accounts for their wider distribution compared to Skylark, although higher ground may be deserted.

It has been known since the 1800s that great numbers visited west Cork in the autumn, and remained for the winter. There is a significant movement at this season; large numbers occur in coastal districts, and birds have been recorded at Fastnet Rock during September to November. Passage is often substantial, and peaks during September and October. There have been at least four counts of 1,000 birds at west Cork sites in September and October. One count at Ballycotton recorded 2,206 coming in over the sea from the south-east and flying west in four hours and twenty minutes on 2 October 1977. Peak counts at other headlands in September and October may reach 500–600 birds.

Little is known of natal dispersal in Irish and British nestlings since few have been recovered, but the median distance travelled by those that have is less than 1 km. Although there are few ring recoveries in support, it is likely that many birds involved in autumn movements in Ireland have an origin in Iceland, Faeroes, and south-west Norway (*A.p. pratensis*), but there is no evidence that foreign Meadow Pipits winter in Ireland.

Red-throated Pipit

Anthus cervinus | Riabhóg phíbrua | A

(monotypic)

World	Arctic regions of Europe and Asia from north Norway east to Alaska
Ireland	60 individuals
Cork	34 individuals
Conservation	Not globally threatened

Red-throated Pipits have occurred at several sites at and near the coast. Most have been in autumn (32) and two in spring, with October accounting for twenty-eight. All sightings are of single individuals. Most occurred at CCBO (11), Dursey Island (7), Mizen Head (5), Ballycotton (5) and Galley Head (2), with singles at four other sites. East Cork sites have accounted for eight. Most Red-throated Pipits have been seen on only one day (27), but individuals have remained for up to six days with one remaining sixteen days at Ballycotton (mean stay = 1.9 days). Red-throated Pipits have increased in frequency in recent years, especially since 2009, with six in 2011 alone.

1959–68	1969–78	1979–88	1989–98	1999–08	2009–18
0	1	3	7	7	16

Earliest and latest occurrence in spring and autumn (all records): 15 May–2 June and 30 September–8 November.

Rock Pipit

Anthus petrosus | Riabhóg chladaigh | A

(subspecies: *petrosus*; *littoralis*)

World	Coastal Europe from France to Kola Peninsula, also Faeroes
Ireland	Resident breeder
Cork	Resident breeder
Conservation	Green-listed

Rock Pipits (*A.p. petrosus*) are confined to rocky coasts and islands. During the three breeding atlas surveys they were present on the coastline from Dursey Island to Youghal. Density was highest at the western peninsulas and in east Cork. About six pairs were breeding at Bull Rock in the late 1800s. The breeding population at CCBO ranged from thirty-nine to fifty-four pairs during 1965–86. At Sherkin Island, numbers averaged twenty-two pairs during 1977–94. Capel and Ballycotton Islands (lighthouse) each host about ten pairs annually, and one pair was at Sovereign Islands in 1993.

There is little difference between winter and summer distribution, apart from a partial withdrawal from exposed coasts, and a wider range around Cork Harbour, due to their habit of occurring in winter on beaches and estuarine saltmarsh habitats, where few if any breed.

Although the Rock Pipit is sedentary, up to sixty have occurred in a day in September at CCBO. These are likely post-breeding aggregations making local movements. Loose gatherings of ten to fifty birds are common on beaches in winter.

Scandinavian Rock Pipit *A.p. littoralis*: There are twenty-one records (25 individuals) of this subspecies. Most occurred in February to May, peaking in March, with one in November. It has occurred at Beara peninsula (6), Sherkin Island (3), Baltimore (1), Roche's Point (2), Ballycotton (7), Knockadoon Head (2) and Ballymacoda Bay (3), with one inland at the Gearagh. Most (18) have occurred on a single day, while others stayed for up to seven days, and one for twenty-four days (mean = 2.8 days). Most records are of singles, but two have occurred together on four occasions. Repeated occurrences at the same site, such as Ballycotton in 2007, 2009, 2010 and 2011, and Eyeries in 2013, 2015 and 2016, suggest possible returning of the same individuals in subsequent years.

Water Pipit

Anthus spinoletta | Riabhóg uisce | A

(subspecies: *spinoletta*)

World	Southern and central Europe, and central Asia
Ireland	Rare winter visitor
Cork	28 individuals (including presumed returning)
Conservation	Green-listed

This species was probably previously overlooked. It is now a winter visitor to the coast. Some (3) arrive in October, but most in November (16). Some apparently new birds continue to appear until April, but some of those seen for the first time after November may have arrived earlier and gone undetected. Most have occurred around Ballycotton (10) and Ballymacoda (9) Bays. Two occurred elsewhere in east Cork, and six at west Cork sites from Lough Beg to Sherkin Island. One was seen inland at the Gearagh. Most records are of single individuals, but two occurred together at Ballycotton and four at Pilmore Strand. Twelve birds occurred on one day, and four remained up to seven days. Twelve birds remained for longer periods, one for 135 days (mean stay = 25.6 days). Long-staying birds indicate wintering behaviour over passage migration, and occurrences at Ballycotton and Ballymacoda Bays in successive winters indicates the return of some of the same individuals.

1959–68	1969–78	1979–88	1989–98	1999–08	2009–18
0	0	0	0	5	23

Earliest and latest occurrence in autumn and spring (all records): 25 October–3 April.

Buff-bellied Pipit

Anthus rubescens | Riabhóg tharr-dhonnbhuí | A

(subspecies: *rubescens*)

World	Central and eastern Asia, North America, and Greenland
Ireland	23 individuals
Cork	7 individuals (6 records)
Conservation	Not globally threatened

One at Lissagriffin on 5–21 October 2007, joined by a second bird on 7–21 October. One at Ballycotton on 31 October–10 November 2007. One at Redbarn Strand on 25 November 2007–21 March 2008. First-year at Pilmore Strand on 12 October 2010. One at Ballycotton on 5–12 November 2011. One at CCBO on 21–4 September 2012.

Chaffinch

Fringilla coelebs | Rí rua | A

(subspecies: *gengleri*; *coelebs*)

World	Europe, Asia, north Africa, and islands, introduced elsewhere
Ireland	Common resident breeder, winter visitor, passage migrant
Cork	Resident breeder, winter visitor, passage migrant
Conservation	Green-listed

The Chaffinch (*F.c. gengleri*) was common across the county during all three breeding atlas surveys. Breeding data are available only for Sherkin Island where thirteen pairs in 1977 had declined to two or three by 1994. Two pairs bred for the first time at CCBO in 2016.

The winter range is identical to the breeding season, with evidence of higher density around Cork Harbour, with no apparent change over time. Winter flocks of 50–200 occur commonly, but this number is sometimes exceeded and up to 450 have been recorded.

A large immigration on the coast in autumn was described in the late 1800s. Many have been observed and obtained at Fastnet Rock in October, and as late as 9 December. Small numbers of migrants appear from mid-September and passage peaks in late October. The autumn maximum at CCBO is typically 50–100, but huge numbers occur in some years, over 9,500 passing south-west on 29–30 October 1959. Migrants also occur at Dursey Island and Beara peninsula, but usually do not exceed 150 birds. Migrants are mainly the northern subspecies (*F.c. coelebs*), which are larger than residents, and females predominate as they migrate further than males.[328]

Irish- and British-bred Chaffinches are sedentary with median movements of less than 1 km at all seasons, but some wander further. Winter numbers are augmented by birds mainly from Scandinavia. One adult ringed at Ovens on 10 July 1974 was found dead at Ballincollig on 28 March 1982, so was at least nine years old.

Brambling

Fringilla montifringilla | Breacán | A

(monotypic)

World	Northern Europe and northern Asia
Ireland	Winter visitor
Cork	Winter visitor in varying numbers
Conservation	SPEC 3; Amber-listed

The Brambling occasionally occurred in considerable numbers in winter in the 1800s, and the severe winter of 1837/8 was the first time it visited the neighbourhood of Cork city. There were four records from lighthouses, one from Old Head of Kinsale (October 1888) and three from Fastnet Rock (November 1891, 1899 and 1905), and one found dead at Buttevant (March 1900). It was described as common in winter in the first half of the 1900s, with about fifty near Fermoy in January 1920.

Winter atlas surveys gave the first account of distribution across the county. The 1981–4 survey showed it was scarce, with one main cluster of occupied 10 km squares around Cork city and harbour. Elsewhere, there was a wide, but thin, scatter of records. Brambling numbers can vary greatly between winters depending on factors such as weather severity, food availability, and a tendency for individuals to winter in different places in different winters.[329]

Small numbers occur at coastal sites from early October, especially at islands and headlands. Autumn counts at CCBO rarely exceed fifteen birds, but 1975 was exceptional with counts of over thirty on six dates and a peak of 200 on 24 October. Other coastal sites from Dursey Island to Knockadoon Head typically record small flocks in autumn, but not annually.

Most Bramblings occur on arable farmland in the south and east, often associating with Chaffinches. Large flocks are rare and only 3 per cent of records are of over fifty birds. Eight winters since the 1970s produced flocks of at least twenty-five birds (Table 7.20). Although some probably occur annually, numbers in recent years have been low, and the cold winters of 2009/10 and 2010/11 did not produce high numbers. Most wintering Bramblings depart in March, and the latest record is of two at CCBO on 10 May 2000. One ringed at CCBO in October 1975 was recovered at Oviedo (Spain) in February 1976, suggesting that some birds in that exceptional winter moved further south.

Table 7.20 Numbers of Brambling in County Cork, 1975/76–2003/04.

Winter	Site	Peak number
1975/76	Ballintubbrid	100
1975/76	Bishopstown	60–70
1977/78	Ballymacoda	25
1981/82	Fota	50
1981/82	Ballymacoda	25
1983/84	Cobh	350
1983/84	Lough Beg	35
1983/84	Carrigrohane	30
1986/87	Ballymacoda	100
1986/87	Douglas	90
1992/93	Kilcolman	90
1992/93	Brinny	60
1994/95	Cloyne	30
2003/04	Oysterhaven	32
2003/04	Grange	27

Serin

Serinus serinus | Seirín | A

(monotypic)

World	Europe, south-west Asia, north Africa, and Canary Islands
Ireland	10 individuals
Cork	6 individuals
Conservation	Not globally threatened

One at CCBO on 11–12 November 1975. One at Galley Head on 2 October 1979. One at CCBO on 11 October 1988. First-year male at CCBO on 27 October–1 November 2002. One at Sherkin Island on 5 November 2005. One at Mizen Head on 4 October 2013.

Greenfinch

Chloris chloris | Glasán darach | A

(subspecies: *chloris*)

World	Europe, Asia, north Africa, and islands, introduced elsewhere
Ireland	Common resident breeder
Cork	Common resident breeder
Conservation	Amber-listed

The Greenfinch was one of our commonest residents in the 1800s. It was breeding throughout the county during 1968–72, even at the western peninsulas. There was little change during subsequent surveys, and while density was low, it was highest in agricultural areas in south and east Cork.

Greenfinches may have increased at CCBO in the early 1960s, and there were about fifty pairs in 1965 and forty-six pairs in 1986. There is also a breeding population at Sherkin Island, where it has averaged thirty-one pairs during 1977–94. Two or three pairs have bred annually at Dursey Island since at least 2007.

An increase was said to occur in winter in west Cork in the late 1800s. Winter surveys in 1981–4 and 2007–11 showed it was widely distributed but was scarce in coastal districts in the west. Higher densities were recorded around Cork Harbour. Flocks of over 100 are uncommon, and the highest count is of 240 at Ballintemple (Cork city) in February 1965.

Greenfinches have suffered from trichomonosis since 2005, a disease caused by a parasitic organism. This disease spreads among birds at feeding stations and water troughs in gardens, and numbers have declined locally.

A marked movement took place on the Cork coast in April 1945, birds remaining for one or two days, followed by another movement towards the end of May, and noticeable as far inland as Fermoy. All had passed on by mid-June. More recently, a spring movement has been noted at CCBO, peaking in April and May, although there is little evidence this is widespread.

Autumn movements have been recorded at Fastnet Rock. Passage at CCBO occurs from August to October, peaking at up to 240 birds. By contrast, observations at Dursey Island and the Beara peninsula indicate that Greenfinches are scarce there, with maximum counts of under fifty birds. It is not known whether these movements involve birds of continental or British origin, or whether they are local. There is evidence from ringing that some birds from Britain spend the winter in Ireland (November to April), and there have been recoveries in Cork of birds ringed in England and Wales.[330]

The cones of noble fir may be a locally important winter food for Greenfinches.[331] About fifty birds in December and about eighty in January were feeding on the seeds of this fir in woodland in Cork.

Siskin

Spinus spinus | Siscín | A

(monotypic)

World	Europe and Asia
Ireland	Common resident breeder, common winter visitor
Cork	Resident breeder and winter visitor
Conservation	Green-listed

Siskins occurred about Cork city in January 1844 and in large numbers in winters 1847/8 and 1849/50. Breeding probably commenced some time before 1900.[332] Glengarriff was a known breeding haunt by 1950, but the full distribution of the Siskin at that time is unknown. During the 1950s and 1960s it was known mostly as an autumn passage migrant at CCBO.

Siskins were breeding in conifer plantations at the western peninsulas and on the border of Cork and Kerry during 1968–72. Breeding extended to the Ballyhoura Mountains, Mitchelstown, Coachford and Mallow. Numbers increased thereafter, and by 2008–11 there was a continuous population across the county, although it was absent from a coastal band from Castlefreke to east Cork and from areas north of the River Blackwater.

The winter atlas survey in 1981–4 showed that Siskins were thinly distributed, mainly in a band across the centre of the county, including Cork Harbour and east Cork, but there were hardly any in south-west Cork. The situation had changed by 2007–11 when distribution covered the entire county. Siskins occur in winter in woodlands with alder, often associating with Lesser Redpolls, and flocks of up to 200 have been recorded. They increasingly occur at garden feeding stations in suburban areas.

The Siskin occurs at Fastnet Rock in October and November. It is an autumn passage migrant, principally in October, to coastal headlands and islands. It is strongly irruptive, and while it occurs every year, numbers in some are very large, while in others few are seen. More occur at CCBO than at other sites, and peak counts of 500 birds are regularly obtained; on 15 October 2005 a count of 1,050 was recorded. High numbers also migrate through Dursey Island and Beara peninsula and as many as 650 have been recorded.

It is unclear if birds involved in autumn movements are continental migrants or part of the Irish population moving south. Some British and Irish birds migrate south to the continent for the winter, while continental birds arrive from the north, but the latter group appear to visit southern England more than Ireland. However, there are winter and early spring exchanges of Siskins between Cork, England, Scotland and Belgium.

Goldfinch

Goldfinch

Carduelis carduelis | Lasair choille | A

(subspecies: *britannica*)

World	Europe, Asia, north Africa and its islands, introduced elsewhere
Ireland	Common resident breeder
Cork	Common resident breeder
Conservation	Green-listed

The Goldfinch was common in the late 1800s except where it was persecuted by bird-catchers, and it occurred in numbers in the bare country about Schull. It was reported to have increased following the passing of the Wild Birds Protection Act, 1930.

The Goldfinch was breeding throughout the county during each of the three breeding atlas surveys. Although it was not breeding at CCBO in the 1960s, it was doing so by the early 2000s. It bred also at Sherkin Island, but the six pairs in 1977 had declined to three pairs by 1992–3, and none bred in 1994.

Winter surveys showed the Goldfinch was present across the county, but was scarce in the west, with highest density centred on Cork Harbour. Its range expanded later, and areas in the west and north where it was previously absent were occupied in 2007–11.

The Goldfinch is most noticeable in autumn when flocks occur throughout the county, especially in weed-infested fields and waste ground, and flocks of 200–50 containing many juveniles are not uncommon. Goldfinches are usually scarce at CCBO in autumn and peak numbers occur in late October, e.g. 250 in October 2016. Counts of 240 and 200, respectively, have been obtained at Dursey Island and Beara peninsula in October. Goldfinches have also occurred at Fastnet Rock in September, October and November.

Many British Goldfinches go south to winter on the continent. One ringed as a juvenile at CCBO on 16 July 2017 was recovered at Alderney (Channel Islands) on 30 October 2017, indicating that some Irish birds may go there too. A British origin for some wintering birds is suggested by recoveries at Rochestown, Kinsale, Leap and CCBO from Wiltshire, Dorset and Devon.

Linnet

Linaria cannabina | Gleoiseach | A

(subspecies: *cannabina*)

World	Europe, Asia, north Africa, and Madeira and Canary Islands
Ireland	Common resident breeder
Cork	Common resident breeder
Conservation	SPEC 2; Amber-listed

The Linnet was a common resident in the 1800s, except in wooded districts. It was breeding widely during 1968–72. Although there was a contraction of range in 1988–91, by 2008–11 all apparent losses had been recovered, and it was present throughout the county. Areas of highest density were at coastal headlands and in agricultural land centred on Cork Harbour.

The Linnet is mainly a summer visitor to CCBO, the breeding population arriving in March, and most leaving by late October. Breeding numbers do not appear to have changed since 1959 with a population of thirty pairs in

1965 and twenty-six pairs in 1986. The breeding population of about sixty pairs at Sherkin Island during the 1980s declined to about twenty pairs by 1994.

Both winter atlas surveys showed the Linnet was widely distributed, it being scarcest in the south-west and the uplands, and commonest in coastal districts from Galley Head eastwards.

Linnet numbers increased in west Cork at migration times in spring and autumn, and in winter in the 1800s. There is a peak at CCBO in early October when up to 2,000 birds have occurred. Elsewhere on the coast, and rarely inland, autumn passage is recorded when flocks of up to 550 occur between September and November, occasionally to December, at sites from Dursey Island to Ballymacoda. Large flocks may continue to be present into January as at Cobh (580) and Cloyne (600), and a flock of 2,000 was recorded at Cloyne in March 1995.

Spring passage is most noticeable in April when small flocks (up to 50) occur at coastal sites, with some arriving from the south and continuing north overland. Many Linnets from Britain and Ireland migrate to France and Spain for the winter, but Irish birds are under-represented due to fewer being ringed.

Twite

Linaria flavirostris | Gleoiseach sléibhe | A

(subspecies: *pipilans*)

World	North-west Europe, south-west and central Asia
Ireland	Scarce resident breeder, recent decrease
Cork	Former breeder, now rare visitor
Conservation	Red-listed

The Twite was probably a common breeding bird in the 1700s and 1800s. Robert Warren found a nest beside a public road at Carrigaline, and the species was noted at Cobh.[333] It was common at the end of the 1800s, frequenting mountains, and large flocks occurred in winter in west Cork. Two were obtained at Bull Rock in February 1903.

There is no indication in the literature of a change in status up to the mid-1960s. However, comparison between the counties in which breeding took place in 1900 and 1968–72 shows that a considerable contraction of range had occurred. Although its former range in Cork is unknown, by 1968–72 it was breeding in only two 10 km squares (Dursey Island and Toe Head) and was present in four others, three in coastal areas in the west and one inland. These, in retrospect, represent the last documented records of breeding in Cork. Two subsequent surveys recorded Twites in three and one 10 km squares in the west, respectively, but none was breeding.

Winter atlas surveys recorded Twites in three 10 km squares near Bantry and one near Garryvoe in 1981–4, and in three squares on the coast from Dursey Island to Ballycotton in 2007–11.

Since the foundation of CCBO in 1959, Twite have occurred there most often in autumn, twenty being recorded in a day in October 1964, but with fewer records in the last three decades. Non-annual records have occurred at coastal sites from Dursey Island to Ballymacoda since the 1960s.

Data for the last three decades (since breeding ceased) are analysed in the decadal table. While most of the records are of single birds, Dursey Island had forty in September 2015, thirteen in January to April 2010, and nine in October 1993, and CCBO had twenty in September 1997, ten in September

1989–98	1999–08	2009–18
35	21	84

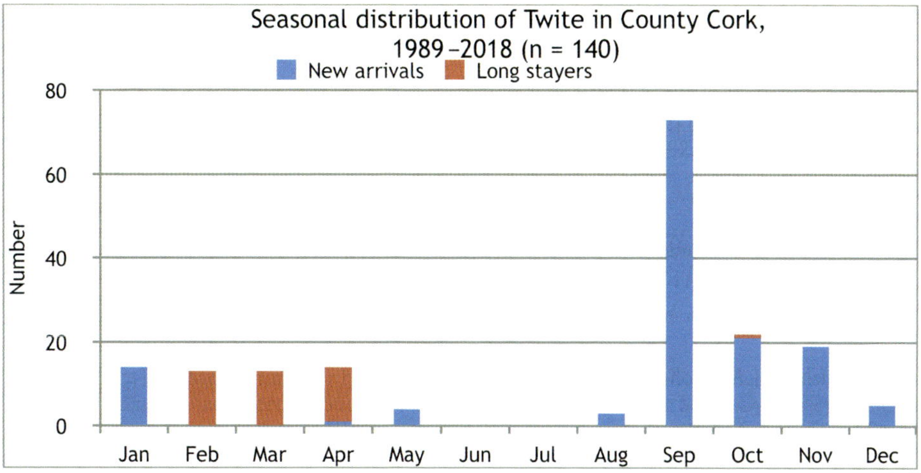

Fig. 7.48 Twite.

2001, and nine in November 2016. The number recorded during the three decades is skewed by these few high counts. Most have occurred at Dursey Island (69), CCBO (46) and Ballycotton (12). Seven other coastal sites had a combined total of twelve birds, with one inland at the Gearagh in January 2013.

Lesser Redpoll

Acanthis cabaret | Miondeargéadan | A

(monotypic)

World	North-west and central Europe, introduced elsewhere
Ireland	Common resident breeder, winter visitor
Cork	Resident breeder and winter visitor
Conservation	Green-listed

The Lesser Redpoll was a rare winter visitor in the mid-1800s, but its occurrence was probably underestimated. It appeared around Cork city in winter 1847/8, and it was increasing and bred in suitable districts near the end of the 1800s and occurred in flocks in winter. It was a widespread breeder during 1968–72, including at the western peninsulas. Losses in some areas during 1988–91 had largely been made good by 2008–11.

After the breeding season Redpolls gather into flocks of usually fewer than 100 birds and they can be found in river valleys feeding among alder and birch. Winter atlas surveys showed it was more widespread at that season, with the main populations present around Cork Harbour and in north-east and east Cork.

Its occurrence at CCBO in the 1960s was unpredictable, but it has since increased and become more regular, and it has bred. Most occurrences are in autumn, especially between August and November, and numbers rarely exceed fifty birds. It is scarce in autumn at Dursey Island and the Beara peninsula and does not occur every year. There is no evidence that Irish birds leave for the winter, like some British birds, although it is suspected that some do.

Common Redpoll

Acanthis flammea | Deargéadan | A

(subspecies: *flammea*; *rostrata*)

World	Arctic regions of Europe and North America, Greenland, and Iceland
Ireland	Rare visitor, mainly in autumn and winter
Cork	7 individuals
Conservation	Not globally threatened

One at Dursey Island on 10–12 October 1978. One at Dursey Island on 3 June 1998. One at Dursey Island on 13 October 2000. One at Dursey Island on 22 October 2005. One at CCBO on 28 October 2007. One at Garinish on 16 October 2011. One at Dursey Island on 19 October 2015. The first, third, fourth, fifth and sixth records refer to the subspecies *flammea* (Mealy Redpoll), while the second and seventh records refer to the subspecies *rostrata* (Greenland Redpoll).

Arctic Redpoll

Acanthis hornemanni | Deargéadan Artach | A

(subspecies: *exilipes*)

World	Arctic regions of North America, also Greenland, Europe, and Asia
Ireland	12 individuals
Cork	3 individuals
Conservation	Not globally threatened

One at Dursey Island on 4–10 October 1999. One at Dursey Island on 1–2 May 2010. One at Dursey Island on 5–10 June 2014. The subspecies was undetermined for the 1999 and 2014 individuals, but the 2010 individual was assigned to *exilipes*.

Common Crossbill

Loxia curvirostra | Crosghob | A

(subspecies: *curvirostra*)

World	North America, Europe, Asia, and north Africa
Ireland	Rare or scarce resident breeder, highly irruptive
Cork	Scarce resident breeder, highly irruptive
Conservation	Green-listed

The Crossbill was rare but occurred in Cork in the 1700s. Vast numbers arrived in the severe winter of 1801/2, the first in late August. Robert Ball knew of it occurring once in the south about 1812, when it devastated orchards. Large numbers appeared in the south of Ireland (Cork not mentioned) in 1838/9, some remaining the whole year, while a few were seen near Cork city in summer 1844.

Hundreds were observed at Monkstown in November 1887, and they remained to the following summer. There was a sudden increase in Ireland in 1888 and thirteen nests were observed at Monkstown.[334] Smith also saw Crossbills at Ballybricken, Carrigaline, Coolmore and Currabinny, and at other locations. Birds were at Glandore in 1887 and 1888, with several obtained near Cloyne, Doneraile and Mallow in January and February 1888.[335] A pair bred at Skibbereen in 1900. Ussher believed the increase was due to the maturing of seed-bearing conifer plantations. Crossbills were recorded in fifteen counties in 1889, sixteen in 1890, seventeen in 1891 and eighteen in 1894, but Cork was not mentioned, and they were said to be established in most of the large conifer plantations in the 1890s.

Many irruptions took place into Ireland in the 1800s and 1900s and into the 2000s, but Cork was not specifically mentioned during most: 1801, 1807, 1838, 1868, 1881, 1887–1890s, 1909, 1927, 1929, 1935, 1953, 1956, 1958, 1959, 1962, 1966, 1972, 1984, 1986, 1990, 1991, 1997 and 2002. In July to September 1929, and again in 1935, flocks visited Glengarriff, but breeding was not suspected.[336] There were few records during the 1950s, 1960s and 1970s, but this may reflect an absence of observation.

Crossbills were not recorded in Cork during the breeding seasons of 1968–72 or the winters of 1981–4. They were present in a handful of 10 km squares in the south, west and north during 1988–91, and breeding was proved in two squares. These resulted from an influx in 1986 when a total of almost 100 birds was recorded at three locations: Inchigeelagh, Doneraile and Doolieve Wood. These remained to the next winter, and juveniles were seen. More occurred in 1987, 1988, 1990 and 1991, and three pairs bred in 1992. Throughout the 1990s and into the 2000s Crossbills could be seen at conifer plantations in many parts of Cork (15 plantations in south, mid and north Cork in 1990, with reports of up to 60 birds). There were records of Crossbills across the county in 2008–11, although there was no significant concentration, and breeding was proved in two squares.

The Crossbill is highly irruptive and irregularly visits CCBO, mostly in October, occurrences often coinciding with influx years. It also occurs irregularly at Dursey Island, again mostly in October, but it seldom occurs at headlands on the Cork coast east of CCBO.

Common Rosefinch

Carpodacus erythrinus | Rósghlasán coiteann | A
(subspecies: undetermined)

World	Europe and Asia
Ireland	Rare spring and autumn passage migrant
Cork	143 individuals
Conservation	Not globally threatened

The Common Rosefinch is an autumn migrant (125) to coastal headlands and islands, but it also occurs in spring (18). Most occur at CCBO (79), Beara peninsula and Dursey Island (31), and Mizen peninsula (21). Away from these it is rare, and only four other sites in west Cork have had records: Sherkin Island (1), Toe Head (1), Galley Head (2) and Old Head of Kinsale (5). Three birds occurred in east Cork, at Cobh, Ballycotton and Knockadoon Head. Most birds (98) stayed for only one day, while thirty-five stayed two to five days, and ten stayed longer, extremes being four for twenty-eight days each and one for thirty-eight days (mean stay = 2.8 days). Of those aged, seventy-four were first-year and four were adult birds.

Thirteen have been sexed as male but only two as female, it being more difficult to identify females because of their duller plumage. Single birds are the norm, but two have occurred on seven occasions, three twice and four once. There has been a sustained increase each decade since the first record in 1971.

1959–68	1969–78	1979–88	1989–98	1999–08	2009–18
0	7	15	34	37	50

Earliest and latest occurrence in spring and autumn (all records): 26 May–15 June and (4 July) 24 August–23 November.

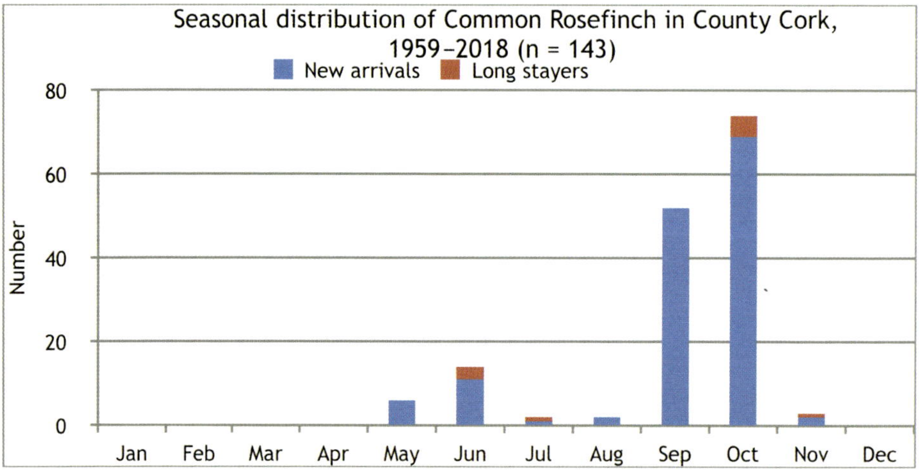

Fig. 7.49 Common Rosefinch.

Bullfinch

Pyrrhula pyrrhula | Corcrán coille | A

(subspecies: *pileata*; *pyrrhula*)

World	Europe and Asia
Ireland	Common resident breeder
Cork	Common and widespread resident breeder
Conservation	Green-listed

The Bullfinch (*P.p. pileata*) was common and increasing in cultivated and wooded districts in the late 1800s but was scarce along the coast.

It was present across the county during all three breeding atlas surveys, and density was uniform, except for the western peninsulas, where it was lowest. Winter distribution mirrored the breeding season, but it was again scarcer at the western peninsulas. The Bullfinch is met with in pairs or family parties, but small flocks of twenty or fewer occur in favoured locations during the winter months.

The Bullfinch was a rare visitor to CCBO during the 1960s, but there has been an increase since about 1980, and it bred for the first time in 2016 when a pair reared two broods. It breeds at Sherkin Island and in good years up to six pairs occur. It rarely visits Dursey Island and does not breed, presumably due to a lack of suitable habitat. It is also rare in October near the tip of the Beara peninsula. There is no evidence that Bullfinches breeding in Ireland are migratory.

Northern Bullfinch *P.p. pyrrhula*: There is one record of this subspecies. Adult male at Dursey Island on 15–18 June 2017.

Hawfinch

Coccothraustes coccothraustes | Glasán gobmhór | A

(subspecies: *coccothraustes*)

World	Europe, Asia, and north Africa
Ireland	Rare autumn and winter visitor, occasionally irruptive
Cork	Minimum of 95 individuals
Conservation	Not globally threatened

Five Hawfinches were obtained before 1850, at Glengarriff (2), near Cork city (2) and near Youghal (1). Dated records (3) occurred in winter, including one in January. Ussher knew of eighty-six records from twenty-three counties, usually singles, but sometimes of many birds. Cork was third in frequency after Dublin and Tipperary, but numbers for each county were not given. Ussher's unpublished notes record one at Doneraile Park (1866), two at unknown locations (winter 1878/9), and one at Castlemartyr (winter

1879/80). Two occurred in the early 1900s (Roancarrigmore lighthouse, November 1913 and Castletownshend, January 1920).

The first occurrence in the 1959–2018 period was one seen at CCBO in October 1969. Most occurrences have been in October (69), with five in other autumn months, while ten have occurred in March, April and May. Occurrences have been characterised by three major influxes in October 1988 (12), October 2005 (25) and October and November 2017 (25). In each year birds were observed in coastal districts and on headlands and islands, and few stayed for longer than one or two days before moving to unknown locations.

Most have occurred at CCBO (44), Dursey Island and Beara peninsula (19), and Mizen peninsula (9). The remaining twelve occurred at eleven sites, most of which were at or near the coast, from Schull eastwards to Fountainstown, with one record at Midleton. Two have occurred inland, at Kilcolman (1997) and Taur (2017). Most birds remained for only one day (50), a few remained up to seven days and five birds were present at CCBO for twelve days in October and November 1988 (mean stay = 2.3 days). Most records were of single individuals apart from flocks of ten and eleven at CCBO in 2005 and 2017, respectively. The number occurring over six decades has increased, but it has been recorded in only eighteen of the sixty years, and in non-influx years the numbers occurring have exceeded one on only three occasions (2, 3 and 5).

1959–68	1969–78	1979–88	1989–98	1999–08	2009–18
0	3	14	4	26	37

Earliest and latest occurrence in spring and autumn (1959–2018): 30 March–30 May and (15 August) 1 October–6 December.

Snow Bunting

Plectrophenax nivalis | Gealóg shneachta | A

(subspecies: *nivalis*; *nivalis*)

World	Arctic regions of North America, also Greenland, Iceland, Europe, and Asia

Ireland	Uncommon winter visitor and passage migrant
Cork	Scarce passage migrant, rare in winter and spring
Conservation	Green-listed

In the 1800s, Snow Buntings were scarce and irregular winter visitors, only one adult having been seen. There were records from Fastnet Rock (October 1886) and Old Head of Kinsale for May, September, November and December between 1889 and 1911.

Snow Buntings occur in autumn from September to November, small numbers stay the winter, and there is a light spring passage. Most occur on or near the coast, and there are some records for Cork Harbour, including one for Cork city. Winter atlas surveys revealed no inland occurrences. Upland areas, where some may winter, are rarely visited by birdwatchers at that season, and it may occur more regularly if searched for. Individuals were seen on the Caha Mountains in October and January in the 1930s, and it has been reported on upland moors in recent decades. There are records in November from Nadd Bog (Boggeragh Mountains) and Halfway (near Inishannon), but these may have been migrants.

Records of migrants are mainly from Dursey Island and CCBO, with small numbers at most headlands, and sometimes at beaches. Migrants at these islands seldom remain for long, most passing on within one or two days. Numbers are usually low, with most counts being of one or two, and rarely reaching ten. The highest count has been at Dursey Island where up to twenty-four were present in October 1987. Snow Buntings have rarely occurred at Dursey Island and CCBO in winter, most such records coming from east Cork beaches, e.g. Ballycotton, Ballymacoda and Youghal. Numbers are always low, the highest count being of four which spent over four months at Ballycotton. Other wintering birds have made shorter stays. Spring migrants occur at several sites such as CCBO, Dursey and Sherkin Islands, but also Seven Heads and other sites east to Youghal. Birds in adult male plumage are rare and are usually recorded in spring.

Assessment of occurrences over time suggests little change in the last four decades, although numbers at some sites vary between years. At Dursey Island, few (or none) occur in some years. It is often difficult to determine the number of individuals involved over one autumn as some move on quickly while others linger, requiring judgement when calculating totals. Migrant and wintering birds at east Cork beaches are rarer now than in the 1960s–1980s period.

Most British ring recoveries involve Icelandic birds (*insulae*), with some from Scandinavia, Greenland and Newfoundland (*nivalis*), but there are no recoveries to positively indicate the origin of Irish birds.[337]

1959–68	1969–78	1979–88	1989–98	1999–08	2009–18
33	55	85	113	97	98

Earliest and latest occurrence in autumn and spring (1959–2018): 10 September –29 May.

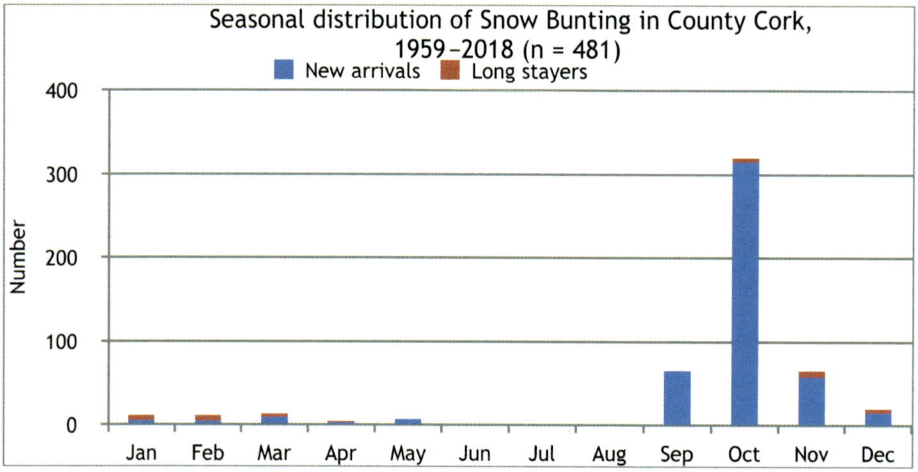

Fig. 7.50 Snow Bunting.

Lapland Bunting

Calcarius lapponicus | Gealóg Laplannach | A

(subspecies: *lapponicus*)

World	Arctic regions of North America, also Greenland, Europe, and Asia
Ireland	Scarce autumn passage migrant, rare in winter and spring
Cork	Scarce passage migrant, rare in winter and spring
Conservation	Green-listed

The first Lapland Bunting recorded in the county (and in Ireland) was a female found dead at Fastnet Rock on 16 October 1887. There were no further records until CCBO opened in 1959.

During 1959–2018 Lapland Buntings occurred in all but eight of the sixty years, the last year without a record being 1983. Most occur as autumn migrants from late August, and only five have been recorded in spring. Small numbers occur in winter, but only twice have some remained throughout the winter (December 1986–March 1987) when fourteen and eleven birds were at Ballymacoda and Ballycotton, respectively. Apart from one at Minane Bridge in February, Lapland Buntings have occurred in winter only at these two sites. Migrants are present for only a day or two and stays of more than five days are unusual.

Lapland Buntings occur at coastal sites (islands, headlands and beaches) from Dursey Island to Ballymacoda Bay, with only one inland record, three birds at Macroom in October 2011. Most have been seen at Dursey Island and CCBO, both sites where there is a regular autumn passage. Smaller numbers occur at headlands between CCBO and Cork Harbour, with few at sites in between. Most of those recorded in east Cork have been at Ballycotton and Ballymacoda Bays.

Lapland Buntings generally occur in numbers of from one to three, but occasionally in flocks of up to ten birds. The largest flock recorded was fourteen at Ballymacoda (see above). The few birds that have been aged or sexed have been males in breeding plumage on migration in April and May. Numbers occurring annually have varied from high in some years, such as 1986 (29), 1989 (20), 1992 (31), 2006 (21) and 2010 (minimum of 64), to fewer than ten, and sometimes one or two in other years. The distribution of autumn migrants in 2010 was different to other years when fourteen of at least forty-five birds in a September influx were in east Cork, while in all other years few have occurred outside west Cork at this season.

Overall, the number of birds recorded has been higher in the last four decades than in the two previous ones. This may be, at least partly, related to an increase in observer effort, especially at Dursey Island where often up to ten birds are seen per annum (29 in 1992 and 39 in 2010).

There are no ring recoveries to positively indicate the origin of Irish birds, but arrival patterns and other data suggest they may have a Greenlandic origin, so they could be (or include) subspecies subcalcaratus.[338]

1959–68	1969–78	1979–88	1989–98	1999–08	2009–18
12	9	62	116	67	126

Earliest and latest occurrence in autumn and spring (1959–2018): 28 August–27 May.

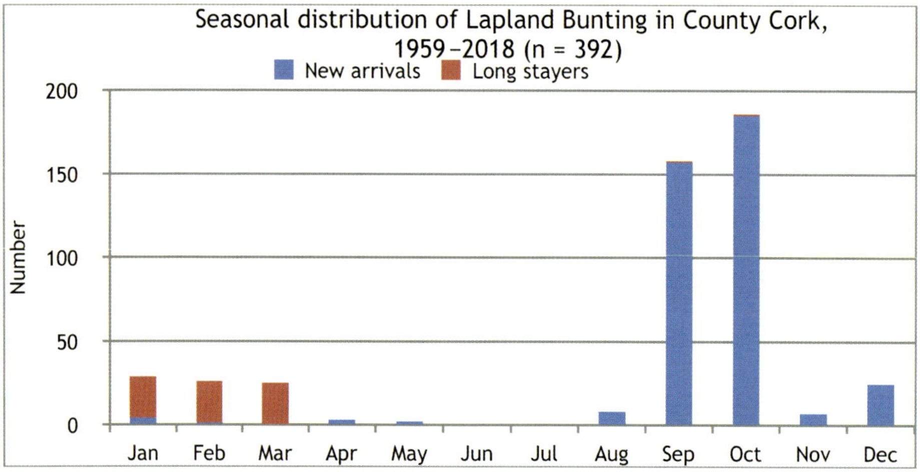

Fig. 7.51 Lapland Bunting.

Scarlet Tanager

Piranga olivacea | Tanagair scarlóideach | A

(monotypic)

World	Eastern North America
Ireland	6 individuals
Cork	5 individuals
Conservation	Not globally threatened

First-year female at Firkeel on 12–14 October 1985. Adult male at Firkeel on 18 October 1985. First-year male at Garinish on 7–11 October 2008. First-year at Mizen Head on 3–6 October 2017. First-year male at CCBO on 16–21 October 2018.

Rose-breasted Grosbeak

Pheucticus ludovicianus | Gobach mór broinnrósach | A

(monotypic)

World	North America
Ireland	9 individuals
Cork	7 individuals
Conservation	Not globally threatened

First-year male at CCBO on 7–8 October 1962. First-year female at CCBO on 11–12 October 1979. First-year male at CCBO on 9–15 October 1983. First-year male at Bull Rock on 10 October 1983. First-year male at Firkeel on 19–20 October 1983. First-year female at CCBO on 7–24 October 1987. Female at Garinish on 29 September–10 October 2016.

Indigo Bunting

Passerina cyanea | Gealóg phlúiríneach | A

(monotypic)

World	North America
Ireland	1 individual
Cork	1 individual
Conservation	Not globally threatened

One at CCBO on 9–19 October 1985.

White-crowned Sparrow

Zonotrichia leucophrys | Gealbhan bán-choróin | A

(subspecies: undetermined)

World	North America
Ireland	1 individual
Cork	1 individual
Conservation	Not globally threatened

Male at Ballynacarriga (Beara) on 20–7 May 2003.

White-throated Sparrow

Zonotrichia albicollis | Gealbhan bán-scornach | A

(monotypic)

World	North America
Ireland	3 individuals
Cork	2 individuals
Conservation	Not globally threatened

One at CCBO on 3 April 1967. First-year at CCBO on 12–18 October 2008.

Dark-eyed Junco

Junco hyemalis | Luachairín shúildubh | A

(subspecies: undetermined)

World	North America
Ireland	4 individuals
Cork	1 individual
Conservation	Not globally threatened

Adult male at Dursey Island on 9 June 2015.

Yellowhammer

Emberiza citrinella | Buíóg | A

(subspecies: *caliginosa*)

World	Europe to central Asia, introduced elsewhere
Ireland	Common resident breeder, recent decrease
Cork	Resident breeder, recent decrease
Conservation	SPEC 2; Red-listed

The Yellowhammer was a common resident in the 1700s and 1800s. A local decrease was noted in Cork between 1934 and 1937, but with numbers apparently returning to normal by 1945. During 1965 to at least 1968, it was again noted as scarce or absent from parts of the county, particularly the east.

Notwithstanding the reported decline, Yellowhammers bred throughout the county during 1968–72. By the late 1980s a significant range contraction had taken place, especially in the west and north-west. The range contracted further over the next two decades and by 2011 ended abruptly around Galley Head and skirted around the eastern edge of the Boggeragh Mountains. South and east Cork held the main populations, together with an area in north-east Cork. The entire range was then present in the most intensively farmed parts of the county. There has been no obvious change in status of the population in east Cork since 1970, although the distribution is rather patchy. There is also a significant population in south Cork between Carrigaline and Kinsale, which appears not to have changed in the last twenty years.

A population of about thirty-five pairs formerly bred at CCBO. This number decreased through the 1970s and 1980s, and 1990 was the last year in which breeding took place. It was last recorded breeding at Sherkin Island in 1985, following a sharp decline from twenty pairs in 1977. The pattern of breeding followed by decline in the 1980s also took place at Dursey Island, and although several pairs were breeding in 1993, the last year of breeding was 1994. There was a significant decline in small-scale arable farming at these islands between the 1970s and the early 1990s.

The winter range closely mirrored the breeding season at different times. Winter flocking is common, and flock size occasionally exceeds 100 birds in areas of weedy winter stubble. The decline of the Yellowhammer, which obtains most of its food on arable land, is consistent with changes in, and intensification of, agricultural practice in Ireland over recent decades.[339] Changes involve a loss of hedgerows as nesting habitat, reduction of insects as food for nestlings and loss of weed seeds in winter due to herbicide use and the switch from spring-sown to winter-sown cereals.

Although the species is sedentary in nature, one was recorded at Fastnet Rock on 6 May 1889, and another was obtained at the same place in November 1905.

Cirl Bunting

Emberiza cirlus | Cirlghealóg | A

(monotypic)

World	Southern Europe and north-west Africa, introduced elsewhere
Ireland	1 individual
Cork	1 individual
Conservation	Not globally threatened

Female at Mizen Head on 8–9 May 2006.

Ortolan Bunting

Emberiza hortulana | Gealóg gharraí | A

(monotypic)

World	Europe to central Asia and north Africa
Ireland	Rare passage migrant, mainly in autumn
Cork	93 individuals
Conservation	Not globally threatened

The Ortolan Bunting is an autumn passage migrant, but with a few spring occurrences. Most birds occur at CCBO (61), Dursey Island (13) and Mizen peninsula (9), with the remainder at Skibbereen (1), Galley Head (5) and Old Head of Kinsale (3). There is only one record for east Cork, one at Knockadoon Head in 2016. The first bird recorded in the county was in 1959, and it has occurred in thirty-nine of the sixty years to 2018, with a maximum of six in any year (1960 and 1976). Where ages were recorded, eleven were adult and twenty-one were first-year birds, and there was an even distribution of males and females with seven of each. Most birds were seen on only one day (53), although some were present for up to eleven days with one remaining for eighteen days at CCBO (mean stay = 2.6 days). There has been no discernible change in status over the six decades of this analysis.

1959–68	1969–78	1979–88	1989–98	1999–08	2009–18
19	15	11	16	14	18

Earliest and latest occurrence in spring and autumn (all records): 5 May–9 May and 23 August–20 October.

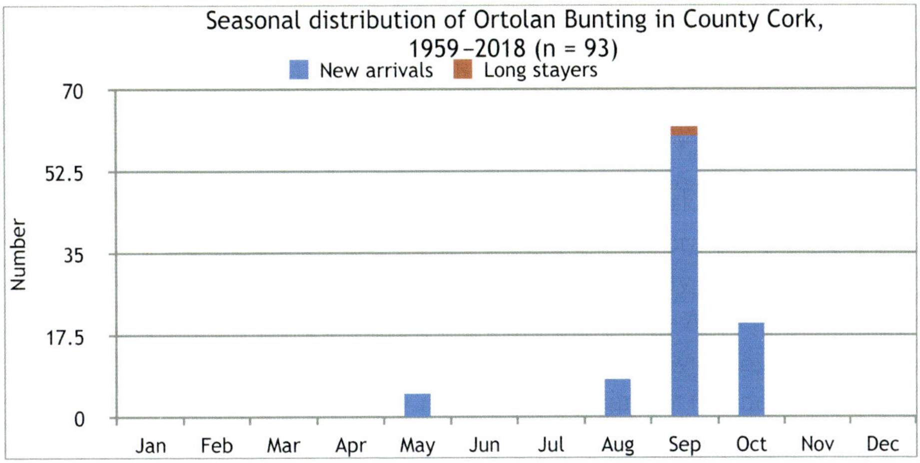

Fig. 7.52 Ortolan Bunting.

Rustic Bunting

Emberiza rustica | Gealóg thuathúil | A

(subspecies: undetermined)

World	Northern Europe and northern Asia
Ireland	21 individuals
Cork	19 individuals
Conservation	Not globally threatened

The Rustic Bunting is a late autumn (all 15 in October) and spring (4) migrant to coastal districts. All occurred singly, most of them at CCBO (11) and Dursey Island (4), with two at Old Head of Kinsale and one each at Toe Head and Sherkin Island. Most have been seen on only one day (13), but a few have stayed for several, and one each for seven and nine days, respectively (mean stay = 2.4 days). This species has become more frequent since the late 1980s, and although it is not annual, it occurred each year between 1988 and 1990, and again each year between 1997 and 2002.

1959–68	1969–78	1979–88	1989–98	1999–08	2009–18
1	0	1	9	6	2

Earliest and latest occurrence in spring and autumn (all records): 26 March–6 June and 5 October–28 October.

Little Bunting

Emberiza pusilla | Gealóg bheag | A

(monotypic)

World	Northern Europe and northern Asia
Ireland	59 individuals
Cork	35 individuals
Conservation	Not globally threatened

The Little Bunting is a late autumn migrant to coastal districts. Most birds have been recorded in October (32), with two in November, and one in spring (April). Most records refer to single birds, although two have occurred together twice at CCBO in different locations in the same year (2016) and once at Dursey Island (1989). Overall, it has occurred mostly at CCBO (16), and also Dursey Island (9), Mizen Head (4), Galley Head (3) and Beara peninsula (2), with one at Old Head of Kinsale. Most have stayed for one or a few days, but singles have remained for up to nine days with one remaining for twelve days at CCBO (mean stay = 2.7 days). An increase in occurrences has taken place over the decades with numbers rising from zero in 1959–68 to seventeen in 2009–18. Fifteen (43 per cent) have occurred in the most recent four-year period (2015–18).

1959–68	1969–78	1979–88	1989–98	1999–08	2009–18
0	2	4	3	9	17

Earliest and latest occurrence in spring and autumn (all records): 21 April and 3 October–13 November.

Yellow-breasted Bunting

Emberiza aureola | Gealóg bhroinnbhuí | A

(subspecies: undetermined)

World	Northern Europe and northern Asia
Ireland	5 individuals
Cork	3 individuals
Conservation	Vulnerable

First-year at CCBO on 11–20 September 1983. Female or first-year at CCBO on 18 September 1985. Female or first-year at Dursey Island on 9 October 2010.

Reed Bunting

Emberiza schoeniclus | Gealóg ghiolcaí | A

(subspecies: *schoeniclus*)

World	Europe, Asia, and Morocco
Ireland	Common resident breeder
Cork	Resident, breeding widely
Conservation	Green-listed

The Reed Bunting was common in low and wet boggy ground in the 1800s. It was present throughout the county as a breeding species during 1968–72. Despite considerable apparent losses during the next survey, the range once again extended across the entire county in 2008–11. Density was greatest around and east of Cork Harbour, and in some coastal areas in the south.

Reed Buntings breed mostly in wet and marshy areas, such as Ballyvergan Marsh and Ballycotton, but some occupy drier farmland habitats and young conifer plantations. They breed at least on some of the larger islands where suitable habitat exists. At CCBO, twenty-three pairs were breeding in 1965 but this was probably exceptional; the numbers decreased thereafter, and the usual population was said to be three to seven pairs. In the 1970s and up to at least the mid-1980s the population at Sherkin Island was about

Reed Bunting

five pairs, but this increased to seventeen to twenty-three pairs in 1991. Increasing scrub might be partly responsible for this increase, but decreased interspecific competition with the Yellowhammer, which ceased to breed on the island after 1985, may be another factor allowing the Reed Bunting to increase.[340] A pair or two also breeds at Dursey Island.

Many Reed Buntings leave their breeding habitats to winter in flocks on farmland and sometimes in gardens with finches and other buntings. During winter atlas surveys they were sparsely distributed throughout the county, but with some gaps in the west, south and north. The densest population was from Old Head of Kinsale north to the River Lee extending into east Cork. This patchiness may be, at least partly, due to the propensity of the species to form flocks of up to 100 birds at this season.

There is no evidence of migration on the south Irish coast involving native or foreign birds. Nevertheless, it is mainly a summer visitor at CCBO, and in autumn flocks of sixty or more often occur at coastal headlands and islands. These flocks may be of birds in the process of moving to farmland and other dry habitats as they join other bunting and finch feeding flocks.

Black-headed Bunting

Emberiza melanocephala | Gealóg cheanndubh | A

(monotypic)

World	South-east Europe and south-west Asia
Ireland	10 individuals
Cork	2 individuals
Conservation	Not globally threatened

Male at Skibbereen on 30 May 2005. First-year at Galley Head on 11–15 September 2009.

Unidentified bunting species

One bunting, either Black-headed or Red-headed (*Emberiza melanocephala/bruniceps*), has been recorded. First-year female at CCBO on 3–4 October 1992; it was killed by a Common Kestrel before it could be specifically identified.

Corn Bunting

Emberiza calandra | Gealóg bhuachair | A

(subspecies: *calandra*)

World	Europe, south-west Asia, north Africa, and Canary Islands
Ireland	Former breeder, now very rare visitor
Cork	Former resident, extinct as breeding species, vagrant since mid-1970s
Conservation	SPEC 2

The Corn Bunting was well known in the 1700s and 1800s, and it was common where woods were absent, especially in coastal districts where cultivation existed. However, by 1900 although recorded as a breeding species, it was decreasing in several districts in Ireland, including an area west of Bandon, although it became more numerous in winter near Bantry, where the climate was milder.

By the early 1950s there had been a major contraction of range throughout Ireland, and it was almost confined to coastal districts. The distribution in Cork was extremely local.[341] It was scarce or absent in coastal areas east of Cork Harbour in 1936, but in 1946 a few pairs were still breeding at Knockadoon Head and near Ballycotton. It was absent west of Cork Harbour until CCBO was reached. Scattered pairs occurred at Mizen Head, and it was found around Crookhaven. Although the population was small, Dursey Island and the mainland of the Beara peninsula had more breeding Corn Buntings than anywhere else in Cork. There was a marked decrease since 1948 at south-west headlands. It bred at CCBO in the 1940s and early 1950s, but there was no record of it during the period 1959–69, and the first record since the bird observatory was established was in 1972.

It was scarce in Cork by the late 1950s, but small numbers continued to breed at coastal sites throughout the 1960s, and it was plentiful at Knockadoon Head in November and December 1962. It was absent from some headlands in the south-west in 1963, notably Mizen Head and around Roaringwater Bay. One was seen at Dunmanway in July 1964, but without evidence of breeding. About six pairs in the vicinity of Roche's Point in 1966 were presumably breeding, while one pair was at Galley Head in July of the same year. Some birds were at Ballycotton in May and June 1967 and 1968, but were not proved to breed, and in 1969 five pairs were holding territory between Galley Head and Ballycotton, but their distribution was not outlined. One bird was at Douglas in 1967 or 1968, a pair at Roaringwater Bay in 1969, and seven birds at Ballycotton in November 1969.

Breeding was proved in one 10 km square at the Mizen peninsula in the 1968–72 breeding atlas, with records of probable breeding at Galley Head, Seven Heads and Ballycotton. The Mizen peninsula record was probably the last case of proved breeding in Cork, and the only sightings during the rest of the 1970s appear to be singles in 1972 (August), 1975 (October) and 1977 (2 in July) at CCBO and singles at Knockadoon Head (June) and Old Head of Kinsale (September) in 1976. During the breeding seasons of 1988–91 there were records from Dursey Island and Sheep's Head, but none involved proved breeding. It is unlikely that a population persisted in the west of the county into the early 1990s as tentatively suggested.[342]

Since the 1980s, only ten records have been documented in Cork, all single birds apart from one involving two, exclusive of the records from Dursey Island and Sheep's Head already mentioned, the most recent being in 2005. All have occurred in coastal districts from Dursey Island to Ballycotton, all former breeding haunts. All records are dated between February and July with a small peak in April, but all without evidence of breeding.

Studies have suggested that changes in agricultural practices are responsible for the decline of the Corn Bunting, especially the switch from spring-sown to winter-sown cereals, and from hay to silage. Important factors are believed to be a general decline in mixed farming and the loss of temporary grasslands and overwinter cereal stubble. The use of herbicides may also have reduced invertebrates and weed seeds.[343;344] Many of these agricultural changes have taken place in Ireland and they may be implicated in the decline and extinction of the species here. Intensification of agricultural practices has accelerated since Ireland's accession to the European Economic Community (later the European Union) in 1973, and this is known to have caused declines in several farmland bird species across Ireland, Britain and elsewhere in Europe.[345]

Bobolink

Dolichonyx oryzivorus | Bobóilinc | A

(monotypic)

World	North America
Ireland	3 individuals
Cork	2 individuals
Conservation	Not globally threatened

One at CCBO on 13–24 September 1982. One at CCBO on 10 October 2003.

Baltimore Oriole

Icterus galbula | Óiréal tuaisceartach | A

(monotypic)

World	North America
Ireland	3 individuals
Cork	2 individuals
Conservation	Not globally threatened

First-year male at Baltimore on 7–8 October 2001. First-year at CCBO on 12–19 October 2006.

Black-and-white Warbler

Black-and-white Warbler

Mniotilta varia | Ceolaire dubh is bán | A

(monotypic)

World	North America
Ireland	2 individuals
Cork	1 individual
Conservation	Not globally threatened

One at CCBO on 18 October 1978.

Blue-winged Warbler

Vermivora cyanoptera | Ceolaire gormeiteach | A

(monotypic)

World	South-east North America
Ireland	1 individual
Cork	1 individual
Conservation	Not globally threatened

First-year male at CCBO on 4–10 October 2000.

Northern Parula Warbler

Setophaga americana | Parúl tuaisceartach | A

(monotypic)

World	South-east North America
Ireland	3 individuals
Cork	2 individuals
Conservation	Not globally threatened

First-year male at Firkeel on 19–24 October 1983. Female at Dursey Island on 25 September 1989.

American Yellow Warbler

Setophaga petechia | Ceolaire buí | A

(subspecies: undetermined)

World	North America, Caribbean islands, and north-east South America
Ireland	5 individuals
Cork	3 individuals
Conservation	Not globally threatened

First-year at CCBO on 24–30 August 2008. First-year at Mizen Head on 26–8 August 2008. One at Mizen Head on 21–2 August 2017.

Yellow-rumped Warbler

Setophaga coronata | Ceolaire buíphrompach | A

(subspecies: undetermined)

World	North America
Ireland	17 individuals
Cork	11 individuals (10 records)
Conservation	Not globally threatened

One at CCBO on 7–8 October 1976; feathers found on 9 October indicated it may have been killed by a Sparrowhawk. First-year at CCBO on 19–20 October 1982. Two first-years at CCBO on 10 October 1983, one remained to 19 October. One at CCBO on 5–7 October 1985. Male at CCBO on 8–9 October 1987. First-year at CCBO on 7–15 October 1993. One at CCBO on 2–6 October 2001. First-year at CCBO on 30–1 October 2005. First-year at CCBO on 5–13 October 2010. First-year at Dursey Island on 3–6 October 2012.

Blackpoll Warbler

Setophaga striata | Ceolaire dubhéadanach | A

(monotypic)

World	North America
Ireland	12 individuals
Cork	5 individuals
Conservation	Not globally threatened

One at CCBO on 6–12 October 1976. First-year at CCBO on 24–31 October 1982. First-year at CCBO on 6 October 1984. One at Dursey Island on 10 October 2006. First-year at Garinish on 11–20 October 2009.

American Redstart

Setophaga ruticilla | Earrdeargán Meiriceánach | A

(monotypic)

World	North America
Ireland	2 individuals
Cork	2 individuals
Conservation	Not globally threatened

Male at CCBO on 13–14 October 1968. One at Galley Head on 13–15 October 1985.

Ovenbird

Seiurus aurocapilla | Éan oighinn | A

(subspecies: undetermined)

World	North America
Ireland	3 individuals
Cork	2 individuals
Conservation	Not globally threatened

First-year at Dursey Island on 24–5 September 1990. One at Mizen Head on 27 September 2014.

Northern Waterthrush

Parkesia noveboracensis | Smólach uisce tuaisceartach | A

(monotypic)

World	North America
Ireland	2 individuals
Cork	2 individuals
Conservation	Not globally threatened

One at CCBO on 10–11 September 1983. First-year at CCBO on 27–30 August 2008.

Wilson's Warbler

Cardellina pusilla | Ceolaire Wilson | A

(subspecies: undetermined)

World	North America
Ireland	1 individual
Cork	1 individual
Conservation	Not globally threatened

First-year male at Dursey Island on 18–21 September 2013.

Potential first records for Cork pending acceptance by the Irish Rare Birds Committee (1 October 2021)

Northern Harrier *Circus hudsonius* Cromán Mheiriceánach
Barry's Head: 2–15 October 2020.

Collared Pratincole *Glareola pratincola* Pratancól muinceach
Gearagh: 18–20 April 2020.

Semipalmated Plover *Charadrius semipalmatus* Feadóg mionbhosach
Lissagriffin: 2 July and 1 August 2021; same at Galley Cove on 1 October 2021.

Greater Sand Plover *Charadrius leschenaultii* Feadóg ghainimh
Schull: 21 April 2021.

Short-billed Dowitcher *Limnodromus griseus* Guilbnín gobghearr
Clonakilty: 3–27 July 2021.

Belted Kingfisher *Megaceryle alcyon* Cruidín creasa
Beara peninsula: 9 November 2020–29 April 2021.

APPENDIX 1

Category D and E Species

Species and records assigned to categories D and E do not form part of the main Cork List. This appendix is likely to under-represent the full range of species that have escaped or been released from captivity. The authors would welcome additions to the list, including new information on the species mentioned, so that population changes in the future can be tracked. Records of wildfowl, gamebirds, birds of prey, parrots and ship-assisted species would be particularly valuable. The origin of many species in and near Cork Harbour is suspected (or known) to be Fota Wildlife Park. The reader is referred to Chapter 7 for further information on introductions and escapes under Mute Swan, Pink-footed Goose, Greylag Goose, Barnacle Goose, Mallard, Red Grouse, Capercaillie, Grey Partridge, Quail, Pheasant, Red Kite, White-tailed Eagle, Golden Eagle, Rock Dove/Feral Pigeon.

Black Swan *Cygnus atratus* **E**

Male bred with female Mute Swan, Castlemartyr, 1843. One, Courtmacsherry, 12–15 January 1992. One, Clonakilty, 18 March 2000. One, CCBO, 25 March–3 April 2003 when found dead. Black Swans were present at two 10 km squares during 2007–11 (W37, River Lee, near Macroom; W67, Cork Lough), but breeding was not proved. These may be the same occurrences referred to by Lever, although no locations were given. Black Swans are known to have occurred at other sites in and around Cork Harbour, and elsewhere, in recent years, but details are lacking.

Swan Goose *Anser cygnoides* **E**

Swan Geese were present at one 10 km square during 2007–11 (W67, Cork Lough), but breeding was not proved.

Bar-headed Goose *Anser indicus* **E**

Two, Fota area, 1991.

Emperor Goose *Anser canagicus* **E**

Two, Fota area, 1991.

Canada Goose *Branta canadensis* D4

The earliest records relate to the winters of 1844/5 and 1845/6 when a flock of six occurred near Inishannon. Another was shot at the River Lee (near Cork city) in January 1854. These had probably escaped from private waterfowl collections.

There was no further mention of this goose until 1962. In October of that year several species of non-native waterfowl, including three pairs of Canada Geese (believed to have been transported from Dublin Zoo) were released at Cork Lough. The geese prospered, and by 1969 about fifty-five were present.

Moult migration flights have become established elsewhere in introduced Canada Goose populations. Cork Lough birds quickly developed a short (25 km) annual summer migration flight to Ballycotton where peak numbers (55 birds) occurred from July to October. In some years a few birds wintered there, but in most years a majority returned to Cork Lough after October. This migration was well established by the late 1960s. These geese also occasionally visited other sites around the east of the county, especially Douglas estuary, Lough Aderry and Ballymacoda, and more recently Rostellan, Slatty Bridge, Cuskinny, and sometimes other sites at Cork Harbour.

The total Cork population had grown only slightly by 1994 to sixty-three adults (and six goslings) at three sites; only two sites, including Cork Lough, held breeding birds. Expansion of range extended only a few kilometres south and west of Cork city, possibly the result of introductions since 1969. Numbers at Cork Lough decreased thereafter. It is highly unlikely the Cork population was ever self-sustaining given its small size, slow growth rate, recent decline, and the considerable food input (at Cork Lough) from local people. Greatest population growth and stability occurred during the period from the late 1960s to the mid-1990s.

Canada Geese have not managed to colonise any site outside of the Cork city area, and records outside of that area usually include only one or two birds at sites such as Timoleague, Dunmanway, the Gearagh and Kilcolman. Six flew over CCBO on 3 October 1999, and one flew past Dursey Island on 25 June 2002. Numbers have declined further in recent years and the peak at Cork Lough was eleven in 2005, and nine in 2006.

Cackling Goose *Branta hutchinsii* D1

One, Lissagriffin, 31 March–7 April 1985.

Upland Goose *Chloephaga picta* **E**
One pair was released at Cork Lough in October 1962.

Egyptian Goose *Alopochen aegyptiacus* **E**
One obtained, near Inishannon, 1 March 1877. Egyptian Geese were present at two 10 km squares during 2007–11 (W44, Timoleague; W65, Belgooly), but breeding was not proved.

Ruddy Shelduck *Tadorna ferruginea* **D1**
One, Ballycotton, 15–25 October 1995. One, Ballycotton, 23–29 April 1996. One, Galley Head and Rosscarbery, 24 July 2005. The Irish Rare Birds Committee considers all recent Irish records of the Ruddy Shelduck to involve birds of doubtful origin. Therefore, no records after 1946 have been accepted as referring to wild birds.

Cape Shelduck *Tadorna cana* **E**
One, Inchydoney, 7–28 January 1986.

Wood Duck *Aix sponsa* **E**
Robert Ball knew of this species kept on ponds in and about Cork, and two were obtained at the River Owenboy near its confluence with Cork Harbour on 10 October 1845. Three pairs were released at Cork Lough in October 1962.

Mandarin Duck *Aix galericulata* **E**
Five (one obtained), River Bandon, 1878. One obtained, River Owennacurra, 11 December 1879. One pair released, Cork Lough, October 1962; two pairs, Cork Lough, 1987; one or two pairs, Cork Lough, late 2000s. Pair displaying, near Kealkill, 7 March 1987. Two pairs, Farran Wood, 1987. One, near Fota, March 1995. Male, Belvelly, 13 November 1995. One, Lee Fields, 23 July 2014. One, the Gearagh, September 2018; two, the Gearagh, February 2021. Mandarin Ducks were present at one 10 km square during 2007–11 (W45, Bandon), but breeding was not proved.

Chiloé Wigeon *Anas sibilatrix* **E**
One pair was released at Cork Lough in October 1962.

Red-crested Pochard *Netta rufina* **D1**
Male, Ballyhonock Lake, 5–14 April, and Ballycotton, 20 May–7 June

1998. Male, Rostellan, 3 August–13 September 1998. Male, Atlantic Pond, 15 November 1998–2004. Male, Atlantic Pond, 15 September 2006. It is possible all these sightings refer to a single individual.

Rosy-billed Duck *Netta peposaca* E

One, Ballycotton, 14 June 1987; one, Belvelly, August 1987–8.

Red-legged Partridge *Alectoris rufa* E

Many attempts at introduction of this species to Ireland have been made by game shooting interests. Several of these appear to have been unsuccessful, except perhaps for some in the eastern half of Ireland. Introductions have been made by gun clubs in Cork (e.g. Rathcormack). However, it is unlikely that any introduction has led to self-sustaining populations. Most introductions appear to be made on a put-and-take basis for shooting. Records are absent from all but one of the atlas projects. Birds were recorded present in winter at Castletownroche in one 10 km square during 2007–11. Individuals occur from time to time, for example one at Garrettstown in October 1994. One was reported near Ladysbridge in June 2014. Singles were seen several times between Ballymacoda and Garryvoe during June 2017 to 2019. These records could represent local escapes or wandering birds from more distant introductions.

Golden Pheasant *Chrysolophus pictus* E

Male, Hollymount Wood, Lee Road, Cork city, 9 April 2011.

Great White Pelican *Pelecanus onocrotalus* E

Two flew across the N25 at Brown Island in mid-October 2018; they were seen frequently at various sites around Cork Harbour thereafter, and one was killed by a wind turbine near Ringaskiddy. One flew across the N25 near Castlemartyr on 28 October 2018.

Brown Booby *Sula leucogaster* D3

Adult dead at tideline, Owenahincha, 2 January 2016.

Sacred Ibis *Threskionis aethiopicus* E

During the 1980s and 1990s the Sacred Ibis bred, at least occasionally, at Cork Lough and in a heronry at Little Island. During this period up to three birds were frequently seen at various sites around Cork Harbour, such as Carrigaline, Kinsale Road landfill, Douglas estuary, Atlantic Pond and Belvelly.

Greater Flamingo *Phoenicopterus roseus* D1

One, near Timoleague, 1 May–4 June 1938. This individual may have been the same as that present in Londonderry in April 1938 and Limerick in November 1938.

Chilean Flamingo *Phoenicopterus chilensis* E

One, Ballycotton, 13–14 October 1983.

Goshawk *Accipiter gentilis* E

An escaped falconer's bird was reported in the early 2000s.

Red-tailed Hawk *Buteo jamaicensis* E

One, Ballymaloe, 12 December 1999 and 2 January 2000.

Indian Peafowl *Pavo cristatus* E

Indian Peafowl were present at two 10 km squares during 2007–11 (W59, Mallow; W67, Blarney), but breeding was not proved.

Helmeted Guineafowl *Numida meleagris* E

Helmeted Guineafowl were present at two 10 km squares during 2007–11 (W24, Drinagh; W67, Blarney), but breeding was not proved.

Kentish Plover *Charadrius alexandrinus* E

One, Redbarn Strand, 2 December 2007–23 January 2008. This bird had been reared in captivity (illegally) in Germany. It was subsequently ringed and released at Greetsiel on the northwest German coast before turning up at Redbarn Strand.

African Collared Dove *Streptopelia roseogrisea* E

One domestic type often referred to as *Streptopelia risoria* was at CCBO on 20 October 1960.

Cockatiel *Nymphicus hollandicus* E

One, Cork city, 28 April 1984.

Ring-necked Parakeet *Psittacula krameri* **E**

One or two, Carrigaline, November 1999, February 2000, March and September 2003, February, March and December 2006. Ring-necked Parakeets were present at two 10 km squares during 2007–11 (W75, Minane Bridge; W76, Carrigaline), but breeding was not proved.

Tawny Owl *Strix aluco* **E**

Two captured near Cork, one in 1884, the other in 1892.

Northern Flicker *Colaptes auratus* **D2**

A male was on board RMS *Mauretania* at Cork Harbour on 13 October 1962; it flew ashore to Roche's Point when the vessel anchored.

House Crow *Corvus splendens* **D2**

One, Cobh, 5 September 2010–21 July 2012.

Song Sparrow *Melospiza melodia* **D2**

Two were on board RMS *Mauretania* at Cork Harbour on 13 October 1962.

White-throated Sparrow *Zonotrichia albicollis* **D2**

Three were on board RMS *Mauretania* at Cork Harbour on 13 October 1962.

Field Sparrow *Spizella pusilla* **D2**

One was on board RMS *Mauretania* at Cork Harbour on 13 October 1962.

Dark-eyed Junco *Junco hyemalis* **D2**

One was on board RMS *Mauretania* at Cork Harbour on 13 October 1962.

Red-headed Bunting *Emberiza bruniceps* **D1**

Male, CCBO, 26 July 1962. Male, Ballycotton, 23–4 June 1964. Male, CCBO, 14 September 1964. Male, CCBO, 26–7 August 1968. Male, CCBO, 14 June 1969. Male, CCBO, 26 May 1978.

Baltimore Oriole *Icterus galbula* **D2**

One was on board RMS *Mauretania* at Cork Harbour on 13 October 1962.

APPENDIX 2

First Records for Cork

First records of the 427 species on the Cork List to 31 December 2018. Categories (A, B, C) are shown, and species marked with an asterisk represent first records for Ireland. For vagrant species that remained for longer than one day, only the first date of occurrence is given. For some records, particularly the earlier ones, the source reference given is not necessarily the first mention in the literature of the record concerned but is usually the first mention of the species where the complete (or near complete) record is given either in the standard references for the county or the country. For records since 1953 the source reference given is almost always to the annual *Irish Bird Report* or the journal *Irish Birds*, regardless of whether the record had been published elsewhere earlier.

Species	**Place**	**Date**	**Source**
Mute Swan (A)	County Cork	Pre 1750	Smith 1750
Bewick's Swan (A)	River Bandon	Early 1879	Ussher 1894
Whooper Swan (A)	Ahanesk	28 January 1879	Payne-Gallwey 1882
Taiga Bean Goose (A)	County Cork	Pre 1845	Harvey 1845
Tundra Bean Goose (A)	Kilcolman	3 March 2006	*Irish Birds* 8: 396
Pink-footed Goose (A)	Ballycotton	2 November 1974	*Irish Bird Report* 22: 10
Greater White-fronted Goose (A)	County Cork	Pre 1845	Harvey 1845
Greylag Goose (A)	County Cork	Pre 1750	Smith 1750
Barnacle Goose (A)	County Cork	Pre 1845	Harvey 1845
Brent Goose (A)	County Cork	Pre 1750	Smith 1750
Ruddy Shelduck (B)	Clonakilty	January 1871	Harvey 1875
Common Shelduck (A)	County Cork	Pre 1750	Smith 1750
Common Wigeon (A)	County Cork	Pre 1750	Smith 1750
American Wigeon (A)	Ballycotton	8 April 1985	*Irish Birds* 3: 299
Gadwall (A)	River Lee	December 1849	Harvey 1875
Common Teal (A)	County Cork	Pre 1750	Smith 1750
Green-winged Teal (A)	Midleton	16 March 1968	*Irish Bird Report* 16: 17–8
Mallard (A)	County Cork	Pre 1750	Smith 1750

Species	Place	Date	Source
American Black Duck (A)	Ballycotton	26 January 1993	*Irish Birds* 5: 331
Pintail (A)	County Cork	Pre 1845	Harvey 1845
Garganey (A)	County Cork	Pre 1845	Harvey 1845
Blue-winged Teal (A) *	Ballycotton	9 September 1910	Kennedy et al. 1954
Shoveler (A)	County Cork	Pre 1845	Harvey 1845
Red-crested Pochard (A)	Reenydonagan Lake	29 December 1927	Kennedy et al. 1954
Common Pochard (A)	County Cork	Pre 1750	Smith 1750
Redhead (A) *	CCBO	12 July 2003	*Irish Birds* 7: 552
Ring-necked Duck (A)	Lough Skahanagh	10 February 1974	*Irish Bird Report* 22: 8
Ferruginous Duck (A)	Kilcolman	20 January 1991	*Irish Birds* 4: 581
Tufted Duck (A)	County Cork	Pre 1845	Harvey 1845
Greater Scaup (A)	County Cork	Pre 1845	Harvey 1845
Lesser Scaup (A)	Dooniskey	4 January 2004	*Irish Birds* 8: 109
Common Eider (A)	Cork Harbour	December 1878	Payne-Gallwey 1882
King Eider (A)	Baltimore	29 January 1959	*Irish Bird Report* 7: 7
Long-tailed Duck (A)	Cork Harbour	January 1878	Payne-Gallwey 1882
Common Scoter (A)	County Cork	Pre 1845	Harvey 1845
Surf Scoter (A)	Crookhaven Bay	5 November 1888	Barrington 1900
Velvet Scoter (A)	Youghal	1 March 1850	Thompson 1851
Bufflehead (A) *	Gearagh	18 January 1998	*Irish Birds* 6: 386
Common Goldeneye (A)	County Cork	Pre 1845	Harvey 1845
Hooded Merganser (B) *	East Ferry	December 1878	Payne-Gallwey 1882
Smew (A)	Enniskean	1 March 1895	Ussher and Warren 1900
Red-breasted Merganser (A)	County Cork	Pre 1845	Harvey 1845
Goosander (A)	County Cork	Pre 1845	Harvey 1845
Ruddy Duck (C)	Ballycotton	7 January 1979	*Irish Birds* 3: 463–4
Red Grouse (A)	County Cork	Pre 1750	Smith 1750
Capercaillie (B)	County Cork	Pre 1750	Smith 1750
Grey Partridge (B)	County Cork	Pre 1750	Smith 1750
Quail (A)	County Cork	Pre 1750	Smith 1750
Pheasant (C)	County Cork	Pre 1750	Smith 1750
Red-throated Diver (A)	County Cork	Pre 1845	Harvey 1845

Species	Place	Date	Source
Black-throated Diver (A)	CCBO	9 April 1962	*Irish Bird Report* 10: 6
Pacific Diver (A)	Crookhaven	18 January 2018	*Irish Birds* 42: 74
Great Northern Diver (A)	County Cork	Pre 1750	Smith 1750
White-billed Diver (A) *	Lough Hyne	3 February 1974	*Irish Bird Report* 22: 5
Black-browed Albatross (A) *	CCBO	24 September 1963	*Irish Bird Report* 11: 8
Fulmar (A) *	Inchydoney	1832	Thompson 1851
Bulwer's Petrel (A) *	Galley Head	1 August 2013	*Irish Birds* 10: 397
Cory's Shearwater (A)	CCBO	10 August 1962	*Irish Bird Report* 10: 7
Great Shearwater (A)	Off Youghal	December 1854	Ussher and Warren 1900
Sooty Shearwater (A)	Off Cork Harbour	24 August 1849	Ussher and Warren 1900
Manx Shearwater (A)	County Cork	Pre 1750	Smith 1750
Balearic Shearwater (A)	CCBO	30 March 1961	*Irish Bird Report* 9: 5
Barolo Shearwater (A) *	Near Bull Rock	6 May 1853	Ussher and Warren 1900
Wilson's Storm Petrel (A)	13 km off Mizen Head	29 August 1987	*Irish Birds* 3: 613
European Storm Petrel (A)	County Cork	Pre 1845	Harvey 1845
Leach's Storm Petrel (A) *	County Cork	September 1818	Thompson 1851
Gannet (A)	County Cork	Pre 1750	Smith 1750
Great Cormorant (A)	County Cork	Pre 1750	Smith 1750
Shag (A)	County Cork	Pre 1750	Smith 1750
Great Bittern (A)	County Cork	Pre 1750	Smith 1750
American Bittern (A)	Myross Wood	Early October 1875	Payne-Gallwey 1882
Little Bittern (A)	Carrigrohane	Summer 1842	Harvey 1845
Night Heron (A)	Castlefreke	Pre 1845	Harvey 1845
Green Heron (A) *	Schull	11 October 2005	*Irish Birds* 8: 377–8
Squacco Heron (A) *	Killeagh	26 May 1849	Thompson 1850
Cattle Egret (A) *	Kilkerran Lake	7 March 1976	*Irish Birds* 1: 73
Little Egret (A) *	Skibbereen	26 May 1940	Kennedy et al. 1954
Great Egret (A)	CCBO	26 October 1997	*Irish Birds* 6: 291
Grey Heron (A)	County Cork	Pre 1750	Smith 1750
Purple Heron (A)	CCBO	2 May 1965	*Irish Bird Report* 13: 14–5
White Stork (A) *	Fermoy	Late May 1846	Thompson 1850
Glossy Ibis (A)	County Cork	Pre 1850	Thompson 1850

Species	Place	Date	Source
Spoonbill (A)	Youghal	Autumn 1829	Thompson 1850
Pied-billed Grebe (A)	Rostellan	1 February 1997	*Irish Birds* 6: 287
Little Grebe (A)	County Cork	Pre 1750	Smith 1750
Great Crested Grebe (A)	County Cork	Pre 1845	Harvey 1845
Red-necked Grebe (A)	Glengarriff	December 1842	Thompson 1851
Slavonian Grebe (A)	Cork Harbour	Early 1879	Ussher 1894
Black-necked Grebe (A)	Barony of Muskerry	1847	Thompson 1851
Honey Buzzard (A)	CCBO	23 August 1972	*Irish Bird Report* 20: 13
Black Kite (A) *	Garryvoe	20 April 1980	*Irish Birds* 6: 71
Red Kite (A)	County Cork	Pre 1750	Smith 1750
White-tailed Eagle (B)	County Cork	Pre 1750	Smith 1750
Griffon Vulture (B) *	Cork Harbour	Spring 1843	Harvey 1845
Marsh Harrier (A)	County Cork	Pre 1845	Harvey 1845
Hen Harrier (A)	County Cork	Pre 1845	Harvey 1845
Pallid Harrier (A) *	Ballyvergan Marsh	22 April 2011	*Irish Birds* 9: 459
Montagu's Harrier (A)	County Cork	Summer 1957	Hutchinson 1989
Goshawk (A)	County Cork	Pre 1750	Smith 1750
Sparrowhawk (A)	County Cork	Pre 1750	Smith 1750
Common Buzzard (A)	County Cork	Pre 1750	Smith 1750
Rough-legged Buzzard (A)	Mitchelstown	18 November 1906	Kennedy et al. 1954
Greater Spotted Eagle (B) *	Castlemartyr-Claycastle	January 1845	Thompson 1849
Golden Eagle (B)	County Cork	Pre 1845	Harvey 1845
Osprey (A)	Blackrock	14 October 1848	Thompson 1851
Common Kestrel (A)	County Cork	Pre 1750	Smith 1750
Red-footed Falcon (A)	CCBO	31 May 1991	*Irish Birds* 4: 584
Merlin (A)	County Cork	Pre 1845	Harvey 1845
Hobby (A) *	Carrigrohane	Summer 1822 (?)	Thompson 1849
Gyr Falcon (A)	Blackrock	23 November 1883	Ussher and Warren 1900
Peregrine Falcon (A)	County Cork	Pre 1750	Smith 1750
Water Rail (A)	County Cork	Pre 1750	Smith 1750
Spotted Crake (A)	Claycastle	October 1843	Thompson 1850
Baillon's Crake (B) *	Claycastle	30 October 1845	Thompson 1850

Species	Place	Date	Source
Corncrake (A)	County Cork	Pre 1750	Smith 1750
Moorhen (A)	County Cork	Pre 1750	Smith 1750
Common Coot (A)	County Cork	Pre 1750	Smith 1750
American Coot (A) *	Ballycotton	7 February 1981	*Irish Birds* 2: 209
Common Crane (A)	County Cork	Pre 1750	Smith 1750
Sandhill Crane (B) *	Castlefreke	14 September 1905	Kennedy et al. 1954
Little Bustard (B)	Ballycotton	24 December 1860	Ussher and Warren 1900
Great Bustard (B)	Castletownbere	9 December 1925	Kennedy et al. 1954
Oystercatcher (A)	County Cork	Pre 1750	Smith 1750
Black-winged Stilt (A) *	Youghal	Winter 1823 or 1824	Thompson 1850
Avocet (A)	Ringaskiddy	Pre 1830	Ussher and Warren 1900
Stone Curlew (A)	Castletownshend	24 February 1913	Kennedy et al. 1954
Little Ringed Plover (A)	Ballycotton	22 September 1968	*Irish Bird Report* 16: 25
Common Ringed Plover (A)	County Cork	Pre 1750	Smith 1750
Killdeer (A)	Crookhaven	30 November 1938	Kennedy et al. 1954
Kentish Plover (A)	Ballycotton	23 April 1970	*Irish Bird Report* 18: 29
Lesser Sand Plover (A) *	Ballymacoda	27 July 2013	*Irish Birds* 10: 81–2
Dotterel (A)	Near Cork city	September 1844	Thompson 1850
American Golden Plover (A)	Ballycotton	10 September 1971	*Irish Birds* 10: 244
Pacific Golden Plover (A)	Kinsale Marsh	5 October 1991	*Irish Birds* 5: 88
European Golden Plover (A)	County Cork	Pre 1750	Smith 1750
Grey Plover (A)	County Cork	Pre 1750	Smith 1750
Northern Lapwing (A)	County Cork	Pre 1750	Smith 1750
Red Knot (A)	Cork Harbour	Pre 1850	Thompson 1850
Sanderling (A)	Claycastle	Pre 1850	Thompson 1850
Semipalmated Sandpiper (A) *	Ballycotton	16 October 1966	*Irish Bird Report* 14: 36–8
Western Sandpiper (A)	Ballydehob	1 September 1999	*Irish Birds* 6: 553
Red-necked Stint (A) *	Ballycotton	2 July 1998	*Irish Birds* 6: 389
Little Stint (A)	Crookhaven	6 September 1932	Kennedy et al. 1954
Temminck's Stint (A)	Ballycotton	15 September 1981	*Irish Birds* 2: 212
Long-toed Stint (A) *	Ballycotton	15 June 1996	*Irish Birds* 6: 73
Least Sandpiper (A)	Clonakilty	13 September 1966	*Irish Bird Report* 14: 32–4

Species	Place	Date	Source
White-rumped Sandpiper (A)	CCBO	8 September 1966	*Irish Bird Report* 14: 35–6
Baird's Sandpiper (A)	Ballycotton	24 September 1966	*Irish Bird Report* 14: 34–5
Pectoral Sandpiper (A)	Ballycotton	2 October 1963	*Irish Bird Report* 11: 21–2
Sharp-tailed Sandpiper (A) *	Ballycotton	1 July 1971	*Irish Birds* 10: 558–9
Curlew Sandpiper (A)	Cork Harbour	29 October 1847	Thompson 1850
Stilt Sandpiper (A)	Ballycotton	14 July 1979	*Irish Birds* 1: 564
Purple Sandpiper (A)	Youghal	Pre 1850	Thompson 1850
Dunlin (A)	County Cork	Pre 1845	Harvey 1845
Broad-billed Sandpiper (A)	Charleville	2 June 1978	*Irish Birds* 3: 623
Buff-breasted Sandpiper (A)	CCBO	27 August 1960	*Irish Bird Report* 8: 15
Ruff (A)	Buttevant	8 October 1889	Ussher 1894
Jack Snipe (A)	County Cork	Pre 1845	Harvey 1845
Common Snipe (A)	County Cork	Pre 1750	Smith 1750
Great Snipe (A)	Clonakilty	17 November 1879	Ussher and Warren 1900
Long-billed Dowitcher (A)	Ballycotton	8 October 1966	*Irish Birds* 3: 650
Woodcock (A)	County Cork	Pre 1750	Smith 1750
Black-tailed Godwit (A)	Fota Island	September 1839	Ussher 1894
Bar-tailed Godwit (A)	County Cork	Pre 1750	Smith 1750
Common Whimbrel (A)	County Cork	Pre 1750	Smith 1750
Hudsonian Whimbrel (A)	Mizen Head	20 September 2011	*Irish Birds* 9: 470
Common Curlew (A)	County Cork	Pre 1750	Smith 1750
Upland Sandpiper (A)	Newcestown	4 September 1894	Ussher and Warren 1900
Common Sandpiper (A)	County Cork	Pre 1845	Harvey 1845
Spotted Sandpiper (A)	Union Hall	28 September 1978	*Irish Birds* 3: 310
Green Sandpiper (A)	Youghal	October 1822 (?)	Thompson 1850
Solitary Sandpiper (A)	Lissagriffin	5 September 1971	*Irish Bird Report* 19: 36–7
Spotted Redshank (A)	Fota Island	26 December 1898	Ussher and Warren 1900
Greater Yellowlegs (A) *	Aghadown	21 January 1940	Kennedy et al. 1954
Greenshank (A)	County Cork	Pre 1845	Harvey 1845
Lesser Yellowlegs (A)	Ballycotton	17 September 1967	*Irish Bird Report* 15: 29
Marsh Sandpiper (A)	Great Island	20 August 1999	*Irish Birds* 7: 397
Wood Sandpiper (A)	CCBO	4 August 1960	*Irish Bird Report* 8: 11–12

Species	Place	Date	Source
Common Redshank (A)	County Cork	Pre 1750	Smith 1750
Turnstone (A)	County Cork	Pre 1845	Harvey 1845
Wilson's Phalarope (A)	Ballycotton	10 September 1967	*Irish Bird Report* 15: 31–2
Red-necked Phalarope (A)	CCBO	16 September 1960	*Irish Bird Report* 8: 16
Grey Phalarope (A)	Youghal (?)	Winter 1841/2	Harvey 1845
Pomarine Skua (A) *	Off Youghal	12 October 1834	Thompson 1851
Arctic Skua (A)	County Cork	Pre 1750	Smith 1750
Long-tailed Skua (A)	Kilbrittain-Kinsale	22 August 1960	*Irish Bird Report* 8: 16
Great Skua (A)	Whiddy Island	Winter 1845/6	Thompson 1851
Ivory Gull (A)	Bantry Bay	31 January 1852	Harvey 1875
Sabine's Gull (A)	Crosshaven	31 August 1962	*Irish Bird Report* 10: 19
Kittiwake (A)	County Cork	Pre 1845	Harvey 1845
Bonaparte's Gull (A)	Dunkettle-Blackrock	15 April 1997	*Irish Birds* 6: 302
Black-headed Gull (A)	County Cork	Pre 1845	Harvey 1845
Little Gull (A)	Cobh	23 February 1888	*Zoologist* (3) 13: 22–3
Ross's Gull (A)	Cobh	24 February 1985	*Irish Birds* 3: 316
Laughing Gull (A) *	Tivoli	12 August 1968	*Irish Bird Report* 16: 36–7
Franklin's Gull (A)	Kinsale	11 January 1999	*Irish Birds* 6: 560
Mediterranean Gull (A)	Old Head of Kinsale	5 October 1964	*Irish Bird Report* 12: 27
Common Gull (A)	County Cork	Pre 1750	Smith 1750
Ring-billed Gull (A)	Ballycotton	20 September 1981	*Irish Birds* 2: 218
Lesser Black-backed Gull (A)	County Cork	Pre 1845	Harvey 1845
European Herring Gull (A)	County Cork	Pre 1750	Smith 1750
Yellow-legged Gull (A)	Rosscarbery	12 October 1955	*Irish Bird Report* 3: 17
Caspian Gull (A)	Youghal	24 February 2007	*Irish Birds* 8: 597–9
American Herring Gull (A) *	Cobh	16 November 1986	*Irish Birds* 4: 246
Iceland Gull (A)	Cork Harbour	25 January 1849	Thompson 1851
Thayer's Gull (A) *	Cork Lough	21 February 1990	*Irish Birds* 7: 231
Glaucous-winged Gull (A) *	Castletownbere	2 January 2016	*Irish Birds* 10: 561
Glaucous Gull (A)	Youghal	Autumn 1833	Thompson 1851
Great Black-backed Gull (A)	County Cork	Pre 1750	Smith 1750
Little Tern (A)	Cork Harbour	Pre 1851	Thompson 1851

Species	Place	Date	Source
Gull-billed Tern (A)	Ballycotton	29 April 1993	*Irish Birds* 5: 223
Caspian Tern (A)	Ballycotton	7 August 1988	*Irish Birds* 4: 100
Whiskered Tern (A)	Ballycotton	18 May 1968	*Irish Bird Report* 16: 37–8
Black Tern (A) *	Roxborough (Midleton)	July 1819 (?)	Thompson 1851
White-winged Tern (A)	Glandore	31 July 1936	Kennedy et al. 1954
Elegant Tern (A)	Ballymacoda	1 August 1982	*Irish Birds* 3: 632
Sandwich Tern (A)	Monkstown	October 1852	Ussher 1894
Royal Tern (A) *	Clonakilty	7 June 2009	*Irish Birds* 9: 268
Lesser Crested Tern (A) *	Ballycotton	7 August 1996	*Irish Birds* 6: 81
Forster's Tern (A)	Ballycotton	11 January 2006	*Irish Birds* 8: 404
Common Tern (A)	County Cork	Pre 1750	Smith 1750
Roseate Tern (A)	Roancarrigmore	Summer 1955	Kennedy 1961
Arctic Tern (A)	County Cork	Pre 1845	Harvey 1845
Common Guillemot (A)	County Cork	Pre 1845	Harvey 1845
Razorbill (A)	County Cork	Pre 1750	Smith 1750
Black Guillemot (A)	County Cork	Pre 1845	Harvey 1845
Little Auk (A)	Cork Harbour	Pre 1845	Harvey 1845
Puffin (A)	County Cork	Pre 1750	Smith 1750
Pallas's Sandgrouse (B)	Mallow	June 1888	Ussher and Warren 1900
Rock Dove and Feral Pigeon (A)	County Cork	Pre 1750	Smith 1750
Stock Dove (A)	Timoleague	1925	Kennedy et al. 1954
Woodpigeon (A)	County Cork	Pre 1750	Smith 1750
Collared Dove (A)	Ballinacurra	July 1962	Hudson 1965
Turtle Dove (A)	County Cork	Pre 1845	Harvey 1845
Mourning Dove (A)	Garinish	25 October 2009	*Irish Birds* 9: 270
Great Spotted Cuckoo (A)	Ringaskiddy	15 February 2009	*Irish Birds* 9: 270
Common Cuckoo (A)	County Cork	Pre 1750	Smith 1750
Yellow-billed Cuckoo (A) *	Youghal	Autumn 1825	Harvey 1845
Barn Owl (A)	County Cork	Pre 1750	Smith 1750
Scops Owl (A)	Fastnet Rock	6 May 1907	Kennedy et al. 1954
Snowy Owl (A)	Inchigeelagh	September 1827	Harvey 1845
Long-eared Owl (A)	County Cork	Pre 1750	Smith 1750

Species	Place	Date	Source
Short-eared Owl (A)	County Cork	Pre 1845	Harvey 1845
Nightjar (A)	County Cork	Pre 1845	Harvey 1845
Common Nighthawk (A) *	Ballydonegan	24 October 1999	*Irish Birds* 6: 563
Chimney Swift (A) *	CCBO	23 October 1999	*Irish Birds* 6: 563–4
Needletail Swift (A) *	CCBO	20 June 1964	*Irish Bird Report* 12: 30
Common Swift (A)	County Cork	Pre 1750	Smith 1750
Little Swift (A) *	CCBO	12 June 1967	*Irish Bird Report* 15: 36–7
Alpine Swift (A) *	15 km off CCBO	Summer 1829	*Irish Birds* 10: 251
Common Kingfisher (A)	County Cork	Pre 1750	Smith 1750
Bee-eater (A)	Trabolgan-Whitegate	April/May 1888	More 1890
Roller (A)	Dunmanway	September 1851	Harvey 1875
Hoopoe (A)	County Cork	Pre 1750	Smith 1750
Wryneck (A)	Fastnet Rock	17 September 1898	Barrington 1900
Yellow-bellied Sapsucker (A) *	CCBO	16 October 1988	*Irish Birds* 4: 102–3
Great Spotted Woodpecker (A)	County Cork	1862	Ussher 1894
Philadelphia Vireo (A) *	Galley Head	6 October 1985	*Irish Birds* 3: 327
Red-eyed Vireo (A)	CCBO	6 October 1967	*Irish Bird Report* 15: 41–2
Golden Oriole (A) *	Midleton-Castlemartyr	Summer 1817 (?)	Thompson 1849
Isabelline Shrike (A)	Old Head of Kinsale	17 October 2006	*Irish Birds* 8: 414
Red-backed Shrike (A)	Fastnet Rock	26 September 1910	Kennedy et al. 1954
Lesser Grey Shrike (A)	Ballycotton	6 September 1985	*Irish Birds* 3: 643
Great Grey Shrike (A)	Near Cork city	1834	Harvey 1845
Woodchat Shrike (A)	Whitegate	11 May 1951	Kennedy et al. 1954
Chough (A)	County Cork	Pre 1750	Smith 1750
Magpie (A)	County Cork	Pre 1750	Smith 1750
Jay (A)	County Cork	Pre 1750	Smith 1750
Jackdaw (A)	County Cork	Pre 1750	Smith 1750
Rook (A)	County Cork	Pre 1750	Smith 1750
Carrion Crow (A)	County Cork	Pre 1750	Smith 1750
Hooded Crow (A)	County Cork	Pre 1750	Smith 1750
Raven (A)	County Cork	Pre 1750	Smith 1750
Goldcrest (A)	County Cork	Pre 1845	Harvey 1845

Species	Place	Date	Source
Firecrest (A) *	Glengarriff	7 December 1943	Kennedy et al. 1954
Ruby-crowned Kinglet (A) *	CCBO	27 October 2013	*Irish Birds* 10: 92–3
Blue Tit (A)	County Cork	Pre 1750	Smith 1750
Great Tit (A)	County Cork	Pre 1750	Smith 1750
Coal Tit (A)	County Cork	Pre 1750	Smith 1750
Bearded Tit (A)	CCBO	13 October 1972	*Irish Bird Report* 20: 20–3
Greater Short-toed Lark (A)	CCBO	14 September 1962	*Irish Bird Report* 10: 21–2
Woodlark (A)	County Cork	Pre 1750	Smith 1750
Skylark (A)	County Cork	Pre 1750	Smith 1750
Shore Lark (A)	CCBO	13 April 2007	*Irish Birds* 8: 602
Sand Martin (A)	County Cork	Pre 1750	Smith 1750
Barn Swallow (A)	County Cork	Pre 1750	Smith 1750
House Martin (A)	County Cork	Pre 1750	Smith 1750
Red-rumped Swallow (A)	Old Head of Kinsale	26 April 1987	*Irish Birds* 3: 635
Cetti's Warbler (A) *	Ballymacoda	21 May 2001	*Irish Birds* 7: 407
Long-tailed Tit (A)	County Cork	Pre 1750	Smith 1750
Greenish Warbler (A)	CCBO	29 August 1961	*Irish Birds* 3: 651
Arctic Warbler (A)	CCBO	8 September 1968	*Irish Bird Report* 16: 44
Pallas's Leaf Warbler (A) *	CCBO	23 October 1968	*Irish Bird Report* 16: 45
Yellow-browed Warbler (A)	CCBO	19 September 1959	*Irish Bird Report* 7: 20
Hume's Leaf Warbler (A) *	Knockadoon Head	18 December 2003	*Irish Birds* 7: 567
Radde's Warbler (A)	CCBO	23 October 1988	*Irish Birds* 4: 107
Dusky Warbler (A)	Old Head of Kinsale	31 October 1987	*Irish Birds* 3: 640–1
Western Bonelli's Warbler (A) *	CCBO	2 September 1961	*Irish Bird Report* 9: 23
Wood Warbler (A)	Glengarriff	8 June 1938	Kennedy et al. 1954
Common Chiffchaff (A)	County Cork	Pre 1845	Harvey 1845
Willow Warbler (A)	County Cork	Pre 1845	Harvey 1845
Blackcap (A)	Youghal (?)	Winter 1833	Harvey 1845
Garden Warbler (A)	Sunday's Well	Pre 1849	Thompson 1849
Barred Warbler (A)	CCBO	6 September 1960	*Irish Bird Report* 8: 22
Lesser Whitethroat (A)	CCBO	11 September 1959	*Irish Bird Report* 7: 20
Common Whitethroat (A)	County Cork	Pre 1845	Harvey 1845

Species	**Place**	**Date**	**Source**
Dartford Warbler (A)	CCBO	27 October 1968	*Irish Bird Report* 16: 43
Western Subalpine Warbler (A)	CCBO	6 October 1962	*Irish Bird Report* 10: 25
Sardinian Warbler (A) *	CCBO	10 April 1993	*Irish Birds* 5: 226
Pallas's Grasshopper Warbler (A)	CCBO	8 October 1990	*Irish Birds* 4: 454
Common Grasshopper Warbler (A)	Youghal	Pre 1849	Thompson 1849
Savi's Warbler (A)	Ballyvergan Marsh	17 June 1985	*Irish Birds* 3: 322
Eastern Olivaceous Warbler (A) *	Dursey Island	16 September 1977	*Irish Birds* 1: 267
Booted Warbler (A)	Ballycotton	2 September 2004	*Irish Birds* 8: 118
Sykes's Warbler (A) *	CCBO	17 October 1990	*Irish Birds* 7: 235
Icterine Warbler (A)	CCBO	26 August 1959	*Irish Bird Report* 7: 19
Melodious Warbler (A) *	Old Head of Kinsale	23 September 1905	Kennedy et al. 1954
Aquatic Warbler (A) *	Bull Rock	20 September 1903	Kennedy et al. 1954
Sedge Warbler (A)	County Cork	Pre 1845	Harvey 1845
Paddyfield Warbler (A)	Galley Head	13 October 1991	*Irish Birds* 5: 225
Blyth's Reed Warbler (A) *	CCBO	20 October 2006	*Irish Birds* 8: 604–5
Marsh Warbler (A) *	Ballyvergan Marsh	5 August 1991	*Irish Birds* 5: 469
Common Reed Warbler (A)	CCBO	27 September 1960	*Irish Bird Report* 8: 21
Great Reed Warbler (A) *	Castletownshend	16 May 1920	Kennedy et al. 1954
Fan-tailed Warbler (A) *	CCBO	23 April 1962	*Irish Bird Report* 17: 52
Bohemian Waxwing (A)	Castlemartyr	1820	Harvey 1845
Treecreeper (A)	County Cork	Pre 1845	Harvey 1845
Wren (A)	County Cork	Pre 1750	Smith 1750
Grey Catbird (A) *	CCBO	4 November 1986	*Irish Birds* 3: 480
Common Starling (A)	County Cork	Pre 1750	Smith 1750
Rosy Starling (A)	CCBO	31 August 1961	*Irish Bird Report* 9: 24
Dipper (A)	County Cork	Pre 1750	Smith 1750
White's Thrush (B) *	Bandon	December 1842	Harvey 1845
Siberian Thrush (A) *	CCBO	18 October 1985	*Irish Birds* 3: 322
Hermit Thrush (A) *	Galley Head	25 October 1998	*Irish Birds* 6: 398
Swainson's Thrush (A)	CCBO	14 October 1968	*Irish Bird Report* 16: 41
Grey-cheeked Thrush (A) *	CCBO	19 October 1982	*Irish Birds* 2: 403
Veery (A) *	CCBO	17 October 2018	*Irish Birds* 42: 88–9

Species	Place	Date	Source
Ring Ouzel (A)	County Cork	Pre 1750	Smith 1750
Blackbird (A)	County Cork	Pre 1750	Smith 1750
Fieldfare (A)	County Cork	Pre 1750	Smith 1750
Song Thrush (A)	County Cork	Pre 1750	Smith 1750
Redwing (A)	County Cork	Pre 1750	Smith 1750
Mistle Thrush (A)	Youghal (?)	1818	Thompson 1849
American Robin (A)	Glengarriff	16 January 1977	*Irish Birds* 1: 265
Spotted Flycatcher (A)	County Cork	Pre 1845	Harvey 1845
Rufous Bush Robin (A) *	Old Head of Kinsale	23 September 1876	Ussher and Warren 1900
European Robin (A)	County Cork	Pre 1750	Smith 1750
Thrush Nightingale (A) *	CCBO	29 October 1989	*Irish Birds* 4: 249
Common Nightingale (A)	CCBO	16 April 1968	*Irish Bird Report* 16: 42
Bluethroat (A)	CCBO	17 September 1959	*Irish Bird Report* 7: 19
Red-flanked Bluetail (A) *	Dursey Island	10 November 2009	*Irish Birds* 9: 276
Black Redstart (A) *	Youghal	Autumn 1818	Harvey 1845
Common Redstart (A)	Fastnet Rock	5 October 1887	Barrington 1900
Whinchat (A)	County Cork	Pre 1845	Harvey 1845
European Stonechat (A)	County Cork	Pre 1845	Harvey 1845
Siberian Stonechat (A)	Old Head of Kinsale	27 October 1988	*Irish Birds* 4: 105
Isabelline Wheatear (A) *	Mizen Head	10 October 1992	*Irish Birds* 5: 97
Northern Wheatear (A)	County Cork	Pre 1845	Harvey 1845
Pied Wheatear (A) *	Knockadoon Head	8 November 1980	*Irish Birds* 2: 113
Black-eared Wheatear (A)	CCBO	26 May 1992	*Irish Birds* 5: 97
Desert Wheatear (A)	Red Strand	27 October 1990	*Irish Birds* 4: 453
Red-breasted Flycatcher (A)	Bull Rock	18 November 1903	Kennedy et al. 1954
Taiga Flycatcher (A) *	Galley Head	21 October 2018	*Irish Birds* 42: 90
Pied Flycatcher (A)	Fastnet Rock	5 October 1886	Barrington 1900
Dunnock (A)	County Cork	Pre 1750	Smith 1750
House Sparrow (A)	County Cork	Pre 1750	Smith 1750
Tree Sparrow (A)	Fastnet Rock	2 October 1911	Nichols 1920
Western Yellow Wagtail (A)	Old Head of Kinsale	29 August 1909	Nichols 1920
Eastern Yellow Wagtail (A)	Dursey Island	3 November 2017	*Irish Birds* 43: 97

Species	**Place**	**Date**	**Source**
Citrine Wagtail (A) *	Ballycotton	15 October 1968	*Irish Birds* 2: 403
Grey Wagtail (A)	County Cork	Pre 1750	Smith 1750
Pied Wagtail (A)	County Cork	Pre 1750	Smith 1750
Richard's Pipit (A)	CCBO	14 October 1959	*Irish Bird Report* 20: 29
Tawny Pipit (A)	CCBO	9 October 1959	*Irish Bird Report* 7: 21
Olive-backed Pipit (A)	CCBO	13 October 1990	*Irish Birds* 4: 451
Tree Pipit (A)	CCBO	11 September 1959	*Irish Bird Report* 7: 21
Pechora Pipit (A) *	Garinish	27 September 1990	*Irish Birds* 6: 83
Meadow Pipit (A)	County Cork	Pre 1750	Smith 1750
Red-throated Pipit (A)	CCBO	9 October 1975	*Irish Bird Report* 23: 27
Rock Pipit (A)	County Cork	Pre 1845	Harvey 1845
Water Pipit (A)	Ballycotton	11 November 2007	*Irish Birds* 8: 602
Buff-bellied Pipit (A)	Lissagriffin	5 October 2007	*Irish Birds* 8: 602
Chaffinch (A)	County Cork	Pre 1750	Smith 1750
Brambling (A)	Near Cork city	Winter 1837/8	Thompson 1849
Serin (A)	CCBO	11 November 1975	*Irish Bird Report* 23: 27
Greenfinch (A)	County Cork	Pre 1750	Smith 1750
Siskin (A)	County Cork	Pre 1845	Harvey 1845
Goldfinch (A)	County Cork	Pre 1750	Smith 1750
Linnet (A)	County Cork	Pre 1750	Smith 1750
Twite (A)	County Cork	Pre 1750	Smith 1750
Lesser Redpoll (A)	County Cork	Pre 1750	Smith 1750
Common Redpoll (A)	Dursey Island	10 October 1978	*Cork Bird Report* 1978: 21
Arctic Redpoll (A) *	Dursey Island	4 October 1999	*Irish Birds* 6: 567
Common Crossbill (A)	County Cork	Pre 1750	Smith 1750
Common Rosefinch (A)	CCBO	7 October 1971	*Irish Bird Report* 19: 54
Bullfinch (A)	County Cork	Pre 1750	Smith 1750
Hawfinch (A)	Glengarriff	20 January 1844	Thompson 1851
Snow Bunting (A)	County Cork	Pre 1845	Harvey 1845
Lapland Bunting (A) *	Fastnet Rock	16 October 1887	Barrington 1900
Scarlet Tanager (A)	Firkeel	12 October 1985	*Irish Birds* 3: 329
Rose-breasted Grosbeak (A) *	CCBO	7 October 1962	*Irish Bird Report* 10: 28

Species	Place	Date	Source
Indigo Bunting (A) *	CCBO	9 October 1985	*Irish Birds* 3: 331
White-crowned Sparrow (A) *	Ballynacarriga (Beara)	20 May 2003	*Irish Birds* 7: 572–3
White-throated Sparrow (A) *	CCBO	3 April 1967	*Irish Bird Report* 15: 42–3
Dark-eyed Junco (A)	Dursey Island	9 June 2015	*Irish Birds* 10: 423–4
Yellowhammer (A)	County Cork	Pre 1750	Smith 1750
Cirl Bunting (A) *	Mizen Head	8 May 2006	*Irish Birds* 8: 414
Ortolan Bunting (A)	CCBO	30 August 1959	*Irish Bird Report* 7: 23
Rustic Bunting (A) *	CCBO	9 October 1959	*Irish Bird Report* 7: 23
Little Bunting (A)	CCBO	11 October 1973	*Irish Bird Report* 21: 30
Yellow-breasted Bunting (A)	CCBO	11 September 1983	*Irish Birds* 2: 577
Reed Bunting (A)	County Cork	Pre 1845	Harvey 1845
Black-headed Bunting (A)	Skibbereen	30 May 2005	*Irish Birds* 8: 393
Corn Bunting (A)	County Cork	Pre 1750	Smith 1750
Bobolink (A)	CCBO	13 September 1982	*Irish Birds* 2: 408
Baltimore Oriole (A) *	Baltimore	7 October 2001	*Irish Birds* 7: 238
Black-and-white Warbler (A) *	CCBO	18 October 1978	*Irish Birds* 1: 446
Blue-winged Warbler (A) *	CCBO	4 October 2000	*Irish Birds* 7: 107
Northern Parula Warbler (A) *	Firkeel	19 October 1983	*Irish Birds* 2: 576
American Yellow Warbler (A)	CCBO	24 August 2008	*Irish Birds* 9: 103
Yellow-rumped Warbler (A) *	CCBO	7 October 1976	*Irish Birds* 1: 95
Blackpoll Warbler (A) *	CCBO	6 October 1976	*Irish Birds* 1: 95
American Redstart (A) *	CCBO	13 October 1968	*Irish Bird Report* 16: 46
Ovenbird (A)	Dursey Island	24 September 1990	*Irish Birds* 4: 459–60
Northern Waterthrush (A) *	CCBO	10 September 1983	*Irish Birds* 2: 577
Wilson's Warbler (A) *	Dursey Island	18 September 2013	*Irish Birds* 10: 99

APPENDIX 3
Duck Decoys

Duck decoys were invented by the Dutch as a means of catching large numbers of wildfowl for food. Decoys usually consisted of a shallow pond or lake of about 1 ha where a series of curved tunnels, or pipes, covered in netting radiated from the centre. These pipes tapered towards the end and terminated in a catching area into which the birds were enticed by the provision of food, and trained dogs were sometimes used to assist in moving the birds towards the catching area.

Two duck decoys are known to have existed in Cork, one at Doneraile Court and one at Longueville House. The decoy at Doneraile was mentioned by Smith and Payne-Gallwey, and its history has recently been recounted by Jim Fox. Its location was on the right bank of the River Awbeg. The outline of it can still be traced but it holds no water, and it is overgrown with trees and shrubs. The Doneraile decoy is remarkably similar to one at Beaulieu, near Drogheda (Louth), which exists today unchanged. It was possibly built between 1727 and 1733 by the second Viscount Doneraile. No records exist to show that this decoy was ever put to active use.

More information is available on the Longueville decoy. It was reputed to have been built in 1750 by John Longfield, but today the decoy pond is overgrown and partially drained. Longueville decoy was not worked with a dog, but by feeding birds up the pipes, and the average take was said to be about 300 ducks per season. This decoy is thought to have been the last one in Ireland to operate, and it apparently caught ducks into the 1930s. Information on the species and catch was given by Thompson. He states that from November to March 1840/1, 216 Common Teal, 100 Mallard and one Common Wigeon were captured, the greatest number on any one occasion being thirty-five Teal and six Mallard. In the same months of 1841/2 only 150 Teal and thirty-two Mallard were obtained. In the season of 1845/6 about 730 head of duck were taken, more than half being Teal, about 300 Mallard, about twenty-five Wigeon and four Shoveler. In the season 1849/50 no birds were taken, but the number of ducks present at the decoy was greater than had been recorded for some years, the winter being unusually severe.

Common Wigeon

Following repair and maintenance work in 1865 and 1875, the annual catch averaged 250 to 300 ducks. On one occasion in 1883/4, forty-four Mallard were taken at one drive. The ducks caught were Teal, Mallard and Wigeon with a few Pintail and Gadwall. During 1893 to 1897 some Gadwall were taken every winter. At least thirty male Pintail were seen at one time and some Shoveler also frequented the decoy. The decoy was not worked during the First World War. Catching resumed after the war, but the Game Preservation Act of 1930 made it illegal 'to kill or take any game by means of any trap, snare, or net', except under licence. The decoy gradually fell into disuse and was dismantled completely by 1938.

APPENDIX 4
Designated Sites

Sites are designated in County Cork as Special Protection Areas (SPA) and Special Areas of Conservation (SAC) entirely or principally for their bird interest. Other designations are Natural Heritage Areas (8), proposed Natural Heritage Areas (103), Nature Reserves (5), Refuges for Fauna (3) and Ramsar Wetlands (4). Some sites hold several designations, and some of the Natural Heritage Areas have no bird interest. Full information about each site is available on the National Parks and Wildlife Service website (www.npws.ie) regarding species, habitats and boundaries (note that some sites extend into adjoining counties).

Site name	SPA/SAC	Site code
Ballycotton Bay	SPA	004022
Ballyhoura Mountains	SAC	002036
Ballymacoda Bay	SPA	004023
Ballymacoda (Clonpriest and Pilmore)	SAC	000077
Bandon River	SAC	002171
Barley Cove to Ballyrisode Point	SAC	001040
Beara Peninsula	SPA	004155
Blackwater Callows	SPA	004094
Blackwater Estuary	SPA	004028
Blackwater River	SAC	002170
Caha Mountains	SAC	000093
Carrigeenamronety Hill	SAC	002037
Castletownshend	SAC	001547
Cleanderry Wood	SAC	001043
Clonakilty Bay	SAC	000091
Clonakilty Bay	SPA	004081
Cork Harbour	SPA	004030
Courtmacsherry Bay	SPA	004219
Courtmacsherry Estuary	SAC	001230

Site name	SPA/SAC	Site code
Derryclogher (Knockboy) Bog	SAC	001873
Dunbeacon Shingle	SAC	002280
Farranamanagh Lough	SAC	002189
Galley Head to Duneen Point	SPA	004190
Glanmore Bog	SAC	001879
Glengarriff Harbour and Woodland	SAC	000090
Great Island Channel	SAC	001058
Kilcolman Bog	SPA	004095
Kilkerran Lake and Castlefreke Dunes	SAC	001061
Lough Hyne Nature Reserve and Environs	SAC	000097
Mullaghanish Bog	SAC	001890
Mullaghanish to Musheramore Mountains	SPA	004162
Mullaghareirk Mountains	SPA	004161
Myross Wood	SAC	001070
Old Head of Kinsale	SAC	004021
Reen Point Shingle	SAC	002281
Roaringwater Bay and Islands	SAC	000101
Seven Heads	SPA	004191
Sheep's Head	SAC	000102
Sheep's Head to Toe Head	SPA	004156
Sovereign Islands	SPA	004124
St Gobnet's Wood	SAC	000106
The Bull and The Cow Rocks	SPA	004066
The Gearagh	SAC	000108
The Gearagh	SPA	004109
Three Castle Head to Mizen Head	SAC	000109

APPENDIX 5

Names of Places

Names and grid references of places in County Cork mentioned in text.

Place	Grid reference
Abisdealy Lake (Lissard)	W1331
Adrigole	V8050
Aghada	W8565
Aghadown	W0334
Aghern	W8992
Aghinagh	W3871
Allihies	V5845
Annagh Bog (Churchtown)	R4916
Annagh Bog (Inishannon)	W5760
Ardfield	W3735
Ardgroom	V6955
Ardnahinch Strand	W9966
Atlantic Pond	W7071
Ballinacurra	W8871
Ballinadee	W5651
Ballinagree	W3680
Ballinamona Lake	W9865
Ballincollig	W5970
Ballincurrig	W8481
Ballindangan Lough	R7509
Ballinhassig	W6262
Ballintotis	W9373
Ballintubbrid	W8470
Ballinwilling	X0168
Ballyandreen	W9662
Ballybraher	W9764

Place	Grid reference
Ballybranagan Strand	W9061
Ballybricken	W7764
Ballybutler Lake	W9272
Ballycotton	W9865
Ballycotton Island (lighthouse)	X0163
Ballycotton Island (small)	X0063
Ballydehob	V9835
Ballydesmond	R1503
Ballydonegan	V5744
Ballygarvan	W6863
Ballyhea	R5417
Ballyhonock Lake	W9973
Ballyhooly	W7299
Ballyhoura Mountains	R6115
Ballyleary Bog	W7868
Ballylicky	W0053
Ballymacoda	X0672
Ballymacrown	W0726
Ballymakeery	W2176
Ballymaloe	W9467
Ballynacarriga (Beara)	V5241
Ballynacarriga Lake	W2850
Ballynoe	W9389
Ballyshane Strand	W9060
Ballyvergan Marsh	X0875
Ballywilliam	W9362

Place	Grid reference
Baltimore	W0426
Bandon	W4954
Bandon estuary	W6049
Banduff Bog	W7074
Banteer	W3897
Bantry	V9948
Bantry Bay	V9048
Barley Cove	V7624
Barley Lake	V8756
Barryroe	W4739
Barry's Head	W7250
Bateman's Lake	W4046
Beara peninsula	V6344
Belgooly	W6653
Belvelly	W7970
Benduff	W2639
Bere Island	V7043
Berehaven lighthouse	V6742
Big Doon	W7048
Bishopstown	W6370
Blackrock (Cork Harbour)	W7271
Blarney	W6175
Blarney Lake	W6074
Bogaghard Ponds	W7684
Boggeragh Mountains	W4287
Brinny	W5159
Broadstrand Bay	W5140
Brooklodge	W7475
Brow Head	V7723
Brown Island	W7972
Bull Rock	V4040
Buttevant	R5409

Place	Grid reference
Caha Mountains	V8155
Caherbarnagh	W1885
Cahermore	V5740
Calf Island East	V9626
Calf Island Middle	V9526
Calf Island West	V9425
Calf Rock	V4437
Cape Clear Island (Clear Island) (CCBO)	V9521
Capel Island	X1069
Careysville	W8599
Carrigacrump	W9066
Carrigadrohid	W4172
Carrigaline	W7362
Carrigaloe	W7767
Carrigmore Island	V9823
Carrignafoy	W8067
Carrignavar	W6781
Carrigrohane	W6171
Carrigtwohill	W8273
Carrigviglash Island	W0131
Castle Warren	W7763
Castlefreke	W3334
Castlehaven	W1830
Castlehyde	W7798
Castlelyons	W8393
Castlemartyr	W9673
Castlemary	W8967
Castlemore	W4466
Castlenalact Lake	W4859
Castlepook	R6111
Castletownbere	V6746

Place	Grid reference
Castletownroche	R6802
Castletownshend	W1831
Cecilstown	R4602
Charleville	R5322
Charleville lagoons	R5325
Churchtown South	W9463
Classis (Classes)	W5670
Claycastle	X1076
Clonakilty	W3841
Clondulane	W8498
Clonmult	W9282
Clonpriest	X0573
Cloyne	W9167
Coachford	W4573
Cobh	W8066
Comeenatrush	W2485
Conna	W9293
Connonagh	W2338
Convamore	W7199
Cool Mountain	W1760
Coolcower	W3571
Coolim Cliffs	W5239
Coolmain Point	W5442
Coolmore	W7662
Coomhola Mountain	V9959
Cork airport	W6665
Cork city	W6771
Cork Harbour	W8265
Cork Lough	W6670
Cork University Hospital	W6470
Corkbeg Island	W8263
Corran Lake	W2239

Place	Grid reference
Courtmacsherry	W5042
Cow Rock	V4239
Creagh	W0730
Cregg	W7798
Croagh Bay	V8930
Crookhaven	V8025
Crossbarry	W5561
Crosshaven	W7961
Crow Head/Island	V5038
Currabinny Wood	W7961
Curraghalicky Lake	W2346
Cuskinny	W8167
Daunt's Rock lightship	W7952
Derrylahan	W1273
Derrynasaggart Mountains	W1680
Doneraile	R6007
Doolieve Wood	W6959
Dooniskey (Dunisky)	W3669
Douglas estuary	W7069
Dower	W9772
Drimoleague	W1245
Drinagh	W2044
Dripsey	W4973
Drishane (Millstreet)	W2892
Dunboy Castle	V6643
Dungourney	W9379
Dunkettle	W7372
Dunmanus Bay	V8034
Dunmanway	W2352
Dunowen Head	W3632
Dunworly	W4737
Durrus	V9442

Place	Grid reference
Dursey Island	V4740
Dursey Sound	V5041
East Ferry	W8568
Enniskean	W3554
Eyeries	V6450
Eyeries Island	V6351
Farnanes	W4467
Farran Wood	W4969
Fastnet Rock	V8816
Fennel's Bay	W8059
Fermoy	W8198
Firkeel	V5241
Flat Head	W7551
Flaxfort Marsh	W5144
Fota Wildlife Park	W8071
Fountainstown	W7858
Frower Point	W6746
Gallanes Lake	W3943
Galley Cove	V7824
Galley Head	W3331
Garinish	V5241
Garrettstown	W5943
Garryhesta	W5268
Garrylucas Marsh	W6043
Garryvoe	X0067
Gearagh (The)	W3170
Glandore	W2235
Glanmire	W7274
Glanworth	R7503
Glen Lough	V8454
Glengarriff	V9256
Goat Island	V8827

Place	Grid reference
Gokane	W1227
Goleen	V8128
Gougane Barra	W0965
Grange	W6968
Great Island	W8168
Great Island Channel	W8369
Grenagh	W5884
Gyleen	W8660
Halfway	W6061
Harbour View	W5344
Harper's Island	W7872
Harty's Quay	W7269
Healy Pass	V7853
High Island	W2229
Hop Island	W7369
Hornet Rock	V7045
Hungry Hill	V7549
Ightermurragh Castle	W9972
Ileclash	W8299
Inch Strand/Bay	W8760
Inchigaggin Marsh	W6371
Inchigeelagh	W2265
Inchydoney	W3938
Inishannon	W5457
Inishcarra	W5572
Iveleary	W2265
Kanturk	R3803
Kealkill	W0456
Kilbrittain	W5246
Kilcatherine	V6153
Kilcolman	R5810
Kilcrea	W5068

Place	Grid reference
Kildorrery	R7010
Kilkerran Lake	W3334
Killavullen	W6499
Killeagh	X0076
Kilmacsimon	W5653
Kilmichael	W2864
Kilworth	R8302
Kilworth Mountains	R8308
Kinsale	W6350
Kinsale Marsh	W6249
Knockadoon Head	X0970
Ladysbridge	W9771
Laght (Lough Pole)	R7404
Lakelands	W7370
Leamlara	W8179
Leap	W2037
Lee Fields	W6271
Lisgoold	W8580
Lissagriffin Lake	V7726
Lissard	W1331
Little Island	W7571
Lombardstown	W4696
Long Strand	W3234
Longueville (duck decoy)	W4998
Lough Aderry	W9373
Lough Allua	W1965
Lough Beg	W7863
Lough Gal	W4075
Lough Hyne	W0928
Lough Mahon	W7470
Lough Nambrackderg	W1561
Lough Rhue	W8561

Place	Grid reference
Macroom	W3472
Mallow	W5598
Mallow lagoons	W5097
Mallow racecourse	W5297
Man-of-War Roads	W8164
Manch	W2952
Marino Point	W7870
Meelin	R2912
Midleton	W8873
Millstreet	W2690
Minane Bridge	W7456
Mitchelstown	R8112
Mizen Head	V7323
Mogeely	W9675
Monkstown	W7666
Moorepark	R8201
Mount Gabriel	V9334
Mourneabbey	W5791
Mullaganish	W2181
Mullaghareirk Mountains	R2318
Mushera Mountain	W3284
Myross peninsula	W2030
Myross Wood (Glandore)	W2036
Myrtleville	W7958
Nadd	W4291
Nagles Mountains	W6894
Newcastle	W5780
Newcestown	W4059
Newfoundland Bay	W7150
Newmarket	R3107
Newtownshandrum	R4721
Nohoval	W7351

Place	Grid reference
Old Head of Kinsale	W6239
Ovens	W5469
Owenahincha	W3035
Oysterhaven	W6949
Passage West	W7668
Pilmore Strand	X0773
Power Head	W8859
Rathbarry	W3335
Rathcormack	W8091
Rathcoursey	W8769
Rathduff	W5984
Reanie's Bay	W7451
Red Strand	W3533
Redbarn Strand	X0874
Reenydonagan Lake	V9950
Ring Strand (Ballymacoda)	X0671
Ringabella	W7857
Ringaskiddy	W7864
River Araglin	R8602
River Argideen	W4745
River Awbeg	R5508
River Bandon	W4654
River Blackwater (callows)	W9198
River Bride (north)	W9493
River Bride (south)	W4868
River Douglas	R8402
River Funshion	R7702
River Glashaboy	W7273
River Ilen	W0833
River Kiltha	W9576
River Leamlara	W8377
River Lee (lakes and reservoirs)	W5172

Place	Grid reference
River Owenboy	W7262
River Owennacurra	W8676
River Tourig	X0280
River Womanagh	X0374
Riverstick	W6557
Riverstown	W7375
Roancarrigmore lighthouse	V7945
Roaringwater Bay	W0030
Robert's Cove	W7854
Robert's Head	W7853
Roche's Point	W8260
Rochestown	W7369
Rocky Bay	W7753
Rosscarbery	W2836
Rosslague	W8169
Rossmore (Great Island Channel)	W8270
Rossmore (west Cork)	W3146
Rostellan	W8765
Roxborough	W9074
Rushbrooke	W7866
Ryecourt Wood	W4567
Saleen	W8767
Sands Cove	W3733
Sandy Cove Island	W6347
Schull	V9231
Seven Heads	W4935
Shanagarry	W9766
Shanballymore	R6707
Sheep's Head	V7233
Shehy Mountains	W0663
Shepperton Lake	W1835
Sherkin Island	W0124

Place	Grid reference
Shippool	W5654
Shreelane Lakes	W1735
Skeagh	W0637
Skibbereen	W1233
Skiddy Island	W1829
Slatty Bridge	W8072
Slieve Miskish Mountains	V6447
Snave	V9954
Sovereign Islands	W6847
Spanish Island	W0327
Spike Island	W8064
Spit Bank	W8065
Stags Rock (Stags of Castlehaven)	W1524
Tanner's Pond (Ballincollig)	W5770
Taur	R2309
Templenacarriga	W8579
Timoleague	W4643
Tivoli	W7072
Toe Head	W1426
Toormore	V8530
Trabolgan	W8360
Tracton	W7356
Union Hall	W2034
Upton	W5359
Watergrasshill	W7684
Wellington Bridge	W6571
Whiddy Island	V9549
White Bay	W8261
Whitegate	W8364
Wood Point	W5242
Youghal	X1078

NOTES

Chapter 1: Introduction

1. T. Gittings, 'Cork Harbour I-WeBS Summary Reports 2011/12–2019/20', unpublished report, 2020.

2. I. Newton, *Farming and Birds* (London: Collins, 2017).

3. G. Gilbert, A. Stanbury and L. Lewis, 'Birds of Conservation Concern in Ireland 4: 2020–2026', *Irish Birds*, vol. 43, 2021, pp. 1–22.

Chapter 2: Cork Environment

1. G.A.L. Johnson, 'Geographical Change in Britain During the Carboniferous Period', *Proceedings of the Yorkshire Geological Society*, vol. 44, 1982, pp. 181–203.

2. M.R. Leeder, 'Recent Developments in Carboniferous Geology: A critical review with implications for the British Isles and N.W. Europe', *Proceedings of the Geologists' Association*, vol. 99, 1988, pp. 73–100.

3. I.A.J. MacCarthy, 'Alluvial Sedimentation Patterns in the Munster Basin, Ireland', *Sedimentology*, vol. 37, 1990, pp. 685–712.

4. T.N. George, G.A.L. Johnson, M. Mitchell, J.E. Prentice, W.H.C. Ramsbottom, G.D. Sevastopulo and R.B. Wilson, 'A Correlation of Dinantian Rocks in the British Isles', *Geological Society of London Special Report* 7, 1976.

5. R.G.W. Heselden, 'Sedimentology and Stratigraphy of the Courceyan-Asbian Limestones (Dinantian, Lower Carboniferous) of the Cork Harbour Area, Southern Ireland', unpublished PhD thesis, University College Cork, vol. 1, 1992.

6. D.D.C. Pochin-Mould, *Discovering Cork* (Cork: Brandon Books, 1991).

7. Ibid.

8. F. Mitchell, *Shell Guide to Reading the Irish Landscape* (Dublin: Country House, 1986).

9. K. Corcoran, *Saving Eden: The Gearagh and Irish nature* (Macroom: Gearagh Press, 2021).

Chapter 3: Ornithology and Ornithologists

1. P. Smiddy, 'Breeding Birds in Ireland: Success and failure among colonists', in D.P. Sleeman, J. Carlsson and J.E.L. Carlsson (eds), *Mind the Gap II: New insights into the Irish postglacial* (Belfast: Irish Naturalists' Journal, 2014), pp. 89–99.

2. Giraldus Cambrensis (Gerald of Wales), *The History and Topography of Ireland (Topographia Hiberniae)*, ed. J.J. O'Meara (Dublin: Dolmen Press, 1982).

3. J.S. Fairley, *An Irish Beast Book* (Belfast: Blackstaff Press, 1984), p. 69.

4. R. Payne-Gallwey, *The Fowler in Ireland* (London: Van Voorst, 1882; Southampton: Ashford Press, 1985), p. 12.

5. D. Ó Drisceoil and D. Ó Drisceoil, *Beamish and Crawford: The history of an Irish brewery* (Cork: Collins Press, 2015).

6. J.A. Murphy, *The College: A history of Queen's/University College Cork, 1845–1995* (Cork: Cork University Press, 1995), pp. 7–8.

7. J.T.R. Sharrock (ed.), *The Natural History of Cape Clear Island* (Berkhamstead: Poyser, 1973), pp. 186–91.

8. W.J. O'Flynn, 'Bird Watching in Cork 1960–1990: Some personal reminiscences', *Cork Bird Report 1989*, 1990, pp. 99–102.

9. Ibid.

10. C.D. Hutchinson, *Birds in Ireland* (Calton: Poyser, 1989), p. 34.

11. M. Ridgway and C.D. Hutchinson (eds), *The Natural History of Kilcolman* (Dublin: O'Brien Printing, 1990).

12. F.J. O'Rourke, *The Fauna of Ireland: An introduction to the land vertebrates* (Cork: Mercier Press, 1970).

13. M. Murphy and S. Murphy (eds), *Ireland's Bird Life: A world of beauty* (Sherkin Island: Sherkin Island Marine Station, 1994).

14. A.A. Myers, C. Little, M.J. Costello and J.C. Partridge (eds), *The Ecology of Lough Hyne* (Dublin: Royal Irish Academy, 1991).

15. J. O'Halloran, T.C. Kelly, J.L. Quinn, S. Irwin, D. Fernández-Bellon, A. Caravaggi and P. Smiddy, 'Current Ornithological Research in Ireland: Seventh Ornithological Research Conference, UCC, November 2017', *Irish Birds*, vol. 10, 2017, pp. 598–638.

16. S. Wing, *The Natural History of Cape Clear 1959–2019* (Skibbereen: privately published, 2020).

17. Cambrensis, *The History and Topography of Ireland (Topographia Hiberniae).*

18. D.C. O'Sullivan (ed.), *The Natural History of Ireland* (Cork: Cork University Press, 2009), pp. 23–5.

19. E. Magennis, '"A Land of Milk and Honey": The Physico-Historical Society, improvement and the surveys of mid-eighteenth-century Ireland', *Proceedings of the Royal Irish Academy*, vol. 102C, 2002, pp. 199–217.

20. C. Smith, *The Antient and Present State of the County and City of Cork*, vol. 2 (Dublin: privately published, 1750), pp. 320–49.

21. R.L. Praeger, *Some Irish Naturalists* (Dundalk: Dundalgan Press, 1949).

22. G. D'Arcy, *Ireland's Lost Birds* (Dublin: Four Courts Press, 1999).

23. G.E. Hutchinson, 'The Harp that Once ... A Note on the Discovery of Stridulation in the Corixid Water-bugs', *Irish Naturalists' Journal*, vol. 20, 1982, pp. 457–66.

24. P.N. Wyse Jackson, 'Robert Ball (1802–1857): Naturalist', *Irish Naturalists' Journal*, vol. 30, 2009, pp. 15–18.

25. Praeger, *Some Irish Naturalists.*

26. J.R. Harvey, 'Memoranda towards a Fauna of the County of Cork, Div.: Vertebrata', in J.R. Harvey, J.D. Humphreys and T. Power (eds), *Contributions towards a Fauna and Flora of the County of Cork* (London: John Van Voorst, and Cork: George Purcell, 1845), pp. 1–24.

27. J.R. Harvey, 'The Fauna of the County Cork', in M.F. Cusack (ed.), *A History of the City and County of Cork* (Dublin: McGlashan & Gill, and Cork: Francis Guy, 1875), pp. 454–66.

28. R. Warren, 'The Harvey Collection of Irish Birds', *Irish Naturalist*, vol. 12, 1903, p. 55.

29. J.P. Cullinane, 'Joshua Reubens Harvey', *Irish Naturalists' Journal*, vol. 17, 1972, pp. 223–5.

30. M. Anglesea (ed.), *Birds of Ireland by Richard Dunscombe Parker* (Belfast: Blackstaff Press, 1984).

31. M. Anglesea, 'The Art of Nature Illustration', in J.W. Foster (ed.), *Nature in Ireland: A scientific and cultural history* (Dublin: Lilliput Press, 1997), pp. 497–523.

32. W. Thompson, *The Natural History of Ireland*, vols 1–3 (London: Reeve, Benham & Reeve, 1849–51).

33. C.B. Moffat, 'Robert Warren', *Irish Naturalist*, vol. 25, 1916, pp. 33–44.

34. C.B. Moffat, 'Robert Warren', *British Birds*, vol. 9, 1916, pp. 295–7.

35. R.J. Ussher, 'Birds of County Cork', *Journal of the Cork Historical and Archaeological Society*, series A, vol. 1, 1892, pp. 230–1, 250–1; vol. 2, 1893, pp. 20, 40, 59–60, 80, 103–4, 123–4, 148, 171–2, 191–2, 211–12, 235–6, 266–8; vol. 3, 1894, pp. 23–4, 42–4.

36. R.J. Ussher, 'A Catalogue of the Birds Observed in this County', in R. Day and W.A. Copinger (eds), *The Ancient and Present State of the County and City of Cork*, vol. 2 (Cork: Guy, 1894), pp. 238–62.

37. R.M. Barrington, 'Richard John Ussher', *Irish Naturalist*, vol. 22, 1913, pp. 221–7.

38. R.M. Barrington, 'Richard John Ussher, DL, MRIA: A memoir', *British Birds*, vol. 7, 1914, pp. 182–5.

39. R.J. O'Connor, 'Book Review: *Birds in Ireland*', *Irish Birds*, vol. 4, 1989, pp. 144–5.

40. C.D. Hutchinson (ed.), *The Birds of Dublin and Wicklow* (Dublin: Irish Wildbird Conservancy, 1975).

41. J.M. Rochford, 'Personalities: C.D. Hutchinson', *British Birds*, vol. 71, 1978, pp. 305–7.

42. T.C. Kelly and J. O'Halloran, 'Obituary: Clive Desmond Hutchinson, 1949–1998', *Irish Birds*, vol. 6, 1998, pp. 329–31.

43. J. O'Halloran and J.T.R. Sharrock, 'Obituary: Clive Desmond Hutchinson, BA (1949–1998)', *British Birds*, vol. 91, 1998, pp. 269–70.

44 R.G.W. Nairn, 'Obituary: Clive D. Hutchinson (1949–1998)', *Irish Naturalists' Journal*, vol. 26, 1999, pp. 145–8.

Chapter 4: Research and Monitoring

1. D.E. Balmer, S. Gillings, B.J. Caffrey, R.L. Swann, I.S. Downie and R.J. Fuller, *Bird Atlas 2007–11: The breeding and wintering birds of Britain and Ireland* (Thetford: BTO Books, 2013).

2. S.T. Cummins, C. Lauder, A. Lauder and T.D. Tierney, 'The Status of Ireland's Breeding Seabirds: Birds Directive Article 12 Reporting 2013–2018', *Irish Wildlife Manuals, No. 114* (Dublin: National Parks and Wildlife Service, 2019).

3. L.J. Lewis, B. Burke, N. Fitzgerald, T.D. Tierney and S. Kelly, 'Irish Wetland Bird Survey: Waterbird status and distribution 2009/10–2015/16', *Irish Wildlife Manuals, No. 106* (Dublin: National Parks and Wildlife Service, 2019).

4. O. Crowe, G.E. Austin, K. Colhoun, P.A. Cranswick, M. Kershaw and A.J. Musgrove, 'Estimates and Trends of Waterbird Numbers Wintering in Ireland, 1994/95 to 2003/04', *Bird Study*, vol. 55, 2008, pp. 66–77.

5. I.M.D. Maclean, G.E. Austin, M.M. Rehfisch, J. Blew, O. Crowe, S. Delany, K. Devos, B. Deceuninck, K. Günther, K. Laursen, M. van Roomen and J. Wahl, 'Global Warming Causes Rapid Changes in the Distribution and Abundance of Birds in Winter', *Global Change Biology*, vol. 14, 2008, pp. 2489–500.

6. L.J. Lewis, G. Austin, H. Boland, T. Frost, O. Crowe and T.D. Tierney, 'Waterbird Populations on Non-estuarine Coasts in Ireland: Results of the 2015/16 Non-Estuarine Coastal Waterbird Survey (NEWS-III)', *Irish Birds*, vol. 10, 2017, pp. 511–22.

7. L.J. Lewis, D. Coombes, B. Burke, J. O'Halloran, A. Walsh, T.D. Tierney and S. Cummins, 'Countryside Bird Survey: Status and trends of common and widespread breeding birds 1998–2016', *Irish Wildlife Manuals, No. 115* (Dublin: National Parks and Wildlife Service, 2019).

8. I. Newton, *Farming and Birds* (London: Collins, 2017).

9. O. Crowe, 'The Garden Bird Survey: Monitoring birds of Irish gardens during winters between 1994/95 and 2003/04', *Irish Birds*, vol. 7, 2005, pp. 475–82.

10. R.M. Barrington, *The Migration of Birds* (London: Porter, and Dublin: Ponsonby, 1900).

11. J.J.D. Greenwood, '100 Years of Ringing in Britain and Ireland', *Ringing and Migration*, vol. 24, 2009, pp. 147–53.

12. I. Newton, *Bird Migration* (London: Collins, 2010).

13. F. Bairlein, 'The Study of Bird Migrations: Some future perspectives', *Bird Study*, vol. 50, 2003, pp. 243–53.

Chapter 5: Recent Changes of Status

1. I. Newton, *Farming and Birds* (London: Collins, 2017).

2. D.J. Pain and M.W. Pienkowski (eds), *Farming and Birds in Europe: The Common Agricultural Policy and its implications for bird conservation* (London: Academic Press, 1997).

3. L.J. Lewis, B. Burke, N. Fitzgerald, T.D. Tierney and S. Kelly, 'Irish Wetland Bird Survey: Waterbird status and distribution 2009/10–2015/16', *Irish Wildlife Manuals, No. 106* (Dublin: National Parks and Wildlife Service, 2019).

4. G. Gilbert, A. Stanbury and L. Lewis, 'Birds of Conservation Concern in Ireland 4: 2020–2026', *Irish Birds*, vol. 43, 2021, pp. 1–22.

5. I. Newton, *Bird Populations* (London: Collins, 2013).

6. V. Keller, S. Herrando, P. Voříšek, M. Franch, M. Kipson, P. Milanesi, D. Martí, M. Anton, A. Klvaňová, M.V. Kalyakin, H.-G. Bauer and R.P.B. Foppen, *European Breeding Bird Atlas 2: Distribution, abundance and change* (Barcelona: Lynx Edicions, 2020).

Chapter 6: Introduction to Systematic List

1. IRBC (Irish Rare Birds Committee), *Checklist of the Birds of Ireland* (Dublin: BirdWatch Ireland, 1998).

2. Ibid.

3. D.T. Parkin, 'Birding and DNA: Species for the new millennium', *Bird Study*, vol. 50, 2003, pp. 223–42.

4. D.T. Parkin and A.G. Knox, *The Status of Birds in Britain and Ireland* (London: Helm, 2010).

5. J. Hobbs, *A List of Irish Birds* (Version 10.2) (www.southdublinbirds.com) (Dublin: South Dublin Branch of BirdWatch Ireland, 2021).

6. G. Anderson, *Birds of Ireland: Facts, folklore and history* (Cork: Collins Press, 2008).

7. IRBC (Irish Rare Birds Committee), *Checklist of the Birds of Ireland.*

8. J. del Hoyo, A. Elliott and J. Sargatal (eds), *Handbook of the Birds of the World* (vols 1–7); J. del Hoyo, A. Elliott and D.A. Christie (eds) (vols 8–16) (Barcelona: Lynx Edicions, 1992–2011).

9. S. Cramp and K.E.L. Simmons (eds), *Handbook of the Birds of Europe, the Middle East and North Africa: The birds of the Western Palearctic* (vols 1–3); S. Cramp (ed.) (vols 4–6); S. Cramp and C.M. Perrins (eds) (vols 7–9) (Oxford: Oxford University Press, 1977–94).

10. Parkin and Knox, *The Status of Birds in Britain and Ireland.*

11. C.D. Hutchinson, *Birds in Ireland* (Calton: Poyser, 1989).

12. IRBC (Irish Rare Birds Committee), *Checklist of the Birds of Ireland.*

13. Hobbs, *A List of Irish Birds.*

14. G. Gilbert, A. Stanbury and L. Lewis, 'Birds of Conservation Concern in Ireland 4: 2020–2026', *Irish Birds*, vol. 43, 2021, pp. 1–22.

15. V. Keller, S. Herrando, P. Voříšek, M. Franch, M. Kipson, P. Milanesi, D. Martí, M. Anton, A. Klvaňová, M.V. Kalyakin, H.-G. Bauer and R.P.B. Foppen, *European Breeding Bird Atlas 2: Distribution, abundance and change* (Barcelona: Lynx Edicions, 2020).

16. A. Staneva and I. Burfield, *European Birds of Conservation Concern: Populations, trends and national responsibilities* (Cambridge: BirdLife International, 2017).

17. del Hoyo, Elliott and Sargatal (eds), *Handbook of the Birds of the World.*

18. IRBC (Irish Rare Birds Committee), *Checklist of the Birds of Ireland.*

19. C.D. Hutchinson, 'Bird Study in Ireland', in J.W. Foster (ed.), *Nature in Ireland: A scientific and cultural history* (Dublin: Lilliput Press, 1997), pp. 262–82.

20. J.P. Hillis, 'Rare Irish Breeding Birds, 1992–2001', *Irish Birds*, vol. 7, 2003, pp. 157–72.

Chapter 7: Systematic List of the Birds of County Cork

1. D.B. Cabot, *Wildfowl* (London: Collins, 2009), p. 41.

2. F. Kelly, *Early Irish Farming: A study based mainly on the law-texts of the 7th and 8th centuries AD* (Dublin: Dublin Institute for Advanced Studies, 1997), p. 299.

3. P.D. O'Donoghue, J. O'Halloran, P.J. Bacon, P. Smiddy and T.F. Cross, 'The Population Genetics of the Mute Swan *Cygnus olor* in Ireland', *Wildfowl*, vol. 43, 1992, pp. 5–11.

4. C. Smith, *The Antient and Present State of the County and City of Cork*, vol. 2 (Dublin: privately published, 1750), pp. 320–49.

5. T.C. Kelly, 'The Status of the Lough. A report compiled on behalf of Cork Corporation' (unpublished report) (Cork: Department of Zoology, University College Cork, 1985).

6. R. Patterson and N.H. Foster, 'Irish Field Club Union: Report of the Fifth Triennial Conference and Excursion, held at Cork, July 11th to 16th, 1907. Vertebrata', *Irish Naturalist*, vol. 16, 1907, pp. 269–72.

7. I. Forsyth, 'A Breeding Census of Mute Swans in Ireland in 1978', *Irish Birds*, vol. 1, 1980, pp. 492–501.

8. P. Smiddy and J. O'Halloran, 'The Breeding Biology of Mute Swans *Cygnus olor* in Southeast Cork, Ireland', *Wildfowl*, vol. 42, 1991, pp. 12–16.

9. P. Smiddy and J. O'Halloran, 'The January 1991 Swan Census in Co. Cork', *Cork Bird Report 1990*, 1991, pp. 85–9.

10. J. O'Halloran, A.A. Myers and P.F. Duggan, 'Some Sub-lethal Effects of Lead on Mute Swan *Cygnus olor*', *Journal of Zoology*, vol. 218, 1989, pp. 627–32.

11. M.M. O'Connell, P. Smiddy and J. O'Halloran, 'Lead Poisoning in Mute Swans (*Cygnus olor*) in Ireland: Recent changes', *Biology and Environment: Proceedings of the Royal Irish Academy*, vol. 109B, 2009, pp. 53–60.

12. J. O'Halloran, P. Smiddy and S. Irwin, 'Movements of Mute Swans in South-west Ireland', *Irish Birds*, vol. 5, 1995, pp. 295–8.

13. R. Payne-Gallwey, *The Fowler in Ireland* (London: Van Voorst, 1882; Southampton: Ashford Press, 1985).

14. E. Rees, *Bewick's Swan* (London: Poyser, 2006).

15. R.F. Ruttledge, 'Winter Distribution of Whooper and Bewick's Swans in Ireland', *Bird Study*, vol. 21, 1974, pp. 141–5.

16. O. Crowe, *Ireland's Wetlands and their Waterbirds: Status and distribution* (Newcastle: BirdWatch Ireland, 2005).

17. B. Burke, J.G. McElwaine, N. Fitzgerald, S.B.A. Kelly, N. McCulloch, A.J. Walsh and L.J. Lewis, 'Population Size, Breeding Success and Habitat Use of Whooper Swan *Cygnus cygnus* and Bewick's Swan *Cygnus columbianus bewickii* in Ireland: Results of the 2020 International Swan Census', *Irish Birds*, vol. 43, 2021, pp. 57–70.

18. O.J. Merne, 'The Changing Status and Distribution of the Bewick's Swan in Ireland', *Irish Birds*, vol. 1, 1977, pp. 3–15.

19. R.J. Ussher and R. Warren, *The Birds of Ireland* (London: Gurney & Jackson, 1900).

20. Merne, 'The Changing Status and Distribution of the Bewick's Swan in Ireland'.

21. J. O'Halloran, M. Ridgway and C.D. Hutchinson, 'A Whooper Swan *Cygnus cygnus* Population Wintering at Kilcolman Wildfowl Refuge, Co. Cork, Ireland: Trends over 20 years', *Wildfowl*, vol. 44, 1993, pp. 1–6.

22. Ruttledge, 'Winter Distribution of Whooper and Bewick's Swans in Ireland'.

23. M. Brazil, *The Whooper Swan* (London: Poyser, 2003).

24. J.G. McElwaine, J.H. Wells and J.M. Bowler, 'Winter Movements of Whooper Swans Visiting Ireland: Preliminary results', *Irish Birds*, vol. 5, 1995, pp. 265–78.

25. J. Hobbs, *A List of Irish Birds* (Version 10.2) (www.southdublinbirds.com) (Dublin: South Dublin Branch of BirdWatch Ireland, 2021).

26. R.F. Ruttledge and M.A. Ogilvie, 'The Past and Current Status of the Greenland White-fronted Goose in Ireland and Britain', *Irish Birds*, vol. 1, 1979, pp. 293–363.

27. R.F. Ruttledge and R. Hall Watt, 'The Distribution and Status of Wild Geese in Ireland', *Bird Study*, vol. 5, 1958, pp. 22–33.

28. Ibid.

29. A.M. Browne, J. O'Halloran and P. Smiddy, 'Introduced Canada *Branta canadensis* and Greylag Goose *Anser anser* Populations in Ireland, 1994', *Irish Birds*, vol. 6, 1998, pp. 233–6.

30. Ruttledge and Hall Watt, 'The Distribution and Status of Wild Geese in Ireland'.

31. P.G. Kennedy, *A List of the Birds of Ireland* (Dublin: Stationery Office, 1961).

32. D.B. Cabot, 'Movements and Migration of the Mallard in Ireland', *Irish Birds*, vol. 1, 1977, pp. 37–45.

33. S. Holloway, *The Historical Atlas of Breeding Birds in Britain and Ireland, 1875–1900* (London: Poyser, 1996).

34. R.F. Ruttledge, 'Winter Distribution and Numbers of Scaup, Long-tailed Duck and Common Scoter in Ireland', *Bird Study*, vol. 17, 1970, pp. 241–6.

35. Ibid.

36. Payne-Gallwey, *The Fowler in Ireland.*

37. Ruttledge, 'Winter Distribution and Numbers of Scaup, Long-tailed Duck and Common Scoter in Ireland'.

38. J.H. Wells and P. Smiddy, 'The Status of the Ruddy Duck in Ireland', *Irish Birds*, vol. 5, 1995, pp. 279–84.

39. S.T. Cummins, A. Bleasdale, C. Douglas, S.F. Newton, J. O'Halloran and H.J. Wilson, 'Densities and Population Estimates of Red Grouse *Lagopus lagopus scotica* in Ireland Based on the 2006–2008 National Survey', *Irish Birds*, vol. 10, 2015, pp. 197–210.

40. Ibid.

41. A. Mee, 'A Survey of Nightjar in the Ballyhoura Hills', Cork and Limerick County Councils, unpublished report, 2020.

42. J. Freeland, D. Allen and S. Anderson, *DNA Analysis of Red Grouse: An analysis of taxonomy and genetic diversity* (Belfast: Environment and Heritage Service, 2006).

43. Smith, *The Antient and Present State of the County and City of Cork.*

44. C. Lever, *The Naturalized Animals of the British Isles* (London: Hutchinson, 1977).

45. G. D'Arcy, *Ireland's Lost Birds* (Dublin: Four Courts Press, 1999).

46. C.D. Deane, 'The Capercaillie as an Irish Species', *Irish Birds*, vol. 1, 1979, pp. 364–9.

47. J.J. Hall, 'The Cock of the Wood', *Irish Birds*, vol. 2, 1981, pp. 38–47.

48. D'Arcy, *Ireland's Lost Birds.*

49. R.E. Longfield, 'Southern Notes', *Irish Naturalists' Journal*, vol. 1, 1926, p. 84.

50. P.G. Kennedy, 'Birds of the Countryside: Waders', *Studies*, vol. 33, 1944, pp. 393–9.

51. Longfield, 'Southern Notes'.

52. Holloway, *The Historical Atlas of Breeding Birds in Britain and Ireland: 1875–1900.*

53. C.B. Moffat, 'The Quail in Ireland: Its present and recent visits', *Irish Naturalist*, vol. 5, 1896, pp. 203–7.

54. C. Lever, *The Naturalized Animals of Britain and Ireland* (London: New Holland, 2009).

55. Kelly, *Early Irish Farming: A study based mainly on the law-texts of the 7th and 8th centuries AD*, p. 300.

56. D.W. Yalden, 'A Distant History of Irish birds', *Irish Birds*, vol. 9, 2011, pp. 225–8.

57. P.A. Robertson and J. Whelan, 'The Ecology and Management of Wild and Hand reared Pheasants in Ireland', *Irish Birds*, vol. 3, 1987, pp. 427–40.

58. P. Smiddy, 'Pheasant Densities in East Cork', *Irish Birds*, vol. 3, 1988, pp. 605–6.

59. J. Fisher, *The Fulmar* (London: Collins, 1952).

60. S.T. Cummins, C. Lauder, A. Lauder and T.D. Tierney, 'The Status of Ireland's Breeding Seabirds: Birds Directive Article 12 Reporting 2013–2018', *Irish Wildlife Manuals, No. 114* (Dublin: National Parks and Wildlife Service, 2019).

61. C.M. Pollock, 'Observations on the Distribution of Seabirds off South-west Ireland', *Irish Birds*, vol. 5, 1994, pp. 173–82.

62. M. Mackey, O. Ó Cadhla, T.C. Kelly, N. Aguilar de Soto and N. Connolly, *Cetaceans and Seabirds of Ireland's Atlantic Margin: Seabird distribution, density and abundance*, vol. 1 (Cork: Coastal and Marine Resource Centre, 2004).

63. C.D. Hutchinson, *Birds in Ireland* (Calton: Poyser, 1989).

64. C.D. Hutchinson, 'Cape Clear Bird Observatory 1970–1980', *Irish Birds*, vol. 2, 1981, pp. 60–72.

65. R.J. Ussher, 'The Great and Sooty Shearwaters on the South Coast', *Irish Naturalist*, vol. 10, 1901, pp. 42–3.

66. R.J. Ussher, 'Great Shearwaters and Sooty Shearwaters in 1901', *Irish Naturalist*, vol. 14, 1905, p. 43.

67. J.E. Flynn, 'Some South-west Irish Seabird Colonies', *Seabird Bulletin*, vol. 2, 1966, pp. 52–5.

68. R.G. Newell, 'Influx of Great Shearwaters in Autumn 1965', *British Birds*, vol. 61, 1968, pp. 145–59.

69. Hutchinson, 'Cape Clear Bird Observatory 1970–1980'.

70. Ussher, 'Great Shearwaters and Sooty Shearwaters in 1901'.

71. C.J. Stone, A. Webb and M.L. Tasker, 'The Distribution of Manx Shearwaters *Puffinus puffinus* in North-west European Waters', *Bird Study*, vol. 41, 1994, pp. 170–80.

72. P.S. Watson, 'Seabirds in the Celtic Sea and off Co. Cork in April 1980', *Irish Naturalists' Journal*, vol. 20, 1982, pp. 388–93.

73. P.G.H. Evans and R.R. Lovegrove, 'The Birds of the South-west Irish Islands', *Irish Bird Report*, vol. 21, 1974, pp. 33–64.

74. T.C. Guilford, J. Meade, R. Freeman, D. Biro, T. Evans, F. Bonadonna, D. Boyle, S. Roberts and C.M. Perrins, 'GPS Tracking of the Foraging Movements of Manx Shearwaters *Puffinus puffinus* Breeding on Skomer Island, Wales', *Ibis*, vol. 150, 2008, pp. 462–73.

75. Hutchinson, *Birds in Ireland.*

76. Anon., 'The Stormy Petrel–*Procellaria pelagica* Commonly Called Mother Carey's Chicken', *Dublin Penny Journal*, vol. 2, 1833, p. 160.

77. R.M. Lockley, 'Southwesternmost Island', *The Countryman*, vol. 37, 1948, pp. 90–3.

78. Kennedy, *A List of the Birds of Ireland.*

79. Evans and Lovegrove, 'The Birds of the South-west Irish Islands'.

80. A.R. Mainwood, 'The Movements of Storm Petrels as Shown by Ringing', *Ringing and Migration*, vol. 1, 1976, pp. 98–104.

81. R.W. Furness and S.R. Baillie, 'Factors Affecting Capture Rate and Biometrics of Storm Petrels on St Kilda', *Ringing and Migration*, vol. 3, 1981, pp. 137–47.

82. Ibid.

83. J. D'Elbee and G. Hemery, 'Diet and Foraging Behaviour of the British Storm Petrel *Hydrobates pelagicus* in the Bay of Biscay during Summer', *Ardea*, vol. 86, 1997, pp. 1–10.

84. H. Boyd, 'The "Wreck" of Leach's Petrels in the Autumn of 1952', *British Birds*, vol. 47, 1954, pp. 137–63.

85. Evans and Lovegrove, 'The Birds of the South-west Irish Islands'.

86. J.B. Nelson, *The Gannet* (Berkhamstead: Poyser, 1978).

87. S.F. Newton, M.P. Harris and S. Murray, 'Census of Gannet *Morus bassanus* Colonies in Ireland in 2013–2014', *Irish Birds*, vol. 10, 2015, pp. 215–20.

88. Ibid.

89. J.B. Nelson, *The Atlantic Gannet* (Norfolk: Fenix Books, 2002).

90. Patterson and Foster, 'Irish Field Club Union: Report of the Fifth Triennial Conference and Excursion, held at Cork, July 11th to 16th, 1907. Vertebrata'.

91. A.G. More, *A List of Irish Birds* (Dublin: Science and Art Museum, 1890).

92. Cummins, Lauder, Lauder and Tierney, 'The Status of Ireland's Breeding Seabirds: Birds Directive Article 12 Reporting 2013–2018'.

93. R.A. Macdonald, 'The Breeding Population and Distribution of the Cormorant in Ireland', *Irish Birds*, vol. 3, 1987, pp. 405–16.

94. T. Gittings, 'Cork Harbour I-WeBS Summary Reports 2011/12–2019/20', unpublished report, 2020.

95. Cummins, Lauder, Lauder and Tierney, 'The Status of Ireland's Breeding Seabirds: Birds Directive Article 12 Reporting 2013–2018'.

96. Evans and Lovegrove, 'The Birds of the South-west Irish Islands'.

97. Payne-Gallwey, *The Fowler in Ireland.*

98. P. Smiddy and O. O'Sullivan, 'The Status of the Little Egret *Egretta garzetta* in Ireland', *Irish Birds*, vol. 6, 1998, pp. 201–6.

99. P. Smiddy, 'Breeding of the Little Egret *Egretta garzetta* in Ireland, 1997–2001', *Irish Birds*, vol. 7, 2002, pp. 57–60.

100. S.T. Ronayne, 'Diet of the Little Egret *Egretta garzetta* in Southern Ireland', *Irish Birds*, vol. 9, 2011, pp. 329–30.

101. C. Cullen, T.C. Kelly and J. O'Halloran, 'Late Summer Foraging of the Little Egret *Egretta garzetta* in East County Cork', *Irish Birds*, vol. 7, 2005, pp. 523–8.

102. Ussher and Warren, *The Birds of Ireland.*

103. P.G. Kennedy, R.F. Ruttledge and C.F. Scroope, *The Birds of Ireland* (Edinburgh and London: Oliver & Boyd, 1954).

104. R.J. O'Sullivan, 'Food Provisioning of Little Grebe *Tachybaptus ruficollis* Chicks', *Irish Birds*, vol. 10, 2015, pp. 175–8.

105. T. Gittings, 'Nocturnal Communal Roosting Behaviour in Great Crested Grebes *Podiceps cristatus*', *Irish Birds*, vol. 10, 2017, pp. 483–92.

106. D'Arcy, *Ireland's Lost Birds.*

107. R. Price and J.A. Robinson, 'The Persecution of Kites and other Species in 18th Century Co. Antrim', *Irish Naturalists' Journal*, vol. 29, 2008, pp. 1–6.

108. R.J. Evans, L. O'Toole and D.P. Whitfield, 'The History of Eagles in Britain and Ireland: An ecological review of placename and documentary evidence from the last 1500 years', *Bird Study*, vol. 59, 2012, pp. 335–49.

109. A. Mee, D. Breen, D. Clarke, C. Heardman, J. Lyden, F. McMahon, P. O'Sullivan and L. O'Toole, 'Reintroduction of White-tailed Eagles *Haliaeetus albicilla* to Ireland', *Irish Birds*, vol. 10, 2016, pp. 301–14.

110. Kennedy, Ruttledge and Scroope, *The Birds of Ireland.*

111. W.J. O'Flynn, 'Population Changes of the Hen Harrier in Ireland', *Irish Birds*, vol. 2, 1983, pp. 337–43.

112. T. Nagle, 'The Status of Birds of Prey and Owls in County Cork', *Cork Bird Report 1996–2004*, 2006, pp. 285–308.

113. M.W. Wilson, S. Irwin, D.W. Norriss, S.F. Newton, K. Collins, T.C. Kelly and J. O'Halloran, 'The Importance of Pre-thicket Conifer Plantations for Nesting Hen Harriers *Circus cyaneus* in Ireland', *Ibis*, vol. 151, 2009, pp. 332–43.

114. M.W. Wilson, D. Fernández-Bellon, S. Irwin and J. O'Halloran, 'Hen Harrier *Circus cyaneus* Population Trends in Relation to Wind Farms', *Bird Study*, vol. 64, 2017, pp. 20–9.

115. B.G. O'Donoghue, 'Duhallow Hen Harriers *Circus cyaneus*: From stronghold to just holding on', *Irish Birds*, vol. 9, 2012, pp. 349–56.

116. B.G. O'Donoghue, 'Hen Harrier *Circus cyaneus* Ecology and Conservation during the Non-breeding Season in Ireland', *Bird Study*, vol. 67, 2020, pp. 344–59.

117. P.A.C. Dawson, 'Hen Harrier *Circus cyaneus* Utilisation of a Limestone Fen in North County Cork', *Irish Birds*, vol. 7, 2005, pp. 503–10.

118. P. Smiddy and C. Cullen, 'Winter Diet of the Hen Harrier *Circus cyaneus* in Coastal East County Cork', *Irish Birds*, vol. 10, 2017, pp. 523–6.

119. D'Arcy, *Ireland's Lost Birds.*

120. IRSG (Irish Raptor Study Group), *Birds of Prey and Owls in Ireland: Restoration* (Clonmel: Irish Raptor Study Group, 2006).

121. I. Newton, *The Sparrowhawk* (Calton: Poyser, 1986).

122. Hutchinson, *Birds in Ireland.*

123. Nagle, 'The Status of Birds of Prey and Owls in County Cork'.

124. C. Kelleher, 'Observation of Bat Predation at a Soprano Pipistrelle (*Pipistrellus pygmaeus* (Leach 1825)) Roost by a Sparrowhawk (*Accipiter nisus* L. 1758) in West Cork', *Irish Naturalists' Journal*, vol. 29, 2008, p. 57.

125. D.W. Norriss, 'The Status of the Buzzard as a Breeding Species in the Republic of Ireland, 1977–1991', *Irish Birds*, vol. 4, 1991, pp. 291–8.

126. T. Nagle, A. Mee, M.W. Wilson, J. O'Halloran and P. Smiddy, 'Habitat and Diet of Re-colonising Common Buzzards *Buteo buteo* in County Cork', *Irish Birds*, vol. 10, 2014, pp. 47–58.

127. Ibid.

128. *Cork Examiner*, 12 May 2020.

129. Evans, O'Toole and Whitfield, 'The History of Eagles in Britain and Ireland: An ecological review of placename and documentary evidence from the last 1500 years'.

130. D'Arcy, *Ireland's Lost Birds.*

131. W. Thompson, *The Natural History of Ireland*, vols 1–3 (London: Reeve, Benham & Reeve, 1849–51).

132. D'Arcy, *Ireland's Lost Birds.*

133. Nagle, 'The Status of Birds of Prey and Owls in County Cork'.

134. P. Smiddy, 'Diet of the Common Kestrel *Falco tinnunculus* in East Cork and West Waterford: An insight into the dynamics of invasive mammal species', *Biology and Environment: Proceedings of the Royal Irish Academy*, vol. 117B, 2017, pp. 131–8.

135. Nagle, 'The Status of Birds of Prey and Owls in County Cork'.

136. Smith, *The Antient and Present State of the County and City of Cork.*

137. Irish Wildbird Conservancy (unpublished data).

138. J.T. Lang, 'Peregrine Survey – Republic of Ireland 1967–68', *Irish Bird Report*, vol. 16, 1969, pp. 8–12.

139. D. Ratcliffe, *The Peregrine Falcon* (Calton: Poyser, 1980).

140. Nagle, 'The Status of Birds of Prey and Owls in County Cork'.

141. Hutchinson, *Birds in Ireland.*

142. D.W. Norriss, H.J. Wilson and D. Browne, 'The Breeding Population of the Peregrine Falcon in Ireland in 1981', *Irish Birds*, vol. 2, 1982, pp. 145–52.

143. D.W. Norriss, 'The 1991 Survey and Weather Impacts on the Peregrine *Falco peregrinus* Breeding Population in the Republic of Ireland', *Bird Study*, vol. 42, 1995, pp. 20–30.

144. Nagle, 'The Status of Birds of Prey and Owls in County Cork'.

145. B. Madden, J. Hunt and D.W. Norriss, 'The 2002 Survey of the Peregrine *Falco peregrinus* Breeding Population in the Republic of Ireland', *Irish Birds*, vol. 8, 2009, pp. 543–8.

146. Nagle, 'The Status of Birds of Prey and Owls in County Cork'.

147. N.P. Moore, P.F. Kelly, F.A. Lang, J.M. Lynch and S.D. Langton, 'The Peregrine *Falco peregrinus* in Quarries: Current status and factors influencing occupancy in the Republic of Ireland', *Bird Study*, vol. 44, 1997, pp. 176–81.

148. Madden, Hunt and Norriss, 'The 2002 Survey of the Peregrine *Falco peregrinus* Breeding Population in the Republic of Ireland'.

149. Ussher and Warren, *The Birds of Ireland.*

150. C.A. Norris, 'Summary of a Report on the Distribution and Status of the Corn-crake (*Crex crex*)', *British Birds*, vol. 38, 1945, pp. 162–8.

151. J.E. Flynn, 'Distribution Map of Breeding Corncrakes in the Glengarriff Area', unpublished report, 1966.

152. M. O'Meara, 'Distribution and Numbers of Corncrakes in Ireland in 1978', *Irish Birds*, vol. 1, 1979, pp. 381–405.

153. Ibid.

154. M. O'Meara, 'Corncrake Declines in Seven Areas, 1978–1985', *Irish Birds*, vol. 3, 1986, pp. 237–44.

155. E. Mayes and T. Stowe, 'The Status and Distribution of the Corncrake in Ireland, 1988', *Irish Birds*, vol. 4, 1989, pp. 1–12.

156. R. Sheppard and R.E. Green, 'Status of the Corncrake in Ireland in 1993', *Irish Birds*, vol. 5, 1994, pp. 125–38.

157. C. Casey, 'Distribution and Conservation of the Corncrake in Ireland, 1993–1998', *Irish Birds*, vol. 6, 1998, pp. 159–76.

158. S. Boisseau and D.W. Yalden, 'The Former Status of the Crane *Grus grus* in Britain', *Ibis*, vol. 140, 1998, pp. 482–500.

159. J. Fisher, *The Shell Bird Book* (London: Ebury Press & Michael Joseph, 1966).

160. Kelly, *Early Irish Farming: A study based mainly on the law-texts of the 7th and 8th centuries AD*, p. 127.

161. K. Corcoran, *Saving Eden: The Gearagh and Irish nature* (Macroom: Gearagh Press, 2021).

162. A. Brown, 'One Hundred Years of Notable Avian Events in *British Birds*', *British Birds*, vol. 100, 2007, pp. 214–43.

163. A.G. Knox, 'Taxonomic Status of "Lesser Golden Plovers"', *British Birds*, vol. 80, 1987, pp. 482–7.

164. P.G. Connors, B.J. McCaffery and J.L. Maron, 'Speciation in Golden Plovers, *Pluvialis dominica* and *Pluvialis fulva*: Evidence from their breeding grounds', *Auk*, vol. 110, 1993, pp. 9–20.

165. R. Sheppard, *Ireland's Wetland Wealth* (Dublin: Irish Wildbird Conservancy, 1993).

166. D.T. Parkin and A.G. Knox, *The Status of Birds in Britain and Ireland* (London: Helm, 2010).

167. P.I. Stanley and C.D.T. Minton, 'The Unprecedented Westward Migration of Curlew Sandpipers in Autumn 1969', *British Birds*, vol. 65, 1972, pp. 365–80.

168. J.R. Wilson, M.A. Czajkowski and M.W. Pienkowski, 'The Migration through Europe and Wintering in West Africa of Curlew Sandpipers', *Wildfowl*, vol. 31, 1980, pp. 107–22.

169. A.R. Hardy and C.D.T. Minton, 'Dunlin Migration in Britain and Ireland', *Bird Study*, vol. 27, 1980, pp. 81–92.

170. J.G. Greenwood, 'Migration of Dunlin *Calidris alpina*: A world-wide overview', *Ringing and Migration*, vol. 5, 1984, pp. 35–9.

171. J.G. Greenwood, 'Geographical Variation and Taxonomy of the Dunlin *Calidris alpina* (L.)', *Bulletin of the British Ornithologists' Club*, vol. 106, 1986, pp. 43–56.

172. C.J. Patten, 'The Natural History of the Ruff', *Irish Naturalist*, vol. 9, 1900, pp. 187–209.

173. A.J. Prater, 'The Wintering Population of Ruffs in Britain and Ireland', *Bird Study*, vol. 20, 1973, pp. 245–50.

174. C.D. Hutchinson, 'Dowitcher Identification', *Irish Birds*, vol. 1, 1980, pp. 526–9.

175. Payne-Gallwey, *The Fowler in Ireland*.

176. G.R. Humphreys, *A List of Irish Birds* (Dublin: Stationery Office, 1937).

177. H.J. Wilson, *The Breeding and Wintering Ecology of the Woodcock* Scolopax rusticola *in Ireland*, Forest and Wildlife Service, Bray, unpublished report, 1982.

178. Ibid.

179. A.N. Hoodless and J.C. Coulson, 'Survival Rates and Movements of British and Continental Woodcock *Scolopax rusticola* in the British Isles', *Bird Study*, vol. 41, 1994, pp. 48–60.

180. A.J. Prater, 'The Wintering Population of the Black-tailed Godwit', *Bird Study*, vol. 22, 1975, pp. 169–76.

181. Ibid.

182 C.D. Hutchinson and J. O'Halloran, 'The Ecology of Black-tailed Godwits at an Irish South Coast Estuary', *Irish Birds*, vol. 5, 1994, pp. 165–72.

183. J. Wilson, 'Operation Godwit', *Cork Bird Report 2005–2006*, 2009, pp. 177–81.

184. S. Pierce and J. Wilson, 'Spring Migration of Whimbrels over Cork Harbour', *Irish Birds*, vol. 1, 1980, pp. 514–16.

185. J.A. Alves, M.P. Dias, V. Méndez, B. Katrínardóttir and T.G. Gunnarsson, 'Very Rapid Long-distance Sea Crossing by a Migratory Bird', *Scientific Reports*, vol. 6, 2016, doi 10.1038/srep38154.

186. Ibid.

187. I.P. Bainbridge and C.D.T. Minton, 'The Migration and Mortality of the Curlew in Britain and Ireland', *Bird Study*, vol. 25, 1978, pp. 39–50.

188. R. Summers, N. Christian, B. Etheridge, S. Rae, I. Cleasby and S. Pálsson, 'Scottish Breeding Greenshanks *Tringa nebularia* Do Not Migrate Far', *Bird Study*, vol. 67, 2020, pp. 1–7.

189. N.J.B.A. Branson, E.D. Ponting and C.D.T. Minton, 'Turnstone Migrations in Britain and Europe', *Bird Study*, vol. 25, 1978, pp. 181–7.

190. D.L. Davenport, 'The Spring Passage of the Pomarine Skua on British and Irish Coasts', *British Birds*, vol. 68, 1975, pp. 456–62.

191. D.L. Davenport, 'The Spring Passage of Pomarine and Long-tailed Skuas off the South and West Coasts of Britain and Ireland', *Irish Birds*, vol. 2, 1981, pp. 73–9.

192. D.L. Davenport, 'Large Passage of Skuas off Scotland and Ireland in May 1982 and 1983', *Irish Birds*, vol. 2, 1984, pp. 515–20.

193. Hutchinson, *Birds in Ireland.*

194. R.W. Furness, *The Skuas* (Calton: Poyser, 1987).

195. E. Magnusdottir, E.H.K. Leat, S. Bourgeon, H. Strøm, A Petersen, R.A. Phillips, S.A. Hanssen, J.O. Bustnes, P. Hersteinsson and R.W. Furness, 'Wintering Areas of Great Skuas *Stercorarius skua* Breeding in Scotland, Iceland and Norway', *Bird Study*, vol. 59, 2012, pp. 1–9.

196. J.T.R. Sharrock, *Scarce Migrant Birds in Britain and Ireland* (Berkhamstead: Poyser, 1974).

197. Cummins, Lauder, Lauder and Tierney, 'The Status of Ireland's Breeding Seabirds: Birds Directive Article 12 Reporting 2013–2018'.

198. J. Higginbotham, 'Gannets on the Bull Rock', *Irish Naturalist*, vol. 8, 1899, p. 251.

199. N. Horton, T. Brough, M.R. Fletcher, J.B.A. Rochard and P.I. Stanley, 'The Winter Distribution of Foreign Black-headed Gulls in the British Isles', *Bird Study*, vol. 31, 1984, pp. 171–86.

200. G.E. MacKinnon and J.C. Coulson, 'The Temporal and Geographical Distribution of Continental Black-headed Gulls *Larus ridibundus* in the British Isles', *Bird Study*, vol. 34, 1987, pp. 1–9.

201. Hutchinson, *Birds in Ireland.*

202 B. Madden and R.F. Ruttledge, 'Little Gulls in Ireland, 1970–1991', *Irish Birds*, vol. 5, 1993, pp. 23–34.

203. C.D. Hutchinson, 'The Changing Status of the Little Gull *Larus minutus* in Ireland', *Irish Bird Report*, vol. 19, 1972, pp. 11–21.

204. B. Madden, 'The Mediterranean Gull in Ireland, 1956–1985', *Irish Birds*, vol. 3, 1987, pp. 363–76.

205. N.J. Buckley, 'Kleptoparasitism of Black-headed Gulls *Larus ridibundus* by Common Gulls *Larus canus* at a Refuse Dump', *Bird Study*, vol. 34, 1987, pp. 10–11.

206. Cummins, Lauder, Lauder and Tierney, 'The Status of Ireland's Breeding Seabirds: Birds Directive Article 12 Reporting 2013–2018'.

207. B. Burke, P. Manley and S. Bayley, 'Migration and Wintering of First-year Lesser Black-backed Gulls *Larus fuscus graellsii* from Two Irish Colour-ringing Projects', *Irish Birds*, vol. 42, 2020, pp. 112–16.

208. G.A. Crème, P.M. Walsh, M. O'Callaghan and T.C. Kelly, 'The Changing Status of the Lesser Black-backed Gull *Larus fuscus* in Ireland', *Biology and Environment: Proceedings of the Royal Irish Academy*, vol. 97B, 1997, pp. 149–56.

209. J.C. Coulson, *Gulls* (London: Collins, 2019).

210. Cummins, Lauder, Lauder and Tierney, 'The Status of Ireland's Breeding Seabirds: Birds Directive Article 12 Reporting 2013–2018'.

211. Coulson, *Gulls*.

212. IRBC (Irish Rare Birds Committee), *Checklist of the Birds of Ireland*.

213. C. Cronin, C. Barton, H. Hussey and M. Carmody (eds), *Cork Bird Report 1996–2004* (Cork: Cork Bird Report Editorial Team, 2006).

214. C. Cronin, C. Barton, H. Hussey and M. Carmody (eds), *Cork Bird Report 2005–2006* (Cork: Cork Bird Report Editorial Team, 2009).

215. Cummins, Lauder, Lauder and Tierney, 'The Status of Ireland's Breeding Seabirds: Birds Directive Article 12 Reporting 2013–2018'.

216. Ibid.

217. N.J. Buckley and T.C. Kelly, 'Breeding Biology of Great Black-backed Gulls *Larus marinus* at a Declining Colony: Cape Clear Island, Co. Cork', *Irish Naturalists' Journal*, vol. 24, 1994, pp. 388–92.

218. B. Burke, N. Fitzgerald, H. Boland, T. Murray, T. Gittings and T.D. Tierney, 'Results from the First Three Years of Monitoring Post-breeding Tern Aggregations in Ireland', *Irish Birds*, vol. 42, 2020, pp. 35–44.

219. Cummins, Lauder, Lauder and Tierney, 'The Status of Ireland's Breeding Seabirds: Birds Directive Article 12 Reporting 2013–2018'.

220. B. O'Mahony and P. Smiddy, 'Breeding of the Common Tern *Sterna hirundo* in Cork Harbour, 1983–2017', *Irish Birds*, vol. 10, 2017, pp. 535–40.

221. Hutchinson, *Birds in Ireland*.

222. Kennedy, *A List of the Birds of Ireland*.

223. Cummins, Lauder, Lauder and Tierney, 'The Status of Ireland's Breeding Seabirds: Birds Directive Article 12 Reporting 2013–2018'.

224. Ibid.

225. J. Fisher and R.M. Lockley, *Sea-birds* (London: Collins, 1954).

226. A. Whilde, 'Auks Trapped in Salmon Drift Nets', *Irish Birds*, vol. 1, 1979, pp. 370–6.

227. Hutchinson, *Birds in Ireland*.

228. Cummins, Lauder, Lauder and Tierney, 'The Status of Ireland's Breeding Seabirds: Birds Directive Article 12 Reporting 2013–2018'.

229. P. Smiddy, 'Auks (Alcidae) Drowned in Fishing Nets in East Cork in January and February 1983', *Irish Naturalists' Journal*, vol. 26, 2001, pp. 414–19.

230. D. Roycroft, T.C. Kelly and L.J. Lewis, 'Birds, Seals and the Suspension Culture of Mussels in Bantry Bay, a Non-seaduck Area in Southwest Ireland', *Estuarine, Coastal and Shelf Science*, vol. 61, 2004, pp. 703–12.

231. P. Smiddy, 'Northern Razorbills on the South Irish Coast', *Irish Birds*, vol. 3, 1987, pp. 451–2.

232. D.E. Sergeant, 'Little Auks in Britain, 1948 to 1951', *British Birds*, vol. 45, 1952, pp. 122–33.

233. Smith, *The Antient and Present State of the County and City of Cork*.

234. Patterson and Foster, 'Irish Field Club Union: Report of the Fifth Triennial Conference and Excursion, held at Cork, July 11th to 16th, 1907. Vertebrata'.

235. Kennedy, Ruttledge and Scroope, *The Birds of Ireland*.

236. R. Hudson, 'The Spread of the Collared Dove in Britain and Ireland', *British Birds*, vol. 58, 1965, pp. 105–39.

237. R. Hudson, 'Collared Doves in Britain and Ireland during 1965–70', *British Birds*, vol. 65, 1972, pp. 139–55.

238. N. Davies, *Cuckoo: Cheating by nature* (London: Bloomsbury, 2015).

239. S.G. Sealy, J. O'Halloran and P. Smiddy, 'Cuckoo Hosts in Ireland', *Irish Birds*, vol. 5, 1996, pp. 381–90.

240. N. Ockendon, C.M. Hewson, A. Johnston and P.W. Atkinson, 'Declines in British-breeding Populations of Afro-Palearctic Migrant Birds are Linked to Bioclimatic Wintering Zone in Africa, Possibly via Constraints on Arrival Time Advancement', *Bird Study*, vol. 59, 2012, pp. 111–25.

241. Nagle, 'The Status of Birds of Prey and Owls in County Cork'.

242. J. Lusby, A. McCarthy, M. O'Clery, Á. Lynch, S. Bayley, T. Nagle, C. Forkan, M. Stanley and B. Nolan, *Barn Owl Monitoring Report 2020* (Kilcoole: BirdWatch Ireland, 2021).

243. Ibid.

244. P. Smiddy, D.P. Sleeman and J. O'Halloran, 'Barn Owl *Tyto alba* Diet in Ireland: A review', *Irish Birds*, vol. 41, 2018, pp. 39–48.

245. P. Smiddy, 'Dominance of Invasive Small Mammals in the Diet of the Barn Owl *Tyto alba* in County Cork, Ireland', *Biology and Environment: Proceedings of the Royal Irish Academy*, vol. 118B, 2018, pp. 49–53.

246. Nagle, 'The Status of Birds of Prey and Owls in County Cork'.

247. P. Smiddy and D.P. Sleeman, 'Diet of Long-eared Owl *Asio otus* and Short-eared Owl *Asio flammeus* in Ireland: A review', *Irish Birds*, vol. 42, 2020, pp. 27–34.

248. Ibid.

249. J. Stafford, 'Nightjar Enquiry, 1957–58', *Bird Study*, vol. 9, 1962, pp. 104–15.

250. A. Mee, 'A Survey of Nightjar in the Ballyhoura Hills', Cork and Limerick County Councils, unpublished report, 2020.

251. O. Crowe, S. Cummins, N. Gilligan, P. Smiddy and T.D. Tierney, 'An Assessment of the Current Distribution and Status of the Kingfisher *Alcedo atthis* in Ireland', *Irish Birds*, vol. 9, 2010, pp. 41–54.

252. A.D. McDevitt, Ł. Kajtoch, T.D. Mazgajski, R.F. Carden, I. Coscia, C. Osthoff, R.H. Coombes and F. Wilson, 'The Origins of Great Spotted Woodpeckers *Dendrocopos major* Colonizing Ireland Revealed by Mitochondrial DNA', *Bird Study*, vol. 58, 2011, pp. 361–4.

253. C.D. Hutchinson, 'Scarce Passerine Migrants in Ireland', *Irish Birds*, vol. 1, 1980, pp. 502–14.

254. D.B. Cabot, 'The Status and Distribution of the Chough, *Pyrrhocorax pyrrhocorax* (L.) in Ireland, 1960–65', *Irish Naturalists' Journal*, vol. 15, 1965, pp. 95–100.

255. R. Rolfe, 'The Status of the Chough in the British Isles', *Bird Study*, vol. 13, 1966, pp. 221–36.

256. J.B. Fox, 'A Transcript of Records of Jay *Garrulus glandarius* from the Books of a Dublin Taxidermist, 1881–1912', *Irish Birds*, vol. 10, 2017, pp. 469–74.

257. D.P. Sleeman and J.B. Fox, 'Did Jays *Garrulus glandarius* Go Extinct Locally Due to Killing for Fishing Flies or Habitat Destruction in 19th Century North Cork?', *Irish Birds*, vol. 42, 2020, pp. 110–11.

258. S. Wing, *The Natural History of Cape Clear 1959–2019* (Skibbereen: privately published, 2020).

259. P.D. O'Donoghue, T.F. Cross and J. O'Halloran, 'Carrion Crows in Ireland, 1969–1993', *Irish Birds*, vol. 5, 1996, pp. 399–406.

260. S.D. Berrow, T.C. Kelly and A.A. Myers, 'The Impact of Hooded Crows *Corvus corone cornix* L. on Populations of Intertidal Molluscs', in A.A. Myers, C. Little, M.J. Costello and J.C. Partridge (eds), *The Ecology of Lough Hyne* (Dublin: Royal Irish Academy, 1991), pp. 79–88.

261. S.D. Berrow, T.C. Kelly and A.A. Myers, 'Crows on Estuaries: Distribution and feeding behaviour of the Corvidae on four estuaries in southwest Ireland', *Irish Birds*, vol. 4, 1991, pp. 393–412.

262. S.D. Berrow, T.C. Kelly and A.A. Myers, 'The Diet of Coastal Breeding Hooded Crows *Corvus corone cornix*', *Ecography*, vol. 15, 1992, pp. 337–46.

263. S.D. Berrow, T.C. Kelly and A.A. Myers, 'The Mussel Caching Behaviour of Hooded Crows *Corvus corone cornix*', *Bird Study*, vol. 39, 1992, pp. 115–19.

264. S.D. Berrow, 'The Diet of Coastal Breeding Raven in Co. Cork', *Irish Birds*, vol. 4, 1992, pp. 555–8.

265. P. Smiddy, 'Diet of Coastal Breeding Ravens *Corvus corax* in East County Cork', *Irish Birds*, vol. 10, 2016, pp. 335–8.

266. P. Smiddy, 'Aspects of the Breeding Biology of Blue Tits (*Cyanistes caeruleus*) and Great Tits (*Parus major*) in County Cork', *Irish Naturalists' Journal*, vol. 37, 2021, pp. 97–101.

267. Ibid.

268. A. Whilde, *Threatened Mammals, Birds, Amphibians and Fish in Ireland. Irish Red Data Book 2: Vertebrates* (Belfast: Her Majesty's Stationery Office, 1993).

269. E. Dempsey, *Birdwatching in Ireland* (Dublin: Gill & Macmillan, 2008).

270. P.F. Donald, *The Skylark* (London: Poyser, 2004).

271. D.M. Bryant and G. Jones, 'Morphological Changes in a Population of Sand Martins *Riparia riparia* Associated with Fluctuations in Population Size', *Bird Study*, vol. 42, 1995, pp. 57–65.

272. C.J. Mead and J.D. Harrison, 'Sand Martin Movements within Britain and Ireland', *Bird Study*, vol. 26, 1979, pp. 73–86.

273. C.J. Mead and J.D. Harrison, 'Overseas Movements of British and Irish Sand Martins', *Bird Study*, vol. 26, 1979, pp. 87–98.

274. P. Smiddy, C. Cullen and J. O'Halloran, 'Autumn Use of a Reedbed by Barn Swallows *Hirundo rustica* and Sand Martins *Riparia riparia* in County Cork', *Irish Birds*, vol. 8, 2007, pp. 243–8.

275. P. Smiddy and J. O'Halloran, 'Breeding Biology of Barn Swallows *Hirundo rustica* in Counties Cork and Waterford, Ireland', *Bird Study*, vol. 57, 2010, pp. 256–60.

276. P. Smiddy, C. Cullen and J. O'Halloran, 'Time of Roosting of Barn Swallows *Hirundo rustica* at an Irish Reedbed during Autumn Migration', *Ringing and Migration*, vol. 23, 2007, pp. 228–30.

277. Smiddy, Cullen and O'Halloran, 'Autumn Use of a Reedbed by Barn Swallows *Hirundo rustica* and Sand Martins *Riparia riparia* in County Cork'.

278. S.J. Ormerod, 'Pre-migratory and Migratory Movements of Swallows *Hirundo rustica* in Britain and Ireland', *Bird Study*, vol. 38, 1991, pp. 170–8.

279. P. Smiddy, 'Winter Flock Size in the Long-tailed Tit *Aegithalos caudatus*, and Possible Effects of the Cold Winters of 2009/10 and 2010/11', *Irish Birds*, vol. 43, 2021, pp. 39–44.

280. J.K. Baker and G.P. Catley, 'Yellow-browed Warblers in Britain and Ireland, 1968–85', *British Birds*, vol. 80, 1987, pp. 93–109.

281. K. Thorup, 'Vagrancy of Yellow-browed Warbler *Phylloscopus inornatus* and Pallas's Warbler *Ph. proregulus* in North-west Europe: Misorientation on great circles?', *Ringing and Migration*, vol. 17, 1998, pp. 7–12.

282. J. Phillips, 'Autumn Vagrancy: "Reverse migration" and migratory orientation', *Ringing and Migration*, vol. 20, 2000, pp. 35–8.

283. J. Rabøl, 'Reversed Migration as the Cause of Westward Vagrancy by four *Phylloscopus* Warblers', *British Birds*, vol. 62, 1969, pp. 89–92.

284. Holloway, *The Historical Atlas of Breeding Birds in Britain and Ireland, 1875–1900.*

285. J.T.R. Sharrock, 'Migration Seasons of the *Sylvia* Warblers at Cape Clear Bird Observatory', *Bird Study*, vol. 15, 1968, pp. 99–103.

286. D.R. Langslow, 'Recent Increases of Blackcaps at Bird Observatories', *British Birds*, vol. 71, 1978, pp. 345–54.

287. R.M. Barrington, *The Migration of Birds* (London: Porter, and Dublin: Ponsonby, 1900).

288. D.R. Langslow, 'Movements of Blackcaps Ringed in Britain and Ireland', *Bird Study*, vol. 26, 1979, pp. 239–52.

289. P. Berthold and S.B. Terrill, 'Migratory Behaviour and Population Growth of Blackcaps Wintering in Britain and Ireland: Some hypotheses', *Ringing and Migration*, vol. 9, 1988, pp. 153–9.

290. A.G. More, *A List of Irish Birds* (Dublin: Science and Art Museum, 1885).

291. J.C. Smith, 'The Garden Warbler in Ireland', *Irish Naturalist*, vol. 3, 1894, pp. 46–7.

292. P. Smiddy, 'Breeding Birds in Ireland: Success and failure among colonists', in D.P. Sleeman, J. Carlsson and J.E.L. Carlsson (eds), *Mind the Gap II: New insights into the Irish postglacial* (Belfast: Irish Naturalists' Journal, 2014), pp. 89–99.

293. D. Winstanley, R. Spencer and K. Williamson, 'Where Have All the Whitethroats Gone?', *Bird Study*, vol. 21, 1974, pp. 1–14.

294. Sharrock, 'Migration Seasons of the *Sylvia* Warblers at Cape Clear Bird Observatory'.

295. Hobbs, *A List of Irish Birds.*

296. D.W. Gibbons, J.B. Reid and R.A. Chapman (eds), *The New Atlas of Breeding Birds in Britain and Ireland* (London: Poyser, 1993).

297. P. Smiddy and B. O'Mahony, 'The Status of the Reed Warbler *Acrocephalus scirpaceus* in Ireland', *Irish Birds*, vol. 6, 1997, pp. 23–8.

298. Ibid.

299. H. Insley and R.C. Boswell, 'The Timing of Arrivals of Reed and Sedge Warblers at South Coast Ringing Sites during Autumn Passage', *Ringing and Migration*, vol. 2, 1978, pp. 1–9.

300. Hutchinson, 'Scarce Passerine Migrants in Ireland'.

301. N. MacCoitir, *Ireland's Birds: Myths, legends and folklore* (Cork: Collins Press, 2015).

302. G. Anderson, *Birds of Ireland: Facts, folklore and history* (Cork: Collins Press, 2008).

303. Ibid.

304. P. Smiddy and R. Nairn, 'Rivers and Canals', in R. Nairn and J. O'Halloran (eds), *Bird Habitats in Ireland* (Cork: Collins Press, 2012), pp. 24–43.

305. O. Crowe, P. Smiddy, R. Whelan and A. Copland, 'Birds of Irish Rivers', in M. Kelly-Quinn and J. Reynolds (eds), *Ireland's Rivers* (Dublin: UCD Press, 2020), pp. 263–85.

306. P. Smiddy, J. O'Halloran, B. O'Mahony and A.J. Taylor, 'The Breeding Biology of the Dipper *Cinclus cinclus* in South-west Ireland', *Bird Study*, vol. 42, 1995, pp. 76–81.

307. J. O'Halloran, P. Smiddy and B. O'Mahony, 'Movements of Dippers *Cinclus cinclus* in Southwest Ireland', *Ringing and Migration*, vol. 20, 2000, pp. 147–51.

308. A.J. Taylor and J. O'Halloran, 'The Diet of the Dipper *Cinclus cinclus* as Represented by Faecal and Regurgitate Pellets: A comparison', *Bird Study*, vol. 44, 1997, pp. 338–47.

309. A.J. Taylor and J. O'Halloran, 'Diet of Dippers *Cinclus cinclus* during an Early Winter Spate and the Possible Implications for Dipper Populations Subjected to Climate Change', *Bird Study*, vol. 48, 2001, pp. 173–9.

310. Hutchinson, *Birds in Ireland.*

311. R.F. Ruttledge, *Ireland's Birds* (Dublin: Witherby, 1966).

312. K.M. Kelleher and J. O'Halloran, 'Breeding Biology of the Song Thrush *Turdus philomelos* in an Island Population', *Bird Study*, vol. 53, 2006, pp. 142–55.

313. K.M. Kelleher and J. O'Halloran, 'Influence of Nesting Habitat on Breeding Song Thrushes *Turdus philomelos*', *Bird Study*, vol. 54, 2007, pp. 221–9.

314. R.D.P. Milwright, 'Redwing *Turdus iliacus* Migration and Wintering Areas as Shown by Recoveries of Birds Ringed in the Breeding Season in Fennoscandia, Poland, the Baltic Republics, Russia, Siberia and Iceland', *Ringing and Migration*, vol. 21, 2002, pp. 5–15.

315. G.J. Fennessy and T.C. Kelly, 'Breeding Densities of Robin *Erithacus rubecula* in different habitats: The importance of hedgerow structure', *Bird Study*, vol. 53, 2006, pp. 97–104.

316. P. Holt, 'A Study of the Passerine Community of Hedgerows on Farmland Adjacent to Kilcolman National Nature Reserve, Co. Cork, Ireland', unpublished MSc thesis, University College Cork, 1996.

317. Hutchinson, *Birds in Ireland.*

318. Corcoran, *Saving Eden: The Gearagh and Irish nature.*

319. Hutchinson, *Birds in Ireland.*

320. Ibid.

321. S. Cummins and J. O'Halloran, 'The Breeding Biology of the Stonechat *Saxicola torquata* in Southwest Ireland', *Irish Birds*, vol. 7, 2003, pp. 177–86.

322. S. Cummins and J. O'Halloran, 'An Assessment of the Diet of Nestling Stonechats *Saxicola torquata* Using Compositional Analysis', *Bird Study*, vol. 49, 2002, pp. 139–45.

323. J.D. Summers-Smith, *The Tree Sparrow* (Cleveland: privately published, 1995).

324. H.M. Dobinson and A.J. Richards, 'The Effects of the Severe Winter of 1962/63 on Birds in Britain', *British Birds*, vol. 57, 1964, pp. 374–434.

325. P. Smiddy and J. O'Halloran, 'Breeding Biology of the Grey Wagtail *Motacilla cinerea* in Southwest Ireland', *Bird Study*, vol. 45, 1998, pp. 331–6.

326. J.T.R. Sharrock, 'Grey Wagtail Passage and Population Fluctuations in 1956–67', *Bird Study*, vol. 16, 1969, pp. 17–34.

327. P. Moore, 'A Tree Pipit *Anthus trivialis* Overwintering in Co. Cork', *Irish Birds*, vol. 41, 2018, p. 112.

328. I. Newton, *Finches* (London: Collins, 1972).

329. Ibid.

330. I.G. Main, 'Overseas Movements to and from Britain by Greenfinches *Carduelis chloris*', *Ringing and Migration*, vol. 19, 1999, pp. 191–9.

331. P.M. Walsh, J. O'Halloran, P.S. Giller and T.C. Kelly, 'Greenfinches *Carduelis chloris* and Other Bird Species Feeding on Cones of Noble Fir *Abies procera*', *Bird Study*, vol. 46, 1999, pp. 119–21.

332. Holloway, *The Historical Atlas of Breeding Birds in Britain and Ireland, 1875–1900.*

333. J.J. Watters, *The Natural History of the Birds of Ireland* (Dublin: McGlashan, 1853).

334 J.C. Smith, 'Crossbills (*Loxia curvirostra*) in Co. Cork', *Irish Naturalist*, vol. 3, 1894, p. 47.

335. R.J. Ussher, 'The Crossbill (*Loxia curvirostra*, L.) in Ireland', *Irish Naturalist*, vol. 1, 1892, pp. 6–9, 28–31.

336. Kennedy, *A List of the Birds of Ireland.*

337. Parkin and Knox, *The Status of Birds in Britain and Ireland.*

338. Ibid.

339. I. Newton, *Farming and Birds* (London: Collins, 2017).

340. R.D. Bell, 'Some Thoughts on the Apparent Ecological Expansion of the Reed Bunting', *British Birds*, vol. 62, 1969, pp. 209–18.

341. Hutchinson, *Birds in Ireland.*

342. Whilde, *Threatened Mammals, Birds, Amphibians and Fish in Ireland.*

343. R.J. O'Connor and M. Shrubb, *Farming and Birds* (Cambridge: Cambridge University Press, 1986).

344. A.J. Taylor and J. O'Halloran, 'The Decline of the Corn Bunting, *Miliaria calandra*, in the Republic of Ireland', *Biology and Environment: Proceedings of the Royal Irish Academy*, vol. 102B, 2002, pp. 165–75.

345. Newton, *Farming and Birds.*

BIBLIOGRAPHY

Alves, J.A., Dias, M.P., Méndez, V., Katrínardóttir, B. and Gunnarsson, T.G., 'Very Rapid Long-distance Sea Crossing by a Migratory Bird', *Scientific Reports*, vol. 6, 2016, doi 10.1038/srep38154

Anderson, G., *Birds of Ireland: Facts, folklore and history* (Cork: Collins Press, 2008)

Anglesea, M. (ed.), *Birds of Ireland by Richard Dunscombe Parker* (Belfast: Blackstaff Press, 1984)

——, 'The Art of Nature Illustration', in J.W. Foster (ed.), *Nature in Ireland: A scientific and cultural history* (Dublin: Lilliput Press, 1997), pp. 497–523

Anon., 'The Stormy Petrel–*Procellaria pelagica* Commonly Called Mother Carey's Chicken', *Dublin Penny Journal*, vol. 2, 1833, p. 160

——, 'Survey of the South-west Islands of Ireland', Royal Irish Academy, Dublin, unpublished report, 1955

Bainbridge, I.P. and Minton, C.D.T., 'The Migration and Mortality of the Curlew in Britain and Ireland', *Bird Study*, vol. 25, 1978, pp. 39–50

Bairlein, F., 'The Study of Bird Migrations: Some future perspectives', *Bird Study*, vol. 50, 2003, pp. 243–53

Baker, J.K. and Catley, G.P., 'Yellow-browed Warblers in Britain and Ireland, 1968–85', *British Birds*, vol. 80, 1987, pp. 93–109

Balmer, D.E., Gillings, S., Caffrey, B.J., Swann, R.L., Downie, I.S. and Fuller, R.J., *Bird Atlas 2007–11: The breeding and wintering birds of Britain and Ireland* (Thetford: BTO Books, 2013)

Barrington, R.M., *The Migration of Birds* (London: Porter, and Dublin: Ponsonby, 1900)

——, 'Bird Records from Irish Lighthouses', *Irish Naturalist*, vol. 19, 1910, p. 104

——, 'The Great Rush of Birds on the Night of March 29th–30th, as Observed in Ireland', *Irish Naturalist*, vol. 20, 1911, pp. 97–110

——, 'Richard John Ussher', *Irish Naturalist*, vol. 22, 1913, pp. 221–7

——, 'Richard John Ussher, DL, MRIA: A memoir', *British Birds*, vol. 7, 1914, pp. 182–5

——, 'Bird Rushes and Wrens', *Irish Naturalist*, vol. 23, 1914, pp. 241–7

—— and Ussher, R.J., 'Irish Breeding-stations of the Gannet, *Sula bassana*', *Zoologist*, series 3, vol. 8, 1884, pp. 473–82

Bell, J. and Watson, M., *A History of Irish Farming 1750–1950* (Dublin: Four Courts Press, 2008)

Bell, R.D., 'Some Thoughts on the Apparent Ecological Expansion of the Reed Bunting', *British Birds*, vol. 62, 1969, pp. 209–18

Berrow, S.D., 'Predation by the Hooded Crow *Corvus corone cornix* on Freshwater Pearl Mussels *Margaritifera margaritifera*', *Irish Naturalists' Journal*, vol. 23, 1991, pp. 492–3

——, 'The Diet of Coastal Breeding Ravens in Co. Cork', *Irish Birds*, vol. 4, 1992, pp. 555–8

——, Kelly, T.C. and Myers, A.A., 'The Impact of Hooded Crows *Corvus corone cornix* L. on Populations of Intertidal Molluscs', in A.A. Myers, C. Little, M.J. Costello and J.C. Partridge (eds), *The Ecology of Lough Hyne* (Dublin: Royal Irish Academy, 1991), pp. 79–88

——, Kelly, T.C. and Myers, A.A., 'Crows on Estuaries: Distribution and feeding behaviour of the Corvidae on four estuaries in southwest Ireland', *Irish Birds*, vol. 4, 1991, pp. 393–412

——, Kelly, T.C. and Myers, A.A., 'The Diet of Coastal Breeding Hooded Crows *Corvus corone cornix*', *Ecography*, vol. 15, 1992, pp. 337–46

——, Kelly, T.C. and Myers, A.A., 'The Mussel Caching Behaviour of Hooded Crows *Corvus corone cornix*', *Bird Study*, vol. 39, 1992, pp. 115–19

——, Mackie, K.L., O'Sullivan, O., Shepherd, K.B., Mellon, C. and Coveney, J.A., 'The Second International Chough Survey in Ireland, 1992', *Irish Birds*, vol. 5, 1993, pp. 1–10

Berthold, P. and Terrill, S.B., 'Migratory Behaviour and Population Growth of Blackcaps Wintering in Britain and Ireland: Some hypotheses', *Ringing and Migration*, vol. 9, 1988, pp. 153–9

Birkhead, M. and Perrins, C., *The Mute Swan* (London: Croom Helm, 1986)

Boisseau, S. and Yalden, D.W., 'The Former Status of the Crane *Grus grus* in Britain', *Ibis*, vol. 140, 1998, pp. 482–500

Boland, H. and Crowe, O., *Irish Wetland Bird Survey: Waterbird status and distribution 2001/02–2008/09* (Newcastle: BirdWatch Ireland, 2012)

——, McElwaine, J.G., Henderson, G., Hall, C., Walsh, A. and Crowe, O., 'Whooper *Cygnus cygnus* and Bewick's *C. columbianus bewickii* Swans in Ireland: Results of the International Swan Census, January 2010', *Irish Birds*, vol. 9, 2010, pp. 1–10

Boyd, H., 'The "Wreck" of Leach's Petrels in the Autumn of 1952', *British Birds*, vol. 47, 1954, pp. 137–63

Branson, N.J.B.A., Ponting, E.D. and Minton, C.D.T., 'Turnstone Migrations in Britain and Europe', *Bird Study*, vol. 25, 1978, pp. 181–7

Brazil, M., *The Whooper Swan* (London: Poyser, 2003)

Brooke, M., *The Manx Shearwater* (London: Poyser, 1990)

Brown, A., 'One Hundred Years of Notable Avian Events in *British Birds*', *British Birds*, vol. 100, 2007, pp. 214–43

Browne, A.M., O'Halloran, J. and Smiddy, P., 'Introduced Canada *Branta canadensis* and Greylag Goose *Anser anser* Populations in Ireland, 1994', *Irish Birds*, vol. 6, 1998, pp. 233–6

Bryant, D.M. and Jones, G., 'Morphological Changes in a Population of Sand Martins *Riparia riparia* Associated with Fluctuations in Population Size', *Bird Study*, vol. 42, 1995, pp. 57–65

Buckley, N.J., 'Kleptoparasitism of Black-headed Gulls *Larus ridibundus* by Common Gulls *Larus canus* at a Refuse Dump', *Bird Study*, vol. 34, 1987, pp. 10–11

—— and Kelly, T.C., 'Breeding Biology of Great Black-backed Gulls *Larus marinus* at a Declining Colony: Cape Clear Island, Co. Cork', *Irish Naturalists' Journal*, vol. 24, 1994, pp. 388–92

Bullock, I.D., Drewett, D.R. and Mickleburgh, S.P., 'The Chough in Ireland', *Irish Birds*, vol. 2, 1983, pp. 257–71

Bunn, D.S., Warburton, A.B. and Wilson, R.D.S., *The Barn Owl* (Calton: Poyser, 1982)

Burke, B., Fitzgerald, N., Boland, H., Murray, T., Gittings, T. and Tierney, T.D., 'Results from the First Three Years of Monitoring Post-breeding Tern Aggregations in Ireland', *Irish Birds*, vol. 42, 2020, pp. 35–44

——, Manley, P. and Bayley, S., 'Migration and Wintering of First-year Lesser Black-backed Gulls *Larus fuscus graellsii* from Two Irish Colour-ringing Projects', *Irish Birds*, vol. 42, 2020, pp. 112–16

——, McElwaine, J.G., Fitzgerald, N., Kelly, S.B.A., McCulloch, N., Walsh, A.J. and Lewis, L.J., 'Population Size, Breeding Success and Habitat Use of Whooper Swan *Cygnus cygnus* and Bewick's Swan *Cygnus columbianus bewickii* in Ireland: Results of the 2020 International Swan Census', *Irish Birds*, vol. 43, 2021, pp. 57–70

Cabot, D.B., 'The Status and Distribution of the Chough, *Pyrrhocorax pyrrhocorax* (L.) in Ireland, 1960–65', *Irish Naturalists' Journal*, vol. 15, 1965, pp. 95–100

——, 'Movements and Migration of the Mallard in Ireland', *Irish Birds*, vol. 1, 1977, pp. 37–45

——, *Wildfowl* (London: Collins, 2009)

Cambrensis, Giraldus (Gerald of Wales), *The History and Topography of Ireland (Topographia Hiberniae)*, ed. J.J. O'Meara (Dublin: Dolmen Press, 1982)

Campbell, H.F., 'Hoopoe Killed by Hawk in Co. Cork', *Irish Naturalists' Journal*, vol. 9, 1948, p. 148

Casey, C., 'Distribution and Conservation of the Corncrake in Ireland, 1993–1998', *Irish Birds*, vol. 6, 1998, pp. 159–76

Clarke, R. and Watson, D., 'The Hen Harrier *Circus cyaneus* Winter Roost Survey in Britain and Ireland', *Bird Study*, vol. 37, 1990, pp. 84–100

Colhoun, K., McElwaine, J.G., Cranswick, P.A., Enlander, I. and Merne, O.J., 'Numbers and Distribution of Whooper *Cygnus cygnus* and Bewick's *C. columbianus bewickii* Swans in Ireland: Results of the International Swan Census, January 2000', *Irish Birds*, vol. 6, 2000, pp. 485–94

—— and Newton, S.F., 'Winter Waterbird Populations on Non-estuarine Coasts in the Republic of Ireland: Results of the 1997/98 Non-Estuarine Coastal Waterfowl Survey (NEWS)', *Irish Birds*, vol. 6, 2000, pp. 527–42

Collins, R. and Whelan, J., 'Mute Swan Herds in Dublin and Wicklow', *Irish Birds*, vol. 5, 1993, pp. 11–22

——, 'Movements in an Irish Mute Swan *Cygnus olor* Population', *Ringing and Migration*, vol. 15, 1994, pp. 40–9

Connors, P.G., McCaffery, B.J. and Maron, J.L., 'Speciation in Golden Plovers, *Pluvialis dominica* and *Pluvialis fulva*: Evidence from their breeding grounds', *Auk*, vol. 110, 1993, pp. 9–20

Coombes, R.H. and Wilson, F.R., 'Colonisation and Breeding Status of the Great Spotted Woodpecker *Dendrocopos major* in the Republic of Ireland', *Irish Birds*, vol. 10, 2015, pp. 183–96

Corcoran, K., *Saving Eden: The Gearagh and Irish nature* (Macroom: Gearagh Press, 2021)

Corkery, I., Irwin, S., Quinn, J.L., Keating, U., Lusby, J. and O'Halloran, J., 'Changes in Forest Cover Result in a Shift in Bird Community Composition', *Journal of Zoology*, vol. 310, 2020, pp. 306–14

Coulson, J.C., *Gulls* (London: Collins, 2019)

Cowley, E., 'Sand Martin Population Trends in Britain, 1965–1978', *Bird Study*, vol. 26, 1979, pp. 113–16

Cramp, S., Bourne, W.R.P. and Saunders, D., *The Seabirds of Britain and Ireland* (London: Collins, 1974)

Cramp, S. and Simmons, K.E.L. (eds), *Handbook of the Birds of Europe, the Middle East and North Africa: The birds of the Western Palearctic* (vols 1–3); Cramp, S. (ed.) (vols 4–6); Cramp, S. and Perrins, C.M. (eds) (vols 7–9) (Oxford: Oxford University Press, 1977–94)

Cranswick, P.A., Bowler, J.M., Delany, S.N., Einarsson, O., Gardarsson, A., McElwaine, J.G., Merne, O.J., Rees, E.C. and Wells, J.H., 'Numbers of Whooper

Swans *Cygnus cygnus* in Iceland, Ireland and Britain in January 1995: Results of the international Whooper Swan census', *Wildfowl*, vol. 47, 1996, pp. 17–30

Crème, G.A., Walsh, P.M., O'Callaghan, M. and Kelly, T.C., 'The Changing Status of the Lesser Black-backed Gull *Larus fuscus* in Ireland', *Biology and Environment: Proceedings of the Royal Irish Academy*, vol. 97B, 1997, pp. 149–56

Cronin, C., Barton, C., Hussey, H. and Carmody, M. (eds), *Cork Bird Report 1996–2004* (Cork: Cork Bird Report Editorial Team, 2006)

——, *Cork Bird Report 2005–2006* (Cork: Cork Bird Report Editorial Team, 2009)

Crowe, O., *Ireland's Wetlands and their Waterbirds: Status and distribution* (Newcastle: BirdWatch Ireland, 2005)

——, 'The Garden Bird Survey: Monitoring birds of Irish gardens during winters between 1994/95 and 2003/04', *Irish Birds*, vol. 7, 2005, pp. 475–82

——, Austin, G.E. and Boland, H., 'Waterbird Populations on Non-estuarine Coasts in Ireland: Results of the 2006/07 Non-Estuarine Coastal Waterbird Survey (NEWS)', *Irish Birds*, vol. 9, 2012, pp. 385–96

——, Austin, G.E., Colhoun, K., Cranswick, P.A., Kershaw, M. and Musgrove, A.J., 'Estimates and Trends of Waterbird Numbers Wintering in Ireland, 1994/95 to 2003/04', *Bird Study*, vol. 55, 2008, pp. 66–77

——, Coombes, R.H., Lysaght, L., O'Brien, C., Choudhury, K.R., Walsh, A.J., Wilson, H.J. and O'Halloran, J., 'Population Trends of Widespread Breeding Birds in the Republic of Ireland 1998–2008', *Bird Study*, vol. 57, 2010, pp. 267–80

——, Coombes, R.H. and O'Halloran, J., 'Estimates and Trends of Common Breeding Birds in the Republic of Ireland', *Irish Birds*, vol. 10, 2014, pp. 23–32

——, Cummins, S., Gilligan, N., Smiddy, P. and Tierney, T.D., 'An Assessment of the Current Distribution and Status of the Kingfisher *Alcedo atthis* in Ireland', *Irish Birds*, vol. 9, 2010, pp. 41–54

——, McElwaine, J.G., Boland, H. and Enlander, I.J., 'Whooper *Cygnus cygnus* and Bewick's *C. columbianus bewickii* Swans in Ireland: Results of the International Swan Census, January 2015', *Irish Birds*, vol. 10, 2015, pp. 151–8

——, McElwaine, J.G., Worden, J., Watson, G.A., Walsh, A.J. and Boland, H., 'Whooper *Cygnus cygnus* and Bewick's *C. columbianus bewickii* Swans in Ireland: Results of the International Swan Census, January 2005', *Irish Birds*, vol. 7, 2005, pp. 483–8

——, Musgrove, A.J. and O'Halloran, J., 'Generating Population Estimates for Common and Widespread Breeding Birds in Ireland', *Bird Study*, vol. 61, 2014, pp. 82–90

——, Smiddy, P., Whelan, R. and Copland, A., 'Birds of Irish Rivers', in M. Kelly-Quinn and J. Reynolds (eds), *Ireland's Rivers* (Dublin: UCD Press, 2020), pp. 263–85

Cullen, C., Kelly, T.C. and O'Halloran, J., 'Late Summer Foraging of the Little Egret *Egretta garzetta* in East County Cork', *Irish Birds*, vol. 7, 2005, pp. 523–8

—— and Smiddy, P., 'Spring and Summer Use of a Reedbed by Barn Swallows (*Hirundo rustica*) and Sand Martins (*Riparia riparia*) in Co. Cork', *Irish Naturalists' Journal*, vol. 29, 2008, pp. 126–8

Cullinane, J.P., 'Joshua Reubens Harvey', *Irish Naturalists' Journal*, vol. 17, 1972, pp. 223–5

Cummins, S.T., Bleasdale, A., Douglas, C., Newton, S.F., O'Halloran, J. and Wilson, H.J., 'Densities and Population Estimates of Red Grouse *Lagopus lagopus scotica* in Ireland Based on the 2006–2008 National Survey', *Irish Birds*, vol. 10, 2015, pp. 197–210

——, Lauder, C., Lauder, A. and Tierney, T.D., 'The Status of Ireland's Breeding Seabirds: Birds Directive Article 12 Reporting 2013–2018', *Irish Wildlife Manuals, No. 114* (Dublin: National Parks and Wildlife Service, 2019)

—— and O'Halloran, J., 'An Assessment of the Diet of Nestling Stonechats *Saxicola torquata* Using Compositional Analysis', *Bird Study*, vol. 49, 2002, pp. 139–45

—— and O'Halloran, J., 'The Breeding Biology of the Stonechat *Saxicola torquata* in Southwest Ireland', *Irish Birds*, vol. 7, 2003, pp. 177–86

D'Arcy, G., *Ireland's Lost Birds* (Dublin: Four Courts Press, 1999)

Davenport, D.L., 'The Spring Passage of the Pomarine Skua on British and Irish Coasts', *British Birds*, vol. 68, 1975, pp. 456–62

——, 'The Spring Passage of Pomarine and Long-tailed Skuas off the South and West Coasts of Britain and Ireland', *Irish Birds*, vol. 2, 1981, pp. 73–9

——, 'Large Passage of Skuas off Scotland and Ireland in May 1982 and 1983', *Irish Birds*, vol. 2, 1984, pp. 515–20

Davies, N., *Cuckoo: Cheating by nature* (London: Bloomsbury, 2015)

Dawson, P.A.C., 'Hen Harrier *Circus cyaneus* Utilisation of a Limestone Fen in North County Cork', *Irish Birds*, vol. 7, 2005, pp. 503–10

Deane, C.D., 'The Capercaillie as an Irish Species', *Irish Birds*, vol. 1, 1979, pp. 364–9

del Hoyo, J., Elliott, A. and Sargatal, J. (eds), *Handbook of the Birds of the World* (vols 1–7); del Hoyo, J., Elliott, A. and Christie, D.A. (eds) (vols 8–16) (Barcelona: Lynx Edicions, 1992–2011)

D'Elbee, J. and Hemery, G., 'Diet and Foraging Behaviour of the British Storm Petrel *Hydrobates pelagicus* in the Bay of Biscay during Summer', *Ardea*, vol. 86, 1997, pp. 1–10

Dempsey, E., *Birdwatching in Ireland* (Dublin: Gill & Macmillan, 2008)

Dobinson, H.M. and Richards, A.J., 'The Effects of the Severe Winter of 1962/63 on Birds in Britain', *British Birds*, vol. 57, 1964, pp. 374–434

Donald, P.F., *The Skylark* (London: Poyser, 2004)

Donovan, G.E., 'Goosander (*Mergus merganser*) in Co. Cork', *Irish Naturalist*, vol. 2, 1893, p. 86

Doran, C.G., 'The Lough of Cork', *Journal of the Cork Historical and Archaeological Society*, series A, vol. 2, 1893, pp. 193–8, 213–19, 237–44

Durand, A.L., 'A Remarkable Fall of American Land-birds on the *Mauretania*, New York to Southampton, October 1962', *British Birds*, vol. 56, 1963, pp. 157–64

——, 'Landbirds over the North Atlantic: Unpublished records 1961–65 and thoughts a decade later', *British Birds*, vol. 65, 1972, pp. 428–42

Evans, P.G.H. and Lovegrove, R.R., 'The Birds of the South West Irish Islands', *Irish Bird Report*, vol. 21, 1974, pp. 33–64

Evans, R.J., O'Toole, L. and Whitfield, D.P., 'The History of Eagles in Britain and Ireland: An ecological review of placename and documentary evidence from the last 1500 years', *Bird Study*, vol. 59, 2012, pp. 335–49

Fairley, J.S., *An Irish Beast Book* (Belfast: Blackstaff Press, 1984)

Fennessy, G.J. and Kelly, T.C., 'Breeding Densities of Robin *Erithacus rubecula* in Different Habitats: The importance of hedgerow structure', *Bird Study*, vol. 53, 2006, pp. 97–104

Fisher, J., *The Fulmar* (London: Collins, 1952)

——, *The Shell Bird Book* (London: Ebury Press & Michael Joseph, 1966)

——, 'The Fulmar Population of Britain and Ireland, 1959', *Bird Study*, vol. 13, 1966, pp. 5–76

—— and Lockley, R.M., *Sea-birds* (London: Collins, 1954)

Flynn, J.E., 'Bird Notes from West Cork', *Irish Naturalists' Journal*, vol. 2, 1929, pp. 201–2

——, 'Snow Bunting in County Cork', *Irish Naturalists' Journal*, vol. 3, 1930, p. 38

——, 'The Diver Family at Glengarriff', *Irish Naturalists' Journal*, vol. 3, 1930, p. 39

——, 'Bird Notes from West Cork', *Irish Naturalists' Journal*, vol. 5, 1935, p. 229

——, 'Whimbrel in Winter and Snow Bunting in County Cork', *Irish Naturalists' Journal*, vol. 6, 1936, p. 74

——, 'Little Egret in Co. Cork', *British Birds*, vol. 34, 1941, pp. 243–4

——, 'Some South-west Irish Seabird Colonies', *Seabird Bulletin*, vol. 2, 1966, pp. 52–5

——, 'Distribution Map of Breeding Corncrakes in the Glengarriff Area', unpublished report, 1966

Forsyth, I., 'A Breeding Census of Mute Swans in Ireland in 1978', *Irish Birds*, vol. 1, 1980, pp. 492–501

Fox, J.B., 'Duck Decoys in North Co. Cork', *Mallow Field Club Journal*, vol. 2, 1984, pp. 111–20

——, 'A Transcript of Records of Jay *Garrulus glandarius* from the Books of a Dublin Taxidermist, 1881–1912', *Irish Birds*, vol. 10, 2017, pp. 469–74

Freeland, J., Allen, D. and Anderson, S., *DNA Analysis of Red Grouse: An analysis of taxonomy and genetic diversity* (Belfast: Environment and Heritage Service, 2006)

Furness, R.W., *The Skuas* (Calton: Poyser, 1987)

—— and Baillie, S.R., 'Factors Affecting Capture Rate and Biometrics of Storm Petrels on St Kilda', *Ringing and Migration*, vol. 3, 1981, pp. 137–47

George, T.N., Johnson, G.A.L., Mitchell, M., Prentice, J.E., Ramsbottom, W.H.C., Sevastopulo, G.D. and Wilson, R.B., 'A Correlation of Dinantian Rocks in the British Isles', *Geological Society of London Special Report* 7, 1976

Gibbons, D.W., Reid, J.B. and Chapman, R.A. (eds), *The New Atlas of Breeding Birds in Britain and Ireland* (London: Poyser, 1993)

Gilbert, G., Stanbury, A. and Lewis, L., 'Birds of Conservation Concern in Ireland 4: 2020–2026', *Irish Birds*, vol. 43, 2021, pp. 1–22

Gittings, T., 'Waterbird Monitoring of Cork Harbour: 1994/95–2002/03', *Cork Bird Report 1996–2004*, 2006, pp. 319–39

——, 'Nocturnal Communal Roosting Behaviour in Great Crested Grebes *Podiceps cristatus*', *Irish Birds*, vol. 10, 2017, pp. 483–92

——, 'Cork Harbour I-WeBS Summary Reports 2011/12–2019/20', unpublished report, 2020

Gray, N., Thomas, G., Trewby, M. and Newton, S.F., 'The Status and Distribution of Choughs *Pyrrhocorax pyrrhocorax* in the Republic of Ireland 2002/03', *Irish Birds*, vol. 7, 2003, pp. 147–56

Green, R.E. and Stowe, T.J., 'The Decline of the Corncrake *Crex crex* in Britain and Ireland in Relation to Habitat Change', *Journal of Applied Ecology*, vol. 30, 1993, pp. 689–95

Greenwood, J.G., 'Migration of Dunlin *Calidris alpina*: A world-wide overview', *Ringing and Migration*, vol. 5, 1984, pp. 35–9

——, 'Geographical Variation and Taxonomy of the Dunlin *Calidris alpina* (L.)', *Bulletin of the British Ornithologists' Club*, vol. 106, 1986, pp. 43–56

Greenwood, J.J.D., '100 Years of Ringing in Britain and Ireland', *Ringing and Migration*, vol. 24, 2009, pp. 147–53

Grimmett, R.F.A. and Jones, T.A., *Important Bird Areas in Europe* (Cambridge: International Council for Bird Preservation, 1989)

Guilford, T.C., Meade, J., Freeman, R., Biro, D., Evans, T., Bonadonna, F., Boyle, D., Roberts, S. and Perrins, C.M., 'GPS Tracking of the Foraging Movements of Manx Shearwaters *Puffinus puffinus* Breeding on Skomer Island, Wales', *Ibis*, vol. 150, 2008, pp. 462–73

Hall, J.J., 'The Cock of the Wood', *Irish Birds*, vol. 2, 1981, pp. 38–47

Hannon, C., Berrow, S.D. and Newton, S.F., 'The Status and Distribution of Breeding Sandwich *Sterna sandvicensis*, Roseate *S. dougallii*, Common *S. hirundo*, Arctic *S. paradisaea* and Little Terns *S. albifrons* in Ireland in 1995', *Irish Birds*, vol. 6, 1997, pp. 1–22

Hardy, A.R. and Minton, C.D.T., 'Dunlin Migration in Britain and Ireland', *Bird Study*, vol. 27, 1980, pp. 81–92

Harvey, J.R., 'Memoranda towards a Fauna of the County of Cork, Div.: Vertebrata', in J.R. Harvey, J.D. Humphreys and T. Power (eds), *Contributions towards a Fauna and Flora of the County of Cork* (London: John Van Voorst, and Cork: George Purcell, 1845), pp. 1–24

——, 'The Fauna of the County Cork', in M.F. Cusack (ed.), *A History of the City and County of Cork* (Dublin: McGlashan & Gill, and Cork: Francis Guy, 1875), pp. 454–66

——, Humphreys, J.D. and Power, T. (eds), *Contributions towards a Fauna and Flora of the County of Cork* (London: John Van Voorst, and Cork: George Purcell, 1845)

Heardman, C., 'Biodiversity: Cork's place in Europe', in M. Hallinan, C. Nelligan and D.P. Sleeman (eds), *Europe and the County of Cork: A heritage perspective* (Cork: Cork County Council, 2018), pp. 104–25, 189–200

Heselden, R.G.W., 'Sedimentology and Stratigraphy of the Courceyan-Asbian Limestones (Dinantian, Lower Carboniferous) of the Cork Harbour Area, Southern Ireland', unpublished PhD thesis, University College Cork, vol. 1, 1992

Higginbotham, J., 'Gannets on the Bull Rock', *Irish Naturalist*, vol. 8, 1899, p. 251

Hillis, J.P., 'Rare Irish Breeding Birds, 1992–2001', *Irish Birds*, vol. 7, 2003, pp. 157–72

Hirst, M., 'Treecreepers *Certhia familiaris* Foraging on Man-made Structures at Kilcolman National Nature Reserve, 2006–2011', *Irish Birds*, vol. 9, 2011, p. 327

Hobbs, J., *A List of Irish Birds* (Version 10.2) (www.southdublinbirds.com) (Dublin: South Dublin Branch of BirdWatch Ireland, 2021)

Holloway, S., *The Historical Atlas of Breeding Birds in Britain and Ireland: 1875–1900* (London: Poyser, 1996)

Holt, P., 'A Study of the Passerine Community of Hedgerows on Farmland Adjacent to Kilcolman National Nature Reserve, Co. Cork, Ireland', unpublished MSc thesis, University College Cork, 1996

Holyoak, D.T. and Ratcliffe, D.A., 'The Distribution of the Raven in Britain and Ireland', *Bird Study*, vol. 15, 1968, pp. 191–7

Hoodless, A.N. and Coulson, J.C., 'Survival Rates and Movements of British and Continental Woodcock *Scolopax rusticola* in the British Isles', *Bird Study*, vol. 41, 1994, pp. 48–60

Horton, N., Brough, T., Fletcher, M.R., Rochard, J.B.A. and Stanley, P.I., 'The Winter Distribution of Foreign Black-headed Gulls in the British Isles', *Bird Study*, vol. 31, 1984, pp. 171–86

Hudson, R., 'The Spread of the Collared Dove in Britain and Ireland', *British Birds*, vol. 58, 1965, pp. 105–39

——, 'Collared Doves in Britain and Ireland during 1965–70', *British Birds*, vol. 65, 1972, pp. 139–55

Humphreys, G.R., *A List of Irish Birds* (Dublin: Stationery Office, 1937)

Hutchinson, C.D., 'The Changing Status of the Little Gull *Larus minutus* in Ireland', *Irish Bird Report*, vol. 19, 1972, pp. 11–21

—— (ed.), *The Birds of Dublin and Wicklow* (Dublin: Irish Wildbird Conservancy, 1975)

——, *Ireland's Wetlands and their Birds* (Dublin: Irish Wildbird Conservancy, 1979)

——, 'Scarce Passerine Migrants in Ireland', *Irish Birds*, vol. 1, 1980, pp. 502–14

——, 'Dowitcher Identification', *Irish Birds*, vol. 1, 1980, pp. 526–9

——, 'Cape Clear Bird Observatory 1970–1980', *Irish Birds*, vol. 2, 1981, pp. 60–72

——, *Birds in Ireland* (Calton: Poyser, 1989)

——, 'Bird Study in Ireland', in J.W. Foster (ed.), *Nature in Ireland: A scientific and cultural history* (Dublin: Lilliput Press, 1997), pp. 262–82

—— and Neath, B., 'Little Gulls in Britain and Ireland', *British Birds*, vol. 71, 1978, pp. 563–82

—— and O'Halloran, J., 'The Waterfowl of Cork Harbour', *Irish Birds*, vol. 2, 1984, pp. 445–56

—— and O'Halloran, J., 'The Ecology of Black-tailed Godwits at an Irish South Coast Estuary', *Irish Birds*, vol. 5, 1994, pp. 165–72

Hutchinson, G.E., 'The Harp that Once … A Note on the Discovery of Stridulation in the Corixid Water-bugs', *Irish Naturalists' Journal*, vol. 20, 1982, pp. 457–66

Insley, H. and Boswell, R.C., 'The Timing of Arrivals of Reed and Sedge Warblers at South Coast Ringing Sites during Autumn Passage', *Ringing and Migration*, vol. 2, 1978, pp. 1–9

IRBC (Irish Rare Birds Committee), *Checklist of the Birds of Ireland* (Dublin: BirdWatch Ireland, 1998)

IRSG (Irish Raptor Study Group), *Birds of Prey and Owls in Ireland: Restoration* (Clonmel: Irish Raptor Study Group, 2006)

Johnson, G.A.L., 'Geographical Change in Britain during the Carboniferous Period', *Proceedings of the Yorkshire Geological Society*, vol. 44, 1982, pp. 181–203

Jones, G., 'Selection against Large Size in the Sand Martin *Riparia riparia* during a Dramatic Population Crash', *Ibis*, vol. 129, 1987, pp. 274–80

Kelleher, C., 'Observation of Bat Predation at a Soprano Pipistrelle (*Pipistrellus pygmaeus* (Leach 1825)) Roost by a Sparrowhawk (*Accipiter nisus* L. 1758) in West Cork', *Irish Naturalists' Journal*, vol. 29, 2008, p. 57

Kelleher, K.M. and O'Halloran, J., 'Breeding Biology of the Song Thrush *Turdus philomelos* in an Island Population', *Bird Study*, vol. 53, 2006, pp. 142–55

—— and O'Halloran, J., 'Influence of Nesting Habitat on Breeding Song Thrushes *Turdus philomelos*', *Bird Study*, vol. 54, 2007, pp. 221–9

Keller, V., Herrando, S., Voříšek, P., Franch, M., Kipson, M., Milanesi, P., Martí, D., Anton, M., Klvaňová, A., Kalyakin, M.V., Bauer, H.-G. and Foppen, R.P.B., *European Breeding Bird Atlas 2: Distribution, abundance and change* (Barcelona: Lynx Edicions, 2020)

Kelly, F., *Early Irish Farming: A study based mainly on the law-texts of the 7th and 8th centuries AD* (Dublin: Dublin Institute for Advanced Studies, 1997), p. 299

Kelly, T.C., 'The Status of the Lough. A report compiled on behalf of Cork Corporation' (unpublished report) (Cork: Department of Zoology, University College Cork, 1985)

——, 'The Origin of the Avifauna of Ireland', in J.L. Davenport, D.P. Sleeman and P.C. Woodman (eds), *Mind the Gap: Postglacial colonization of Ireland* (Belfast: Irish Naturalists' Journal, 2008), pp. 97–107

—— and O'Halloran, J., 'Obituary: Clive Desmond Hutchinson, 1949–1998', *Irish Birds*, vol. 6, 1998, pp. 329–31

Kennedy, P.G., 'Birds of the Countryside: Waders', *Studies*, vol. 33, 1944, pp. 393–9

——, *A List of the Birds of Ireland* (Dublin: Stationery Office, 1961)

——, Ruttledge, R.F. and Scroope, C.F., *The Birds of Ireland* (Edinburgh and London: Oliver & Boyd, 1954)

Kirby, J.S., Rees, E.C., Merne, O.J. and Gardarsson, A., 'International Census of Whooper Swans *Cygnus cygnus* in Britain, Ireland and Iceland: January 1991', *Wildfowl*, vol. 43, 1992, pp. 20–6

Knox, A.G., 'Taxonomic Status of "Lesser Golden Plovers"', *British Birds*, vol. 80, 1987, pp. 482–7

Lack, P. (ed.), *The Atlas of Wintering Birds in Britain and Ireland* (Calton: Poyser, 1986)

Lang, J.T., 'Peregrine Survey – Republic of Ireland 1967–68', *Irish Bird Report*, vol. 16, 1969, pp. 8–12

Langslow, D.R., 'Recent Increases of Blackcaps at Bird Observatories', *British Birds*, vol. 71, 1978, pp. 345–54

——, 'Movements of Blackcaps Ringed in Britain and Ireland', *Bird Study*, vol. 26, 1979, pp. 239–52

Leach, I.H., 'Wintering Blackcaps in Britain and Ireland', *Bird Study*, vol. 28, 1981, pp. 5–14

Leeder, M.R., 'Recent Developments in Carboniferous Geology: A critical review with implications for the British Isles and N.W. Europe', *Proceedings of the Geologist's Association*, vol. 99, 1988, pp. 73–100

Lever, C., *The Naturalized Animals of the British Isles* (London: Hutchinson, 1977)

——, *The Naturalized Animals of Britain and Ireland* (London: New Holland, 2009)

Lewis, L.J., Austin, G., Boland, H., Frost, T., Crowe, O. and Tierney, T.D., 'Waterbird Populations on Non-Estuarine Coasts in Ireland: Results of the 2015/16 Non-Estuarine Coastal Waterbird Survey (NEWS-III)', *Irish Birds*, vol. 10, 2017, pp. 511–22

——, Burke, B., Fitzgerald, N., Tierney, T.D. and Kelly, S., 'Irish Wetland Bird Survey: Waterbird status and distribution 2009/10–2015/16', *Irish Wildlife Manuals, No. 106* (Dublin: National Parks and Wildlife Service, 2019)

——, Coombes, D., Burke, B., O'Halloran, J., Walsh, A., Tierney, T.D. and Cummins, S., 'Countryside Bird Survey: Status and trends of common and widespread breeding birds 1998–2016', *Irish Wildlife Manuals, No. 115* (Dublin: National Parks and Wildlife Service, 2019)

—— and Kelly, T.C., 'A Short-term Study of the Effects of Algal Mats on the Distribution and Behavioural Ecology of Estuarine Birds', *Bird Study*, vol. 48, 2001, pp. 354–60

—— and Kelly, T.C., 'Aspects of the Spatial Ecology of Waders along an Estuarine Gradient', *Irish Birds*, vol. 9, 2012, pp. 375–84

Lloyd, C.S., 'Inventory of Seabird Breeding Colonies in the Republic of Ireland', Forest and Wildlife Service, Bray, unpublished report, 1982

——, Tasker, M.L. and Partridge, K., *The Status of Seabirds in Britain and Ireland* (London: Poyser, 1991)

Lockley, R.M., 'Southwesternmost Island', *The Countryman*, vol. 37, 1948, pp. 90–3

Longfield, R.E., 'Southern Notes', *Irish Naturalists' Journal*, vol. 1, 1926, p. 84

——, 'Wandering Flocks of Jays in Co. Cork', *Irish Naturalists' Journal*, vol. 1, 1927, p. 177

Lusby, J., McCarthy, A., O'Clery, M., Lynch, Á., Bayley, S., Nagle, T., Forkan, C., Stanley, M. and Nolan, B., *Barn Owl Monitoring Report 2020* (Kilcoole: BirdWatch Ireland, 2021)

—— and O'Clery, M., *Barn Owls in Ireland: Information on the ecology of Barn Owls and their conservation in Ireland* (Kilcoole: BirdWatch Ireland, 2014)

MacCarthy, I.A.J., 'Alluvial Sedimentation Patterns in the Munster Basin, Ireland', *Sedimentology*, vol. 37, 1990, pp. 685–712

MacCoitir, N., *Ireland's Birds: Myths, legends and folklore* (Cork: Collins Press, 2015)

Macdonald, R.A., 'The Breeding Population and Distribution of the Cormorant in Ireland', *Irish Birds*, vol. 3, 1987, pp. 405–16

Mackey, M., Ó Cadhla, O., Kelly, T.C., Aguilar de Soto, N. and Connolly, N., *Cetaceans and Seabirds of Ireland's Atlantic Margin: Seabird distribution, density and abundance*, vol. 1 (Cork: Coastal and Marine Resource Centre, 2004)

MacKinnon, G.E. and Coulson, J.C., 'The Temporal and Geographical Distribution of Continental Black-headed Gulls *Larus ridibundus* in the British Isles', *Bird Study*, vol. 34, 1987, pp. 1–9

Maclean, I.M.D, Austin, G.E., Rehfisch, M.M., Blew, J., Crowe, O., Delany, S., Devos, K., Deceuninck, B., Günther, K., Laursen, K., van Roomen, M. and Wahl, J., 'Global Warming Causes Rapid Changes in the Distribution and Abundance of Birds in Winter', *Global Change Biology*, vol. 14, 2008, pp. 2,489–500

Madden, B., 'The Mediterranean Gull in Ireland, 1956–1985', *Irish Birds*, vol. 3, 1987, pp. 363–76

——, Hunt, J. and Norriss, D., 'The 2002 Survey of the Peregrine *Falco peregrinus* Breeding Population in the Republic of Ireland', *Irish Birds*, vol. 8, 2009, pp. 543–8

—— and Ruttledge, R.F., 'Little Gulls in Ireland, 1970–1991', *Irish Birds*, vol. 5, 1993, pp. 23–34

Magennis, E., '"A Land of Milk and Money": The Physico-Historical Society, improvement and the surveys of mid-eighteenth-century Ireland', *Proceedings of the Royal Irish Academy*, vol. 102C, 2002, pp. 199–217

Magnusdottir, E., Leat, E.H.K., Bourgeon, S., Strøm, H., Petersen, A., Phillips, R.A., Hanssen, S.A., Bustnes, J.O., Hersteinsson, P. and Furness, R.W., 'Wintering Areas of Great Skuas *Stercorarius skua* Breeding in Scotland, Iceland and Norway', *Bird Study*, vol. 59, 2012, pp. 1–9

Main, I.G., 'Overseas Movements to and from Britain by Greenfinches *Carduelis chloris*', *Ringing and Migration*, vol. 19, 1999, pp. 191–9

Mainwood, A.R., 'The Movements of Storm Petrels as Shown by Ringing', *Ringing and Migration*, vol. 1, 1976, pp. 98–104

Mayes, E. and Stowe, T., 'The Status and Distribution of the Corncrake in Ireland, 1988', *Irish Birds*, vol. 4, 1989, pp. 1–12

McDevitt, A.D., Kajtoch, Ł., Mazgajski, T.D., Carden, R.F., Coscia, I., Osthoff, C., Coombes, R.H. and Wilson, F., 'The Origins of Great Spotted Woodpeckers *Dendrocopos major* Colonizing Ireland Revealed by Mitochondrial DNA', *Bird Study*, vol. 58, 2011, pp. 361–4

McElwaine, J.G., Wells, J.H. and Bowler, J.M., 'Winter Movements of Whooper Swans Visiting Ireland: Preliminary results', *Irish Birds*, vol. 5, 1995, pp. 265–78

McGrath, D. and Walsh, P.M., 'The Breeding Population of Kittiwakes on the South Coast of Ireland, 1985–95', *Irish Birds*, vol. 5, 1996, pp. 375–80

Mead, C.J. and Harrison, J.D., 'Sand Martin Movements within Britain and Ireland', *Bird Study*, vol. 26, 1979, pp. 73–86

—— and Harrison, J.D., 'Overseas Movements of British and Irish Sand Martins', *Bird Study*, vol. 26, 1979, pp. 87–98

Mee, A., 'A Survey of Nightjar in the Ballyhoura Hills', Cork and Limerick County Councils, unpublished report, 2020

——, Breen, D., Clarke, D., Heardman, C., Lyden, J., McMahon, F., O'Sullivan, P. and O'Toole, L., 'Reintroduction of White-tailed Eagles *Haliaeetus albicilla* to Ireland', *Irish Birds*, vol. 10, 2016, pp. 301–14

Merne, O.J., 'The Status of the Canada Goose in Ireland', *Irish Bird Report*, vol. 17, 1970, pp. 12–17

——, 'The Changing Status and Distribution of the Bewick's Swan in Ireland', *Irish Birds*, vol. 1, 1977, pp. 3–15

——, 'Greylag Geese in Ireland, March 1986', *Irish Birds*, vol. 3, 1986, pp. 207–14

—— and Murphy, C.W., 'Whooper Swans in Ireland, January 1986', *Irish Birds*, vol. 3, 1986, pp. 199–206

Milwright, R.D.P., 'Redwing *Turdus iliacus* Migration and Wintering Areas as Shown by Recoveries of Birds Ringed in the Breeding Season in Fennoscandia, Poland, the Baltic Republics, Russia, Siberia and Iceland', *Ringing and Migration*, vol. 21, 2002, pp. 5–15

Mitchell, F., *Shell Guide to Reading the Irish Landscape* (Dublin: Country House, 1986)

Mitchell, P.I., Newton, S.F., Ratcliffe, N. and Dunn, T.E., *Seabird Populations of Britain and Ireland* (London: Poyser, 2004)

Moffat, C.B., 'The Quail in Ireland: Its present and recent visits', *Irish Naturalist*, vol. 5, 1896, pp. 203–7

——, 'Robert Warren', *Irish Naturalist*, vol. 25, 1916, pp. 33–44

——, 'Robert Warren', *British Birds*, vol. 9, 1916, pp. 295–7

Moore, N.P., Kelly, P.F., Lang, F.A., Lynch, J.M. and Langton, S.D., 'The Peregrine *Falco peregrinus* in Quarries: Current status and factors influencing occupancy in the Republic of Ireland', *Bird Study*, vol. 44, 1997, pp. 176–81

Moore, N.W., 'The Past and Present Status of the Buzzard in the British Isles', *British Birds*, vol. 50, 1957, pp. 173–97

Moore, P., 'A Tree Pipit *Anthus trivialis* Overwintering in Co. Cork', *Irish Birds*, vol. 41, 2018, p. 112

More, A.G., *A List of Irish Birds* (Dublin: Science and Art Museum, 1885) (annotated copy housed in the Natural History Museum, Dublin)

——, *A List of Irish Birds* (Dublin: Science and Art Museum, 1890) (annotated copy housed in the Natural History Museum, Dublin)

Mundy, R., 'Lesser Black-backed Gull showing the Characters of the South Scandinavian Race *Larus fuscus intermedius* Breeding in Co. Cork', *Irish Birds*, vol. 4, 1990, p. 230

Murphy, J.A., *The College: A history of Queen's/University College Cork, 1845–1995* (Cork: Cork University Press, 1995)

Murphy, M. and Murphy, S. (eds), *Ireland's Bird Life: A world of beauty* (Sherkin Island: Sherkin Island Marine Station, 1994)

Murphy, S., Lewis, L.J. and Kelly, T.C., 'The Spatial Ecology of Wildfowl in Courtmacsherry Bay, Southern Ireland, with Particular Reference to Shelduck *Tadorna tadorna*', *Irish Birds*, vol. 8, 2006, pp. 51–8

Myers, A.A., Little, C., Costello, M.J. and Partridge, J.C. (eds), *The Ecology of Lough Hyne* (Dublin: Royal Irish Academy, 1991)

Nagle, T., 'The Status of Birds of Prey and Owls in County Cork', *Cork Bird Report 1996–2004*, 2006, pp. 285–308

——, Mee, A., Wilson, M.W., O'Halloran, J. and Smiddy, P., 'Habitat and Diet of Re-colonising Common Buzzards *Buteo buteo* in County Cork', *Irish Birds*, vol. 10, 2014, pp. 47–58

Nairn, R.G.W., 'Obituary: Clive D. Hutchinson (1949–1998)', *Irish Naturalists' Journal*, vol. 26, 1999, pp. 145–8

——, *Ireland's Coastline: Exploring its nature and heritage* (Cork: Collins Press, 2005)

Nelson, J.B., *The Gannet* (Berkhamstead: Poyser, 1978)

——, *The Atlantic Gannet* (Norfolk: Fenix Books, 2002)

Nesbitt, L., 'Summary of Bird Movements at Irish Bird Observatories 1963', *Irish Naturalists' Journal*, vol. 15, 1965, pp. 50–3

Newell, R.G., 'Influx of Great Shearwaters in Autumn 1965', *British Birds*, vol. 61, 1968, pp. 145–59

Newton, I., *Finches* (London: Collins, 1972)

——, *The Sparrowhawk* (Calton: Poyser, 1986)

——, *Bird Migration* (London: Collins, 2010)

——, *Bird Populations* (London: Collins, 2013)

——, *Farming and Birds* (London: Collins, 2017)

Newton, S.F., Harris, M.P. and Murray, S., 'Census of Gannet *Morus bassanus* Colonies in Ireland in 2013–2014', *Irish Birds*, vol. 10, 2015, pp. 215–20

Ní Lamhna, É., Nairn, R., Benson, L. and Kelly, T.C., 'Urban Habitats', in R. Nairn and J. O'Halloran (eds), *Bird Habitats in Ireland* (Cork: Collins Press, 2012), pp. 196–212

Nichols, A.R., *Catalogue of the Birds in the Barrington Collection, National Museum, Dublin* (Dublin: John Faulkner, 1920)

——, *A List of Irish Birds* (Dublin: Stationery Office, 1924)

Norris, C.A., 'Summary of a Report on the Distribution and Status of the Corncrake (*Crex crex*)', *British Birds*, vol. 38, 1945, pp. 162–8

Norriss, D.W., 'The Status of the Buzzard as a Breeding Species in the Republic of Ireland, 1977–1991', *Irish Birds*, vol. 4, 1991, pp. 291–8

——, 'The 1991 Survey and Weather Impacts on the Peregrine *Falco peregrinus* Breeding Population in the Republic of Ireland', *Bird Study*, vol. 42, 1995, pp. 20–30

——, Wilson, H.J. and Browne, D., 'The Breeding Population of the Peregrine Falcon in Ireland in 1981', *Irish Birds*, vol. 2, 1982, pp. 145–52

Ockendon, N., Hewson, C.M., Johnston, A. and Atkinson, P.W., 'Declines in British-breeding Populations of Afro-Palearctic Migrant Birds are Linked to Bioclimatic Wintering Zone in Africa, Possibly via Constraints on Arrival Time Advancement', *Bird Study*, vol. 59, 2012, pp. 111–25

O'Clery, M., McDonald, B. and Lusby, J., *Duhallow Raptor Conservation Project, 2012–13, Final Report* (Kilcoole: BirdWatch Ireland and IRD Duhallow, 2013)

O'Connell, M.M., Smiddy, P. and O'Halloran, J., 'Lead Poisoning in Mute Swans (*Cygnus olor*) in Ireland: Recent changes', *Biology and Environment: Proceedings of the Royal Irish Academy*, vol. 109B, 2009, pp. 53–60

O'Connor, R.J., 'Book Review: *Birds in Ireland*', *Irish Birds*, vol. 4, 1989, pp. 144–5

—— and Shrubb, M., *Farming and Birds* (Cambridge: Cambridge University Press, 1986)

O'Donoghue, B.G., 'Duhallow Hen Harriers *Circus cyaneus*: From stronghold to just holding on', *Irish Birds*, vol. 9, 2012, pp. 349–56

——, 'Hen Harrier *Circus cyaneus* Ecology and Conservation during the Non-breeding Season in Ireland', *Bird Study*, vol. 67, 2020, pp. 344–59

O'Donoghue, P.D., Cross, T.F. and O'Halloran, J., 'Carrion Crows in Ireland, 1969–1993', *Irish Birds*, vol. 5, 1996, pp. 399–406

——, O'Halloran, J., Bacon, P.J., Smiddy, P. and Cross, T.F., 'The Population Genetics of the Mute Swan *Cygnus olor* in Ireland', *Wildfowl*, vol. 43, 1992, pp. 5–11

Ó Drisceoil, D. and Ó Drisceoil, D., *Beamish and Crawford: The history of an Irish brewery* (Cork: Collins Press, 2015)

O'Flynn, W.J., 'Population Changes of the Hen Harrier in Ireland', *Irish Birds*, vol. 2, 1983, pp. 337–43

——, 'Bird Watching in Cork 1960–1990: Some personal reminiscences', *Cork Bird Report 1989*, 1990, pp. 99–102

O'Halloran, J., Kelly, T.C., Quinn, J.L., Irwin, S., Fernández-Bellon, D., Caravaggi, A. and Smiddy, P., 'Current Ornithological Research in Ireland: Seventh Ornithological Research Conference, UCC, November 2017', *Irish Birds*, vol. 10, 2017, pp. 598–638

——, Myers, A.A. and Duggan, P.F., 'Lead Poisoning in Swans and Sources of Contamination in Ireland', *Journal of Zoology*, vol. 216, 1988, pp. 211–23

——, Myers, A.A. and Duggan, P.F., 'Some Sub-lethal Effects of Lead on Mute Swan *Cygnus olor*', *Journal of Zoology*, vol. 218, 1989, pp. 627–32

—— Ridgway, M. and Hutchinson, C.D., 'A Whooper Swan *Cygnus cygnus* Population Wintering at Kilcolman Wildfowl Refuge, Co. Cork, Ireland: Trends over 20 years', *Wildfowl*, vol. 44, 1993, pp. 1–6

—— and Sharrock, J.T.R., 'Obituary: Clive Desmond Hutchinson, BA (1949–1998)', *British Birds*, vol. 91, 1998, pp. 269–70

——, Smiddy, P. and Irwin, S., 'Movements of Mute Swans in South-west Ireland', *Irish Birds*, vol. 5, 1995, pp. 295–8

——, Smiddy, P. and O'Mahony, B., 'Movements of Dippers *Cinclus cinclus* in Southwest Ireland', *Ringing and Migration*, vol. 20, 2000, pp. 147–51

——, Smiddy, P., O'Mahony, B., Taylor, A.J. and O'Donoghue, P.D., 'Aspects of the Population Biology of the Dipper in South-west Ireland', *Irish Birds*, vol. 6, 1999, pp. 359–64

——, Smiddy, P., Quishi, X., O'Leary, R. and Hayes, C., 'Trends in Mute Swan Blood Lead Levels: Evidence of grit reducing lead levels', *Waterbirds*, vol. 25 (special publication 1), 2002, pp. 363–7

O'Mahony, B. and Smiddy, P., 'Breeding of the Common Tern *Sterna hirundo* in Cork Harbour, 1983–2017', *Irish Birds*, vol. 10, 2017, pp. 535–40

O'Meara, M., 'Distribution and Numbers of Corncrakes in Ireland in 1978', *Irish Birds*, vol. 1, 1979, pp. 381–405

——, 'Corncrake Declines in Seven Areas, 1978–1985', *Irish Birds*, vol. 3, 1986, pp. 237–44

Ormerod, S.J., 'Pre-migratory and Migratory Movements of Swallows *Hirundo rustica* in Britain and Ireland', *Bird Study*, vol. 38, 1991, pp. 170–8

O'Rourke, F.J., *The Fauna of Ireland: An introduction to the land vertebrates* (Cork: Mercier Press, 1970)

O'Sullivan, D.C. (ed.), *The Natural History of Ireland* (Cork: Cork University Press, 2009)

O'Sullivan, R.J., 'Food Provisioning of Little Grebe *Tachybaptus ruficollis* Chicks', *Irish Birds*, vol. 10, 2015, pp. 175–8

Pain, D.J. and Pienkowski, M.W. (eds), *Farming and Birds in Europe: The Common Agricultural Policy and its implications for bird conservation* (London: Academic Press, 1997)

Parkin, D.T., 'Birding and DNA: Species for the new millennium', *Bird Study*, vol. 50, 2003, pp. 223–42

—— and Knox, A.G., *The Status of Birds in Britain and Ireland* (London: Helm, 2010)

Patten, C.J., 'The Natural History of the Ruff', *Irish Naturalist*, vol. 9, 1900, pp. 187–209

Patterson, R. and Foster, N.H., 'Irish Field Club Union: Report of the Fifth Triennial Conference and Excursion, held at Cork, July 11th to 16th, 1907. Vertebrata', *Irish Naturalist*, vol. 16, 1907, pp. 269–72

Payne-Gallwey, R., *The Fowler in Ireland* (London: Van Voorst, 1882; Southampton: Ashford Press, 1985)

——, *The Book of Duck Decoys* (London: Van Voorst, 1886)

Phillips, J., 'Autumn Vagrancy: "Reverse migration" and migratory orientation', *Ringing and Migration*, vol. 20, 2000, pp. 35–8

Pierce, S. and Wilson, J., 'Spring Migration of Whimbrels over Cork Harbour', *Irish Birds*, vol. 1, 1980, pp. 514–16

Pochin-Mould, D.D.C., *Discovering Cork* (Cork: Brandon Books, 1991)

Pollock, C.M., 'Observations on the Distribution of Seabirds off South-west Ireland', *Irish Birds*, vol. 5, 1994, pp. 173–82

Praeger, R.L., *Some Irish Naturalists* (Dundalk: Dundalgan Press, 1949)

Prater, A.J., 'The Wintering Population of Ruffs in Britain and Ireland', *Bird Study*, vol. 20, 1973, pp. 245–50

——, 'The Wintering Population of the Black-tailed Godwit', *Bird Study*, vol. 22, 1975, pp. 169–76

——, *Estuary Birds of Britain and Ireland* (Calton: Poyser, 1981)

Preston, K., 'Census of Great Crested Grebes, Summer 1975', *Irish Bird Report*, vol. 23, 1976, pp. 38–43

Price, R. and Robinson, J.A., 'The Persecution of Kites and other Species in 18th century Co. Antrim', *Irish Naturalists' Journal*, vol. 29, 2008, pp. 1–6

Rabøl, J., 'Reversed Migration as the Cause of Westward Vagrancy by Four *Phylloscopus* Warblers', *British Birds*, vol. 62, 1969, pp. 89–92

Ratcliffe, D., *The Peregrine Falcon* (Calton: Poyser, 1980)

Rees, E., *Bewick's Swan* (London: Poyser, 2006)

Ridgway, M. and Hutchinson, C.D. (eds), *The Natural History of Kilcolman* (Dublin: O'Brien Printing, 1990)

Robertson, P.A. and Whelan, J., 'The Ecology and Management of Wild and Hand-reared Pheasants in Ireland', *Irish Birds*, vol. 3, 1987, pp. 427–40

Rochford, J.M., 'Personalities: C.D. Hutchinson', *British Birds*, vol. 71, 1978, pp. 305–7

Rolfe, R., 'The Status of the Chough in the British Isles', *Bird Study*, vol. 13, 1966, pp. 221–36

Ronayne, S.T., 'Breeding Little Egrets *Egretta garzetta* and Grey Herons *Ardea cinerea* at an Urban Habitat, the Atlantic Pond, Cork City', *Irish Birds*, vol. 9, 2011, pp. 327–8

——, 'Diet of the Little Egret *Egretta garzetta* in Southern Ireland', *Irish Birds*, vol. 9, 2011, pp. 329–30

Roycroft, D., Kelly, T.C. and Lewis, L.J., 'Birds, Seals and the Suspension Culture of Mussels in Bantry Bay, a Non-seaduck Area in Southwest Ireland', *Estuarine, Coastal and Shelf Science*, vol. 61, 2004, pp. 703–12

Ruttledge, R.F., *Ireland's Birds* (London: Witherby, 1966)

——, 'The Present Breeding Distribution of the Tree Sparrow in Ireland', *Irish Bird Report*, vol. 14, 1967, pp. 50–4

——, 'The Kingfisher Population', *Irish Bird Report*, vol. 15, 1968, pp. 11–14

——, 'Winter Distribution and Numbers of Scaup, Long-tailed Duck and Common Scoter in Ireland', *Bird Study*, vol. 17, 1970, pp. 241–6

——, 'Winter Distribution of Whooper and Bewick's Swans in Ireland', *Bird Study*, vol. 21, 1974, pp. 141–5

—— and Hall Watt, R., 'The Distribution and Status of Wild Geese in Ireland', *Bird Study*, vol. 5, 1958, pp. 22–33

—— and Ogilvie, M.A., 'The Past and Current Status of the Greenland White-fronted Goose in Ireland and Britain', *Irish Birds*, vol. 1, 1979, pp. 293–363

Sealy, S.G., O'Halloran, J. and Smiddy, P., 'Cuckoo Hosts in Ireland', *Irish Birds*, vol. 5, 1996, pp. 381–90

Sergeant, D.E., 'Little Auks in Britain, 1948 to 1951', *British Birds*, vol. 45, 1952, pp. 122–33

Sharrock, J.T.R., 'Migration Seasons of the *Sylvia* Warblers at Cape Clear Bird Observatory', *Bird Study*, vol. 15, 1968, pp. 99–103

——, 'Grey Wagtail Passage and Population Fluctuations in 1956–67', *Bird Study*, vol. 16, 1969, pp. 17–34

—— (ed.), *The Natural History of Cape Clear Island* (Berkhamstead: Poyser, 1973)

——, *Scarce Migrant Birds in Britain and Ireland* (Berkhamstead: Poyser, 1974)

——, *The Atlas of Breeding Birds in Britain and Ireland* (Tring: British Trust for Ornithology, 1976)

Sheppard, R., *Ireland's Wetland Wealth* (Dublin: Irish Wildbird Conservancy, 1993)

—— and Green, R.E., 'Status of the Corncrake in Ireland in 1993', *Irish Birds*, vol. 5, 1994, pp. 125–38

Sleeman, D.P. and Fox, J.B., 'Did Jays *Garrulus glandarius* Go Extinct Locally Due to Killing for Fishing Flies or Habitat Destruction in 19th Century North Cork?', *Irish Birds*, vol. 42, 2020, pp. 110–11

Smiddy, P., 'Northern Razorbills on the South Irish Coast', *Irish Birds*, vol. 3, 1987, pp. 451–2

——, 'Pheasant Densities in East Cork', *Irish Birds*, vol. 3, 1988, pp. 605–6

——, 'The Waterfowl of Ballymacoda, Co. Cork', *Irish Birds*, vol. 4, 1992, pp. 525–48

——, 'The Effect of the *Kowloon Bridge* Oil Spill in East Cork', *Irish Birds*, vol. 4, 1992, pp. 559–60

——, 'The Waterfowl of the Blackwater Estuary (Youghal Harbour), Cos Waterford and Cork', *Irish Naturalists' Journal*, vol. 25, 1996, pp. 157–65

——, 'Cormorant *Phalacrocorax carbo* Breeding Numbers in Waterford, East Cork and Mid Cork', *Irish Birds*, vol. 6, 1998, pp. 213–16

——, 'The Effect of the Cork Harbour Oil Spill of November 1997 on Birds', *Irish Naturalists' Journal*, vol. 26, 1998, pp. 32–7

——, 'Auks (Alcidae) Drowned in Fishing Nets in East Cork in January and February 1983', *Irish Naturalists' Journal*, vol. 26, 2001, pp. 414–19

——, 'Breeding of the Little Egret *Egretta garzetta* in Ireland, 1997–2001', *Irish Birds*, vol. 7, 2002, pp. 57–60

——, 'Bird and Mammal Mortality on Roads in Counties Cork and Waterford, Ireland', *Bulletin of the Irish Biogeographical Society*, vol. 26, 2002, pp. 29–38

——, 'Breeding Waterfowl at Ballycotton, County Cork, 1960–2004', *Irish Birds*, vol. 7, 2005, pp. 497–502

——, 'The Site Characteristics and Use of Nests of the Barn Swallow (*Hirundo rustica* L.)', *Irish Naturalists' Journal*, vol. 29, 2008 pp. 107–10

——, 'Post-fledging Roosting at the Nest in Juvenile Barn Swallows (*Hirundo rustica*)', *Irish Naturalists' Journal*, vol. 31, 2010, pp. 44–6

——, 'Roosting by Barn Swallows (*Hirundo rustica*) and Sand Martins (*Riparia riparia*) in Fields of Maize (*Zea mays*) in Co. Cork', *Irish Naturalists' Journal*, vol. 31, 2010, pp. 128–30

——, 'Breeding Birds in Ireland: Success and failure among colonists', in D.P. Sleeman, J. Carlsson and J.E.L. Carlsson (eds), *Mind the Gap II: New insights into the Irish postglacial* (Belfast: Irish Naturalists' Journal, 2014), pp. 89–99

——, 'Diet of Coastal Breeding Ravens *Corvus corax* in East County Cork', *Irish Birds*, vol. 10, 2016, pp. 335–8

——, 'Unprecedented Numbers of Black-tailed Godwits *Limosa limosa* at the Blackwater Estuary (Youghal Harbour), Counties Cork and Waterford in January and February 2016', *Irish Birds*, vol. 10, 2016, p. 444

——, 'Diet of the Common Kestrel *Falco tinnunculus* in East Cork and West Waterford: An insight into the dynamics of invasive mammal species', *Biology and Environment: Proceedings of the Royal Irish Academy*, vol. 117B, 2017, pp. 131–8

——, 'Dominance of Invasive Small Mammals in the Diet of the Barn Owl *Tyto alba* in County Cork, Ireland', *Biology and Environment: Proceedings of the Royal Irish Academy*, vol. 118B, 2018, pp. 49–53

——, 'Winter Flock Size in the Long-tailed Tit *Aegithalos caudatus*, and Possible Effects of the Cold Winters of 2009/10 and 2010/11', *Irish Birds*, vol. 43, 2021, pp. 39–44

——, 'Aspects of the Breeding Biology of Blue Tits (*Cyanistes caeruleus*) and Great Tits (*Parus major*) in County Cork', *Irish Naturalists' Journal*, vol. 37, 2021, pp. 97–101

—— and Cullen, C., 'Winter Diet of the Hen Harrier *Circus cyaneus* in Coastal East County Cork', *Irish Birds*, vol. 10, 2017, pp. 523–6

——, Cullen, C. and O'Halloran, J., 'Time of Roosting of Barn Swallows *Hirundo rustica* at an Irish Reedbed during Autumn Migration', *Ringing and Migration*, vol. 23, 2007, pp. 228–30

——, Cullen, C. and O'Halloran, J., 'Autumn Use of a Reedbed by Barn Swallows *Hirundo rustica* and Sand Martins *Riparia riparia* in County Cork', *Irish Birds*, vol. 8, 2007, pp. 243–8

—— and Duffy, B., 'Little Egret *Egretta garzetta*: A new breeding bird for Ireland', *Irish Birds*, vol. 6, 1997, pp. 55–6

—— and Nairn, R., 'Rivers and Canals', in R. Nairn and J. O'Halloran (eds), *Bird Habitats in Ireland* (Cork: Collins Press, 2012), pp. 24–43

—— and O'Halloran, J., 'The Breeding Biology of Mute Swans *Cygnus olor* in Southeast Cork, Ireland', *Wildfowl*, vol. 42, 1991, pp. 12–16

—— and O'Halloran, J., 'The January 1991 Swan Census in Co. Cork', *Cork Bird Report 1990*, 1991, pp. 85–9

—— and O'Halloran, J., 'Breeding Biology of the Grey Wagtail *Motacilla cinerea* in Southwest Ireland', *Bird Study*, vol. 45, 1998, pp. 331–6

—— and O'Halloran, J., 'The Ecology of River Bridges: Their Use by Birds and Mammals', in J. Davenport and J.L. Davenport (eds), *The Effects of Human Transport on Ecosystems: Cars and planes, boats and trains* (Dublin: Royal Irish Academy, 2004), pp. 83–97

—— and O'Halloran, J., 'The Waterfowl of Ballycotton, County Cork: Population change over 35 years, 1970/71–2004/05', *Irish Birds*, vol. 8, 2006, pp. 65–78

—— and O'Halloran, J., 'Ballycotton: Habitat change and loss of wetland avian biodiversity, 1970–2004', *Cork Bird Report 1996–2004*, 2006, pp. 309–18

—— and O'Halloran, J., 'The Distribution of Waterfowl at Ballycotton, County Cork in Relation to Habitat and Habitat Change', *Irish Birds*, vol. 8, 2008, pp. 351–8

—— and O'Halloran, J., 'Breeding Biology of Barn Swallows *Hirundo rustica* in Counties Cork and Waterford, Ireland', *Bird Study*, vol. 57, 2010, pp. 256–60

——, O'Halloran, J., Coveney, J.A., Leonard, P.G. and Shorten, M., 'Winter Waterfowl Populations of Cork Harbour: An update', *Irish Birds*, vol. 5, 1995, pp. 285–94

——, O'Halloran, J., O'Mahony, B. and Taylor, A.J., 'The Breeding Biology of the Dipper *Cinclus cinclus* in South-west Ireland', *Bird Study*, vol. 42, 1995, pp. 76–81

—— and O'Mahony, B., 'The Status of the Reed Warbler *Acrocephalus scirpaceus* in Ireland', *Irish Birds*, vol. 6, 1997, pp. 23–8

——, O'Mahony, B., Fernández-Bellon, D. and O'Halloran, J., 'The Oldest Dipper *Cinclus cinclus* in Ireland and Britain, thus Far', *Irish Birds*, vol. 10, 2016, p. 445

—— and O'Sullivan, O., 'The Status of the Little Egret *Egretta garzetta* in Ireland', *Irish Birds*, vol. 6, 1998, pp. 201–6

—— and Sleeman, D.P., 'Diet of Long-eared Owl *Asio otus* and Short-eared Owl *Asio flammeus* in Ireland: A review', *Irish Birds*, vol. 42, 2020, pp. 27–34

——, Sleeman, D.P. and O'Halloran, J., 'Barn Owl *Tyto alba* Diet in Ireland: A review', *Irish Birds*, vol. 41, 2018, pp. 39–48

Smith, C., *The Antient and Present State of the County and City of Waterford* (Dublin: privately published, 1746)

——, *The Antient and Present State of the County and City of Cork* (Dublin: privately published, vol. 2, 1750), pp. 320–49

——, *The Antient and Present State of the County of Kerry* (Dublin: privately published, 1756)

Smith, J.C., 'The Garden Warbler in Ireland', *Irish Naturalist*, vol. 3, 1894, pp. 46–7

——, 'Crossbills (*Loxia curvirostra*) in Co. Cork', *Irish Naturalist*, vol. 3, 1894, p. 47

Stafford, J., 'The Wintering of Blackcaps in the British Isles', *Bird Study*, vol. 3, 1956, pp. 251–7

——, 'Nightjar Enquiry, 1957–58', *Bird Study*, vol. 9, 1962, pp. 104–15

Staneva, A. and Burfield, I., *European Birds of Conservation Concern: Populations, trends and national responsibilities* (Cambridge: BirdLife International, 2017)

Stanley, P.I. and Minton, C.D.T., 'The Unprecedented Westward Migration of Curlew Sandpipers in Autumn 1969', *British Birds*, vol. 65, 1972, pp. 365–80

Stone, C.J., Webb, A. and Tasker, M.L., 'The Distribution of Manx Shearwaters *Puffinus puffinus* in North-west European Waters', *Bird Study*, vol. 41, 1994, pp. 170–80

Summers, R., Christian, N., Etheridge, B., Rae, S., Cleasby, I. and Pálsson, S., 'Scottish Breeding Greenshanks *Tringa nebularia* Do Not Migrate Far', *Bird Study*, vol. 67, 2020, pp. 1–7

Summers-Smith, J.D., *The Tree Sparrow* (Cleveland: privately published, 1995)

Sweeney, P. and Sweeney, E., 'Greylag Goose (*Anser anser* (L.)) Nesting Activity on a Ruined Castle', *Irish Naturalists' Journal*, vol. 32, 2013, pp. 75–6

Taylor, A.J. and O'Halloran, J., 'The Diet of the Dipper *Cinclus cinclus* as Represented by Faecal and Regurgitate Pellets: A comparison', *Bird Study*, vol. 44, 1997, pp. 338–47

—— and O'Halloran, J., 'Diet of Dippers *Cinclus cinclus* during an Early Winter Spate and the Possible Implications for Dipper Populations Subjected to Climate Change', *Bird Study*, vol. 48, 2001, pp. 173–9

—— and O'Halloran, J., 'The Decline of the Corn Bunting, *Miliaria calandra*, in the Republic of Ireland', *Biology and Environment: Proceedings of the Royal Irish Academy*, vol. 102B, 2002, pp. 165–75

Thompson, P.S., Greenwood, J.J.D. and Greenway, K., 'Birds in European Gardens in the Winter and Spring of 1988–89', *Bird Study*, vol. 40, 1993, pp. 120–34

Thompson, W., *The Natural History of Ireland*, vols 1–3 (London: Reeve, Benham & Reeve, 1849–51)

Thorup, K., 'Vagrancy of Yellow-browed Warbler *Phylloscopus inornatus* and Pallas's Warbler *Ph. proregulus* in North-west Europe: Misorientation on great circles?', *Ringing and Migration*, vol. 17, 1998, pp. 7–12

Toms, M., *Owls* (London: Collins, 2014)

Tubbs, C.R., *The Buzzard* (Newton Abbot: David & Charles, 1974)

Ussher, R.J., 'The Crossbill (*Loxia curvirostra*, L.) in Ireland', *Irish Naturalist*, vol. 1, 1892, pp. 6–9, 28–31

——, 'Birds of County Cork', *Journal of the Cork Historical and Archaeological Society*, series A, vol. 1, 1892, pp. 230–1, 250–1; vol. 2, 1893, pp. 20, 40, 59–60, 80, 103–4, 123–4, 148, 171–2, 191–2, 211–12, 235–6, 266–8; vol. 3, 1894, pp. 23–4, 42–4

——, 'A Catalogue of the Birds Observed in this County', in R. Day and W.A. Copinger (eds), *The Ancient and Present State of the County and City of Cork*, vol. 2 (Cork: Guy, 1894), pp. 238–62

——, 'The Great and Sooty Shearwaters on the South Coast', *Irish Naturalist*, vol. 10, 1901, pp. 42–3

——, 'Great Shearwaters and Sooty Shearwaters in 1901', *Irish Naturalist*, vol. 14, 1905, p. 43

——, 'Great Shearwaters and Fulmars', *Irish Naturalist*, vol. 17, 1908, pp. 81–2

—— and Warren, R., *The Birds of Ireland* (London: Gurney & Jackson, 1900) (includes unpublished notes by Ussher held at the Royal Irish Academy, Dublin)

Vernon, J.D.R., 'Icelandic Black-tailed Godwits in the British Isles', *British Birds*, vol. 56, 1963, pp. 233–7

Viney, M., 'Wild Sports and Stone Guns', in J.W. Foster (ed.), *Nature in Ireland: A scientific and cultural history* (Dublin: Lilliput Press, 1997), pp. 524–48

—— and Viney, E., *Ireland's Ocean: A natural history* (Cork: Collins Press, 2008)

Walsh, P.M., O'Halloran, J., Giller, P.S. and Kelly, T.C., 'Greenfinches *Carduelis chloris* and other Bird Species Feeding on Cones of Noble Fir *Abies procera*', *Bird Study*, vol. 46, 1999, pp. 119–21

Warren, R., 'The Harvey Collection of Irish Birds', *Irish Naturalist*, vol. 12, 1903, p. 55

Watson, D., *The Hen Harrier* (Berkhamstead: Poyser, 1977)

Watson, P.S., 'Seabirds in the Celtic Sea and off Co. Cork in April 1980', *Irish Naturalists' Journal*, vol. 20, 1982, pp. 388–93

Watters, J.J., *The Natural History of the Birds of Ireland* (Dublin: McGlashan, 1853)

Wells, J.H. and Smiddy, P., 'The Status of the Ruddy Duck in Ireland', *Irish Birds*, vol. 5, 1995, pp. 279–84

Wernham, C.V., Toms, M.P., Marchant, J.H., Clark, J.A., Siriwardena, G.M. and Baillie, S.R. (eds), *The Migration Atlas: Movements of the birds of Britain and Ireland* (London: Poyser, 2002)

Whelan, R., Smiddy, P. and O'Clery, M., 'The Occurrence of Nesting Sand Martins *Riparia riparia* in Built Structures during the 2016 Breeding Season', *Irish Birds*, vol. 10, 2016, pp. 442–4

Whilde, A., 'Auks Trapped in Salmon Drift Nets', *Irish Birds*, vol. 1, 1979, pp. 370–6

——, 'The 1984 All Ireland Tern Survey', *Irish Birds*, vol. 3, 1985, pp. 1–32

——, *Threatened Mammals, Birds, Amphibians and Fish in Ireland. Irish Red Data Book 2: Vertebrates* (Belfast: Her Majesty's Stationery Office, 1993)

Whitaker, J., *British Duck Decoys of Today* (London: Burlington, 1918)

Williams, E., 'On the Occurrence of the Greenland and Iceland Falcons in Ireland, during the Spring of 1905', *Irish Naturalist*, vol. 14, 1905, pp. 201–4

Williamson, K. and Ruttledge, R.F., 'Icelandic Black-tailed Godwits Wintering in Ireland', *British Birds*, vol. 50, 1957, pp. 524–6

Wilson, H.J., 'The Breeding and Wintering Ecology of the Woodcock *Scolopax rusticola* in Ireland', Forest and Wildlife Service, Bray, unpublished report, 1982

Wilson, J., 'Operation Godwit', *Cork Bird Report 2005–2006*, 2009, pp. 177–81

Wilson, J.R., Czajkowski, M.A. and Pienkowski, M.W., 'The Migration through Europe and Wintering in West Africa of Curlew Sandpipers', *Wildfowl*, vol. 31, 1980, pp. 107–22

Wilson, M.W., Fernández-Bellon, D., Irwin, S. and O'Halloran, J., 'Hen Harrier *Circus cyaneus* Population Trends in Relation to Wind Farms', *Bird Study*, vol. 64, 2017, pp. 20–9

——, Irwin, S., Norriss, D.W., Newton, S.F., Collins, K., Kelly, T.C. and O'Halloran, J., 'The Importance of Pre-thicket Conifer Plantations for Nesting Hen Harriers *Circus cyaneus* in Ireland', *Ibis*, vol. 151, 2009, pp. 332–43

Wing, S., *The Natural History of Cape Clear 1959–2019* (Skibbereen: privately published, 2020)

Winstanley, D., Spencer, R. and Williamson, K., 'Where Have All the Whitethroats Gone?', *Bird Study*, vol. 21, 1974, pp. 1–14

Wolfe, J.J., 'Crossbills Nesting in South Cork', *Irish Naturalist*, vol. 9, 1900, pp. 293–4

Wynn, R.B. and Yésou, P., 'The Changing Status of Balearic Shearwater in Northwest European Waters', *British Birds*, vol. 100, 2007, pp. 392–406

Wyse Jackson, P.N., 'Robert Ball (1802–1857): Naturalist', *Irish Naturalists' Journal*, vol. 30, 2009, pp. 15–18

Yalden, D.W., 'A Distant History of Irish Birds', *Irish Birds*, vol. 9, 2011, pp. 225–8

INDEX

Note: Page locators in *italics* refer to illustrations, tables and figures. Locators in **bold** refer to the main entry of that particular species.